Preface to the First Edition

The numbers coming into the surveying profession reading full-time for degrees and diplomas increase year by year. Particularly in the case of such students, it is desirable that the teaching approach should be to guide their reading rather than to dispense facts in lectures. Students following correspondence courses need to supplement these by further reading. In the subject of Valuations, beyond the intermediate state, there is comparatively little reading matter available, other than "Modern Methods of Valuation". However, in no way does this book attempt to compete with "Modern Methods of Valuation" which covers principles, law and practice; it is intended to complement the latter. There is said by some to be a gap between the principles of valuation as taught for examinations and practice. The purpose of this book is to provide more reading matter at final year level to bridge this gap.

The form of the book, as originally intended, was to have been a series of case studies with notes by way of explanation but during the course of preparation it became apparent that some subjects were better covered by text, hence the chapters vary widely in form. There are also wide differences of approach but in this subject there is frequently room for more that one opinion and this is a not unimportant factor of which a student should be aware. In expressing my thanks to the authors I should explain here that their opinions are confined to the chapter each has written. It is also appropriate that I should disclaim all responsibility for the views expressed. The idea of the book was mine and the choice of authors; my intention in that connection was to try to balance those engaged with teaching with those in practice, a coming together of "town and gown" which is perhaps an all too rare occurrence for the good of the valuer's profession.

I decided to draw the line at the end of November 1979 as far as legislation is concerned; thus no account is taken of the Local Government, Planning and Land Bill.

W H Rees BSc FRICS
March 1980

Note by W H Rees

As this is the last edition of this book which I shall edit, I decided that it would be prudent to find a co-editor and thus provide continuity for further possible editions. I was delighted when Richard Hayward accepted my invitation to join me.

Preface to the Fifth Edition

Since the last edition, three of the contributors have died and nine others have decided not to continue. We are most grateful to those concerned who have allowed successors to use the matter in the last edition and to others who have provided much useful help.

For the first time in the history of this book valuer-contributors are joined by other professionals, chapter 8, Taxation being written by a chartered accountant and a taxation specialist.

The purpose of the book remains as stated in the Preface to the First Edition, as do the remarks relating to the opinions of the various authors and the disclaimer as to our own views.

As with all such books, the relevant case and statute law may have altered since the chapters were written.

R E H Hayward
W H Rees
November 2000

Authors and Contents

Table of Cases

A

S

Table of Statutes

Table of Statutory Instruments

Chapter 1

Agricultural Properties

Many of the assets within agricultural properties are similar to those found elsewhere. The principles behind their valuation are no different to those for other properties, the essence being the comparison of like with like. For instance, many include vacant and let dwelling-houses, shops and commercial buildings, as well as farmland and woodland. However the variety of such assets even within comparatively small agricultural properties, and the very existence of some such as milk quota and sporting rights, can surprise the less experienced practitioner and lead to valuations of some complexity.

Furthermore, legislation affecting the valuation of agricultural properties is often different and more intrusive than that encountered in the urban world, this being the case particularly when valuations are made for taxation purposes. For instance, the relationship between the landlord and tenant and subtenant of an asset within an agricultural property can be governed by any one or more of numerous statutes including the Agricultural Holdings Act 1986, the Agriculture Act 1986, the Agricultural Tenancies Act 1995, the Rent Act 1977, the Rent (Agriculture) Act 1976, the Housing Acts, which include specific categories for agricultural occupancies, and the Landlord and Tenant Acts 1927 and 1954.

While it may not be relevant as to whether a property is agricultural for simple valuations, the practitioner must be alert always to the differing definitions within different statutes which can result in the same asset being agricultural for one or more purposes but not others. For instance the definition of agriculture in section 96 of the Agricultural Holdings Act 1986 (the 1986 Act) contrasts with that for agricultural premises in Schedule 5 of the Local Government Finance Act 1988. While section 66 of the Town and Country Planning Act 1990 incorporates the definition of agricultural holding from section 1 of the 1986 Act, the definition of agricultural buildings and operations in Part 6 of the Town and Country Planning (General Permitted Development) Order 1995 SI 418 is particularly detailed and restrictive. Indeed section 1 of the

1

Agricultural Tenancies Act 1995 specifically permits the parties to a farm business tenancy to agree by notice to each other that its character is primarily or wholly agricultural even in situations where others might judge otherwise.

Furthermore, whether an agricultural asset qualifies for capital gains tax reliefs depends, not only on the classes listed in section 155 of the Taxation of Chargeable Gains Act 1992, but also on the trading status of the taxpayer. Whereas the definition of agricultural property for the purposes of relief in section 115 of the Inheritance Tax Act 1984 includes only those cottages, farm buildings and farmhouses, together with the land occupied with them, as are of a character appropriate to the property.

Valuations of freeholds mainly with vacant possession

The valuer must take care to identify the various assets within the property by means of an inspection and diligent enquiries and to consider the various markets for those assets. The days when an agricultural property was valued as a whole are long gone, not least because of the risk of a speculative purchaser immediately splitting such a property into lots and selling on at a profit.

In particular the various dwellings need to be valued separately and only after detailed investigation into the rights by which they are occupied and the legislation governing those rights. Typically the valuation of such dwellings is based on evidence from the local dwelling-house market. However, the practitioner needs to be alert to the existence of agricultural occupancy conditions, which have been attached to planning consents for most new agricultural dwellings granted since 1947, as these can reduce vacant possession values by as much as 50%, and more recently by planning agreements although these are less common.

Farmland should also be valued in lots based on numerous factors such as: accessibility; geographical location; soil quality; planning restrictions such as conservation areas and tree preservation orders; designations such as sites of special scientific interest or environmentally sensitive areas; irrigation abstraction licences; covenants in the title; possible bids from near neighbours and the detailed complexities of the Common Agricultural Policy of the European Union regarding set-aside and arable area payments, grant schemes such as the Apple and Pear Grubbing-up Scheme, 1998 as well as the possibility of higher non-agricultural bids whether for garden extensions, pony paddocks or more valuable uses.

Farmland values have fluctuated dramatically since 1970 and the practitioner will need to research with care the local and wider market for values per hectare. When historic values are required, typically for taxation valuations, *Farmland Market* (published by *Farmers Weekly*, in association with the Royal Institution of Chartered Surveyors) can be a useful source of information.

When valuing farmland the practitioner must be alert particularly to areas, however small, which may have been let before September 1 1995 resulting in a tenancy under the Agricultural Holdings Act 1986. Such tenancies, which could have been created orally by the mere payment and acceptance of rent, can continue for two further generations and reduce values by as much as 66% from vacant possession figures.

The valuer of farmland should make enquiries to establish whether any area has been used within the last six years by a milk producer, whether for grazing dairy cows or followers or for the production of hay or silage, as milk quota might then attach to that land under the Dairy Produce Quota Regulations.

Farm buildings are particularly difficult to value and the best, even when new, are worth but a small fraction of their current building cost and only when offered with an appropriate area of farmland. Many traditional buildings which are little used for modern agriculture have no value, unless they have potential for conversion to non-agricultural uses, and others including less sophisticated modern framed buildings are not valued separately but are included within the farmland value per hectare. Good quality modern framed buildings to a high specification but capable of a wide variety of uses may be valued to a spot figure checked against the value thereof per hectare of the farmland included with those buildings.

The valuer should be aware of the limited market for selling disused modern framed buildings for dismantling and re-erection elsewhere.

Since the mid 1980s farmers have been encouraged to diversify with the result that disused farm buildings are sometimes let for non-agricultural purposes. The valuation of the rent from such buildings needs to be approached with caution as planning permission may not have been obtained and the terms of the letting may be unclear, although still attracting the security of tenure provisions of the Landlord and Tenant Act 1954.

When valuing an agricultural property which includes traditional buildings the valuer needs to check whether they are listed as being of architectural or historic interest or are included within a conservation area under the Planning (Listed Buildings and Conservation Areas) Act 1990.

Study 1

Valuation of freehold of agricultural property mainly with vacant possession

This illustrates the valuation of a farm comprising a substantial vacant period house, six further dwellings occupied under a variety of different arrangements and 175 hectares of land and buildings, two of which are let away for non-agricultural uses that do not have planning permission.

DWELLINGS

Name	Description	Occupier	Basis of occupation	(£)	(£)
The Court House	Period detached	Vacant		285,000	
The Farm-house	Modern detached subject to agricultural occupancy condition	Farm manager	Assured shorthold tenancy	60,000	
Oak Cottage	Modern semi-detached	An agricultural employee	Assured agricultural occupancy	35,000	
Ash Cottage	Modern semi-detached	Widow of former employee	Rent (Agriculture) Act 1976	35,000	
1 Court Cottages	Victorian end-terrace	A long standing agricultural employee	Rent (Agriculture) Act 1976	25,000	
2 Court Cottages	Victorian mid-terrace	A retired agricultural employee	Rent (Agriculture) Act 1976	22,500	
3 Court Cottages	Victorian end-terrace	Non-employee	Rent Act 1977	25,000	

Total Valuation of Dwellings   487,500

LAND

Area ha	Description (All eligible for arable area payments except where otherwise indicated)	Value per ha (£)	
65	Grade 1 upland with irrigation abstraction licence	8,000	520,000
15	Grade 2/3 arable marshland	4,900	73,500
20	Grade 3 arable marshland	4,300	86,000
25	Grade 2/3 arable marshland over railway	4,300	107,500
35	Grade 3 arable marshland over railway	3,700	129,500
5	Pasture, ineligible for arable area payments, with vacant possession	3,100	15,500
10	Pasture, ineligible for arable area payments, as let	1,500	15,000
175			

Total valuation of land   947,000

ie £5,410 per hectare

OWNER OCCUPIED BUILDINGS

Total valuation of owner occupied buildings   22,500

ie £130 per hectare

LET BUILDINGS

Name	Use	Rent pa (£)	
Barn A	Storage of caravans	600	
Barn B	Car repairs	1,140	
	Total rent received	1,740	
	YP in perp @ 17.5 % 5.71		
	Total valuation of let buildings		9,935
	Say		10,000

Total valuation   £1,467,000

ie £ 8,383 per hectare

Valuations for rent reviews

For agricultural properties the rent review process and the basis of valuation on review are usually determined by statute rather than the lease or tenancy document. Indeed the inexperienced practitioner should be wary of following any procedures or valuing to any basis in any lease, tenancy or other document agreed between landlord and tenant without checking whether it contravenes the relevant statutory provisions, which are often mandatory regardless of any attempt by the parties to contract out of them.

While the majority of rent reviews will continue to be subject to the Agricultural Holdings Act 1986, all tenancies and leases entered into after September 1 1995 are farm business tenancies which are subject instead to the Agricultural Tenancies Act 1995.

Agricultural Holdings Act 1986

Section 12 and Schedule 2 of the Act provide for rental arbitrations at three yearly intervals subject to 12 to 24 month's notice by either party.

The arbitration procedure is set out elsewhere in the Act. The practitioner should be aware that this procedure differs from that in the Arbitration Act 1996 and that it is subject to certain mandatory time limits which cannot be varied either by the parties or the arbitrator.

The rent properly payable is defined at considerable length in Schedule 2. Broadly this requires a market rent to be awarded, but specifically excludes any element of appreciable scarcity or marriage value from the rents of comparable holdings. Furthermore, it obliges the arbitrator to have regard to all relevant factors some of which are referred to specifically including the productive capacity and related earning capacity of the holding. Beyond that the arbitrator is to disregard the rental value of tenant's improvements, landlord's improvements to the extent that they were funded by government grants, the tenant's occupation of the holding, any system of high farming practised by the tenant and any dilapidations to the holding.

This definition was considered in some detail by the Court of Appeal in *JW Childers* v *Anker* [1996] 1 EGLR 1 and the need for clarity in weighing the relevant factors in *Enfield London Borough Council* v *Pott* [1990] 2 EGLR 7.

More specifically the definition directs the arbitrator to have regard to rents fixed in other arbitrations so the decision in *Land Securities plc* v *Westminster City Council* [1992] 2 EGLR 15 has no application to 1986 Act arbitrations.

In theory, prior to September 1 1995, it was possible to enter in to a lease of agricultural property containing whatever rent review procedure and rent formula the parties might agree. In practice such leases are very rare.

Agricultural Tenancies Act 1995

Section 9 of the Act permits the parties to agree that the rent shall not be reviewed, or that it shall be varied on a specified date or dates by or to a specified amount, or in accordance with a formula which does not preclude a reduction and does not require the exercise of judgment in its application.

Otherwise section 10 permits the parties to agree the frequency of rent reviews but, in the absence of such, provides for a three-year cycle. Either way it is a statutory condition precedent that a rent review notice be served by either party 12 to 24 months before the review date.

The amount of rent properly payable is defined in section 13 broadly as a market rent but disregarding the rental value of any tenant's improvements, the occupation of the tenant and any dilapidations. Any other basis of valuation is prohibited by the Act even if aimed to achieve a lower figure than the section 13 definition.

Section 12 provides that, if the parties agree subsequent to the statutory rent review notice, disputes can be referred for determination to a person acting otherwise than as an arbitrator. In the absence of such an agreement an arbitrator must be appointed, or an application made for such appointment to the President of the Royal Institution of Chartered Surveyors, before the rent review date. Such an arbitrator will act in accordance with the procedures in the Arbitration Act 1996.

Rent Properly Payable

The traditional method of valuation was to value field by field based on comparable evidence. This approach still holds good although for valuations under the Agricultural Holdings Act 1986 it is necessary to consider also a calculation of related earning capacity.

To date it is too early to judge what weight arbitrators under the 1986 Act will give to evidence analysed from farm business tenancies under the 1995 Act.

The comparable approach has been refined with different figures being attributed to the dwelling-houses, the land and in some cases the buildings.

The dwelling-houses need to be considered individually with special attention to the rights by which each are occupied.

Typically the farmhouse will be occupied by the farm tenant himself and often as a condition of the tenancy. The valuer will have to judge the quality and size of the house and consider its rental value against evidence analysed from similar lettings under the appropriate legislation for the rent review, whether the Agricultural Holdings Act 1986 or the Agricultural Tenancies Act 1995. While some regard might be had to evidence of rents of similar dwellings on assured or assured shorthold tenancies, arbitration awards suggest that such evidence carries little weight except possibly on small part time holdings.

Analysis of comparable lettings invariably reveal that farm dwellings occupied by current employees, rent free or subject to the deduction permitted under the Agricultural Wages Order, are valued to lower figures than those occupied by retired workers and their families, or non-employees, in those cases where the occupants are protected either by the Rent (Agriculture) Act 1976 or the Rent Act 1977 and can be charged a Fair Rent. Dwellings sublet on assured shorthold tenancies are analysed at higher figures still, perhaps based on 50% of the rent received or achievable by the agricultural tenant, but the practitioner needs to check whether such sublettings are subject to licences between the landlord and the agricultural tenant which specify how such values are to be calculated.

Agricultural land is valued in blocks based on evidence from similar blocks on comparable lettings. An entire holding may be regarded as one block where it is small or the valuer regards it as appropriate. More normally there will be several blocks based on factors including accessibility, geographical location, soil quality, cropping potential, irrigation abstraction licences and eligibility for arable area payments.

Usually farm buildings are included within the farmland rental value per hectare but the analysis of comparable evidence might reveal good quality modern framed buildings to a high specification, especially if that includes environmental control, being valued separately to a figure per square metre.

In analysing comparable evidence the valuer must have regard to the terms of the tenancies and must make adjustments where they differ from that of the subject holding. Obviously rents would normally be lower on those holdings where the tenant is required to reside in the farmhouse, or where the tenancy is full repairing or insuring rather than requiring the landlord to contribute or, in the case of tenancies under 1986 Act only, incorporates the model clauses (The Agriculture (Maintenance, Repair and Insurance of Fixed Equipment) Regulations SI 1973 No 1473). For farm business tenancies under the 1995 Act only, the valuer must consider whether there is evidence that differences such as the length of term, or the existence of tenant break clauses, or a restrictive user clause should require the comparable evidence to be adjusted.

The valuer must be alert also to holdings where the tenant receives payments which should be reflected in the rental valuation. Examples are those received from management agreements on sites of special scientific interest under the Wildlife and Countryside Act 1981 or from the letting of milk quota.

Study 2

Rental valuation of agricultural property

This illustrates the valuation of the rent properly payable under the Agricultural Holdings Act 1986 of a farm approached over an unusually long drive maintainable by the tenant comprising a farmhouse occupied by the tenant, four cottages in a variety of occupations and 168 hectares of land and buildings, including a grain store erected at the expense of the tenant.

1 Comparable Valuation

DWELLINGS

Name	Description	Occupier	Basis of occupation	Rent properly payable (£ pa)
Farm-house	Modern detached	The tenant		1,500
Cottage No 1	Edwardian detached	Current employee	Rent free	1,000
Cottage No 2	End terrace	Current employee	Rent free	500

Name	Description	Occupier	Basis of occupation	Rent properly payable (£ pa)
Cottage No 3	Mid terrace	Retired employee	Rent (Agriculture) Act	1,000
Cottage No 4	End terrace	Let away	Assured shorthold tenancy	1,500

Rent properly payable for dwellings   5,500

LAND AND BUILDINGS

Area ha	Description (All eligible for arable area payments except where otherwise stated)	per ha (£)	Rent properly payable (£ pa)
142	Grade 1 cereal land	175	24,850
6	Grade 3 A heavy soil	150	900
10	Grade 3 A gravel soil	150	1,500
6	Pasture ineligible for arable area payments	50	300
4	Sundries ineligible for arable area payments		
168			

Rent properly payable for land and buildings
before deductions   27,550
                    ie £164 per hectare per annum

Deductions

	Long drive maintainable by tenant Say	500	
	Tenant's improvement. Grain storage for 158 ha @ £22.5	3,555	
	Total deductions		4,055

Rent properly payable for land and buildings   23,495
                    ie £140 per hectare per annum

RENT PROPERLY PAYABLE   28,995
        Say   29,000
                    ie £173 per hectare per annum

2 Calculation of Related Earning Capacity

		(£)	(£)	(£)	(£)
GROSS MARGINS					
ha			per ha	per crop	
80	Winter wheat		820	65,600	
40	Winter barley		680	27,200	
20	Spring peas		580	11,600	
18	Winter beans		620	11,160	
158	Total gross margin				115,560
FIXED COSTS					
excluding rent and finance					
Labour			95	15,010	
Machinery and power					
Depreciation		85			
Fuel and oil		35			
Repairs, tax and insurance		58			
		178	28,124		
General Overheads			90	14,220	
Total fixed costs excluding rent and finance					57,354
Net farm income before rent and finance					58,206
Sundry income					
Rent from two cottages				5,250	
Other				1,125	
					6,375
					64,581
Related earning capacity					
Allow 50%				32,291	
deduct tenant's improvement					
Grain storage for 158 hectares @ 22.5				3,555	
RENT PROPERLY PAYABLE					28,736
Say					£29,000

Valuations of freehold agricultural properties let on tenancy

Freehold reversions in agricultural properties are offered for sale less frequently than vacant properties with the result that evidence for valuations is often hard to find. This is the case particularly for smaller properties, say of less than 200 hectares, and in those parts of the country where livestock or mixed farming predominates, as agricultural land investors have tended to favour large arable farms.

The majority of sales and valuations will continue to be of properties let on tenancies subject to the Agricultural Holdings Act 1986, which sell to a discount of as much as 66% from vacant values and show a yield of around 3½–6½% on the rent properly payable. However the practitioner will need to be alert to the those subject to the Agricultural Tenancies Act 1995 where market evidence is even more sparse.

Frequently such valuations are required for taxation purposes where the tenancy is between members of the same family and where the rent passing is lower than the rent properly payable.

The practitioner must study the terms of the tenancy agreement, particularly those which impose repairing or other liabilities on the landlord, and will need to consider the level of rent, the earliest date upon which it can be reviewed and the amount of the rent properly payable.

Agricultural investment valuations are often calculated on the gross rent without deductions for repairs, insurance and management but, whether valuing gross or net, the practitioner must ensure that the comparable evidence is analysed and applied on a consistent basis.

It is sensible to check any investment valuation of an agricultural property against evidence from elsewhere on a value per hectare basis and on a comparison with vacant possession value which is usually in the range from 33% to 50%.

Study 3

Valuation of freehold agricultural property let on tenancy

This illustrates the valuation of the freehold reversion in a farm let under a tenancy protected by the Agricultural Holdings Act 1986 where the rent passing is £193 pa and the rent properly payable is £14,915 pa. Valuation date October 25. Term date in tenancy October 11.

	Rental value (£ pa)	Capital value (£)
Rent passing	193	
YP for 1 year @ 4%	0.96	
Value of rent passing		185
Rent properly payable	14,915	
YP in perp deferred 1 year @ 5%	19.05	

Value of rent properly payable	284,131
TOTAL VALUATION	284,316
Say	284,300
ie £ 3258 per ha and 43% of vacant possession value of	£655,000

Valuations of agricultural tenancies

The overwhelming majority of agricultural tenancy agreements contain an absolute prohibition on assignment or subletting. Indeed, in respect of verbal tenancies created before September 1 1995, the Agricultural Holdings Act 1986 provides for such a clause to be incorporated if either party wish the tenancy to be committed to writing. Furthermore, in those few agreements which provide for such alienation subject to landlord's consent, in the absence of specific wording in the agreement, there is no statutory provision against that consent being withheld unreasonably.

As a result agricultural tenancies are sold rarely and any valuations must be based on other evidence.

Typically such valuations are required when an agricultural landlord wishes to buy his tenant out or the tenant wishes to surrender, or for compensation for compulsory purchase under section 48 of the Land Compensation Act 1973, or for taxation where there is a business, family or partnership connection between the landlord and tenant. Different approaches are required for each of these purposes.

The buy out or surrender approach is entirely a matter for the parties and is dependent more upon the relative negotiating strength of each rather than any valuation principles. At one extreme the tenant might be paid no more than his entitlement to compensation under the tenancy agreement and the statutory provisions applicable, whether under the Agricultural Holdings Act 1986 or the Agricultural Tenancies Act 1995. At the other, the tenant might be able to negotiate a percentage of the vacant possession value, particularly if he holds under an agreement granted before July 12 1984 and has family who qualify to succeed to the tenancy. More normally the parties will negotiate for the tenant to receive a proportion of the vacant possession premium, being the difference between the investment and vacant possession values of the property.

Section 48 Land Compensation Act 1973

Where all or part of an agricultural holding is compulsorily acquired the tenant is entitled to compensation for his interest and the Act directs that, in assessing the amount due, there shall be disregarded any right of the landlord to serve a notice to quit and any such notice already served.

In practice such valuations of tenancies protected by the Agricultural Holdings Act 1986 usually are based on a split of the vacant possession premium, although a different method was used in *Wakerley* v *St Edmundsbury Borough Council* [1979] 1 EGLR 19. Typically 50% of the premium is allowed as the value of the tenant's interest but this is dependent upon the facts of the case. For instance a lower percentage might be used if the tenant is elderly and without successors and a higher percentage in those unusual cases where the tenant is a limited company.

Partnership tenancies

The approach to be used in valuations for the purposes of Inheritance Tax or Capital Gains Tax, typically where there is a business or family connection between the landlord and tenant and the tenancy is an asset of such a partnership, depends upon the facts of each case (see *Valuation of Agricultural Tenancies for Taxation Purposes* – Central Association of Agricultural Valuers Numbered Publication 171, September 1997).

It was decided in *Walton's Exors* v *Commissioners of Inland Revenue* [1996] EGLR 189, CA that, where it can be proved that the freeholders were not interested in securing a surrender of the tenancy protected by the Agricultural Holdings Act 1986 nor had finance available to make a significant payment for surrender, such valuations should be based not on a proportion of the vacant possession premium but on four elements. The first is the capitalisation of the bottom slice of profit rent, between the rent passing and the rent properly payable as defined in the Act, over the period until the earliest date upon which the rent could be reviewed, at the relatively secure rate of say 5%. The second is the capitalisation of the top slice of profit rent, between the full rental value or tender rent and the rent properly payable, over three years being the shortest possible period between rent reviews under the Act, at a less secure rate of say 12%. To these figures is added any compensation due for tenant's improvements and a sum for tenant right.

The principle that such tenancies were assets to which value attached had been decided in the earlier Scottish case of *Baird's Exors* v *Commissioners of Inland Revenue* [1991] 1 EGLR 201 LTS when a value equivalent to 25% of the vacant possession premium was awarded.

Study 4

Valuations of agricultural tenancy

This illustrates the two methods of valuing a tenancy protected under the Agricultural Holdings Act 1986 of a farm comprising 56 hectares let at a passing rent of £1,200 per annum where the capital value of tenant's improvements is £5,000, tenant right £25 per hectare, dilapidations £2750, the rent properly payable £9,570 per annum, the open market rental value by tender £13,500 per annum, the vacant possession value £420,000, and the freehold value subject to the tenancy £168,000.

(a) Section 48 Land Compensation Act 1973 method

	(£)	(£)
Vacant possession value	420,000	
Investment value	168,000	
Vacant possession premium		252,000
Value of tenant's interest at 50%		126,000
	ie £ 2250 per hectare	

(b) Walton method	pa	
Rent properly payable	9,570	
Rent passing	1,200	
Profit rent (bottom slice)	8,370	
YP @ 5% for 1.75 years	1.6	
Value of profit rent (bottom slice)		13,392
Additional profit rent		
Tender rent	13,500	
Rent properly payable	9,570	
Profit rent (top slice)	3,930	
YP @ 12% for 3 years	2.4	
Value of profit rent (top slice)		9,432
Tenant's improvements		5,000
Tenant right £25 per hectare		1,400

	29,054
less dilapidations	2,750
Value of tenancy	26,474
Say	26,475
ie £472 per ha	

Valuation of milk quota

Milk quota was introduced by the European Commission on March 31 1984 to control the Community budget. Each milk producer was given a quota based usually on his 1983 level of production, but reduced successively by percentages therefrom, and risks being charged super levy if that quota is exceeded. Although quota is attached to land, and usually cannot be sold separately therefrom, a market for quota has developed with transfers being achieved, prior to September 1 1995, by means of grazing licences or 12 to 24 month tenancies which were outside the Agricultural Holdings Act 1986 and, subsequent to that date, by farm business tenancies under the Agricultural Tenancies Act 1995.

Milk quota attached to freehold agricultural properties with vacant possession

When a dairy farm is valued with vacant possession the milk quota is an important element to be considered and usually is valued separately to the freehold. Such valuations are depend upon the number of litres of quota, the percentage of those litres remaining unused in the current quota year and the butterfat percentage. The last two of these factors can be ascertained readily from the farm milk sale and Intervention Board records and the price per litre from market evidence or from reports in the Central Association of Agricultural Valuers News Letters.

While farm milk sales and Intervention Board records will reveal also the total number of litres of quota attached to the holding as defined in the Community legislation, the valuer needs to satisfy himself as to what volume attaches to the area he is valuing. If, as is invariably the case, other land to that being valued has been used for milk production within the previous five years an apportionment will have to be made, if necessary by an arbitrator, in accordance with the Dairy Produce Regulations then current and the wide definition of areas used for milk production in *Puncknowle Farms Ltd v Kane* [1985] 2 EGLR 8.

Milk quota attached to agricultural properties subject to a tenancy

When considering the value for either the landlord or tenant of a let agricultural holding to which milk quota attaches, the valuer must ascertain and apportion the total number of litres of quota to the area being valued in the same way as if the property was vacant.

While milk quota attaches to the land, and is invariably the property of the landlord, the tenant is nevertheless entitled to compensation for a proportion of that quota in certain circumstances and the valuation must allow for this. The valuer must consider whether the tenant is so entitled and, if so, the amount of that compensation which can fall under one or both of two headings where guidance was given in *Grounds* v *A–G of the Duchy of Lancaster* [1989] 1 EGLR 6, *Surrey County Council* v *Main* [1992] 1 EGLR 26, *Carson* v *Cornwall County Council* [1993] 1 EGLR 21 and *Creear* v *Fearon* [1994] 2 EGLR 12.

The first is for excess standard quota which is calculated on the number of litres by which the actual quota exceeds the standard laid down in the Milk Quota (Calculation of Standard Quota) Regulations then current. The area for calculating standard quota is limited to that used for the feeding of dairy cows.

The second is for the tenant's fraction of standard quota which depends upon the ratio between the rental value of the tenant's improvements and the sum of the rent properly payable for the holding and the rental value of the tenant's improvements all calculated at 1983 values under the Agriculture Act 1986.

Once the number of litres to which the tenant is entitled to compensation has been calculated a price per litre is then applied as for vacant properties.

Study 5

Valuations of milk quota

This illustrates the apportionment and valuations required when the milk producer, who occupies some of the land to which milk quota is registered under a tenancy protected by the Agricultural Holdings Act 1986 (Home Farm) and the remainder (The Marshes) as owner, quits that tenancy and sells the quota attached to his owned land.

(a) Apportionment under current Dairy Produce Quota Regulations

Total quota registered to the producer (litres)	385,576

Areas used by the producer for milk production (ha)

Home Farm	180.01
The Marshes	25.19
Total area	205.2

Therefore the apportionment of the total quota between ownerships pro rata to the areas used for milk production is (litres)

Home Farm	338,243
The Marshes	47,333
Total quota	385,576

(b) Valuation of milk quota attached to land owned with vacant possession

Value of quota attached to The Marshes (£)		
47,333 litres 100% used 4.1% butterfat		
@ per litre (£)	0.47	22,246.51
Say		22,247

(c) Valuation of compensation due to tenant under the Agriculture Act 1986

(i) Calculation of excess standard quota		
Area used for feeding and housing dairy cows in 1983 (ha) (from the Home Farm records)		32.12
Standard Quota (litres) from current from current Milk Quota (Calculation of Standard Quota) Order 7,140 litres per ha		229,337
Excess Standard Quota (litres)		
Quota apportioned to Home Farm	338,243	
less standard quota	229,337	
Excess standard quota		108,906
(ii) Calculation of tenant's fraction		
Rent properly payable for Home Farm in 1983 (£ per ha)	62.5 A	
Rental value of tenant's improvements on Home Farm in 1983		
(£ per hectare)	12.5 B	
therefore tenant's fraction	0.17	$\dfrac{B}{A+B}$

Applying tenant's fraction to standard
quota (litres)   38,987.29
Say   38,987

(iii) Total quota eligible for compensation (litres)

    Excess standard quota   108,906
    Tenant's fraction   38,987

                                                              147,893

(iv) Compensation due to tenant of Home Farm (£)
    147,893 litres 100% used 4.1% butterfat
    @ per litre (£)   0.47   69,509.71
Say   69,510

Valuations for compulsory purchase for highway purposes

The law of compulsory purchase as it applies to agricultural properties is little different to that which applies to property generally. Compulsory powers are frequently used in respect of agricultural properties, more often to acquire part, rather than the whole of the land concerned, typically for a new or widening of a highway: particular valuation problems arise.

While the practitioner will be faced with such valuations of property falling in one or more of three categories, being with vacant possession, subject to a tenancy or on behalf of a tenant, each involves valuations of the land taken, of severance and injurious affection and of disturbance.

The owner occupier's claim for land taken is relatively straight forward. However the calculation of severance and injurious affection to the land remaining can be complex where, post scheme, small areas may be isolated and access impeded, perhaps with the effects thereof being partially offset by accommodation works or betterment. Before and after valuation evidence reflecting such parameters invariably is lacking but the valuer may gain some assistance from compensation agreed elsewhere, usually expressed as a percentage of the before value.

The claim by the owner of land subject to a tenancy is partly dependant upon the reduction in the rent passing as a result of the acquisition, which should be agreed between the landlord, tenant and acquiring authority, and partly upon the reduction in future rental due to the reduced area and the effects of any severance and injurious affection.

The claim by an agricultural tenant is frequently the most complex, partly because of the absence of a market in tenant's interests and the resultant artificial nature of that valuation under section 48 of the Land Compensation Act 1973, but also because of the number of statutory or contractual heads of claim whether under the Agricultural Holdings Act 1986, the Agricultural Tenancies Act 1995 or the tenancy agreement. Once again the effect on the value of the tenant's interest of severance and injurious affection creates particular difficulty.

Significant disturbance damage can be suffered by owner occupiers or tenants during the construction process and such claims require attention to detail on the part of the claimant and his valuer from the earliest stages of the scheme to ensure that all are recorded and reported in writing to the acquiring authority's valuer before the evidence is lost.

Study 6

Valuations for compulsory purchase of part of agricultural property for highway purposes

(a) Freehold agricultural property mainly with vacant possession
1 Value of land taken

		(£)	(£)
Area acquired (ha)	2.01		
Value of land taken 2.01 ha @		5,415	10,884
Say			£10,900

2 Severance and injurious affection
A. Land taken for accommodation road

		(£)	(£)	(£)	(£)
Area lost to cultivation (ha)	2.21				
Value of land lost to cultivation 2.21 ha @				5,415	11,967

B. Land left with reduced access
Area affected (hectares)   142.45

PRE SCHEME VALUE OF LAND LEFT WITH REDUCE ACCESS
142.45 ha @   5,415   771,367

POST SCHEME VALUE OF LAND LEFT WITH REDUCED ACCESS
142.45 hectares @   5,035   717,236

Difference between pre and post
scheme values   54,131

> ie 7% of pre scheme value

C. Old farmhouse occupied by employee protected by Rent (Agriculture)
Act 1976

PRE SCHEME VALUE OF OLD FARMHOUSE   35,000
POST SCHEME VALUE OF OLD FARMHOUSE   29,000

Difference between pre and post scheme values   6,000

> ie 17% of pre scheme value

D. Waterlogging in OS 5454
0.12 ha Allow 40% of 5415 per ha   260

E. Cost of new drainage ditch in OS 1180   2,380

F. Replacing bricks to new culvert   100

G. Repairing accommodation fence   200

H. Regrading banks to stream   450

I. Refixing matting to ditch banks   2,000

Total severance and injurious affection   77,488

Say   £77,500

3 Disturbance   (£)   (£)

A. Loss of potatoes in OS 2584 on entry
   Area affected   1.31 ha
   Yield from remainder of field   34.82 tonnes per ha
   Price (£)   50   per tonne
Loss   2,281

B. Loss of wheat in OS 4398 on entry
   Area affected   0.11 ha
   Yield from remainder of field   7.05 tonnes per ha
   Price (£)   95   per tonne
Loss   74

C. Subsoiling compacted area in OS 4398   50

D. Remedial works to land drains   460

E. Land occupied by ground water pressure meters   50

F. Cost of signs to exclude public from accommodation road 200

G. Losses resulting from delays caused by restricted
access during construction based on *Wye College Farm
Management Pocketbook* 28th ed

Implement	Time lost	Cost per ha (£)	ha per 8 hour day	Total cost
Combine harvester	3	71	14	373

Implement	Time lost	Cost per ha (£)	ha per 8 hour day	Total cost		
FWD tractor with folding discs	3	22.5	20	169		
Tractor with seed drill	2	31.5	10	79		
				621		
H. Cost of redrawing IACS plans				700		
I. Management time spent re-organising affairs 184 hours @ 30				5,520		
Total disturbance					9,956	
Say					£10,000	

4 Surveyors fees on Ryde's Scale (1996)   (£)   (£)

For agricultural property on scale 2.3 table C.

Compensation (£)		10,900	
Cost of accommodation works as agreed (£)		47,670	
		58,570	
On first £	5,000		600
On next £	45,000 @ 2.5%		1,125
On balance of £	8,570 @ 1.875%		160.69

For disturbance, severance and injurious affection
to agricultural property on scale 2.5 table E

Severance and injurious affection (£)		77,500	
Disturbance (£)		10,000	
		87,500	
less for dwelling (£)		6,000	
		81,500	
On first £	5,000		750
On next £	45,000 @ 3.5%		1,575
On balance of £	31,500 @ 2.75%		866.25

For disturbance, severance and injurious affection
to private dwellings on scale 2.6 table F.

Severance and injurious affection (£)		6,000	330
Travelling (112 miles) and out of pocket expenses			70

Total of surveyors fees
   Note: The claimant is registered for value added tax   £5,476.94

5 Interest
To be calculated in accordance with section 52(a) Land Compensation Act
1973 as amended

(b) Freehold agricultural property subject to tenancy
1 Value of land taken   (£)   (£)   (£)
Area acquired (ha)   6.14

Rent passing for land taken (£ pa)
as agreed with tenant and acquiring
authority   475

Rent properly payable for land taken (£ pa)
6.14 ha @ 86.5   531

Freehold value of land taken

Rent passing (£ pa)	475		
YP for 1.5 years at 4%	1.424		
Value of rent passing		676.4	
Rent properly payable (£ pa)	531		
YP in perp deferred 1.5 years at 5%	18.594		
Value of reversion to rent properly payable		9,873.41	
			10,549.81
Value of land taken			
Say			£10,550

ie £1718 per ha

2 Severance and injurious affection   (£)   (£)   (£)

PRE SCHEME VALUE OF LAND RETAINED
Area retained (ha)   14.2

Rent passing for land retained (£ pa) as
 agreed with tenant and acquiring authority
  14.2 ha @ 75.07   1,066

Rent properly payable for land
retained (£ pa)
  14.2 ha @ 78.15   1,110

Freehold value of land retained

Rent passing (£ pa)	1,066		
YP for 1.5 years at 4%	1.424		
Value of rent passing		1,517.98	
Rent properly payable (£ pa)	1,110		
YP in perp deferred 1.5 years at 5%	18.594		
Value of reversion to rent properly payable		20,639.34	
Pre scheme value of land retained			22,157.32

POST SCHEME VALUE OF LAND RETAINED

A. Northern block
Area retained (ha)   2.7
Rent passing for land retained (£ pa)
  2.7 hectares @ 75   203
Rent properly payable for land retained (£ pa)
  2.7 hectares @ 84.3   228

Freehold value of land retained
Rent passing (£ pa)		203
YP for 1.5 years at 4%		1.424
Value of rent passing		289.07
Rent properly payable (£ pa)	228	
YP in perp deferred 1.5 years at 5%	18.594	
Value of reversion to rent properly payable (£)		4,239.43

Post scheme value of northern block	4,528.5

B. Southern block sandwiched between cliff and new road with very poor access

Area retained (ha)	11.5

Rent passing for land retained (£ pa)
11.5 ha @ 75	863

Rent properly payable for land retained(£ pa)
11.5 ha @ 72.15	830

Freehold value of land retained
Rent passing (£ pa)		863
YP for 1.5 years at 5%		1.406
Value of rent passing		1,213.38
Rent properly payable (£ pa)		830
YP in perp deferred 1.5 years at 6%		15.279
Value of reversion to rent properly payable		12,681.57

Post scheme value of southern block	13,894.95
Post scheme value of both blocks	18,423.45
Severance and injurious affection being the difference between pre and post scheme values	3,733.87
Say	£3,730

ie 17% of pre scheme value

3 Surveyors fees on Ryde's Scale (1996) for agricultural property

		(£)	(£)
For land taken on scale 2.3 table C.			
Compensation (£)		10,550	
Fencing agreed as accommodation works (£)		5,000	
		15,550	
On first £	5,000		600
On balance of £	10,550 @ 2.5%		263.75
For severance and injurious affection on scale 2.5 table E.			
Compensation (£)	3,730		720
Travelling (71.5 miles) and out of pocket expenses			45
			1,628.75
Value added tax @ 17.5%			285.03

Note: The claimant is not registered for value added tax.

Total of surveyors fees	£1,913.78

4 Interest
To be calculated in accordance with section 52(a) Land Compensation Act 1973 as amended

(c) Tenant's interest in agricultural property

1 **Value of Unexpired Term or Interest in Land**	(£)	(£)	(£)
S 20 (1) Compulsory Purchase Act 1965. (The 1965 Act) S 48 Land Compensation Act 1973. (The 1973 Act)			

Calculated at 50% of the difference between the value of the land acquired with vacant possession and the value subject to the tenancy.

Area acquired (ha)	2.12		
Value with vacant possession.			
2.12 ha @		4,800	10,176
Value as an investment.			
Rent passing for land taken (£ pa) as agreed with landlord and acquiring authority	160		
Rent properly payable for land taken (£ pa) 2.12 ha @ 85	180		
Investment value of land taken			
Rent passing (£ pa)	160		
YP for 1 year at 4%	0.962		
Value of rent passing		153.92	
Rent properly payable (£ pa)	180		
YP in perp def 1 year at 5%	19.048		
Value of reversion to rent properly payable		3,428.64	
Total investment value			3,582.56
Difference between vacant possession and investment values			6,593.44
Allow 50%			3,296.72
Deduction under s 48 (5) of the 1973 Act. 4 times the rent reduction of	160		640.00
Value of unexpired term or interest			2,656.70
Say			£2,660

2 **Severance and injurious affection**	(£)	(£)	(£)
S 20 (2) of the 1965 Act.			

As a result of the acquisition the southern portion of field OS 8474 was left with substandard access.

Area of OS 8474 severed (ha)	5.03		
VALUE OF UNEXPIRED TERM OR INTEREST PRE SCHEME			
Vacant possession value			
5.03 ha @		4,200	21,126

Rent passing for land retained (£ pa)				
5.03 ha @	75.12	378		
Rent properly payable for land retained (£ pa)				
5.03 ha @	78.15	393		
Investment value of land retained				
Rent passing (£ pa)		378		
YP for 1 year at 4%		0.962		
Value of rent passing			363.64	
Rent properly payable (£ pa)		393		
YP in perp def 1 year at 5%		19.048		
Value of reversion to rent properly payable			7,485.86	
Total investment value				7,849.50
Difference between vacant possession and investment value				13,276.50
Value of unexpired term or interest pre scheme allow 50%				6,638.25

VALUE OF UNEXPIRED TERM OR INTEREST POST SCHEME

Vacant possession value				
5.03 ha @		3,000		15,090
Rent passing for land retained (£ pa)				
5.03 ha @	75.12	378		
Rent properly payable for land retained (£ pa)				
5.03 ha @	60	302		
Investment value of land retained				
Rent passing (£ pa)		378		
YP for 1 year at 4%		0.962		
Value of rent passing			363.64	
Rent properly payable (£ pa)		302		
YP in perp def 1 year at 5%		19.048		
Value of rent properly payable			5,752.50	
Total investment value				6,116.14
Difference between vacant possession and investment values				8,973.86
Value of unexpired term or interest post scheme allow 50%				4,486.93
Total severance and injurious affection being difference between pre and post values				2,151.32
Say				£2,150

ie 34% of the pre scheme value

3 Just allowance by an incoming tenant   (£)   (£)   (£)
S 20 (1) of the 1965 Act.

Sch 8 Agricultural Holdings Act 1986. (The 1986 Act)
The Agriculture (Calculation of Value for Compensation) Regulations SI 1978 No 809.

A. Liming 1988. Para 4 SI 1978 No 809
25 tonnes per ha of chalk at £12.28 per tonne

on 2.12 ha ie costing		650.84	
Three crops off Allow	0.57		370.98

B. Residual value of artificial manures. Para 5 SI 1978 No 809.

1997 Barley crop.
    125 kg per ha 20.10.10 fertiliser on
    2.12 ha ie 2.65 tonnes

10% phosphate @ per tonne One crop off		15.8	41.87
10% potash @ per tonne One crop off		9.2	24.38

1996 Wheat crop.
    100 kg per ha 8.20.16 fertiliser on
    2.12 ha ie 2.12 tonnes

20% phosphate @ per tonne Two crops off		7.9	16.75
16% potash @ per tonne Two crops off		4.6	9.75

1995 Wheat crop.
    250 kg per ha 0.20.20 fertiliser
    on 2.12 ha ie 5.3 tonnes

20% phosphate @ per tonne Three crops off		3.9	20.67

C. Acts of Husbandry. Para 8 SI 1978 No 809.
    2.12 ha Two post harvest cultivations calculated
    on Central Association of Agricultural Valuers

costings		8.25	34.98

D. Labour to farm yard manure. para 7 Sch 8 the 1986 Act.
    35 tonnes per ha on 2.12 ha
    7.42 10 tonne loads each taking 30minutes
    to transport and spread

3.71 hours Driver time @ per hour		6.25	23.19
3.71 hours Tractor time @ per hour		5	18.55
3.71 hours Spreader time @ per hour		5	18.55

Unexhausted manurial values. Para 7 SI 1978 No 809.

74.2 tonnes FYM No crop off say		2.5	185.5
74.2 tonnes FYM One crop off say		1.25	92.75
74.2 tonnes FYM Two crops off say		0.63	46.75

Total of just allowance by an incoming tenant	904.67
Say	£905

4 Any other loss or injury   (£)   (£)   (£)
S 20 (1) of the 1965 Act.

A. Temporary loss of access
The access and gateway to the severed land was not available throughout 1996 and 1997.
Entry for each operation thus involved a long detour

20 operations in all each involving an extra 10 minutes ie 5 minutes each way.			
3.33 hours Driver time @ per hour		6.25	20.81
3.33 hours Tractor time @ per hour		5	16.65
3.33 hours Implement time including combine @ per hour		5	16.65

B. Permanent fencing
It was agreed that the claimant would replace the fence to the old road

255 metres @ per m		5	1,275

C. Damage to internal fences
The DOT made no attempt to secure internal fences when these were cut resulting in them collapsing

West fence in OS 8474 to OS 6274		
35 metres @ per m	4	140
East fence in OS 8474 to OS 3200		
215 metres @ per m	4	860

D. Alternative gateway
The access between OS 8474 and 3200 was in the area acquired immediately beside the old A20.

Allow for new gate, posts and erection	200

E. Damage to temporary fence
The temporary fence erected by the acquiring authority to the old road was substandard and had to be repaired by the claimant

Allow	25
F. Redrawing IACS plans	50

G. Claimant's time
Both the claimant and his son had to spend much time reorganising their affairs

Allow	500
Total of any other loss or injury	3,104.11
Say	£3,104

5 Additional compensation
S 60 (2) of the 1986 Act.

4 times the rent reduction of	160	£640

6 Surveyor's fees on Ryde's Scale (1996) for agricultural property
For land taken on scale 2.3 table C.

Compensation	£2,570	538

Addition for leasehold interest on scale 2.2 table B

Annual Rent	160	50

For severance, injurious affection and disturbance on scale 2.5 table E.

Compensation	£6,259	
First	5,000	750
Balance	1,259 @ %3.5	44.07
Travelling (45 miles) and out of pocket expenses		30
Total surveyors fees		£1,412.07

Note: The claimant is registered for value added tax.

7 Interest
To be calculated in accordance with s52(a) Land Compensation Act 1973 as amended.

Valuations for capital gains tax

On an owner occupied agricultural property, or a freehold reversion where there is no business or family relationship between landlord and tenant, capital gains tax valuations are relatively straight forward.

In those cases where the disposal was not at arm's length a current valuation may be needed of the asset concerned.

For purposes of indexation, valuations will often be required at March 31 1982 if the asset disposed of was originally acquired before that date. Evidence of 1982 values may be difficult to obtain but *Farmland Market* (published by *Farmers Weekly* in association with the Royal Institution of Chartered Surveyors) can be a useful source of such information.

Often capital gains tax valuations of agricultural properties are needed as a result of part disposals from a larger property. In such cases valuations may be required of the property remaining at the date of disposal and at March 31 1982. On large properties this results in a major valuation exercise following a relatively minor property disposal.

Significant complications can arise when such valuations are required of an agricultural property which is let, either currently or at March 31 1982 or at both dates, if the transferor has an interest both in the freehold reversion and the tenancy, particularly if the disposal is to a party connected to the transferor. An example of such a case might involve a transferor, who previously farmed as an owner occupier, granting a tenancy under the Agricultural

Holdings Act 1986 to himself and other members of his family, thereafter farming in partnership with those same people and finally gifting all or part of the freehold to one or more of them.

Study 7

Valuations of agricultural property for capital gains tax

This illustrates the valuations required when a farmhouse, previously part of a large farm owned by a family trust since the 1960s and let since then under a tenancy protected by the Agricultural Holdings Act 1986 to tenants of the family farming partnership, is gifted with vacant possession to one of those partners.

As the gift comprises both the freehold reversion and the agricultural tenancy in the farmhouse, valuations will be required of both at the date of disposal and at March 31 1982.

The trust property comprises the farmhouse, two other dwellings, farm buildings and 63.13 hectares of land let at a rent of £1,656 per annum which has remained unchanged since before 1982. Date of gift November 16. Term date of tenancy October 11.

Rental Valuations

DWELLINGS	@ 31st March 1982		@ disposal date	
	Rent properly payable (£ pa)	Tender rent (£ pa)	Rent properly payable (£pa)	Tender rent (£ pa)
The farmhouse	1,600	3,500	2,500	5,000
The manager's house	600	750	1,000	1,000
The worker's cottage	400	500	750	750
BUILDINGS	500	650	500	650

LAND

All eligible for arable area payments except where otherwise stated

Area (ha)	Description	per ha	per ha	per ha	per ha	per ha	per ha	per ha	per ha
54.37	Grade 1	132.25	7,190	172	9,352	192.5	10,466	230	12,505
1.01	Grade 2	108.25	109	140	141	156.5	158	190	192
6.99	Grade 5 (ineligible)	12	84	16	112	25	175	30	210
0.76	Sundries								

63.13	Total		10,483		15,005		15,549		20,307
	Say	166	10,480	238	15,010	246	15,550	322	20,310

Valuations of freehold @ disposal date (£)

THE FARMHOUSE

Rent passing (£ pa)	25	
apportioned at		
YP @ 5 % for 1.8 years	1.63	41
Value of rent passing		
Rent properly payable (£ pa)	2,500	
YP in perpetuity deferred 1.8 years @ 6.5 %	13.71	34,275
Value of freehold reversion in the farmhouse		34,316
Say		34,320

THE ENTIRE PROPERTY

but excluding the farmhouse from the valuation at the disposal date

	@ 31st March 1982			@ disposal date	
	(£)	(£)		(£)	(£)
Rent passing (£ pa)	1,656			1,631	
YP for 1.5 years @ 4%	1.43		1.8 years @ 5%	1.63	
Value of rent passing		2,368			2,659
Rent properly payable (£ pa)	10,480			13,050	
YP in perpetuity deferred					
1.5 years @ 5%	18.6		1.8 years @ 6.5%	13.71	
Value of rent properly payable		194,928			178,916
Value of freehold reversion in entire property but excluding the farmhouse from the valuation					
at the disposal date		197,296			181,575
Say		197,300			181,580

Valuations of tenancy

THE FARMHOUSE

	@ disposal date	
	(£)	(£)
Rent properly payable (pa)	2,500	
Rent passing (pa)	25	
Profit rent (bottom slice)(pa)	2,475	
YP for 1.5 years @ 5(%)	1.4	
Value of profit rent (bottom slice)		3,465
Tender rent (pa)	5,000	
Rent properly payable (pa)	2,500	
Profit rent (top slice)(pa)	2,500	
YP for 3 years @ 12(%)	2.4	
Value of profit rent (top slice)		6,000
Tenant's improvements		0

Tenant right				0
				9,465
less dilapidations				0
Value of tenancy				9,465
Say				9,470

THE ENTIRE PROPERTY
but excluding the farmhouse from the valuation at the disposal date

	@ 31st March 1982		@ disposal date	
	(£)	(£)	(£)	(£)
Rent properly payable (pa)	10,480		13,050	
Rent passing (pa)	1,656		1,631	
Profit rent (bottom slice)(pa)	8,824		11,419	
YP for 2 years @ 5%	1.84	1.5 years @ 5%	1.4	
Value of profit rent				
(bottom slice)		16,236		15,987
Tender rent (pa)	15,010		15,310	
Rent properly payable (pa)	10,480		13,050	
Profit rent (top slice)(pa)	4,530		2,260	
YP for 3 years @ 12%	2.4	3 years @ 12%	2.4	
Value of profit rent (top slice)		10,872		5,424
Tenant's improvements		0		0
Tenant right say		1,560		1,560
		28,668		22,971
less dilapidations		0		0
Value of tenancy excluding				
the farmhouse from the				
valuation at the disposal date		28,668		22,971
Say		£28,670		£22,970

Valuations for inheritance tax

As for capital gains tax on an owner occupied agricultural property, or a freehold reversion where there is no business or family relationship between landlord and tenant, inheritance tax valuations are relatively straight forward and indeed are free of the complications of requiring 1982 or part disposal values.

However, if agricultural property relief is being claimed, there may be doubts as to whether all the property is occupied for agriculture as defined in section 117 of the Inheritance Tax Act 1984, or is agricultural property as defined in section 115(2) with care being needed in deciding whether the cottages, farm buildings and farmhouses are of a character appropriate to the property being valued, or a discrepancy between the value eligible for relief as defined in section 115(3) and the market value.

Again such valuations become complex when required of an agricultural property which is let if the transferor has an interest both in the freehold reversion and the tenancy.

A brief background to the legislation will be found in the Country Landowners Association *Inheritance Tax Agricultural Property Relief* (2nd ed, CLA T8/95, October 1995).

Study 8

Valuations of agricultural property for inheritance tax

This illustrates the valuations required on the death of an individual who was part owner of the freehold of a farm let on a tenancy, protected under the Agricultural Holdings Act 1986, to himself and others as members of a family farming partnership at a passing rent of £2,000 pa.

The farm comprises the farmhouse, four other dwellings and 88.79 hectares of land and buildings and including planning permission to convert one of those buildings to a dwelling.

Date of death December 13. Term date of tenancy October 11.

Note: The valuations below will need to be adjusted to reflect the deceased's share in the freehold and tenancy.

Rental valuations

DWELLINGS

Name	Description	Occupier	Status of occupation	Rent properly payable (£ pa)	Tender rent (£ pa)
The Farm-house	Period detached	The transferor and spouse	As joint agricultural tenant	2,500	10,000
Willow House	Edwardian detached	A partner	As joint agricultural tenant	1,500	3,500
1 Oak Cottages	Victorian terraced	A farm employee	Rent (Agriculture) Act 1976	750	750

2 Oak Cottages	Victorian terraced	Non employee	Assured shorthold tenancy	1,500	2,000
3 Oak Cottages	Victorian terraced	Non employee	Assured shorthold tenancy	1,500	2,000

LAND

Area ha	Description			
59.72	Grade 1 to 3 arable eligible for arable area payments rent properly payable (£ per ha pa)	100	5,972	
	tender rent (£ per ha pa)	220		13,138
29.07	Pasture ineligible for arable area payments rent properly payable (£ per ha pa)	50	1,454	
	tender rent (£ per ha pa)	125		3,634
88.79	Total		15,176	35,022
	Say		15,180	35,020
	ie per hectare		171	394

Market valuation of freehold	(£)	(£)
Rent passing (pa)		
	2,000	
YP for 1.75 years @ 4%	0.96	
Value of rent passing		1,920
Rent properly payable (pa)	15,180	
YP in perp def 1.75 years @ 5%	13.79	
Value of reversion to rent properly payable		209,332
Value of planning permission for conversion of barn to dwelling		50,000
Market value of freehold		261,252
Say		261,300
ie £2,943 per ha		

Valuation of freehold eligible for agricultural property relief

	(£)	(£)
Rent passing (pa) adjusted to exclude Willow House (being a dwelling in excess of character appropriate to the property) and numbers 3 and 4 Oak Cottages (neither being occupied for the purposes of agriculture) Say	1,400	

YP for 1.75 years @ 4 %	0.96	
Value of rent passing		1,344
Rent properly payable (pa)		
excluding Willow House and nos. 3 and 4		
Oak Cottages	10,680	
YP in perp def 1.75 years @ 5%	13.79	
Value of reversion to rent properly payable		147,277
Value of freehold eligible for agricultural		
property relief		148,621
Say		148,600

ie £1,674 per ha

	(£)	(£)
Market valuation of tenancy		
Rent properly payable (pa)	15,180	
Rent passing (pa)	2,000	
Profit rent (bottom slice) (pa)	13,180	
YP @ 5% for 1.75 years	1.6	
Value of profit rent (bottom slice)		21,088
Tender rent (pa)	35,020	
Rent properly payable (pa)	15,180	
Profit rent (top slice) (pa)	19,840	
YP @ 12% for 3 years	2.4	
Value of profit rent (top slice)		47,616
Tenant's improvements		0
Tenant right at Say £25 per ha		2,220
		70,924
less dilapidations		0
Value of tenancy		70,924
Say		70,900

Valuation of tenancy eligible for agricultural property relief

	(£)	(£)
Rent properly payable (pa)		
excluding Willow House and nos. 3 and 4		
Oak Cottages	10,680	
Rent passing (pa)		
excluding Willow House and nos. 3 and 4		
Oak Cottages	1,400	
Profit rent (bottom slice) (pa)	9,280	
YP @ 5% for 1.75 years	1.6	
Value of profit rent (bottom slice)		14,848
Tender rent (pa)		
excluding Willow House and nos. 3 and 4		
Oak Cottages	27,520	

Rent properly payable (pa)		
excluding Willow House and nos. 3 and 4		
Oak Cottages	10,680	
Profit rent (top slice) (pa)	16,840	
YP @ 12% for 3 years	2.4	
Value of profit rent (top slice)		40,416
Tenant's improvements		0
Tenant right at Say £25 per ha		2,220
		57,484
less dilapidations		0
		57,484
Value of tenancy		
Say		57,500

Further reading

Muir, Watt & Moss *Agricultural Holdings* (14th ed), Sweet & Maxwell, 1998

Scammell and Densham's *Law of Agricultural Holdings*, (8th ed) Butterworths, 1997

Chapter 2

Residential Properties

The long history of legislation in respect of residential property, restricting rent and providing security of tenure, combined with changes in the economic structure of the country, has moved residential property into a category of its own, with a valuation approach entirely different from other types of property. Unlike the market for commercial property, there are two separate and distinct markets, purchase and rental, and there is no fixed correlation between them. All valuations must be carried out with an eye on the vacant possession value of the property and, except in the very rarest cases, this is the maximum value which can be put on a property, no matter how high or how low a yield it would seem to give.

The vacant possession value for the purposes of this chapter may be defined as the price at which the house or flat could be sold to an owner-occupier in the open market, the underlying assumption being that this is the maximum value. It is therefore made on the assumption that the property is in such a state of repair and of such a nature as to be capable of being mortgaged to a building society or similar lending institution. This definition is applied to a single unit of accommodation and not to an estate of houses or to a block or blocks of flats taken as a whole. The reason for this restricted definition is that there should be a distinction both from properties which require emendation (in terms of conversion, repair or change in design) and from properties which require a "wholesale" as opposed to a "retail" consideration.

The Rent Act 1977 (as amended by the Housing Act 1980), the Housing Act 1988 and the Housing Act 1996 are the principal Acts of Parliament giving security of tenure to tenants and limiting the rent payable in respect of residential property, whether furnished or unfurnished (provided that the letting comes within the ambit of the Acts). There are several types of tenancy under the Acts, *viz*:

1. Tenancies created prior to January 1989
   (i) Regulated
   (ii) Secure

2. Tenancies created after January 1989
   (i) Assured Shorthold
   (ii) Assured

Rateable Value (RV) limits are still applicable to Rent Act tenancies but these can only now exist if the letting was created before this system of local taxation was abolished, ie created before January 1989. The limits are £1,500 in Metropolitan London and £750 elsewhere, as at March 31 1990, whether furnished or unfurnished. As rateable values are no longer the basis for local taxation, these RV limits will now remain constant with no future changes. Theoretically, all but the very best houses and flats are capable of being within rent control or having tenants with security of tenure. There will be no new RV limits or any substituted system, because a tenancy which is entered into after the commencement of the Housing Act 1988 cannot be a protected tenancy (except in the special cases set out by section 34 of that Act). The Act came into force in January 1989.

The four types of tenancies referred to above are briefly described as follows:

1. *Regulated*

This is a tenancy to which the Rent Act 1977 applies. It provides security of tenure for the tenant and up to two successors. It also restricts the rent which can be charged by a landlord to a "fair rent", which is fixed by a Rent Officer or, on appeal, by a Rent Assessment Committee; a successor spouse to the tenant has the same protection as the tenant, but a non spouse successor becomes an Assured Tenant (see later). Within the definition of "fair rent" there is the assumption that the supply of residential accommodation within the area is in approximate balance with the demand for such accommodation. This assumption removes the "scarcity factor" from the normal working of the market and thus ensures a rent which is lower than the open market rent, except in locations where no scarcity of rented accommodation exists.

2. *Secure*

A secure tenancy is one which exists in the public sector and is governed by sections 28–61 of the Housing Act 1980 and the

Housing Act 1985. Tenants in the public sector now enjoy security of tenure, subject to certain exceptions. Because this type of property has no effect on the valuation of residential property in the open market no more will be said of it.

3. *Assured Shorthold*

These were the creation of section 20 of the Housing Act 1988 and were tenancies:

(i) granted for a term certain of not less than six months,
(ii) in respect of which there was no power for the landlord to determine the tenancy at any time earlier than six months from its beginning, and
(iii) in respect of which a notice was served by the person who was to be the landlord on the person who was to be the tenant stating that the tenancy was to be a shorthold tenancy.

All residential tenancies granted after February 28 1997 are now Assured Shorthold tenancies unless there is a written agreement which unambiguously creates an Assured Tenancy (Housing Act 1996); thus there is now no need for a pre-tenancy notice. Because of this change, it is necessary to check the date when the tenancy came into existence. If the tenancy was created before February 28 1997, then in the absence of a written agreement, and pre-tenancy notice such a tenancy would either be a Regulated Tenancy or an Assured Tenancy.

In summary therefore all tenancies entered into before January 1989 would be Assured Tenancies unless specifically created as an Assured Shorthold Tenancy and there was a pre-tenancy notice and unless the other conditions given at the beginning of this section were complied with. If the same tenancy was created after 1997, then such a tenancy would automatically be an Assured Shorthold Tenancy unless there was a written agreement specifically making it an Assured Tenancy.

In this type of tenancy the tenant will not have security of tenure and only a limited degree of control of the rent; the latter would not help the tenant very much because he/she could be evicted within a short term.

4. *Assured*

These are the creation of section 1 of the Housing Act 1988 and, briefly stated, are tenancies where the tenant has full security of tenure but there is a limited control on the rent. They resemble commercial tenancies which are within the control of Part II of the Landlord and Tenant Act 1954 and as stated above must now be specifically created.

The various types of tenancies will be considered in greater detail later in this chapter. However, in view of the statement given above that all valuations must be carried out with an eye on the vacant possession value of the property, it is clear that it is the security of tenure provisions which have the greatest impact on the valuation of a property if it is subject to a tenancy. It should also be borne in mind that there are a number of other tenancies which are outside the control of either the Rent Acts or the Housing Act 1988 and these include company lettings, holiday lettings, service tenancies, licences and those where there is a large measure of service.

A further factor which distinguishes residential property from commercial property is the different taxation treatment which is given to it. An owner-occupier is specially favoured from the point of view of capital gains tax, as any increase in the value of his principal home is free from taxation. It is worth noting the expectation of capital growth on the sentiment of persons in the market for residential accommodation. It is usually this which determines whether they will become tenants or owner-occupiers.

The net cost of interest to a basic rate taxpayer of a 100% mortgage on a £100,000 house (interest only) at say 8.5% is £8,500 pa. Assuming that the house increases in value over 10 years by an average of 5% pa compound, the value of the house in 10 years' time will be £162,890. Therefore the total net cost over 10 years to the 22% basic rate taxpayer will be as follows:

	£
Net annual cost	8,500
Amount of £1 pa for 10 years at 8.5%	14.84
Total interest cost	126,140
Less: Capital gain over 10-year period	62,890
True net cost	63,250
Annual equivalent of true net cost (£63,250 ÷ 14.84)	£4,260

A tenant of the same property will not make any gain over the period of the tenancy. Such a tenant should therefore be prepared to pay only a rent for the property which equates to an average of £4,260 pa over the 10-year period, if he is to be in the same position as his neighbour who purchased the house with the aid of a mortgage. Such a rent would be totally unacceptable to an investor, since over the 10-year period he would see an average yield of only 4.26%. This would obviously be unacceptable at times of interest rates at say 6–7%, unless the investor could obtain vacant possession at the end of the 10-year period. If the latter were possible then the investor's average yield over the 10-year period would be 9.26% before tax. The investor's after tax position is complicated by the fact that part of the capital gain will be tax free being that part above the rate of inflation or will be reduced by tapering relief.

Before considering this matter further, some comment must be made on the state of the housing and mortgage market at the time of writing this chapter, which is the spring of 2000, and the historical context in which it is set.

A recession overtook the housing market throughout the United Kingdom in about the spring/summer of 1989. This recession had its beginning on August 1 1988 which is the date the then Conservative government removed multiple income tax relief and substituted a single limited tax relief for each property. This fiscal change was published in March 1988 to come into effect after the end of July 1988 and this in turn led to a "bunching" in demand during that half-year period, resulting in a massive increase in capital values; at the same time, interest rates were falling generally. The "bunching" in demand led to a much quieter housing market in the autumn of 1988 and the government then attempted to reduce inflationary pressures in the economy by increasing interest rates, so that they doubled over a period of less than half a year. These two factors led to fall in capital values which was then exacerbated by the Kuwait War, and was then followed by a general recession in both the national and in the international economies. Capital values continued to fall and, with the exception of one or two false starts, they continued to fall until about the middle of 1996. The fall in values from the peak to the troughs was an average of approximately one-third although in some areas, the fall was much greater; London and the south east were the hardest hit.

As already stated, the housing market began a recovery from about the beginning of 1996 and significant price rises were recorded in the period 1997–2000 (1999 being a boom year). This recovery was due to a combination of very low interest rates and the general recovery in the economy as a whole. Government also became very aware of the importance of a strong housing market for the stability of the general economy, and hence there is now a general political philosophy to the effect that the housing market must be maintained with both main parties wishing to see a gentle but sustainable increase, year on year, of the market as a whole. The government changed in the spring of 1997 and the new Labour government has increased interest rates and increased stamp duty levels; with the latter changes only affecting the higher tier values. This change in fiscal policy has not yet had a material effect on the market and the generally held view is that the recovery will be sustained and that prices will continue to increase at a modest year on year rate. As at the date of writing the market had not only recovered to its pre-recession peak levels but has significantly risen above it; however this is not the case in all areas of the country.

The underlying hypothesis of this chapter is that mortgage interest rates will remain within the range of 5%–9% over the course of the foreseeable future and that house prices will rise at a moderate and sustainable rate. This has been the trend since the second world war during which time the market has seen many peaks and troughs and gyrating interest rates. The rate of increase in value has varied according to location within the country, but the average increase between the end of 1973 and the end of 1997 for all properties in the United Kingdom has been approximately 8% pa compound which is almost exactly in line with the rate of inflation. A more detailed picture is as follows:

	House Prices	(Cost of Living)
1983–1988	15.67% pa compound	(4.89%)
1983–1993	7.06% pa compound	(5.17%)
1993–1997	2.05% pa compound	(3.05%)
1997–1999	8.62% pa compound	(2.22%)
1983–1999	6.10% pa compound	(4.20%)

The figures given in brackets are the increases in the cost of living index over the same period. The house prices are calculated from statistics prepared by the Halifax from the average purchase price per sq ft of superficial floor space (excluding external walls, garages and outbuildings) weighted by region. It should also be noted that in the years 1973–1978 average earnings rose at about the same rate as average inflation, and house prices rose at a slower rate. In the years 1978–1988 average earnings increased by more than the average rate of inflation, leading to house prices rising faster than the rate of inflation, but in the years 1988–1997 average earnings also rose faster than the cost of living but house prices rose at a slower rate. However after 1997 house prices rose considerably faster than both average earnings and prices.

It is clear from the above analysis that the assumption that house prices will continue to rise over the foreseeable future, with perhaps short-term periods of stagnation or even falls, is a reasonable hypothesis, but over the long term there is no guarantee that the rate of increase will outpace the cost of living.

The adoption of a mortgage rate in our calculations of 8.5% is justified by consideration of mortgage rates between 1973 and 2000, when the lowest variable rate was 7.44% and the highest rate was 15.4%.

The percentage of households which owned their own houses, as opposed to renting them, rose continuously from the turn of the century until about 1989 when it got very close to 70%. The reasons for this increase were a combination of the prospects for large capital gains the reduction in both the public and private sector supply of housing to rent and the easy availability of mortgage finance. The second of these was caused by the "right to buy" policy of the Conservative government which reduced the number of local authority units (the proceeds could not by statute be re-invested in the building of new council housing) and the private sector was affected by security of tenure and rent control legislation. Tenants and potential tenants were encouraged to buy by freely available mortgages of up to 100% of the purchase price at low interest rates.

As stated above, the "bubble burst" during 1989 and in the subsequent years this led to the creation of a new term in the housing vocabulary of "negative equity". The high percentage of loan taken by many and rising interest rates, combined with falling prices led to many owner-occupiers having mortgages which were greater than the capital value of their property. Many homes were repossessed by the banks and building societies and many families became disenchanted with the concept of owner occupation. For an unprecedented period of years, many owner-occupiers found themselves unable to move because they could not sell their properties and this led to a restriction on the mobility of labour at a time when unemployment was increasing rapidly.

For the first time in at least two generations (and possibly three), many people, both sophisticated and not sophisticated, saw the rental market as providing a more satisfactory home than the owner-occupier market. Some families will never recover from the trauma of losing their home and their capital, and will hereafter wish to rent rather than buy; others believe that the advantages of mobility and the opportunity of capital growth in other investment markets overrides the disadvantages of renting.

The changes in the law concerning security of tenure and rent control has also led to the expansion of the private sector market for letting, so that the demand for good quality housing can now be met by the supply of such housing, albeit as a short term lettings only. The economics of renting, as opposed to buying, as set out above still apply so that a significant part of the return to the investor will remain as capital gain as opposed to just rent. The likely net result is that the headline rent will be lower than the opportunity cost of capital but the cost of renting over any significant period of time will still be greater than the annualised cost of purchase.

There will also be the social factor, which is that in all but a few cases lettings will be on a short term basis, and thus the tenants will not have security of tenure. This in turn will prevent them from investing in their own homes so that for the average property, there will have to be the social disadvantage of living with somebody else's fixtures and fittings in so far as this applies to kitchens, bathrooms etc. The properly advised investor will not allow tenants to have security of tenure because of their need to retain the ability to make capital gains by disposals, and these will only be fully available by the sale of vacant properties. This situation will only change if residential tenants are prepared to take on long term rental commitments with frequent rent reviews, and where such tenants are of good covenant. In the view of the writer, this is unlikely to happen because the principal advantage of renting, as opposed to buying, is the benefit of mobility which would be lost if a long term commitment had to be taken.

It will therefore be clearly apparent that the rent which can practically and economically be paid by a tenant will be significantly below the amount which an owner-occupier can afford to pay by way of interest on a mortgage. At this level of rent no landlord can afford to let unless he is guaranteed that he can obtain vacant possession after the letting term (in order to secure a capital gain).

The above analysis has demonstrated that if an investor can get vacant possession after, say, 10 years, he will have secured a real rate of return before tax of 9.26% assuming that his tenant equates himself with purchaser. This net yield compares with gilt-edged investments, where a 10-year stock is currently yielding about 6%.

The situation is, however, entirely different if vacant possession is not obtainable at the end of the term. We will assume that a higher rent than £4,260 pa could be extracted from a tenant, notwithstanding that this would make renting more expensive than purchasing if vacant possession values rose by 5% pa compound. For the purposes of this example, we will assume a rent of £5,000 per annum and we will also assume that this rent would rise annually in accordance with increases in the cost of living, which we will take at 3% pa. On this basis, the rental value after 10 years would have increased to £6,720 pa.

The true performance of this investment, provided that vacant possession was available at the end of the 10-year period, would have been an average running yield at a purchase cost of £100,000 at 5.86% plus capital growth of 5% giving a total average yield of 10.86%. However, if vacant possession was not available at the end of the term, all that would be receivable at that time would be the increased rent of £6,720 pa which would not only provide an unacceptably low yield but also, in all probability, a capital loss both in real and in nominal terms. (No allowance has been made so far for landlord's outgoings and the rents utilised have been assumed to be net rents).

Without the ability to sell the property with vacant possession at the end of the first lease, the yield on the investment is quite unacceptable, as the capital growth which was taken into account in estimating the yield when comparing it with a gilt-edged stock would no longer be present. It is true that the rent would increase over the term but this would only mitigate the loss, not reduce it.

There are, therefore, two conclusions from the above analysis:

(a) Security of tenure if given to a tenant is disruptive of capital value, as an acceptable return on the investment can only be obtained if there is a reversion to vacant possession value.

(b) That owing to the existence of capital growth the average cost of occupation over any period is significantly lower than the interest which is actually being paid. The result of this is that tenants should also pay a rent which is considerably lower than what at first view would be a fair return on the value of the property they occupy.

It will be appreciated from the above that for all property which is saleable with vacant possession to an owner-occupier the valuation approach must be geared to the differential between the vacant and investment values; it must thus be concerned with marriage values. If the £100,000 house which was referred to earlier was let at, say, £2,500 pa, it would be illogical to consider that the investment value is only £25,000 (applying, say, a 10% return), because the size of the gap between that figure and the vacant possession figure of £100,000 is an incentive for the parties to merge their interest and share the profit. This will be considered in detail later in this chapter.

Valuation on the basis of vacant possession

Owner-occupation in the UK rose from about 30% of dwellings in 1945 to 65% now at the end of the 1990s. The reason for this growth was a combination of the effects of

(i)   restrictions on private renting (which led landlords to sell off their properties rather than offer them to let),
(ii) greater affluence and
(iii) the appeal of owner-occupation to the young, the unmarried and those with higher incomes who could not obtain rented accommodation from local authorities.

The prices realised in the market place are as a result of

(i)   the demand and supply for housing;
(ii) and (iii) Government policy to encourage the sale of council and Housing Association Property.

There are no exact limits to the category of property which is capable of sale with vacant possession to owner-occupiers because it is a matter which has very much to do with personal taste, and at its extremes includes any property capable of being lived in – from a sprawling country estate to a dilapidated shack. In practice, the lending policies of the financial institutions tend to limit the compass of the category although it does vary for different parts of the country.

Whether a property is indeed saleable to an "ordinary purchaser" and the price it would obtain is a matter for local knowledge. A valuer from outside the area must take extreme care when valuing a property in an area of which he has no intimate knowledge; while there are general trends of value the specific will override the general.

The principal factors for determining the value of a property are the following:

(i)   location;
(ii) accommodation;
(iii) the nature of the housing unit;
(iii) the state of repair, appearance and quality of finish;
(v)   the quality and quantum of the fixtures and fittings;
(vi) the potential of the neighbourhood in terms of general improvement and the potential of the property to be improved and modernised; and
(vii) plot size.

All of the above points are really self explanatory and are well known to those who practise in the house agency sphere. Some locations are more fashionable than others owing frequently to historical or current associations with the rich and famous, also owing to matters such as accessibility to transportation facilities, shopping and recreational facilities and other community advantages.

There is a great deal of snobbishness in the housing market which is reflected in the fact that some postal districts are more fashionable than others, and likewise with telephone exchanges, especially in the past, when the exchanges had names rather than numbers. Other locational factors which influence value are travelling distance from the centre of the nearest major city and the attractiveness of the local townscape and of the local surrounding area. In most towns there is a right side and wrong side "of the tracks" and similar properties will have different values according to which side they are on. The political complexion of the local authority coupled with the level of council tax is also relevant.

The quantum of accommodation is clearly important and usually the larger the property the higher the value. There is, however, no exact correlation between value and size; therefore, if a house has 20% more accommodation it does not mean that it is 20% more valuable. In many areas there is an optimum size of a house or flat which commands the highest price per m^2. For example, a three-room flat with kitchen and bathroom/WC having a gross internal floor area of 60 m² might be worth, say, £60,000 but a four-room flat in the same block having a gross internal area of, say, 70 m² may be worth only £67,000; thus the value has fallen from £1000 per m² to just under £960 per m² as a result of the increase in size. The reason for this is that three-bedroom flats (ie four rooms) are more suitable to family occupation (as opposed to occupation by young married or elderly occupants) but the unit itself is not as suitable as, say, a three-bedroom semi-detached house. Similar results may well be found when comparing three-bedroom houses with four-bedroom houses, but in reverse, so that the former is less valuable per sq.m. than the latter. It is not possible to set out any rules, since the effect of accommodation will vary according to area and housing type.

The nature of the housing unit also causes variation in value per m², so that a three-bedroom semi-detached house is less valuable than a three-bedroom detached house of exactly the same size and in exactly the same location. Thus detached houses have a higher value per unit of area than semi-detached houses, which in turn have a higher value than terraced housing. Other similar influences on value are historical associations, prettiness (in the case of cottages) and appearance of grandeur (in the case of large houses) coupled with the size of the garden which accompanies the house. With flats the influencing factors are whether they are purpose built or conversions and whether they have lifts to the upper floors. The size and grandness of the entrance hall, the extent of services offered and provisions for security also have a large part to play.

As a general rule, it is not possible to value residential property by applying to it a local price per sq.m. Unlike offices which may have a local value of £150 per m² it is not possible to say that flats in the given area have a value of say £120 per m². Houses and flats in the United Kingdom are not sold on this basis and are consequently not devalued on this basis. This is because there is no uniform unit of comparison. Thus offices are valued on a net usable floor space basis and this basis is defined in the RICS code of measuring practice with modern offices being in clear space. Residential units are not in clear space but are divided by non-demountable walls. The value of the unit will be determined by the scale of the rooms, the quantum of bathrooms, the outlook from the rooms and the "general feel" of the unit on first approach. Small variations in size will not affect the value except within the immediate development (being the block of flats or the housing estate). The general unit of comparison is a similar property taken overall and then adjusted to reflect the factors already discussed. However in some markets, particularly in the Far East, housing units are sold on a price per m^2 and this applies when United Kingdom property is sold to that market; this only applies in very particular locations within the United Kingdom market which in this case means within very particular locations within the Central London market. Properties should only be valued in the way that the market approaches valuation so that the "currency" for shops is so much per m^2. Zone A, for offices it is so much per m^2 of usable space and for factories and warehouses, it is so much per m^2 of gross internal space. Thus for houses and flats, the "currency" is so much per unit.

Before most house purchasers actually buy a house they have the property surveyed in order to ascertain its structural condition and its state of repair. Surveyors are asked to quantify the actual costs of necessary repairs and purchasers then seek a deduction from the purchase price to reflect this cost. Where the defect is major, such as the need for a new roof, there is clearly some betterment to the property and the deduction allowed will frequently be less than the actual cost of the works. However, in many cases the reverse will be true because of the degree of upheaval and mess necessitated by the repairs. The major repairing items which affect value are settlement and heave, the extent of any dampness (both rising dampness and penetrating dampness), the extent of any rot (both dry rot and wet rot, although the former is the more relevant) and the condition of the roof and services.

Coupled with the actual condition of the property is the appearance which the property gives on being shown. The lay purchaser is influenced by a house which looks as if it has been well maintained, with fresh decorations and the appearance of having been modernised. If he feels that he can virtually move in without carrying out a great deal of work he will pay a higher price than if the reverse is true. The importance of this aspect is shown by the American market where the agent arranges viewing days (with the owners absent) and where they "improve" the property by putting in flower displays around the house and generally giving it the best "beauty treatment" that they can. The structural survey may then prove that the appearance of being in good repair is superficial only and that there are major defects which require remedying, owing possibly to an excess of DIY zeal.

The first price agreed between vendor and purchaser is of the property as first shown and the final price will depend on the result of the survey. The writer has known many cases where a dream house has been turned into a nightmare because rising damp and dry rot have been concealed behind panelling, walls have been removed without supporting RSJs being provided and extensions constructed without building regulation consent and without adequate foundations. In the latter case the deduction in value following the survey will almost certainly exceed the actual cost of the works to allow for the fact that the house will be uninhabitable while the builders work within the property.

Where works of modernisation and decorations have been correctly carried out the actual standard of these works will influence the final value of the property, although taste has a great deal to do with the final answer; taste and expenditure do not always go hand in hand. Stated at its simplest, a house with central heating will have a higher value than a house without it unless the heating system itself is obsolete and requires replacing. A house with double glazing will have a higher value than a house with single glazing although not necessarily by the amount of the extra cost of this provision. A house or flat with attractive cornices, good quality doors, and high-quality kitchen and bathroom fittings will have a higher value than a house or flat with lower quality and this really speaks for itself. However, if the owner's taste is unusual in choice of colours or materials then this could well nullify the extra costs which have been expended in providing that particular standard of finish.

The aspect of a property is frequently seen as an important factor by purchasers. Thus south facing gardens are often seen as important selling points even though the converse of this is that the rooms on the north side of the house get no sunlight whatsoever. The same considerations with regard to sunlight apply to flats as well as houses and where these have balconies, orientation is of very great importance.

Finally, the potential of the neighbourhood and also of the house itself will be a determining factor in the valuation of the property. Areas change from being unfashionable to fashionable and the chances of such an event happening should be taken into account when valuing the property. The reverse is also true and, consequently, civic proposals often have a great deal to do with value. If there are local authority proposals to build a large comprehensive school on a green field site of high natural beauty this will reduce the value of the houses where views will be altered. Likewise, if there are major road proposals or similar future schemes which the buying public may apprehend as being adverse.

The reverse, of course, is also true and proposed improvements in communications will increase the value of property in an area which might come within a more reasonable commuter distance of a major city, whereas at the present time it is beyond such a reasonable distance. Similarly, the potential of the property itself to be increased in size or improved will also play a part in determining its value.

The one further consideration which is vital in the valuation of a residential property is its mortgageability. Many of the lending societies limit their lending to post-first world war purpose-built houses or flats of standard-built construction in good structural condition with pitched tiled or slated roofs. Some will accept flat roofs or flats provided by conversion or above shops, but the more the property varies from the standard the more difficult it will be to mortgage and therefore to sell, and therefore the lower its relative value. It is absolutely vital that the nature of the construction of the property is such that it will conform with the general policies of the lending institutions. The more unusual or experimental the nature of the construction, the more difficult it will be to mortgage the property and consequently the more difficult it will be to sell it, thus resulting in a lower value. A simple example is provided by houses with thatched roofs, which are more difficult to mortgage than houses with pitched tiled roofs and consequently more difficult to sell; the result is that these have a lower value than adjacent houses of similar size but of more conventional construction unless their "prettiness" moves values in the opposite direction.

The valuation effect of mortgage policy is most clearly seen in connection with ex-council tower blocks. During the boom years of the late 1980's, these flats were saleable because mortgage finance was generally available. This is now no longer the case because some lenders restrict their lending to those blocks where not less than 80% of the flats are in owner-occupation with only a maximum of 20% of the tenants remaining as council tenants. It will easily be seen that where the 80% mark has not been passed, then flats in these blocks will be extremely difficult to sell and the situation will not change in the future because the lending policy creates a self-fulfilling prophecy whereby the 80% mark cannot be passed in the years ahead. While in general house and flat prices have now recovered to their 1988/89 peak, this does not apply to ex-council towers which has seen virtually no recovery except where the watershed mark has been passed; negative equity in these locations still remains a major problem.

The difference between residential property and commercial property is that no profit in economic terms is derived from the use of residential property; it has a functional use in providing shelter but it also has an esoteric function in providing enjoyment of ownership and occupation to its possessor. In view of the latter, it is like any other possession in that it has a subjective value to its owner. Consequently, the more the property can be enjoyed by its possessor the higher its value. The exercise of the function of the valuer is not purely mathematical any more than the valuation of a picture or sculpture is a mathematical exercise.

Insofar as houses are concerned, the usual rule is that the larger the plot (ie the bigger the garden), the more valuable the house. Where the house is set in "grounds" extra acreage will not have as great an effect on the value as the initial acreage. Thus there will be little difference in value between a house set in 10 acres and a house set in 11 acres, but there will be a considerable difference in value between a house set in 1 acre and a house set in 2 acres. This proposition is put on the basis that there is no development value. Where there is development value, then clearly every additional square metre will have value because of the potential to redevelop in due course.

No consideration has been given above to whether the property is freehold or leasehold or the effects of the latter tenure on the value. This will be dealt with later in this chapter as also will the effect of the level of service charges, particularly in relation to flats.

Freehold properties let on Regulated Tenancies

It should be noted that the law concerning succession to a statutory tenancy has been amended by section 39 of and Schedule 4 to the Housing Act 1988. A tenancy which passes to a spouse will continue as a Regulated Tenancy, but otherwise the tenancy which passes is an Assured Tenancy; the occupancy period in order to qualify to succession with the predecessor in title is increased to two years. The questions which must be asked by the valuer when valuing properties let on regulated tenancies are as follows:

1. Is there a single letting of the property or is there more than one tenancy?

2. Is there a single occupier-tenant for the whole of the property so that on the death or departure of that tenant vacant possession will be obtained?
3. What is the age of the tenant(s) and what is his/her (or their) state of health?
4. Can the property be sold in its existing state with vacant possession or will it need modification or improvement? If the later, what is the likely cost?
5. Are there successors to the statutory tenant? If so, will there be a transmission to a spouse (thus continuing the Regulated Tenancy) or to a member of the family (which would therefore be a succession to an Assured Tenancy)?

Study 1

A semi-detached house in a good suburban location with a vacant possession value of £80,000 is let to a Regulated Tenant at £2,200 pa exclusive of rates (just registered).

1 (a) Value of landlord's interest as a pure investment

		£	£
Rent reserved			2,200
Less: Outgoings:	repairs	220	
	insurance	170	
	management	110	500
Net Income			1,700
YP in perp @, say, 10%			10
			17,000

This is the value to an investor on first view, but it is, of course, much too low, for reasons that will follow. The value of £17,000 is the apparent value to an investor and not to a sitting tenant. Between the investor and the sitting tenant there is a considerable marriage value notwithstanding the non-open market saleability of the tenant's interest. The tenant possesses a limited state of irremovability which has the effect of preventing the real value of the property being realised.

1 (b) Consideration of marriage value

	£	£
Vacant possession value	80,000	
Investor's value (on first view)	17,000	
		17,000
Marriage value (MV)	63,000	
Say 50% of MV	× 0.5	31,500
Potential value (60.6% of VP value)		48,500

Net yield 3.5%

The potential value must now be looked at in detail, as its yield is considerably below that possibly expected for what appears to be a fairly poor investment, bearing in mind the commercial banks' minimum lending rates. The disadvantages of the investment are, first, that there are landlord's outgoings which might in some years take a large part (or even all) of the income and, second, that the annual rate of growth of the income is limited, as it is not truly related to economic factors (the new rent is determined by a rent officer or rent assessment committee, who can be influenced by non-economic factors). The advantages are, first, that the income is secure, since its non-payment will give early vacant possession (which would be a windfall), and, second, that there is a long-term rate of capital growth which far outstrips any other normal investment.

Assuming that the underlying value of the property increases over 10 years from £80,000 to £130,000 (an increase of 5% pa compound) and that vacant possession is obtained after 10 years, the capital growth would equate to 10.36% pa compound before tax if the property was purchased for £48,500. If inflation is also running at 3% pa average over the 10-year period, the net return to a 40% tax-paying investor is as follows:

(i)   Initial income yield 60% × 3.5% = 2.1%. This will be subject to increase every two years by reregistration.

(ii) Capital gain after tax = £55,572 (ie sale price of £130,000 less original cost (£48,500) and tax of £25,928*. The annual equivalent of the gain is 7.93%.†

---

* The capital gains tax is calculated by taking 40% of the difference between £130,000 and the indexed value of the purchase price of £48,500 (ie multiplied by

Marriage value usually envisages a deal between the landlord and the tenant. In this case it is taken to mean the effect of a potential deal done immediately. The value achieved by this calculation is thus used to calculate the return to the investor if the deal is not done and he waits 10 years for vacant possession. The true annual yield is the income yield plus the growth yield, which in this example is a minimum of 2.1% + 7.93% = 10.03% after tax which is equivalent to 16.72% before tax for a higher rate taxpayer, (minimum yield because the rent would also increase over the period).

The valuation approach for investments of this kind is therefore to take a percentage of the vacant possession value of the property. The percentage taken will depend on the answers to the questions raised at the beginning of this section. It is rare for the percentage to exceed 60% and the usual percentage taken in average circumstances is 40%–50%, but the valuer must use his own judgment to decide on the appropriate percentage. Reverting back to Study 1(b), this would mean a lower percentage of the ultimate marriage value than 50% will be taken in all but unusual cases. The general rule is that the longer the anticipated delay until vacant possession can be obtained, the lower the percentage of vacant possession value. If the rental yield after taking say 40% of VP is unusually high, then the value would increase until a more usual yield is achieved.

A sale to a sitting tenant would justify a higher price because the true marriage value can be calculated using an investment value of £48,500 and a VP value of £80,000, ie £64,200 (80% of the VP value). The discount to the sitting tenant will be greater the lower the prospect for obtaining vacant possession by natural means.

Freehold properties let on unfurnished tenancies with fair rent and with multiple tenancies

During the inter-war period, and for a decade or so after the second world war, it was quite common for houses to be let in floors with the bathroom and wc shared by the tenants. There was no self-

---

1.03 to the power 10), which at 40% = £25,928. Deducting the tax from the sale price of £130,000 leaves a net realisation of £104,072.

† The increase in the value of the property over the 10-year period is represented by the multiplier 2.146% (ie £104,072 divided by £48,500) and the annual equivalent of this figure is 7.93%.

containment between the floors and cooking facilities were provided in the smallest room on each floor. Thus a standard three-bedroom semi-detached house would often be occupied as two two-room flats each with a kitchen and sharing bathroom, WC and garden. The statistical chance of obtaining full vacant possession is less than for single tenancy houses.

There is a market for part-possession houses, as some owner-occupiers see the let portion as a way of helping to pay for the mortgage and there is, of course, a considerable discount on the full vacant possession value. The extent of the discount is difficult to estimate; it will vary according to location and type of property and the age of the remaining tenant(s).

Such property is of considerable appeal to young, first-time buyers who are prepared to put up with the disadvantages of sharing but can buy at a discount to the full VP value and also benefit from the additional income; the actual discount will depend on the age and type of the tenants and whether they have any children living with them. However, these properties may be difficult to mortgage.

The property is also of interest to an investor who might use the vacant portion to rehouse a tenant from another part-possession property which he owns elsewhere.

Study 2

A three-storey house with vacant possession of the ground and first floors and the second floor let to a regulated tenant at a rent of £2,000 pa (recently registered). The accommodation comprises two rooms and a back-addition room on the ground floor, three rooms and a back addition (containing bathroom and WC) on the first floor and three rooms on the second floor. The second-floor tenant uses the bathroom and wc on the first floor but the back addition could be extended to provide a bathroom and WC for the second floor and thus also provide a self-contained unit (1).

	£	£
Vacant possession value of ground and first floors if made into self-contained maisonette, Say		60,000
Value of second-floor flat (say 40% of assumed VP value of, say, £40,000) (2)		16,000
		76,000

Less:	(i)	cost of improvement work (3) say	9,000	
	(ii)	fees on improvement work	900	
	(iii)	legal and agent's fees	1,800	
	(iv)	interest on (i) and (ii) (4)	233	
	(v)	developer's profit, Say (5)	11,400	
				£23,333

Residue, being value of the house, acquisition costs and interest until sale of improved building		52,667
PV £1 6 months at 9%		0.96
Value of house plus acquisition cost of, Say, 2%		50,570
Therefore value of house = 50,570 ÷ 1.02 = Say		£49,575

Notes
(1) The tenant's consent to improvements is required, but the courts have power to authorise modernisation works if the tenant does not lose accommodation which he reasonably requires. The improvements will increase the fair rent to, say, £2,500 pa.
(2) Some rent rebate may be necessary during the course of the works because of disturbance to the second-floor tenant.
(3) No allowance has been given in this example for improvement grants and the costs are assumed to include VAT.
(4) Short-term finance cost has been taken at 9% averaged over six months.
(5) The developer's profit has been taken as 15% of the final estimated entirety value to reflect his risks and to justify the exercise fully.

Study 3

A conventional three-bedroom semi-detached house has vacant possession of the ground floor and an "old controlled" (1) tenant in the first floor paying a rent of £100 pa exclusive.

	£ pa	£ pa	£
Vacant possession value			50,000
Rent reserved for first floor – £100, but Say		700 (1)	
Fair rent of ground floor, say		700	
		1,400	

Less: repairs	120		
insurance	60		
management	70		
		250	
		1,150	
YP in perp at 10%		10	
Theoretical investment value		11,500	11,500
Vacant possession premium			38,500
(if VP value £50,000), say half vacant			
possession premium		19,250	0.5
Value with part possession (2)		£30,750 (3)	

Notes
(1) Old controlled tenancies no longer exist by virtue of the Housing Act 1980. These tenancies can now be registered without any additional formality at a fair rent. Some "unconverted" controlled tenancies will be found for many years owing to landlord's ignorance or indolence.
(2) This example is one hypothetical method of arriving at a part-possession value. There may be an established market in the area which will give a value by comparison. The alternative to a sale with part possession is to wait until full possession is obtained, but this will entail a loss of income and a positive expense if there is a void rate provision. Consequently, the part-possession value must be dependent on the statistical probability of obtaining vacant possession. This type of property may be difficult to mortgage.
(3) The answer in this example is 61.5% of vacant possession value. This is a little on the low side if there is a good prospect of vacant possession of the let portion, and a higher percentage of vacant possession value could well be paid in such a case. Everything really depends on the age and frailty of the tenant and whether there are any potential successors to the tenancies.

Freehold tenancies let on Assured Shorthold Tenancies

The basis of these tenancies is that the tenant will not have security of tenure. They have to be let for a fixed term of not less than six months, there must be no power for the landlord to determine the tenancy at any time earlier than six months from the beginning of the tenancy and if the tenancy was created before February 28 1997 there must be a notice served in the prescribed form before the tenancy is entered into stating that the tenancy is to be a shorthold tenancy; if these formalities are not followed the tenancy becomes an Assured Tenancy. The situation is reversed for all tenancies created after February 28 1997 and these are therefore Assured Shorthold Tenancies unless specifically agreed otherwise in writing. In the case of a grant of an Assured Shorthold Tenancy, section 21 of the Housing Act 1988 provides that a court shall make an order for possession of the dwelling-house subject to the landlord's giving not less than two months' notice that he requires possession of the dwelling, given before or at any time after the date on which the tenancy was due to come to an end. The tenancy may be allowed to continue without affecting the right of the landlord to obtain possession provided that a minimum of two months' notice is given that possession is required.

The only restriction on rent is provided for by section 22 of the Act, under which a tenant may make an application to a rent assessment committee for a determination of the rent which, in the committee's opinion, the landlord might reasonably be expected to obtain under the Assured Shorthold Tenancy. The rights of the rent assessment committee are limited to amending the rent being paid, but they can only reduce the rent being paid under the Assured Shorthold Tenancy in question if it is significantly higher than the rent which the landlord might reasonably be expected to be able to obtain had it been an Assured Tenancy.

This Rent Assessment Committee function is somewhat academic, since if the rent determined by the committee is unacceptably low the landlord will determine the tenancy at the first available opportunity. The existence of an Assured Shorthold Tenancy is a mandatory ground for possession by the court.

As stated at the very beginning of this chapter, the maximum value of any property in usual circumstances is the vacant possession value. The question therefore to be asked is: "What discount from the vacant possession value is appropriate because of the existence of the Assured Shorthold Tenancy?". The answer lies in the length of the tenancy and the quality of the tenant.

Notwithstanding the nature of the tenancy it is illegal for a landlord to evict a tenant from the premises without a court order. Consequently, delays could be experienced in obtaining possession since no action for possession through the courts can be commenced until after the landlord's notice has expired. It could well take two to three months (longer in some areas) to obtain a hearing date from the court and at least another 28 days for actual possession to be obtained. Usually the court will specify a possession date 28 days after the initial court hearing and a further delay can then be experienced, as a court officer will be required in order actually to obtain possession if a tenant does not go voluntarily.

Because of the above, it is reasonable to allow a minimum period of six to nine months after the expiration date of the tenancy in order to obtain actual possession. During this period rent should be paid by the tenant, but in many cases it might in fact be difficult to obtain the rent.

In view of the above, the maximum value which can be placed on a property let on an assured shorthold tenancy should be 90%–95% of its vacant possession value.

Freehold properties let on Assured Tenancies

The only difference between an Assured Tenancy and a Regulated Tenancy is that the rent payable under the Assured Tenancy is an open market rent and not a fair rent. There is no rent officer or committee to interfere with the freely negotiated rent between the landlord and the tenant and the tenancy can provide for a review of the rent, with any mechanism that the parties desire in order to fix the rent on review.

For the purposes of securing an increase in the rent under an Assured Tenancy, the landlord must serve on the tenant a notice in the prescribed form proposing a new rent to take effect at the beginning of the new period of the tenancy specified in the notice, being a period beginning not earlier than:

(a) In the case of a yearly tenancy – six months; in the case of a tenancy where the period is less than one month – one month; and in any other case a period equal to the period of the tenancy.

(b) Except in the case of a statutory periodic tenancy, the first anniversary of the date on which the first period of the tenancy began.

(c) If the rent under the tenancy had previously been increased by virtue of a notice, the first anniversary of the date on which the increased rent took effect.

Where the parties cannot agree the rent payable, the matter is referred to a Rent Assessment Committee who, after taking due notice of any improvements or dis-improvements effected to the premises by the tenant, shall determine the rent at which they consider that the dwelling-house concerned might reasonably be expected to be let in the open market by a willing landlord under an assured tenancy, ie an open market rent.

It is therefore clear that the rent payable under an assured tenancy should be considerably higher than that payable under a fair rent, unless there was little or no scarcity of accommodation to let with security of tenure, but the prospects of obtaining possession are no different than with a regulated tenancy, except that the higher rent and more readily available alternative accommodation could cause an increase in the lessee's mobility.

As at the time of writing this chapter there has been only limited evidence of the rents which properties let on Assured Tenancies actually fetch in the open market. Much will depend on the length of the term and the provisions for review of the rent, which will in turn influence the prospect of vacant possession. In the years 1989 to 1995 the supply of houses and flats to let was greatly increased by the poor selling market and the number of BES (Business Expansion Schemes) developments which became available. However, most landlords will not voluntarily let on this basis and the majority of cases which come before Rent Assessment Committees are second succession regulated tenancy cases.

In exactly the same way as with regulated tenancies, the greater the prospect of vacant possession the higher the percentage of vacant possession value which will be applicable to the property. If the rent review clause is such that the tenants have to pay a rent which is considerably above the open market rental value, then the probability is that they will not stay at the property and consequently vacant possession can be assumed. This will usually allow a value of between 80% and 90% of the vacant possession value to be taken. If the net yield is higher than the cost of finance the value may be 100% of vacant possession value.

If no precautions have been taken in terms of the form of the rent review, the likelihood is that the tenant will remain in possession and go to the Rent Assessment Committee at the appropriate time, then the value of the property is likely to be something in the region of 60% to 80% of its vacant possession value.

Freehold "break up" valuations

Up to now we have considered single properties with one or more tenants. Many residential investments comprise blocks of purpose-built flats. The valuation of these is a combination of investment analysis as in Study 1 and development appraisal. In a block of any size, a number of flats may become vacant over the course of a year. It is usual either to sell the block unbroken with possession of a number of flats or to "break the block up"; partially broken blocks are more difficult to sell as the statistical probabilities have been altered by the part disposals and consequently the yield requirement on the balance must be higher. The more vacant flats there are, the greater the potential for general block improvement, which will increase the average sales price of the individual flats.

The general considerations already discussed apply to flat break-up as they do to individual units. The flats must be mortgageable and capable of smooth management. There must be an acceptable management structure involving a single company which is responsible for repairs, insurance, services, etc, and with which each lessee covenants in respect of payment for a share of the costs; the landlord must similarly covenant in respect of unsold flats. It should be noted that freehold flats are difficult to mortgage because of the problems of enforcing positive covenants and therefore a break-up is best effected by the grant of long leases (over 60 years). This may be changed shortly as the Government is considering the introduction of a commonhold system which will allow the sale of freehold flats.

The block should be in good repair before the break-up is commenced because the buying public are (and should be) wary of involving themselves in a large service charge commitment for works which should have been carried out years before. Similarly, sitting tenants would not be too happy for their landlords to escape existing repairing obligations by passing these on to them as buyers. It must be borne in mind that if a sitting tenant buys his flat on a mortgage he will from an income point of view be financially worse off for many years than if he had continued renting his flat, ie his interest and service charge expenditure will be considerably greater than his rent at the beginning of his mortgage term. The viability and profitability of a break-up scheme will depend on the number of tenant sales.

The sale of the block as a whole will be affected by the Landlord and Tenant Act 1987, which requires an option to be given to the tenants and any sale can be held up for up to seven months. The Act provides that a sale cannot be made to an outside purchaser for less than the block is offered to the tenants. This Act was honoured more in the breach than otherwise and therefore the Housing Act 1996 has brought in penalties for non-observance. There is also a special scheme for sales by auction where the tenants need not bid but may then pre-empt a purchase at the sales price.

Study 4

A tenant rents a flat at £1,250 p.a. including £200 pa for services. The vacant possession value is £40,000 for a 999-year lease at £25 pa. The sitting-tenant price is £30,000.

		£
Present expenditure	rent	1,050
	services	200
Rent fixed for 2 years		1,250
Proposed expenditure		
Assuming 100% (interest only) mortgage		
at 8.5% mortgage interest		2,550
Less tax relief at 15%		382
Net interest		2,168
Services		200
Share of repairs and external decorations		130
Ground rent		25
Building insurance		45
Annual cost		£2,568

Assuming that rent and all costs rise by 5% pa compound, but that interest charges remain constant, it will be about 20 years before expenditure is equated. The compensation for this is the immediate capital profit of £10,000 and the added psychological advantages of ownership, although this is not without worry in view of rising service charges.

Study 5

A block of 12 self-contained flats on three floors. Three flats are vacant and nine are let to regulated tenants at £1,450 pa each including £350 for services (just registered). The block is in good repair and the vacant flats are worth £44,000 each if sold on 99-year leases with £25 pa ground rents.

	£	£	£
Let flats – 9 × £1,450		13,050	
Less: services – 9 × £350	3,150		
repairs at, say, £150 per flat	1,350		
management at, say, 5%	652		

insurance, say	540		
		5,692	
		7,358	
YP in perp at 6%		16.67	122,658
(approx 30% of vacant possession value)			
Vacant flats			
3 × £44,000		132,000	
3 × £25 × 9 YP		675	
		132,675	
Less			
(i) legal and agent's fees 3% on realisation	3,980		
(ii) profit at 15% on realisation (1)	19,901		
		23,881	
		108,794	
Less interest for, say, 6 months @ 9%		4,895	
			103,899
			£226,557
Say (2)			£226,500

Notes
(1)   The profit is required in the same way as any wholesaler requires a profit because of the risk of obtaining the estimated prices, etc.
(2)   The average price per flat is just under £18,900.
(3)   Assuming subsequent sale of four of let flats to sitting tenants at two-thirds of the vacant possession value, the total realisation including the three vacant flats is about £250,000, leaving five flats at nil cost and a profit of £23,250.

The above example may be made more complex by having a variety of tenants at different rents and the need to estimate the relevant fair rents. Furnished and Assured Shorthold lettings should be considered separately from unfurnished regulated lettings and also from Assured lettings as the appropriate yield will be different. Potential decreases of rent must be considered because of possible registration, together with the cost of putting the block and the individual flats into a saleable state.

Some blocks, because of their location or design, are not capable of being broken up and must be considered as pure investments. If this is the case, then the yield must be much higher because of the absence of the potential to the vacant possession profit. If the state of repair is poor, then the cost of refurbishment and modernisation can be taken as an end-allowance together with the financing costs thereon. Many schemes have failed because insufficient allowance has been made for financing the refurbishment expenditure which might be required, either to the exterior and common parts or to the interior of flats which are going to be sold off immediately. In order to ensure the viability of a break-up, this expenditure must be recovered from the initial sales or the financing cost must be deducted from the general income, or a combination of both.

It is very unusual to consider a modernisation/break-up without a reasonable percentage of vacancies at the beginning, as in Study 5, since it is very risky to assume that large numbers of sitting tenants will buy; they very often do not have the resources and are too old to obtain a mortgage. It is therefore better to wait until there is a good vacant possession base.

In many developments there are several blocks. A break-up might best be effected by modernising and selling one block at a time. Sitting tenants can be moved by court order to suitable alternative accommodation to assist in making individual blocks totally vacant. If a sales campaign to sitting tenants is mounted then the purchasers should be made aware of the developer's future intentions as to expenditure, so that they do not buy cheaply in a "run-down" break-up and then find themselves called upon for further large capital sums to meet their share of the improvement expenditure, which they might not be able to afford. It would be better to effect the improvements first and then to sell to the sitting tenants, who will no doubt need larger mortgages.

In considering the value of any block regard must also be had to its development potential. Can, for example, the roof space be developed or can further units be built in the grounds?

In some break up situations some of the flats have already been sold off and those leases may now be sufficiently short to allow for some assumption as to further income being derived from surrenders and re-grants. Under the provisions of the Leasehold Reform, Housing and Urban Development Act 1993 lessee owner-occupiers have the right at any time once they pass the residence test to require a lease extension of 90 years and the reduction of their ground rent to a peppercorn. The price at which the lease extension will be sold will include not less than half the marriage value (the Act says not less than half but in practice only half the marriage value is usually awarded). There is also a right for collective enfranchisement by long lessees although there are qualifying tests in so far as the relevant percentages are concerned.

This will be considered in the next chapter under Leasehold Reform and will not be considered further here.

Leasehold properties

The entire analysis so far has been concerned with freehold properties, but many residential properties are leasehold. Consequently, the above analysis must be amended to take account of the unexpired term and whether a purchaser of the leasehold interest could exercise rights under the Leasehold Reform Act 1967. If he can, then the leasehold value is in fact the freehold value less the cost of enfranchisement with allowance for a margin to reflect the risks involved.

Chapter 3 deals in detail with the Leasehold Reform Act 1967, its subsequent amendments and the Leasehold Reform, Housing and Urban Development Act 1993, but, of particular concern in the present analysis, is the case of *Bickel* v *Duke of Westminster* [1977] 1 EGLR 27. In that case, the Court of Appeal upheld the lessor's right to refuse consent to an assignment of the lease where the assignee would be in a position, after assignment, to enfranchise and where the assignor was not in that position. Some leases are drawn with an assignment clause which only requires registration of the assignment; in these circumstances, clearly, the decision in the *Bickel* case could not be relevant. Where consent is required to the assignment (and this can be refused), then the *Bickel* decision is totally relevant and, in such a case, the investment nature of the property must continue until the lease expires. The subletting clause is equally important because if no consent to subletting is required, then, provided the lease has more than 21 years to run, it may be possible to create an enfranchiseable situation by the grant of a sublease.

Referring back to the analysis, so far it will be seen that the rate of interest effectively adopted in capitalising the income to arrive at a freehold value must depend on the possibility of early possession. In a leasehold situation it must depend on the probability of obtaining vacant possession before the end of the lease, bearing in mind the right for the tenancy to pass twice. Where there is a distinct probability that possession will be obtained, then the only difference between the freehold and leasehold situations is the cost of enfranchisement, assuming that this cannot be prevented, and Study 6 below illustrates an approach in this situation. When valuing freehold reversions to ground-rented long leases, regard must be had to the potential for enfranchisement by the lessee or his successor in title. A view must be taken as to future enfranchisement possibilities.

It will be necessary to study Chapter 3 in order to calculate the enfranchisement price which is applicable under the different circumstances which apply. With the exception of houses having a rateable value in London of less than £1,000 and elsewhere of less than £500, the enfranchisement price will be the value of the landlord's interest plus half the marriage value and this could amount to a very significant sum. Only in respect of the houses below the rateable value limits will the enfranchisement price be confiscatory so far as the landlord is concerned.

Study 6

Using the same information as in Study 1, where the vacant possession value freehold is £80,000 and the property is let at £2,200 exclusive, but assuming that there is a lease with 60 years unexpired at £10 pa and no consent is required to assign.

The sole question to be asked in connection with this property is whether vacant possession will be obtained within a sufficiently short period of time to allow an owner-occupier purchaser to enfranchise. With 60 years unexpired, this is certainly going to be the case and therefore the market will value this property by reference to a percentage of the vacant possession value of the freehold.

Because there are 60 years unexpired, something in the order of 25%–35% of the freehold vacant possession value would be appropriate if 40% were going to be used if the investment was identical but freehold. If the subtenant was elderly then up to 55% might be paid.

Notes

(1) If vacant possession were obtained immediately and all prices remained constant, an assignee could enfranchise in three years.

Assuming the cost of enfranchisement is £750, including all costs, then the leasehold vacant possession value should be approximately £79,000, which would give the investor a greater profit than he would have had if he had purchased a similar freehold, but the risks are greater.

(2) Taking a mid-point for the valuation, ie 35%, this would give a valuation of £28,000. If the net income after outgoings was, say,

£1,500 pa, this is the equivalent of a yield of 4.33% coupled with a sinking fund of 3% and an income tax adjustment of 40%.

Study 6(a)

Using the same information as in study 6 but assuming a lease with 30 years unexpired. The prospects of possession are still very good but there is greater risk which must be reflected in the yield. Therefore using a yield of say 6%.

	£	
Net income	1,500	
YP 30 years at 6.0% & 3% (tax at 40%)	10.52	
	15,780	
Say		£15,750

Note

The cost of the freehold if enfranchised immediately (including costs) would be about £4,500, but it rises to about £8,500 with 20 years unexpired (at constant prices). If house prices remain constant, vacant possession in 10 years will yield a profit of £55,750 [£80,000 – (£15,750 + £8,500)] if the landlord will sell immediately vacant possession is obtained and excluding costs of sale. However the enfranchisement price may be increased to reflect the fact that the enfranchisee will need to be in occupation for at least 3 years which will reflect on the sale price of the unexpired lease to an owner occupier.

Study 7

Using the same information as in Study 6 but assuming the landlord's consent to assignment can and would be refused.

	£ pa	£ pa
Rent reserved		2,200
Less: outgoings:		
ground rent	10	
repairs	500	
insurance	80	
management	110	
		700

Net income	1,500
YP 60 years at 9% & 3%	
(tax at 40%) (1)	9.98
	14,970
Say	£15,000

Note

(1) The justification for a yield as low as 9% is the probability of vacant possession which will allow negotiations with the landlord for an amalgamation of the two interests.

Study 8

Using the same information as in Study 7 but assuming an unexpired lease of 10 years irrespective of restrictions on assignment.

	£ pa	£ pa
Rent reserved		2,200
Less: outgoings:		
ground rent	10	
repairs	500	
insurance	80	
management	110	
		700
Net income		1500
YP 10 years at 12% & 3% (tax at 40%)		3.77
		5,655
Less: dilapidations, Say		1,500
		£4,155
Say		£4,150

There must be a general probability that vacant possession will not be obtained before the lease expires. A surrender will often be negotiated in lieu of dilapidations before the end of the lease. If the lease were shorter still (below three years), then, even if possession is obtained, no enfranchisement could take place and consequently an even higher yield should be taken. However, there is a "blackmail value" against the landlord: see Study 8(a) below. Studies 7 and 8 apply equally to flats as to houses, since enfranchisement now applies to both.

It will therefore be seen that the length of lease is of paramount importance, since the longer the lease the greater the prospect of obtaining vacant possession and a market usually exists for leases of all lengths for owner-occupation. The value of a leasehold flat with 60 years unexpired reflects the hope of a vacant possession sale without the need for a lease extension, so that the potential unexpired term is relevant to the value to a greater degree than would be the case with a house. However, in considering the situation of flats it has been assumed that "break-ups" are possible and are not prevented by covenants. Where a vacant possession sale to an owner-occupier is not possible at all or only at a very low price (either because it is a house which cannot be enfranchised or it is a flat with too short a lease) there is a second way of looking at its value and that is by considering the effect of a reletting on the landlord.

Study 8(a)

Using the same information as in Study 8 but assuming that vacant possession has been obtained, that there are three years unexpired, that £3,000 pa is the rent which would be payable on an assured tenancy and allowing £500 pa in outgoings. The "fag end" leaseholder could not be prevented from letting to an Assured Tenant who could be young with a young family. This would be very destructive of the landlord's value.

		£	£	£
(i)	Value of leasehold interest as an investment			
	Net income	2,500		
	YP 3 years at 12% & 3%			
	(tax at 40%)	1.52		
			3,800	
Less:	dilapidations		750	
				3,050
(ii)	Value of freehold interest			
	Rent reserved	10		
	YP 3 years at 5%	2.72		
			27	
Reversion to assured rent		3,000		
Less:	outgoings	500		
Net income		2,500		

YP in perp deferred 3 years at 7% (1)	11.66		
		29,150	
Say			29,180
Total of both values (3,050 + 29,180)			32,230
Value if interests merged			80,000
Therefore marriage value			47,770
Splitting this equally			0.5
			23,885
Add: value of leasehold interest as above			3,050
Value of the leasehold interest (2)			£26,935
Say			£27,000

The landlord therefore nets £53,000 immediately (excluding costs of sale) compared with the value of possibly 35% × £80,000 = £28,000.

Notes
(1) A comparatively high yield has been taken, as the "blackmail" assumption is that there could be a letting to a tenant who is most unlikely to give VP in the foreseeable future.
(2) Allowance may be made for the costs of sale including interest until sold and for the risk of obtaining the estimated vacant possession price.

The threat of a reletting once possession is obtained could therefore influence the yield on a leasehold flat or house investment and, provided there is a reasonable presumption of possession before the lease end, a lower yield could be appropriate. Hence, if in Study 8 the tenant was an octogenarian widow living alone in a flat, the yield might be reduced to as low as 8–9 %.

The basic problem with non-enfranchiseable houses and flats is that as the lease gets shorter the vacant possession value gets lower and the prospect of possession before the lease end diminishes. Hence the yield must rise as the term gets shorter.

In study 9 below, the minimum value of a short lease to an owner-occupier is considered, but there is also a special value of a vacant short lease to an investor for rehousing purposes.

Study 8(b)

Using the same information as in study 1, where the freehold value of the house was £80,000 with possession but only £32,000 if let at £2,200 pa (taking 40% of VP value).

Value of possession (£80,000 – £32,000)	£48,000	
Maximum price freeholder prepared to pay for possession, say 50%	0.5	
		£24,000

It will be seen from this study that the landlord of the property referred to in study 1 can pay £24,000 for a lease of any length and this will enable him to make a profit on this first house of approximately £24,000. The profit will not be exactly £24,000 because the landlord would have to bear the cost of the legal proceedings necessary to force the tenant into the alternative accommodation and he will also be liable for the tenant's removal expenses. In addition he will be liable for dilapidations in respect of the property which he has purchased. It will be remembered that one of the grounds for obtaining possession against a statutory tenant under the Rent Acts is the offer of suitable alternative accommodation. This is not a mandatory ground, the court has discretion and will order possession only if it considers it reasonable to do so. Assuming that the alternative accommodation is suitable as to size and location (as to size, the local housing authority's standards are the test) then reasonableness will turn on the general conduct of the landlord, the nature of the alternative accommodation and the personal circumstances of the tenant.

Since Parliament's intention was that this ground for possession should be available, the general presumption must be that the courts will grant possession provided that the alternative accommodation is suitable. However, there is no requirement that the alternative accommodation is owned by the same landlord, merely that the tenant must be given the same security of tenure and that the rent is similar. The alternative accommodation must be suitable to the needs and the means of the tenant.

Flat leases are a normal market commodity, so that the vacant possession value of any lease term can usually be determined fairly accurately by comparison with market transactions. The only problem is one of finance for the vacant property, because a mortgage must usually be repaid many years before the end of its term. However, the minimum value for a lease should be as shown in Study 9 below.

Study 9

A maisonette has an unexpired term of five years at £20 pa on full repairing and insuring terms. A rent of £3,000 pa on internal repairing terms is the estimated rental value under an assured tenancy.

	£ pa	£ pa	£
Assured rental value		3,000	
Less: ground rent	20		
insurance	40		
service charge	250		
		310	
Rent saving		2,690	
YP 5 years at 8% (1) (2)		3.99	
		(3)	10,733
Say			£10,750

Notes
(1) Single rate used, as this gives the annual rent saving for the term, discounted to the present day, and thus no sinking fund at a low rate is required.
(2) The rate of interest reflects the absence of a rent increase after two years and the net cost after tax relief.
(3) There will usually be security of tenure at the end of the lease by virtue of Part I of the Landlord and Tenant Act 1954.

The above must represent a minimum value, since a scarcity element may be added; this would not be an illegal premium under the Rent Acts.

American Style "Lofts"

This is a new style of property which has been recently been coming in to the market following the obsoleteness of some types of commercial property. The basic concept is that a large clear space is sold with planning permission for use as residential accommodation, but that space is not fitted out as a flat as such. The accommodation is serviced by the provision of mains water, sewage, electricity and gas but the purchaser fits out the accommodation to suit his own needs. The usual feature of accommodation of this kind is that the space is large and frequently of above average height which allows for the construction of mezzanine floors or even duplex apartments within the shell.

This type of accommodation is generally seen most in the disused docks areas of the major cities where old multi-storey warehouses have been converted to residential use. This use is now spreading to city centres where office blocks, telephone exchanges, etc, are being converted into residential use.

Because of the popularity of this kind of unit, consideration should be given to the conversion of obsolescent commercial buildings into residential use. The valuation method will of course be the residual method which is discussed in Chapter 13.

Non-owner-occupier-type properties

In general, there are three types of properties in this category: (i) properties in very poor repair or in areas scheduled for redevelopment; (ii) houses in multi-occupation; (iii) artisan dwellings.

(i) Freehold properties in very poor repair or scheduled for redevelopment

There comes a point in the life of a property when the cost of repair may approach the value of the property in good repair. In such a case, the maximum vacant possession value is the site value less the cost of demolition. In addition, a landlord can be forced to do repairs to the property under the Public Health Acts and the Housing Acts and, in most cases, by a civil action for breach of covenant to repair. Consequently, there will be a capital loss situation in any investment. Local authorities are now much more active than they have been and are serving notices under Schedule 9 to the Housing Act 1957 requiring major works of repair to be carried out; if the works are not done then they may enter and do the works themselves, recovering the costs from the rents. Grants are often available for the work.

Study 10

A terrace house is let to a single tenant at £650 pa. It has been badly neglected and will now cost £7,000 to put it into an acceptable state of repair, in which condition it will have a fair rental value of £1,450 pa. The house is let to a couple who are 45 years old.

	£ pa	£ pa	£
Fair rent in repair		1,450	
Less: annual repairs	145		
insurance	46		
management	72		
		263	
		1,187	
YP in perp at 10%		10	11,870
Less: necessary repairs		7,000	
supervision fees at 10% (1)		700	7,700
			£4,170
OR:			
Present fair rent		650	
Less: outgoings:			
repairs	250		
insurance	46		
management	32		
		328	
		322	
YP in perp at 20% (2)		5	
			1,610
Less: essential repairs			1,250
			£360

Notes
(1) The 10% yield reflects the non-probability of early possession.
(2) The 20% yield reflects possibilities of notices under the Public Health Acts, etc.

This valuation could be affected by the availability of local authority grants and up to 90% of the cost of works could be obtained. If this is the case then the works are worth doing and the value of the property will be considerably increased.

If, because of the disrepair, the tenant does not pay his rent, then possession may be obtained and a profit made, but this is a doubtful occurrence as the tenant would be advised to press for the repairs via the local authority or to sue on the landlord's deemed repairing covenants.

For property of this type there is also always the threat of compulsory purchase as a property which is "unfit for human habitation" (see chapter 9).

(ii) Houses in multiple occupation

This is a similar category to Studies 2 and 3, but the properties are not of a sufficient quality or of satisfactory design to allow part owner-occupation either by conversion (into all or part self-contained units) or by sharing or the local planning authority will not give consent for change of use to a single dwelling or for self-containment. Consequently, there may be little logic in leaving part vacant in the hope of full vacant possession (unless this is an early probability), since there is a security risk from squatters and there will be substantial outgoings in the form of repairs, insurance and rates (NB conversion of 2 dwellings into a single dwelling does not require planning permission).

Such property is the poorest possible type for investment. There is very little prospect of capital profit because there is little or no vacant possession market and consequently there is almost no option but to relet when a part becomes vacant. This type of property is also politically sensitive, since the tenants are usually the poorer members of society who need statutory protection. Rent levels are often geared to Social Security payments and therefore suspect to political interference.

Even when there is a vacant possession market for improved properties the probability of obtaining early possession must usually be remote, since the pure mathematical chance of this happening is substantially worse than for a single tenanted property. This opinion is reinforced by the fact that in areas where this type of property exists there is less movement towards owner-occupation by the tenants, and so voluntary moves are rarer except for local authority and housing association rehousing. This type of property usually occurs in Housing Action Areas, where the local authority can compel improvements and repairs. Large grants are available but there is still a residual capital liability.

(iii) Artisan dwellings

This is the name given to purpose-built blocks of houses and flats built around the turn of the century by the predecessors of today's housing associations, usually charitable trusts; good examples are the blocks built by the Guinness and Peabody Trusts. At the time they were built, these properties were a major advance on the "working class" housing then existing.

By today's standards, however, these artisan blocks are old fashioned and lack modern amenities. They were built to a very high density with blocks being close together and often five storeys high, without lifts. They tend to have open landings and staircases. Some of the better blocks can be, or have been modernised, to acceptable standards so that "break-ups" can be effected. Others, by reason of their design or location, can never be made acceptable owner-occupier-type properties although they can be provided with standard amenities. Consequently, they provide a purely investment type of property and the yield must reflect a pure income situation. Regard must be given to the state of repair and the level of outgoings on insurance, services and management. Finally, regard must be had to potential obsolescence, since it is hoped that, in due course, this type of housing will be demolished as not being of a sufficiently high standard in our enlightened age.

Study 11

"Trust Building" comprises an artisan block of 16 flats on four floors, without a lift. The location is such that there is no potential for vacant possession sales. All the flats are let on regulated tenancies. Local evidence indicates an increase in registered rent levels of 20% including services since the rents were last registered. All rents are exclusive of rates with tenants responsible for internal decorations only. Services are limited to cleaning and lighting of common parts for which £20 pa was included in the old registered rent. The present cost of the services is £36 per flat. No application has yet been made for new registrations.

Floor	Flat No	Registered Rent	Date of Registration	Rent Payable
Ground	1–4	£350 pa each	2 years ago	£350 pa
First	5–8	£350 pa each	2 years ago	£350 pa
Second	9–12	£329 pa each	2 years ago	£329 pa
Third	13–16	£300 pa each	2 years ago	£300 pa

Value as at January 1991

	£ pa	£ pa	£
Rent reserved		5,316	
Less: services 16 × £36	576		
repairs 16 × £30	480		
insurance 16 × £15	240		

management at 10%	532		
		1,828	
		3,488	
YP 3 months at 13% (1)		0.23	802
		6,380	
Reversion to new fair rents (2)			
Less: service/repairs/insurance	1,296		
management at 10%	638		
		1,934	
		4,446	
YP in perp at 15%	6.67}		
PV £1 – 0.25 years at 15%	0.97}	6.50	
			28,899
			29,701
Less: allowance for essential repair			1,600
			28,101
Say			28,100

Notes

(1) The effective date of registration is the date of registration and the two-year period runs from this date. The first date on which the higher rent is payable is the first rent day following registration. Hence the above assumes a delay of three months from date of application to date of determination. The actual period will depend on local pressures on the rent officer service.

(2) Assumed modern fair rent £5,316 + 20% = £6,380.

Although a different rate has been taken in capitalising the old rent and the new rent, it is not considered that there is any real difference in the security of the new rent. The differentiation reflects the possibility that the full increase in rent will not be achieved due to The Rent Acts (Maximum Fair Rent) Order 1999, or to an over assessment of the Fair Rental Value.

The high yields reflect the absence of a review procedure to cover the increasing cost of services between registrations and the risk that the estimated rent will not be achieved.

The valuation set out is the same as in a break-up valuation, only the rates of interest are different to reflect the absence of the capital profit.

In this case the rents payable are the same as the old registered rents, but this is not always the case owing to landlords' neglect or personal reasons. There may also be some "old controlled" tenancies or tenancies where legal increases have not been effected.

As and when flats become vacant, they will be relet on either assured or assured shorthold tenancies, which will increase the rents.

Leaseholds

If a freehold interest in this category of property is one of the poorest investments available, then a leasehold interest must be the worst of all. The discussion contained in Studies 6 to 9 applies equally here concerning probabilities to possession, but there is also the problem of dilapidations. The covenants of the lease must be considered in valuing the lease and if there are the usual forms of repairing covenant then the property may be a liability rather than an asset. An appropriate sinking fund must be used.

Furnished property

This type of property does not really warrant separate treatment as there is no substantial difference today between a furnished and an unfurnished letting, since both types are subject to the same rent and security of tenure controls. The maximum value is the vacant possession value but care must be taken in considering whether the rent passing is "fair". Also, allowance must be made for deterioration of furnishings. The prospect of possession will also be influenced in this category by the nature of the tenant and whether he is a temporary resident of this country.

Although in theory there is no difference between furnished and unfurnished lettings, in practice there often is a difference. Many landlords try to find loopholes in the law which allow them to let on a temporary basis other than on Assured Shorthold Tenancies. These loopholes are (1) holiday lettings, (2) company lettings, (3) licences, (4) service tenancies.

(1) Holiday lettings can be legitimate, particularly in seaside or tourist areas. In such a case the tenants will not have a tenancy *per se* but a license. A letting following a holiday letting will be an Assured Shorthold Tenancy. However, sham holiday lettings will be caught as the court will look at intention as opposed to the wording of the agreement.

(2) Companies do not have security of tenure under the Rent Acts. However, most company lettings are sham, with the company formed for the precise purpose of taking the tenancy. These

lettings have not yet been tested in the courts, but the writer considers them to be fraught with danger. What has also not been tested is the relationship between the company and the occupier and whether this could itself create a tenancy. If a tenancy is found to be a sham then it will revert to whatever tenancy would have been created at the time the tenancy was entered into.

It is established that a company may have a tenancy of residential property under the Landlord and Tenant Act 1954 for the purposes of its business (eg a nurses' home). If the premises are not for its business then for what purpose can a company require premises? Hence the sham, as the sole purpose is the housing of an employee or director. Does this therefore lead to a subletting with the landlord's consent or can it create a licence?

(3) Licences have been considered by the House of Lords in *Street* v *Mountford* [1985] 1 EGLR 128: any letting giving exclusive occupation is a lease and the tenant could have security of tenure. Only if there is no exclusive occupation can there be a licence and then it must not be a sham. A letting to a husband and wife or man and mistress by way of two separate agreements with neither given exclusive occupation is a sham, as is one to two cousins simultaneously. A number of licences to a group of friends with each being independently liable, only for a given rent and with no licence being dependent on any other is almost certainly valid and will exclude security of tenure.

(4) Service tenancies will succeed in avoiding security of tenure if there is a significant degree of service. One boiled egg a day delivered to the door is unlikely to be sufficient service to qualify; daily cleaning of the room with breakfast and clean linen regularly provided should be sufficient.

Where tenants do not have permanent security of tenure

In these cases, vacant possession can be obtained by the landlord although this might be delayed for up to nine months. The value is therefore the vacant possession value deferred for the period until it is known that possession will be available, less the cost of obtaining possession.

The method of valuation must be the same as for any other residential property except that the rate of interest must be adjusted to take account of the chances that the rent may be increased or decreased (by the rent officer). The rate of interest must also reflect the percentage of the rent attributable to the furniture and the value of the latter. It is essential, in any sale, that the furniture be conveyed with the property.

When advising an owner as to the value of a furnished property for the purpose of sale, account should always be taken of the attitude of any prospective purchaser to err on the side of caution. Hence, if a tenancy is going to expire within a short period then the valuer should be advised to wait to see if possession will be given up. However, if a valuation for mortgage purposes is being carried out, full consideration should be given to the possibility that possession may not be given at the end of the term.

© E F Shapiro 2000

Chapter 3
Leasehold Enfranchisement

The Leasehold Reform Act 1967 was a measure to protect certain occupiers of houses holding under long leases at low rents and is now extended to cover flats held on long leases. It was the culmination of argument since the latter part of the last century that although the land in such circumstances may belong in equity to the freeholder the buildings should belong to the leaseholder. Leasehold enfranchisement, or the principle that certain residential leaseholders should be able to expand their interest in premises, was the policy of both major political parties in 1966, but the basis for enfranchisement was the subject of controversy and still can be.

The original principle of enfranchisement adopted in the 1967 Act was that the tenant might either take a further 50-year term of ground lease at a so-called "modern ground rent" subject to review after 25 years or purchase the landlord's interest. In the latter case the enfranchisement price payable was the value of the landlord's interest subject to the tenant's unexpired term of lease and right to a further 50-year ground lease at a "modern ground rent". This principle only applies to lower value houses in the now widened setting of enfranchisement.

The 1967 Act was amended by section 82 of the Housing Act 1969, which requires that any additional bid that a tenant might make be excluded from the enfranchisement price. Section 118 of the Housing Act 1974 extended the right to enfranchisement to a further range of interests subject to different rules for assessing prices in the higher rateable value bands. It also made the amendment that for enfranchisement qualification purposes rateable values may be adjusted to discount the value of any tenant's improvements. The Leasehold Reform Act 1979 closed the loophole of creating intermediate interests to increase the cost in the enfranchisement price provisions of the 1967 Act revealed and confirmed by the House of Lords in *Jones* v *Wrotham Park Settled Estates* [1979] 1 All ER 286.

The Housing Act 1980 made further amendments to the 1967 Act, reducing the original five years' residential occupation requirement to three years and correcting and improving the 1974 Act provisions regarding rateable value adjustments. It also included within the protection of the 1967 Act tenancies terminable on death or marriage to close an avoidance loophole, it extended transitional relief to tenants paying modern ground rents on lease extensions and introduced a formula approach for those enfranchising against minor superior tenants. Section 142 provided for the referral of valuation disputes in the first instance to leasehold valuation tribunals drawn from rent assessment panels, with an appeal right to the Lands Tribunal.

The Leasehold Reform, Housing and Urban Development Act 1993 introduced provisions for collective leasehold enfranchisement by tenants of certain flat blocks, amended the enfranchisement rules for houses under the 1967 Act, provided for Estate Management Schemes for newly enfranchisable estates and the right to a management audit of service charge accounts. The 1993 Act enables certain occupying leaseholders to acquire collectively the freehold, and any other superior interests, in a block of flats where they hold long leasehold interests. The individual flat leaseholders also each have the right to be granted a new lease of their flat with an additional 90 years added to the existing lease term free of ground rent, whether or not collective enfranchisements rights have been exercised.

The 1993 Act provisions are intended to help flat leaseholders where the lease terms are getting too short to be saleable or to be suitable security for lending. They are also aimed at helping tenants to improve the management of their blocks, through the rights to service audits and management codes. The audit right enables detailed investigation as to the appropriateness and efficiency of management functions and service charge expenditure. The Secretary of State is given power to approve codes of management practice designed to promote improved standards and while there are no direct sanctions relating to the operation of a code such may be used as evidence in proceedings.

Tenants of houses in higher rateable value bands were excluded from the right to enfranchisement under the Act of 1967 but now the value limits are removed by the 1993 Act, although the right to the alternative of a 50-year lease extension at a section 15 "modern ground" rent is excluded for these newly qualified houses. An additional "low rent" test qualification was added by the 1993 Act whereby at the commencement of the tenancy, or before April 1 1990, the tenancy should be at either no rent or a rent of no more than two thirds of the letting value (if a pre 1963 tenancy) or two thirds of the Rateable Value (for post 1963 tenancies). For such newly qualified houses the valuation assumptions in section 9(1A) of the 1967 Act are to be made requiring the parties to share any marriage value and it is not to be assumed that the tenant has any right to carry on in possession at the end of the tenancy. The 1993 Act also excludes certain houses let by charitable housing trusts from the right to enfranchisement.

It should be noted that residential tenants, whether or not qualifying for enfranchisement, generally still have the alternative, but probably less beneficial, protection of Part I of the Landlord and Tenant Act 1954. This effectively entitles them to a protected tenancy at the expiry of the long lease.

Valuations of landlords' and tenants' interests in residential property must take account of the effect of enfranchisement rights and may be specifically required for the assessment of enfranchisement price or lease extension terms. They must take account of the statutes and the precedents which interact to influence the rights to enfranchisement and the basis and terms for it. There have been numerous references to the Lands Tribunal since 1967 on issues under the Act and, since 1980 many before leasehold valuation tribunals as the initial reference point for disputes. Many cases have been heard by the courts too under the provisions but the basis for the valuations required is still not always clear. The importance of surveyors understanding the consequences of the tests for qualification under the Act, for example, was stressed in a breach of duty of care case. In this the valuer did not remind the client of the consequences under the Act of agreeing to a rateable value in excess of £1,500 and was held to be negligent: see *McIntyre* v *Herring Son & Daw* [1988] 1 EGLR 231.

The principal precedents on enfranchisement terms include: *Kemp* v *Josephine Trust Ltd* (1970) 217 EG 351; *Farr* v *Millersons Investments Ltd* (1971) 218 EG 1177, in which the Lands Tribunal described the generally recognised approaches including three alternative means of calculating the modern ground rent; *Official Custodian for Charities* v *Goldridge* (1973) 227 EG 1467, in which the Court of Appeal disapproved of the "adverse differential" approach; *Norfolk* v *Trinity College, Cambridge* [1976] 1 EGLR 215 and *Lloyd-Jones* v *Church Commissioners for England* [1982] 1 EGLR 209, the first two cases under the additional valuation rules of section 9 (1A) introduced by the 1974 Act; and the Lands Tribunal case of *Hickman* v *Phillimore Kensington Estate Trustees* (1985) 274 EG

1261, upheld in the Court of Appeal in *Mosley* v *Hickman* [1986] 1 EGLR 161, which confirmed that, by first extending a lease, certain leaseholders could enfranchise more cheaply. While the option of extending the leases of houses in the higher rateable value bands may still in some cases be a lower cost approach to expanding an interest for a tenant, the Hickman loophole was closed by section 23 of the Housing and Planning Act 1986.

The decision in *Pearlman* v *Harrow Keepers and Governors* [1978] 2 EGLR 61 dealt with tenant's improvements and qualification for enfranchisement under the 1974 Act amendments. On the issue of what is "a house" for enfranchisement purposes, a majority decision of the House of Lords in *Tandon* v *Spurgeons Homes Trustees* [1982] 2 EGLR 73 held that the tenant of a shop in a parade occupied together with living accommodation above qualified even though the shop part was also a protected business tenancy. The Duke of Westminster has taken a number of cases to the Court of Appeal and one to the House of Lords over the issue of what is a "low rent" in the setting of the qualification provisions of the Act: see studies 1 and 2 (*infra*).

The provisions of the Housing Acts, which granted public sector tenants the right to buy long leases of flats, granted wider rights than those which applied to long-leased flats in the private sector which were originally excluded from the leasehold reform provisions. The Landlord and Tenant Act 1987 conferred on groups of tenants of flats certain rights of first refusal to acquire their landlord's reversion and the Leasehold Reform, Housing and Urban Development Act, 1993 further extended these provisions with the right to collective acquisition of a landlord's freehold reversion on blocks of flats or for individual occupying tenants to have an additional 90 years added to their lease terms.

Sections 89–93 of the Housing Act 1996 introduced amendments to the Landlord and Tenant Act 1987 to deal with contractual arrangements structured to avoid the "rights of first refusal", granted to certain flat tenants in the 1987 Act and made it an offence for a landlord not to comply with the provisions of Part I of the 1987 Act. Schedule 6 to the 1996 Act introduced revised procedures for a landlord to serve offer notices on tenants with rights of enforcement by certain flat tenants against a purchaser.

A proposition was made at (1979) 249 EG 31 – "Even freeholders may have human rights" by Harry Kidd – that the Leasehold Reform Act 1967 was contrary to the European Convention on Human Rights as it can allow the expropriation of property unjustly and without fair compensation, other than in the public interest. Following this line of reasoning the principle of the enfranchisement price provisions was questioned by the European Commission on Human Rights and before the European Court of Human Rights at the instigation of the Trustees of the Duke of Westminster. The application was rejected and no breach of the convention was held to have occurred (1984) 8 CSW 633 and *The Times*, February 22 1986.

Following the abolition of the domestic rating system by the Local Government Finance Act of 1988, the Local Government and Housing Act of 1989 conferred powers on the Secretary of State to make consequential amendments to the 1967 Act. By SI 1990 No 434 and SI 1990 No 701, a new section 1 *(1)(a)* of the 1967 Act was inserted to deal with houses qualifying for enfranchisement after April 1 1990, where the house had no rateable value on March 31 1990. In such cases, and where the lease is after March 31 1990, a formula is to be applied to compute an "R" value, which was not to exceed £25,000 for the house to qualify. These provisions are, to a degree, superseded by the effect of the 1993 Act's amendments to the qualification rules for houses.

Leasehold Valuation Tribunals have jurisdiction to determine the terms of acquisition of interests under the 1993 Act and to approve estate management schemes. This jurisdiction is to be exercised by rent assessment committees acting as leasehold valuation tribunals and the regulations dealing with this are The Rent Assessment Committee (England and Wales) (Leasehold Valuation Tribunal) Regulations 1993 SI 1993 No 2408. Following the conferment of new jurisdiction on leasehold valuation tribunals by the Acts of 1985, 1987 and 1996, with a right of appeal to the Lands Tribunal, the Lands Tribunal, (Amendment) Rules 1997 (SI 1997 No 1965) now require leave to appeal to be obtained from the leasehold valuation tribunal.

Reference to the Act and sections of statute in this chapter are to the Leasehold Reform Act 1967 unless otherwise stated. The *Handbook of Leasehold Reform*, Sweet and Maxwell (1988) but currently out of print, is a loose-leaf manual on the subject. It includes digests of many earlier cases of note, including court decisions, Lands Tribunal determinations and Leasehold Valuation Tribunal determinations, many of the latter of which are unreported. For unreported cases referred to in this chapter which are dealt with in the *Handbook of Leasehold Reform*, references to the digests contained therein are given. HBLR LVT 108, for example, refers to the leasehold valuation tribunal case of *Booer* v *Clinton Devon Estates* (1989) which is found digested as LVT 108 in the handbook. Unreported Lands Tribunal determinations are referred to with the Tribunal reference such as LRA/39/1997.

Study 1

Enfranchisement of a long term of lease

This illustrates enfranchisement qualification and price assessment under the Act for an interest in a house held with a long term of lease unexpired.

Study Facts

A terraced house in the Midlands, built 50 years ago, is held on lease with 75 years' unexpired term at a rent of £15 pa. The rateable value of the house was £140 at March 23 1965. The present tenant has occupied the house for the last seven years as assignee of an earlier tenant. An estate of freehold ground rents in the same locality totalling £100 pa with reversions in 80 years was recently sold as an investment at auction for £950. An enfranchising tenant on the same estate, with a 78 years' unexpired term, recently paid 10.5 YP for his freehold on your advice.

Qualification

Prior to November 1 1993, the Act applied only to houses with a certain rateable value and within other financial limits. There were six rateable value tests to apply in assessing whether interests qualify to enfranchise. Compliance with any one would permit enfranchisement. The tests were:

(a) was the rateable value not more than £200 (£400 in Greater London) on March 23 1965 or at the commencement of the tenancy or

(b) was the rateable value not more than £500 (£1,000) in the valuation list on April 1 1973 or

(c) was the rateable value not more than £750 (£1,500) on April 1 1973 and was the tenancy created before February 18 1966 or

(d) if the rateable value in the above cases was greater than the limits given, would the figure be within the limits were any tenant's improvements to be discounted from the rateable value or

(e) where the provisions of the Act would apply to a tenant of a house, but for the financial limit being exceeded, the same right to acquire the freehold applies if the limits of rateable value of £750 (£1,500 in Greater London) were not exceeded or

(f) if a tenancy was entered into on or after April 1 1990 and the house and premises had no rateable value at March 31 1990 the "R" value does not exceed £25,000 under the following formula:

$$R \quad = \quad \frac{P \times 1}{1 - (1 + I)^{-T}}$$

where P is any premium paid for the lease or 0 if no premium was paid.

I is *0. 06*

T is the term of years of the lease.

Since November 1 1993 the rateable value limits have effectively been abolished for those acquiring a freehold under the Act. They remain relevant for qualifying to have an extended lease or to determine the correct approach to enfranchisement price.

The rateable value at the appropriate day (see Section 1 (4) and (5) of the Act), in this case March 23 1965, was less than £200. The house is held on a long lease of over 21 years at a low rent (1) which was less than two-thirds of the rateable value at the appropriate day. The premises appear to be within the definition of a house (2) and have been occupied (3) by the enfranchising tenant for at least the last three years (4) as his only or main residence. The tenant qualifies to serve notice on the freeholder desiring to have the freehold or an extended lease (5) (6).

If the rateable value was more than £500 (£1,000 in Greater London) and the tenancy created after February 18 1966 or the "R" value for a post-March 31 1990 lease is greater than £16,333 the question of different assumptions under section 9 (1A) and a payment of marriage value could apply in this case: see *Lowther* v *Strandberg* (1985) 274 EG 53. If the tenant had qualified with a house with a rateable value of over £750 (£1,500 in Greater London) or under the 1993 Act's additional qualification test he would only qualify to have the freehold under the assumptions in section 9 (1C) and not an extended lease at a modern ground rent.

Analysis of comparables

The auction result indicates a market investment yield on these ground rents of 10.5%. The enfranchisement settlement shows a discount rate of 9.5%. From this information a reasonable enfranchisement capitalisation rate is assumed as 10%.

Valuation for enfranchisement price (7)	
Term	£ pa
Ground rent	15
YP in perp (8) at 10% (9)	10
Enfranchisement price (10) (11)	£150

Reversion
Of nominal value only (8)
Plus landlords fees & costs (12)

Notes

(1) The full rules on qualification are in sections 1 to 4 of the 1967 Act as amended by section 118 and Schedule 8 to the Housing Act 1974, section 141 and Schedule 21 to the Housing Act 1980 and sections 63 to 68 of the Leasehold Reform, Housing and Urban Development Act 1993. Section 4 defines a low rent in certain cases as being less than two-thirds of the letting value and in *Manson* v *Duke of Westminster* [1981] 2 EGLR 78 the Court of Appeal held that the "letting value" for this purpose should include the annual equivalent of the premium paid: see also *Collin* v *Duke of Westminster* [1985] 1 EGLR 109 and *Johnston* v *Duke of Westminster* [1986] 2 EGLR 87, which both involved the issue of "low rent". Tenancies not at a low rent under the section 4(1) test of the Act may now qualify under the test in section 4(A) introduced by the 1993 Act. This test covers tenancies at zero rent or those at rents satisfying the new tests in relation to rateable values or letting values set out in section 4 (A).

(2) A "house" is defined in section 2 and while it does not include horizontally divided flats themselves it can include a whole house which is converted into flats or part of a building divided vertically as long as there is no part of other premises above or below: see article at (1974) 229 EG 1165 "Leasehold Reform Act 1967: Meaning of 'House'" by W A Leach' see also *Peck* v *Anicar Properties Ltd* (1970) 216 EG 1135, *Wolf* v *Crutchley* (1970) 217 EG 401, *Baron* v *Phillips* [1978] 2 EGLR 59, and *Tandon's* case *(supra)*. In *Cresswell* v *Duke of Westminster* [1985] 2 EGLR 151 a terraced house with accommodation over a side access passage was held to be within the definition of a house

for the purposes of section 2 (2). See also *Malpas* v *St Ermin's Property Ltd* [1992] 1 EGLR 109 where the Court of Appeal confirmed that a house converted horizontally into two flats could be enfranchised where the respondent held the long lease of the whole premises and occupied part.

(3) Section 1 allows enfranchisement rights to a tenant "occupying . . . as his residence". In *Poland* v *Earl Cadogan* [1980] 2 EGLR 75 the tenant was overseas having, without effect, instructed that the house be let. Later, mortgagees took possession for a time. The Court of Appeal held that these were not periods of occupation as his residence: see also *Fowell* v *Radford* (1969) 213 EG 757. In *Duke of Westminster* v *Oddy* [1984] 1 EGLR 83, a tenant who held the leasehold interest in a house as a bare trustee for a company was held not to be a tenant with rights under the Act even though he occupied the house with his family.

(4) For enfranchisement qualification section 1 (1)(b) required occupation for the last five years or for periods amounting to five years in the last 10. Schedule 21, 1(1) to the 1980 Act amended each period of five to three years, ie three out of the last 10.

(5) The form of notice to be used by a tenant under the Act is set out in the Leasehold Reform (Notices) Regulations 1997 (SI 1997 No 640).

(6) In *Oliver* v *Central Estates (Belgravia) Ltd* [1985] 2 EGLR 230 a house in London had a rateable value of £347 on March 23 1965 but £1,347 on April 1 1973. The LVT held that the rateable value at the relevant time was the 1973 figure and the enfranchisement price fell to be assessed under section 9 (1A). In *MacFarquhar* v *Phillimore* [1986] 2 EGLR 89 the Court of Appeal considered cases where the rateable values of two London houses exceeded £1,500 on April 1 1973 but subsequent proposals were agreed to reduce these retrospectively. It held that the altered rateable values were the relevant values for the purposes of the 1967 Act and the tenants qualified to enfranchise. The distinction between the relevance of rateable values for qualification purposes and for the purpose of determining the basis for enfranchisement price assessment should be noted: see *Oliver*'s case (*supra*). This is set out in section 9 of the Act.

(7) The valuation date is the date of the tenant's notice: sections 9 (1) and 37 (1).

(8) The reversion to the modern ground rent in 75 years can be ignored although ones as far distant as 56 years have been valued: see *Gordon v Lady Londesborough's Marriage Settlement Trustees* (1974) 230 EG 509 and *Uziell Hamilton v Hibbert Foy* (1974) 230 EG 509. In *Collins v Howell-Jones* [1981] 2 EGLR 108 a leasehold valuation tribunal capitalised a 50 year term of a £5 pa ground rent and then the reversion to the modern ground rent deferred 50 years both at 7%. Leasehold Valuation Tribunals have valued reversions after long terms of lease, including one after 72 years in *Wright v Hawkins* (1989) HBLR LVT 110 and 79 years in *Jong v Peachey Property Management Co Ltd* (1981) HBLR LVT 11 and *Pemberton v Alders Ltd* (1982) HBLR LVT 19. If the rateable value was over £500 and the approach adopted in the *Lowther* case (*supra*) were to be applied, then the reversion could be sufficiently significant to be included here. In *Cummings v Severn Trent Water Authority* (1985) HBLR LVT 69 a ground lease with reviews to "current market value", the first in 14 and a half years and then after a further 30 and 60 years, was dealt with by reverting in perpetuity to a section 15 rent in 14 and a half years.

(9) The 10% rate follows various Lands Tribunal decisions, but in this case is supported by the all-important comparables.

(10) See similar decisions in *Jenkins v Bevan-Thomas* (1972) 221 EG 640 (10% basis), *Barber v Eltham United Charities Trustees* (1972) 221 EG 1343 (10% basis), *Janering v English Property Corporation Ltd* (1977) 242 EG 388 (11% basis), and *Ugrinic v Shipway (Estates) Ltd* (1977) 244 EG 893 (9% basis). An exception was *Cohen v Metropolitan Property Realisations* [1976] 2 EGLR 182 where 7% was adopted for a 59 year term. In *Yates v Bridgwater Estates Ltd* (1982) 261 EG 1001 (LVT) a leasehold valuation tribunal valued a £3.62 pa ground rent receivable for 971 years at £10, a capitalisation rate of interest of 36.2%. Leasehold Valuation Tribunal determinations for long lease terms in the last few years have typically been at between 5% and 10% discount rate with a number of determinations as high as 15 %, 17.5 %, 19% and even 50% where the rent was very low and the term very long.

(11) In its decision in *Re Castlebeg Investments (Jersey) Ltd's Appeal* [1985] 2 EGLR 209, the Lands Tribunal accepted the landlord's unchallenged evidence of long-term ground rents selling at 16 years' purchase, 6.5% and more, when linked with an obligation on tenants to insure the premises through the

landlord's agency: see *Lynch* v *Castlebeg Investments (Jersey) Ltd* [1988] 1 EGLR 223 in which the LVT capitalised the potential insurance commission at 6.5 years' purchase as against 14.28 years' purchase for the ground rent. In *Wells* v *Hillman* (1987) HBLR LVT 95 insurance commission rights were reflected by an extra two years' purchase, while in *Pabarti* v *Calthorpe Estate Trustees* (1990) HBLR LVT 119 the insurance commission was capitalised at just over three years' purchase. Both these latter cases were Leasehold Valuation Tribunal determinations. In *Divis* v *Middleton* (1983) 268 EG 157, a ground rent with fixed rent increases in 23 and 56 years and to full value in 89 years was capitalised at 7%.

(12) Section 9(4) of the Act requires, *inter alia*, an enfranchising tenant to bear the landlord's costs of any valuation of the house and premises. In *Naiva* v *Covent Garden Group Ltd* [1994] EGCS 174 the Court of Appeal confirmed that the landlord's costs of an LVT reference were not recoverable.

Study 2

A lease extension of 50 years

This further considers enfranchisement qualification together with the assessment of a modern ground rent on a 50-year lease extension to a house held with a short term unexpired. Extensions are rare in practice, as they may not give the tenant such a good deal, but they have to be assumed in enfranchisement cases where a tenant wishes to buy the freehold of a house with a rateable value of less than £500 (£1,000 in London) under the section 9(1) original provisions of the Act. A tenant wanting a 50-year extension and falling in the £500 to £750 rateable value band (£1,000 to £1,500 in London) would also qualify to take such an extension at a modern ground rent under section 15 and not on the less favourable assumptions to be adopted for assessing enfranchisement price under section 9 (1A) which would apply if the tenant wished to take the freehold. For such lessees extension may be more beneficial: see article at (1983) 268 EG 876 at p978 and further in studies 7, 8 and 9.

A 98-year-old house in London is held with one-year unexpired term of lease at a rent of £25 pa. The rateable value of the house at March 23 1965 was £380 and at April 1 1973 was £980. The lessee has held the interest for the last 10 years during which time he has used the ground floor as his main residence for the six years when he was not abroad. A small area of the ground floor has been sublet as a betting office and the upper floors have been sublet as unfurnished flats. Both sublettings contravene the headlease covenants. The house is a "listed building" with an indeterminate but reasonable future life. The adjacent houses do not qualify to enfranchise but the plot might have some redevelopment potential. The plot area is 800 m^2 and evidence shows that the house if improved might sell for £300,000 freehold with vacant possession. Sites in this area are worth about 40% of the freehold vacant possession value of such houses.

Qualification

The house is within the £400 London rateable value limit at March 23 1965, the "appropriate day", in this case under section 1 of the Act and the low rent level. It has been occupied as a main residence for more than three out of the last 10 years. Although the house has been used for other purposes it may still qualify for enfranchisement. (1)

Assessment of modern ground rent for extended lease (2)

	£	£ pa
Standing house approach: (3)		
Entirety value (4) (5)	300,000	
Site value at, say, 40% (6)	120,000	
Section 15 rent at 7% (7)	0.07	
		8,400
Less: factor to reflect possible repossession rights, say 10% (8)		840
Section 15 modern ground rent (9) (10), reviewable after 25 years		£7,560

Notes

(1) See the Court of Appeal cases *Harris* v *Swick Securities Ltd* (1969) 211 EG 1125 and *Lake* v *Bennett* (1969) 213 EG 633. In *Baron* v *Phillips (supra)* the subletting of the ground-floor shop part led to the loss of enfranchisement rights. In *Tandon* v *Spurgeons Homes Trustees* [1982] 2 EGLR 73 a shop in a parade with a flat above occupied by the retailer was held to be a house. In *Methuen-Campbell* v *Walters* [1978] 2 EGLR 58 the Court of Appeal held that the enfranchisement right did not

extend to 1.6 acres of paddock demised with the premises: see also *Gaidowski* v *Gonville & Caius College, Cambridge* [1976] 1 EGLR 72.

(2) Where a tenant requires, and is within the value groups qualifying for, an extended lease the new rent is fixed not earlier than 12 months before the original lease termination date and the tenant bears all costs: section 15 (2) (b). In *Burford Estate & Property Co Ltd* v *Creasey* [1986] 1 EGLR 231 a Leasehold Valuation Tribunal rejected a tenant's claim that acceptance by the landlord of the old ground rent for four years of the extended lease implied agreement of the old rent figure as the revised section 15 rent for the first 25 years of the new lease. Section 15 rents have also been determined on lease extensions reported in *Eckert* v *Burnett* (1987) HBLR LVT 98 and *Duke of Norfolk* v *Brandon* (1988) HBLR LVT 103.

(3) The "standing house" approach is one of the three generally accepted ways of arriving at the section 15 modern ground rent and stems from *Kemp* v *Josephine Trust Ltd* (1970) 217 EG 351 but the Lands Tribunal heavily criticised it in *Miller* v *St John Baptist's College, Oxford* (1977) 243 EG 535 and in *Embling* v *Wells & Campden Charity's Trustees* [1978] 2 EGLR 208. It has stated that the approach should be used only where there is no relevant evidence of a market in residential development land. About 30% of the cases referred to the Lands Tribunal and over 40% of those heard by Leasehold Valuation Tribunals have led to determinations using the standing house approach.

(4) Entirety value has become a way of expressing the full freehold vacant possession value. There may not be evidence on the basis required by the Act but there should normally be evidence to support the entirety value: see *Carthew* v *Estates Governors of Alleyn's College of God's Gift* (1974) 231 EG 809.

(5) As the modern ground rent is to be the "letting value of the site for the uses . . . of the existing tenancy, other than uses which by the terms of the new tenancy are not permitted", (section 15 (2) (a)), there may be some argument here as to the basis for the entirety value: see *Lake* v *Bennett* (1971) 219 EG 945, where the best entirety value for the house was the value as two maisonettes, the most profitable use permitted by the lease. In *Kingdon* v *Bartholomew Estates Ltd* (1971) 221 EG 48 the basis for entirety value was taken to be mixed commercial and residential use. In *Barrable* v *Westminster City Council* (1984) 271 EG 1273, the entirety value was taken at the higher value as if

the house was converted into three flats, even though it was used as a single dwelling at the date of enfranchisement.

(6) The approach may be acceptable where the house is likely to remain standing for the foreseeable future. This figure was adopted in *Carthew*'s case *(supra)* but the percentage accepted will depend on locality, building costs, site attributes, evidence etc. In decided cases the figure has ranged between 11.5% and 50% of entirety value and the percentage to be adopted depends very much on location. In the Kensington and Hampstead areas of London, for example, 40% has often been applied and 50% was determined by the Lands Tribunal in *Cadogan Estates Ltd* v *Hows* (1989) 2 EGLR 216. In North and East London figures of 27.5% to 30% have been used while in South Wales, the Midlands and the North figures of between 20% and 30% are more usual. In Sheffield site value proportions as low as 11.5 % and 13 % have been determined by Leasehold Valuation Tribunals in the cases *Duke of Norfolk* v *Bell* (1982) 263 EG 445 and *Duke of Norfolk* v *Brandon* (1988) HBLR LVT 103. In the *Embling* case *(supra)* the Lands Tribunal was critical of the valuers' arbitrary approach to the percentage for site value.

(7) This percentage will depend on the evidence available and decisions by the Lands Tribunal have ranged between 6% and 8%, but 7% seems to have become the generally accepted figure. In *Windsor Life Assurance Company Ltd* v *Buckley* (1995) LRA/5/1994 the Lands Tribunal applied 7% to a term ground rent of £45 for 70½ years and 6.5% to the Section 15 reversion. The Lands Tribunal in an appeal by Mrs JB Taylor (1998) LRA/10/1998 applied a 6.5% return to compute the section 15 rent for a reversion in 65½ years and capitalised that at 6.5%. The term ground rent was discounted at 7%. Yields of 6% and 6.5% have more recently been applied by the tribunals on the basis of evidence particularly in prime central London areas.

(8) Section 17 reserves to the landlord the future right to possession for redevelopment where a tenant takes an extended lease. In *Carthew*'s case *(supra)* it was accepted that the possibility of repossession might reduce any potential rental bid under section 15.

(9) Section 15 modern ground rents calculated as percentages of "fair rents" were held to be inappropriate in *Carthew*'s case *(supra)*.

(10) An alternative method might be the "new for old approach" see Study 5.

Study   3

Enfranchisement price with a short lease unexpired

This study examines the possible enfranchisement price on the facts assumed in Study 2, as the tenant in that case must look at the alternative merits of buying the freehold rather than taking the 50-year lease extension. The approach adopted takes account of the reversion to the full value of the house and premises after the assumed 50-year extension has expired. This follows the decision of the Lands Tribunal in *Haresign* v *St John the Baptist's College, Oxford* (1980) 255 EG 711, but whether such a reversion is built into the valuation will depend on the significance of it and the strength of evidence to justify such an approach. As the rateable value at the relevant time, the date of the tenant's notice to have the freehold, was not more than £1,000 the price will be under the assumptions required by section 9(1), which are the original ones excluding any marriage value.

Valuation for enfranchisement price assessment

Term		£ pa	£
Rent receivable		25	
YP 1 year at 7% (1)		0.935	23
Reversion to section 15 rent figure			
adopted as assessed in Study 2		7,560	
YP 50 years at 7%	13.8		
PV of £1 in 1 year at 7% (3)	0.935	12.90	£97,524
Reversion to standing house			
value (2) in 51 years		£300,000	
PV £1 in 51 years at 7%		0.032	
			9,600
Enfranchisement price of the			
freehold interest say			£107,147
Say			£107,100

Notes

(1) Although 7% has been adopted in numerous LVT determinations since 1980, in *Haresign's* case (*supra*) 6% was adopted in discounting a term rent for three years and the reversion to the section 15 rent. 7% was adopted only for the reversion to the standing house value at the end of the notional lease extension. However, in the case of *Lowther* v *Strandberg* [1985] 1 EGLR 203, the Lands Tribunal approved the adoption

of 9% throughout the valuation. The figure finally adopted will depend on the relative strength of any supporting evidence. In *Speedwell Estates Ltd* (1998) LRA/070/97 the Lands Tribunal applied 7% to a term rent fixed for 38 years and 6.5% to the reversionary section 15 rent.

(2) In *Haresign's* case (*supra*) it was argued successfully that, as the residue of the contractual term was so short, the three-stage basis of taking into account the landlord's reversion to freehold possession at the end of the lease extension was sufficiently material to be included: see also, for example, *Lowther's* case (*supra*.). In *Ball* v *Johnson* (1973) 226 EG 470 a reversion to a development site after the 50-year lease extension was allowed for where the house had additional land with it not likely to remain as garden indefinitely. In *Griffith* v *Allen* (1989)) HBLR LVT 115 a Leasehold Valuation Tribunal determined a price reflecting a reversion to standing house value after a combined term and 50-year lease extension giving a total period of 84 years. A discount rate of 7% was used for the term and 50 years' extension, while 8% was applied to the *Haresign* reversion.

(3) Dependable evidence is often difficult to obtain. In *Letorbury Properties Ltd* v *Ivens* (1982) 265 EG 51 the sale of a freeholder's interest at auction a few days after the enfranchisement notice was held to be the most dependable market evidence.

Study 4

Enfranchisement against two superior interests and alternative approaches to section 15 rent

This illustrates an approach to assessing enfranchisement price for a house held with a short term of lease but with two superior interests against which to enfranchise. The house has a limited life but the freeholder wishes to impose covenants on the freehold which will restrict the future use and thereby the value of the premises.

The house is on a 0.2 ha site and is held with a five-year unexpired term of sublease at £10 pa from the head leaseholder who holds an eight-year unexpired term from the freeholder at a rent of £5 pa. The occupying subleaseholder qualifies to enfranchise and has recently served valid notices to acquire both superior interests. The rateable value was less than £500 (£1000 in

Greater London) on March 31 1990 and the enfranchisement price falls to be assessed under section 9(1) of the Act. It has been agreed that the freeholder will impose covenants in the conveyance to the effect that not more than one house be erected on the site, that any new design and layout be within certain constraints, and that there be various domestic limitations. The existing house, after much-needed improvements, would have an entirety value of £90,000. Comparable sites have sold for £7.50 per m^2 for flat development, planning permission for which is readily forthcoming in the neighbourhood. Large single house plots with planning permission but subject to restrictive covenants such as those to be imposed are worth about £180,000 per ha.

Valuations for enfranchisement price assessment (1)

Section 15 modern ground rent (2)

	£	£	£ pa
Cleared site approach:			
0.2 ha restricted use value (3)			
0.2 × £180,000 site value		36,000	
Section 15 rent at 6%		0.06	
			2,160
Or			
Standing house approach:			
Entirety value	90,000		
Site value at 30% (4)	0.3		
		27,000	
Section 15 rent at 6%		0.06	
			1,620

Assume that the cleared site approach is the most appropriate in this case (5) and the section 15 modern ground rent is £2,160 pa

Enfranchisement price of the freehold interest

Term	£ pa	£
Rent receivable	5	
YP 8 years at 6%	6.2	
		31
Reversion to section 15 rent	2,160	
YP in perp (6) at 7% deferred 8 years	8.3	
		17,928
		17,959

Enfranchisement price of freeholder's interest

Say	£18,000

Enfranchisement price of the headlessee's interest

Term		£ pa	£
Rent receivable		10	
Rent payable		5	
Net income		5	
YP for 5 years at 7% and 4%			
(Tax at 40%) (7)		2.6	
			13
Reversion to section 15 rent		2,160	
Rent payable		5	
Net income		2,155	
YP 3 years at 8% and 4%			
(Tax at 40%)	1.6		
PV £1 in 5 years at 8%	0.68		
		1.1	
			2,371
			2,384

Enfranchisement price of the headlessee's
interest (8) (9) (10)
Say   £2,400

Notes

(1)   Provisions on enfranchisement by subtenants are in section 5(4) and Schedule 1. The "reversioner", in this case the freeholder, acts for all the superior interests in dividing up the total enfranchisement price: see *Goldsmiths' Company* v *Guardian Assurance Co Ltd* (1970) 216 EG 595; *Hameed* v *Hussain* (1977) 242 EG 1063; *Nash* v *Central Estates (Belgravia) Ltd* (1978) 249 EG 1286; *Burton* v *Kolup Investments Ltd* (1978) 251 EG 1290; *Mortiboys* v *Dennis Fell Co* (1984) HBLR LVT 49 and *Pilgrim* v *Central Estates (Belgravia) Ltd* [1986] 1 EGLR 234. The Leasehold Valuation Tribunal determination in *Booth* v *Bullivant* (1987) HBLR LVT 97 dealt with enfranchisement against the freeholder and head leaseholder by a tenant with a 970-year term. It awarded a nominal £10 to each of the freeholder and head leaseholder for incomes of £1.15 and £2.35 pa receivable for 970 years.

(2   In *Farr* v *Millersons Investments Ltd* (1971) 218 EG 1177 it was suggested that the valuer should use one main approach to the section 15 site value and one of the other accepted approaches

as a check. As the life of the property is unsure, the so-called "new for old approach" may have been appropriate here: see Study 5.

(3) Section 10 deals with the restrictions which may be imposed on the freehold title and in the event of dispute these matters may be referred to a Leasehold Valuation Tribunal with an appeal to the Lands Tribunal. Restrictions on the freehold title to be conveyed may reduce the enfranchisement price but the rights to be assumed in assessing the section 15 rent must be borne in mind: see *Buckley* v *SRL Investments Ltd* (1970) 214 EG 1057; *Peck* v *Hornsey Parochial Charities Trustees* (1970) 216 EG 943; *Hulton* v *Girdlers Company* (1971) 219 EG 175; *Grime* v *Robinson* (1972) 224 EG 815 and *Barrable* v *Westminster City Council* (1984) 271 EG 1273 for examples of covenants being imposed on the freehold titles conveyed: see note (5) to Study 2 (*supra*).

(4) This percentage is assumed but would have to be justified: see note (6) to Study 2 (*supra*).

(5) See the Lands Tribunal's decision in *Farr*'s case (*supra*).

(6) The effect of the notional 25-year rent review and the reversion after 50 years may be ignored in the calculation: see *Farr*'s case (*supra*) but see also *Haresign*'s case (*supra*).

(7) This interest is valued on a conventional dual rate tax adjusted basis as it is a short leasehold interest. Although differential risk rates are applied with 6% to the term and 7% to the reversion in this study, in many tribunal decisions the same rate has been applied to both the term the reversion and in recent cases a lower discount rate to the reversions where these will have review provisions. Wholly net of tax valuations were rejected in *Perrin* v *Ashdale Land & Property Co Ltd* (1971) 218 EG 573. The tax rate is a negotiable issue depending on timing, length of lease unexpired and current rates.

(8) Marriage of the head leasehold and freehold interests prior to enfranchisement can put up the total enfranchisement price a little in this sort of case.

(9) The House of Lords in *Jones* v *Wrotham Park Settled Estates* [1979] 1 All ER 286 confirmed the validity of a freeholder creating an intermediary headlease which increased the total costs of enfranchisement from £300 to £4,000. This loophole was closed by the Leasehold Reform Act 1979, which requires that the enfranchisement price cannot be artificially increased by transactions involving the creation of new intermediary or

similar interests or the alteration in the terms of the lease after
February 15 1979.

(10) In this case had the head leaseholder a profit rent of not more
than £5 pa and a reversion of not more than one month the
enfranchisement price would be on a formula basis as set out
in Schedule 21 (6) to the Housing Act 1980. The formula is:

$$\text{Price (P)} = \frac{\text{Profit Rent (R)}}{2\frac{1}{2}\% \text{ Consols Yield (Y)}} - \frac{R}{Y(1-Y)^n}$$

and it capitalises the term profit rent at the 2.5% Consols yield at
the valuation date. The formula counts any part of a year as a
whole year. There is no appeal to a Leasehold Valuation Tribunal on
the enfranchisement price in such cases: see *Afzal* v *Cressingham
Properties Ltd* (1981) HBLR LVT 4.

Study 5

Enfranchisement price, comparables and the adverse differential issue

This considers further the methods of arriving at site value in the
original "extended lease at a modern ground rent" enfranchisement hypothesis of section 9 (1) of the Act and examines the now
generally discredited concept of the "adverse differential". It also
looks at the role of enfranchisement settlements as evidence and
the use of other risk rate approaches.

Assume a large residential estate, the subject of much
enfranchisement with most of the houses held on ground leases
expiring shortly at rents of £15 pa each. The landlord's valuers have
settled several enfranchisement claims at figures of about £29,000
for 5 year unexpired terms. The frontage, depth and amenity of the
plots vary although the plot areas are all very similar. The
landlord's valuers have established a comprehensive approach to
site values on the estate and have developed techniques to adjust
prices to allow for minor differences between the plots.

The study considers enfranchisement price from both the
landlord's and the tenant's point of view in order to draw out
various matters which might be the subject of negotiation. The
interest concerned qualifies for enfranchisement within the less
than £500 (£1,000 in Greater London) rateable value bands and the
price falls to be assessed under the assumptions of section 9 (1). The
same assumptions apply following the abolition of domestic rating

to new houses which had no rateable value at March 31 1990 and the "R" value calculated under section 1 (1) (*a*) (ii) is no more than £16,333. The lease has an unexpired term of 5 years at £15 pa. The house is agreed to have a limited life and the plot has a frontage of 10 m, a depth of 40 m and an area of 400 m².

Enfranchisement price using arguments to the landlord's advantage

Term	£	£ pa	£
Rent receivable		15	
YP 5 years at 6% (1)		4.2	
			63
Reversion to section 15 rent			
Cleared site approach: (2)			
10 m frontage at landlord's			
adjusted rate per m frontage for			
plots of 400 m², 10 m at £3,000 (3)	30,000		
Adjustments to site value: (4)			
(i) for better			
location/amenity + 15%	4,500		
(ii) 10 m deeper than average			
plot + 10%	3,000		
Site value Say	37,500		
Section 15 rent at 6%	0.06		
Section 15 rentt		2,250	
YP in perp at 6% deferred 5			
years (5) (6)		12. 5	
			28,125
			28,188
Landlord's view of enfranchisement price (7)			
Say			£28,200

Notes
(1) A rate of 6% or 7% was generally adopted for the term following *Farr*'s case *(supra)* and the lower rate favours the landlord.
(2) The "cleared site" approach is used as the house has a limited life.
(3) A similarly scheduled approach to site values on a large estate was used by the landlord's valuers in *Siggs* v *Royal Arsenal Co-operative Society Ltd* (1971) 220 EG 39. The Lands Tribunal was noncommittal, but accepted the approach in that case.

(4) These adjustments are given solely by way of illustration and could only be used in practice with clear and careful justification.

(5) The section 15 rent has been recapitalised at the same 6%. This follows *Official Custodian for Charities* v *Goldridge* (1973) 227 EG 1467 in which the Court of Appeal disapproved of the "adverse differential". This differential was earlier adopted by the Lands Tribunal to take account of section 82 of the Housing Act 1969 and its exclusion of the tenant's bid for the freehold in assessing enfranchisement price: see *Farr's* case (*supra*) and many subsequent cases where site values were decapitalised at about 6% and then recapitalised at 8% giving an adverse differential of 2%. In *Grainger* v *Gunter Estate Trustees* (1977) 246 EG 55 an attempt to use a larger adverse differential was rejected by the Lands Tribunal. In *Wilkes* v *Larkcroft Properties Ltd* [1983] 2 EGLR 94, the Court of Appeal held that evidence of "en bloc" sales of ground rents did not justify a claim that the adverse differential should be applied as a matter of law. The court's decision also provides a useful examination of the nature of evidence required to support a decision by the Tribunal.

(6) In *Siggs* case (*supra*) it was held that the large estate landlord might bid for leaseholds as they came on to the market in order to reap marriage value. This enabled the larger estate to argue against the "adverse differential" and apply one common rate in arriving at the section 15 rent reflecting the "marriage by sale incentive". The Court of Appeal disapproved of this argument in the *Goldridge* case (*supra*).

Enfranchisement settlements should be adjusted before use as evidence to reflect the "*Delaforce* effect". This is the extra amount a tenant might be willing to pay in a negotiated settlement to avoid the worry, risk and costs of litigation: see *Delaforce* v *Evans* (1970) 215 EG 315 and *Ugrinic's* case (*supra*). In this study settlements are given at about £2,900 and it therefore might be argued that there has been a £800 *Delaforce* effect allowance in this price. In *Wilkes's* case (*supra*) the Court of Appeal held that the Lands Tribunal did not err in declining to make a deduction for the *Delaforce* effect.

Enfranchisement price using arguments to the tenant's advantage

Term	£	£ pa	£
Rent payable		15	
YP 5 years at 8% (1)		4	
			60
Reversion to section 15 rent			
New-for-old approach: (2)			
Sale price of new house on site	120,000		
Less: building costs	99,000		
Section 15 site value	21,000		
Or			
Sale price of new house on site	120,000		
Site value at 20%	0.2		
Section 15 site value	24,000		
Adopt lower	21,000		
Section 15 rent at 8% 0.08	1,680		
YP in perp at 10% (3)			
deferred 5 years		6.2	
Tenant's view of enfranchisement price			10,476
Say			£10,500

Notes

(1) The tenant might argue for 8% or higher if interest rates are high. In *Patten* v *Wenrose Investments Ltd* (1976) 241 EG 396 8% was accepted at all stages of the valuation and in *Lowther*'s case (*supra*) 9% was adopted throughout for houses in Holland Park, London.

(2) The new-for-old approach is an alternative where the house has a limited but indeterminate future life: see *Gajewski* v *Anderton* (1971) 217 EG 885 and *Farr*'s case (*supra*). The figures here are assumed and would need to be supported with evidence.

(3) Although the Court of Appeal disapproved of the adverse differential in the *Goldridge* case (*supra*), there have since been cases before the Lands Tribunal when it has been accepted. The Court of Appeal left the matter open if the concept could be justified with evidence and reason: see *Lead* v *J & L Estates Ltd* [1975] 2 EGLR 200 and *Perry* v *Barry Marson Ltd* (1976) 238 EG 793.

Possible form of valuation for a negotiated settlement in this study

Term	£	£ pa	£
Rent reserved		15	
YP 5 years at 7% (1)		4.1	
			62
Reversion to section 15 rent			
Site value Say	28,500		
Section 15 rent at 7% (1)	0.07		
		1,995	
YP in perp at 7% deferred			
5 years		10.19	
			20,329
Enfranchisement price (2)			20,391
Say			£20,400

Notes

(1) For the settlement figure calculation a rate of 7% has been adopted but this would of course need to be supported by dependable market evidence derived from consistent analysis of market transactions and settlements: see *Custins* v *Hearts of Oak Benefit Society* (1969) 209 EG 239.

(2) No reversion to the "standing house" after 55 years has been included as was in *Haresign*'s case (*supra*), as the house has been stated to have a very limited future life and such an approach would be inconsistent with the cleared site and new-for-old approaches adopted.

Study 6

Rateable value adjustment for qualification to discount tenant's improvements

This examines further qualification for enfranchisement or extension under section 1(4A) of the Act introduced by section 118 and Schedule 8 to the Housing Act 1974 as amended by section 141 and Schedule 21 to the Housing Act 1980.

A bungalow outside London is held on the residue of a long lease with a 10-year unexpired term at a ground rent of £12 pa. The premises had a rateable value of £300 at March 23 1965, one of £805 at April 1 1973 and has a gross external area (GEA or "reduced covered area") of 225 m². The present occupier purchased the premises in 1970 and at once built an extension with a GEA of 25 m². The previous owner had installed central heating and erected a double garage in 1968. The value of the bungalow now, freehold and with vacant possession, would be £175,000 but without the tenant's improvements would be only £90,000. Apart from the rateable values the tenant is assumed to be otherwise qualified under the provisions of section 1(1) and 1(5) of the Act, ignoring the 1993 Act extensions. The rateable values after adjustment to discount the tenant's improvements may not exclude qualification.

Qualification for enfranchisement

The tenant did not qualify under the original 1967 rateable value limit and at first sight does not qualify under the amendments of the Housing Act 1974 as the rateable value is over £750. The tenant may, however, use section 1(4A) of the Act, contained in Schedule 21 to the Housing Act 1980, which enables a tenant who is otherwise qualified to claim a reduction (1) in the notional rateable value for enfranchisement purposes to exclude the annual value of any tenant's (2) improvements (3) (4). The tenant might also qualify under the 1993 Act amendment removing rateable value limits but on a less favourable enfranchisement price basis because the 1993 Act basis omits the right to the 50-year extension of the lease.

Adjustment to the rateable value under section 1 (4A) of the Act and Schedule 8 to the Housing Act 1974

		£	£
Rateable value at 1.4.1973 (5)		805	
Gross value at 1.4.1973		1,000	
Analysis of gross value	225 m² at £3.67		825
	Central heating		80
	Double garage		95
	Gross value		1,000
Assessment without tenant's improvements (exclude 25 m² extension)	200 m² at £3.67		734
	No central heating		–
	No garage		–
	Gross value		734
Less: statutory deductions			151
Adjusted rateable value without tenant's improvements (5) (6) (7)			£583

Therefore, the lessee will qualify to enfranchise but the enfranchisement price will be under the additional section 9(1A) introduced in 1974 as the rateable value is over £500 (7) (8). For houses on leases granted after March 1990 and which had no rateable value at March 31 1990 the "R" value under section 1(1)(*a*)(ii) of the Act should be less than £16,333 for qualification under section 9(1). To qualify under section 9(1A) the "R" value should be between £16,333 and £25,000.

Notwithstanding the above process the tenant might alternatively claim qualification under section 1(A) of the Act, introduced by the 1993 Act, which confers in certain circumstances the same rights to have the freehold as apply to those within the earlier financial limits. In such circumstances no right to an extended lease can be claimed and the enfranchisement price is on the additional basis set out in section 9(1C) introduced by the 1993 Act.

Notes

(1) The tenant first serves a notice on the landlord requiring him to agree to the nature of the improvements to be discounted and proposing a figure for the reduced rateable value. The form of notice is set out in Schedule 8 to the Housing Act 1974 and the tenant must specify the improvements and works concerned. The tenant must bear the reasonable costs of the landlord's investigations of the works of improvement claimed on notices, see Schedule 21 (8) to the Housing Act 1980.

   Failing agreement between the parties the county court may determine the extent and nature of the improvements to be taken into account and the valuation officer may be required to determine the reduced rateable value excluding the value of the improvements.

(2) The provisions regarding the adjustment of rateable value apply to the improvements of both the current and any previous tenants: see Schedule 8 (1) (1) to the Housing Act 1974.

(3) The time scale for the procedures under Schedule 8 is generally mandatory and a tenant may only make a single application under them for that reason: see *Pollock* v *Brook-Shepherd* [1983] 1 EGLR 84. In *Arieli* v *Duke of Westminster* [1984] 1 EGLR 81 the Court of Appeal overruled a county court decision to refuse an extension of time to originate an application to the court under

Schedule 8. In *Johnson* v *Duke of Devonshire* (1984) 272 EG 661 a 12-day extension of the time to refer the matter to the court was held to be within the judge's discretion.

(4) Or addition, for example a garage. Whether the construction of a replacement house is an improvement is not clear but the Court of Appeal in *Pearlman* v *Harrow School Keepers and Governors* [1978] 2 EGLR 61 decided that the installation of central heating in this context was an improvement.

(5) There has been some uncertainty as to the date by reference to which rateable values are to be considered and adjusted: see article at (1983) 266 EG 187 – "Leasehold Reform Act – Notional reductions in rateable value" by Nigel T Hague QC – *Pollock*'s case (*supra*) and *Mayhew* v *Free Grammar School of John Lyon* [1991] 2 EGLR 89. It has been held to be the date of agreement or the date of the certificate issued by the District Valuer *John Lyon's Charity* v *Vignaud* [1992] 2 EGLR 122.

(6) The rating valuation approach illustrated is likely to be adopted by the valuation officer where the landlord and tenant cannot agree to the appropriate reduction.

(7) See *Woodruff* v *Hambro* [1991] 1 EGLR 107 where a tenant who had surrendered a long lease and taken a new one was precluded from applying for a reduction in rateable value in order to qualify for enfranchisement.

(8) There is no appeal against a valuation officer's certificate of adjusted rateable value. In *R* v *Valuation Officer for Westminster and District, ex parte Rendall* [1986] 1 EGLR 163 a tenant argued that the valuation officer's certified adjusted rateable value was incorrect, it being more than £1,500, and sought judicial review. This was dismissed by the High Court and the Court of Appeal.

(9) As the rateable value after adjustment is over £500 and under £750 the enfranchisement price must be assessed under the assumptions in section 9(1A) of the Act, introduced by section 118(4) of the Housing Act 1974.

Study 7

Enfranchisement price where the rateable value is over £500 (£1,000 in London) – the "Norfolk" approach

This illustrates the assessment of enfranchisement price under the provisions of section 9(1A), which do not assume a modern ground rent for the notional 50-year lease extension and do not preclude any marriage value from a tenant's bid. Using the facts adopted in Study 6 and with the rateable value reduced to £583 the tenant now qualifies to enfranchise but at a price to be assessed under the less favourable assumptions introduced by section 118 of the Housing Act 1974 and interpreted by the Lands Tribunal in *Norfolk* and developed further in *Lloyd-Jones* (*supra*): see also notes (4) and (5).

Enfranchisement price assessment

Valuation of lessor's interest excluding marriage value (1)

Term	£ pa	£ pa	£
Rent receivable		12	
YP 10 years at 7%		7	
			84

Reversion to a protected tenancy under the Landlord and Tenant Act 1954 (2) Assume "registered rent" 10% of the unimproved (3) standing house value (4) of £90,000	9,000		
Less: outgoings, say 15%	1,350		
		7,650	
YP in perp at 12 % (4) deferred 10 years		2. 7	
			20,655
Value of lessor's interest excluding marriage value			20,739
(see note (5) for an alternative approach)			
Say			£20,750

Valuation of the lessee's interest excluding marriage value

Term	£ pa	£
Annual value of house, say 10% on the full freehold vacant possession value (4) of £175,000	17,500	
Less: rent payable	12	
Profit rent	17,488	
YP 10 years at 12% (4) and 4% (tax 40%) (6)		3.9
		68,203
Value of lessee's interest excluding marriage value, Say		£68,200

(No value has been put on the reversionary right as a reversion in 60 years is too distant to be significant)

Apportionment of marriage value to arrive at enfranchisement price (7)

		£	£	£
(i)	Value of lessor's interest exclusive of marriage value			20,750
(ii)	Assessment of lessor's share of marriage value			
(a)	Freehold vacant possession value	175,000		
	Less: lessee's improvements	85,000		
			90,000	
(b)	Value of lessor's interest exclusive of marriage value		20,750	
(c)	Value of lessee's interest exclusive of marriage value	68,200		
	Less: value to the lessee of his improvements (8)	33,150		
			35,050	
(d)	Total value of freehold and leasehold interests unmarried (b + c)		55,800	
(e)	Gain on the marriage of the interests ignoring the lessee's improvements (a–d)		34,200	
	Lessor's share of the gain Say 50% (9)		0. 5	
				17,100
(iii)	Enfranchisement price (i) + (ii)			£37,850

Notes

(1)   The study follows the apportionment of marriage value basis used by the Lands Tribunal in *Norfolk* v *Trinity College, Cambridge* [1976] 1 EGLR 215. The principle confirmed in that case was that section 9(1A) of the 1967 Act does not exclude the extra value of the "tenant's bid" where the enfranchisement price is assessed under the section 118 provisions of the Housing Act 1974. The price therefore includes any extra amount that a tenant might pay for the value increase arising on the marriage of the freehold and leasehold interests.

(2)   The principle in section 9(1A) (*b*) is that, instead of reverting to a "modern ground rent", enfranchising tenants revert in this enfranchisement hypothesis to a protected tenancy under the Landlord and Tenant Act 1954.

(3)   The valuation must ignore the value of any tenant's improvements, section 9(1A)(d). It seems that the word "improvement"

here has a wider meaning than for the rateable value reduction provisions in Schedule 8 to the 1974 Act but would exclude maintenance and renewals. They must have added to the value.

(4) The assumptions here are illustrative only and clearly will be open to negotiation. In *Lloyd-Jones* v *Church Commissioners for England* [1982] 1 EGLR 209, the second reported case under the different valuation assumptions introduced by the 1974 Act, it was held that such an approach was inappropriate in London as it was almost unknown for tenants of particular types of house to continue in occupation on a protected tenancy. Tenants, it was stated, normally surrender and renew leases or purchase the freehold. In *Lowther* v *Strandberg* [1985] 1 EGLR 203 the effect of the tenant's bid on a Holland Park, London W14, estate was considered and the Lloyd-Jones approach was adopted even though the reversion was not for another 81 years. A discount rate of 9% was adopted throughout the valuation. See also *Vignaud* v *Keepers and Governors of the Free Grammar School of John Lyon* [1996] 2 EGLR 179.

(5) In the *Lloyd-Jones* case (*supra*) the Lands Tribunal accepted the strong settlement evidence of the landlord's valuers and dismissed, as being out of touch with reality, the tenant's view of the reversion to a fair rent. The approach adopted for the landlord's reversion was simply to deduct 10% from the freehold vacant possession value to reflect the risk of a tenant claiming a tenancy under Part I of the Landlord and Tenant Act 1954.

The Lloyd-Jones case approach to the value of the landlord's interest excluding marriage value would be:

Term	£	£
As before		84
Reversion to unimproved vacant possession value	90,000	
Less: for risk of tenant claiming a tenancy under Part I of the Landlord and Tenant Act 1954		
10%	9,000	
	81,000	
× PV of £1 in 10 years 7%	0.5	
		40,500
Say		£40,584

In this study the enfranchisement price would be little different but in the *Lloyd-Jones* case it increased the enfranchisement price by £68,000.

(6) The lessee's term has been valued on a conventional dual rate tax-adjusted basis. No value has been placed here on the lessee's right to a further protected tenancy although as there may be a loss in value to the lessor there is an argument for such a further element in the valuation.

(7) This is the apportionment approach used in *Norfolk*'s case (*supra*).

(8) This has been found by taking an annual value of £8,500 for the improvements and capitalising it on a dual rate tax-adjusted basis for the lessee's term unexpired.

(9) The percentage applied in the *Norfolk* case (*supra*) but valuers must be cautioned against simply following the approaches adopted in other cases. Market evidence is likely to be of equal or even greater relevance.

Study 8

Further rateable value adjustments as a means of reducing enfranchisement price

This study considers the effect of obtaining a further reduction in the rateable value to below the £500 (£1,000) figure in order to enable the enfranchisement price to be assessed under section 9(1) which is more favourable to the tenant.

The facts are as in Studies 6 and 7 except that on detailed survey of the house and premises it becomes apparent that part of the bungalow dates from 1900 and part from 1922. The gross external area of the premises as originally constructed in 1900 can be shown to be 160 m². The value of the freehold interest without these further improvements is assumed to be £75,000.

Qualification for enfranchisement

Adjustment to the rateable value under sections 1(4A) and 9(1B) of the Act

Gross value at 1.4.1973 (as before)	1,000
Gross value without tenant's improvements	
160 m² at £3.67	587
No central heating	–
No double garage	–

Gross value	587
Less: statutory deductions	126
Adjusted rateable value without tenant's improvements (1)	£461

The tenant qualifies to enfranchise but as the rateable value was less than £500 (outside Greater London) the enfranchisement price will be assessed under the original section 9(1) assumptions of the 1967 Act.

Enfranchisement price assessment (2)

Term	£ pa	£
Rent reserved	12	
YP 10 years at 7%	7	
		84

Reversion to section 15 rent			
Standing house approach (3)			
Entirety value	175,000		
Site value at 25%	43,750		
Section 15 rent at 7%	0.07		
		3,063	
YP in perp at 7% deferred 10 years		7.3	
			22,360
Enfranchisement price (4)			22,444
This compares with £37,850 in Study 7 (5) say			£22,400

Notes

(1) In the Leasehold Valuation Tribunal case of *Effra Investments Ltd v Stergios* (1982) 264 EG 449 a house in London had been converted into flats taking the rateable value over £1,000. Although the tribunal gave the tenant time to take action to reduce the rateable value under Schedule 8 to the Housing Act 1974 he did not do so and the tribunal fixed a high enfranchisement price under the provisions of section 9(1A), closely following the principles of the *Norfolk* decision.

(2) In this case the value of any tenant's bid, which might include a marriage value component, is excluded by section 82 of the Housing Act 1969.

(3) This approach was considered more fully in Studies 3 and 4.

(4) It can be seen from this enfranchisement price that it may well be in the lessee's interest to try to ensure that the rateable value is reduced to a level which brings the enfranchisement price

within the original 1967 Act valuation rules and excludes the marriage value component and the resulting higher enfranchisement prices adopted under section 9(1A) in the *Lloyd-Jones* and *Norfolk* cases (*supra*).

(5) The enfranchisement price assumptions for premises in the £500 to £750 (£1,000 to £1,500 in London) rateable value bands or with an "R" value of between £16,333 and £25,000 are those of section 9(1A). If a tenant in that band of value served a notice to take an extended lease instead, he would be entitled to such at a modern ground rent for 50 years notwithstanding the rateable value level. Subject to the repossession rights of section 17 (see Study 12) he might avoid having to pay the marriage value element yet would still obtain a very valuable interest: see article at (1983) 268 EG 876, 978 and Study 9 (*infra*).

Study 9

Houses in the higher rateable value bands: extending a lease as an alternative to, or prior to, buying a freehold

The Lands Tribunal in *Hickman* v *Phillimore Kensington Estate Trustees* [1985] 274 EG 1261, confirmed by the Court of Appeal in *Mosley* v *Hickman* [1986] 1 EGLR 16, determined that a tenant of a house in the higher rateable value band who had first extended the lease for 50 years at a section 15 rent could enfranchise at a price reflecting that extended lease. Section 23 of the Housing and Planning Act 1986 subsequently amended section 9(1A) of the Act to require that no rights to such an extension be assumed and that if a lease has been extended it is to be assumed to terminate at the original term date. The alternative of lease extension was also examined at (1983) 268 EG 876, 978, where it was shown that in some cases it might be potentially better value for a tenant than enfranchising under the assumptions of section 9(1A). This alternative is still available to otherwise qualified tenants in the higher rateable value band who wish to expand their interest under the Act by a lease extension but without any capital outlay.

This study assumes a house in London with an April 1 1973 rateable value of £1,300 held on a ground lease with 10 years to run. The ground rent is £50 pa and the tenant is assumed to qualify under the provisions of the Act, as amended, the house is agreed as having an unencumbered freehold value of £200,000 at the relevant date.

Lease extension

The tenant of the house can claim under section 14 to have an extended lease of a further 50 years in addition to the present term and such a lease would be in substitution for the existing one. This new lease would be for 60 years, the first 10-year term at the existing rent and the further 50 years at the section 15 rent. The section 15 rent would not be fixed until the last year of the original lease, in this case in nine years time. The new lease would normally provide for a rent review, in this case in 35 years time: section 15(2). The tenant will still have the right to serve a notice to have the freehold up to the time when the original lease would have expired, in this case during the next 10 years: section 16(1)(a). Such an arrangement would ensure that the tenant retains a quite valuable interest at quite a modest cost, in present value terms, without having to borrow any capital. However, the nature of the section 15 rent could mean that the interest is less marketable than it would be if freehold.

Valuation of the tenant's interest after extension

	£ pa	£
Term of existing lease		
Full annual value, say 10% on £200,000	20,000	
Less: ground rent	50	
Profit rent	19,950	
YP 10 years at 10 % and 4 % (tax 40 %) (1)	4,187	
		83,531
Reversion to section 15 rent and extended term (2)		
Full annual value as before	20,000	
Section 15 rent based on site value as 30% of the entirety of £200,000 = £60,000		
Section 15 rent at 9% of £60,000 (3)	5,400	
Profit rent	14,600	
YP 50 (4) years at 10% and 4% (tax 40%) deferred 10 years at 10%	3.476	
		50,750
Total value of tenant's extended interest		134,281
Say		£135,000

Notes

(1) All the component figures would need justification by reference to market evidence.

(2) While the extended lease will in due course become a wasting asset it might appeal to a tenant who simply wishes to remain in occupation and has no intention, or wish, to realise the best financial return from a sale of the house.

(3) Extension of the lease would limit the cost of remaining in occupation to the present value of the liability to pay the section 15 rent. This would be about £21,000 in discounted net present cost terms at 10% today, but this should be compared with the present value of the interest obtained and the cost of an outright purchase of the freehold (*supra*).

(4) It must be remembered that a tenant who has taken an extended lease can be the subject of proceedings for possession for redevelopment under section 17, subject to the compensation right for the value of the extended lease: see Study 12.

Study 10

Higher Rateable Value Bands

Enfranchisement price if notice to have the freehold is served after the lease has been extended

If the leaseholder were to serve notice to have the freehold after the lease was extended, but before November 7 1986, the enfranchisement price was assessed "on the assumption that the vendor was selling . . . subject to the tenancy". The tenancy existing would be the extended one at a section 15 "modern ground rent", not the type of tenancy normally to be assumed under section 9(1A) for houses in the higher rateable value band. The tenant was not, however, excluded from the notional market in which the freehold is being sold and a marriage value bid may be included in an enfranchisement price calculation.

Following the amendment of section 9(1A)(a) by the Housing Act of 1986, introduced to overturn part of the effect of the *Hickman* decisions, the assumption to be applied in computing the enfranchisement price is that even if the lease has been extended at section 15 rent it is to be assumed that it will terminate on the original term date. In such circumstances the enfranchisement price might be computed as in Study 7.

Study 11

Enfranchisement Price for a house qualifying beyond the applicable financial limits in section 1 – section 9 (1C) Approach

Under amendments to the Act introduced by sections 63 to 66 of the Leasehold Reform, Housing and Urban Development Act 1993, a new subsection 9(1C) was added which provides that houses now qualifying under the provisions of the new sections 1A, 1AA and 1B the enfranchisement price is computed in accordance with section 9(1A) with the additional assumptions of section 9(1C). In any such case where there is marriage value to be taken into account, the share of marriage value to which the tenant is entitled shall not exceed 50%. The new section 9(A) provides additional elements of compensation to the landlord where a tenant qualifies under sections 1A or 1B to include:

(i)   any reduction in value of any interest of the landlord in other property resulting from the acquisition;

(ii) any loss or damage which results, to the extent that it relates to the landlord's ownership of any interest in other property;

(iii) any loss of development value, which means any increase in value attributable to the possibility of demolishing, re-constructing or carrying out substantial works of construction on the whole or a substantial part of the house or premises.

Facts

A detached Victorian 6-bedroomed house in a north London suburb with a large garden with various tenant's improvements. The house has rateable values of £530 at April 1 1963 and £2,200 on April 1 1973. It is held with 53 years unexpired of a 85-year lease at a ground rent of £200 pa with a fixed uplift of the ground rent every 15 years to the sum by which ¼% of the capital value of the unimproved house exceeds £200 pa. The tenant qualifies under section 1A(1) of the Act by the extension of the qualification rules beyond the previous rateable value limits. The assumptions to apply to assessment of the price are those in Section 9(1A) as amended by 9(1C). The ground rent will be reviewed to £2,500 in three years time. The freehold vacant possession value is £1,250,000, but the tenant's improvements contribute £200,000 to that value. The value of the leaseholder's present lease is agreed to be £725,000 excluding the value of their improvements.

Term		
Current ground rent	£200	
YP 3 years @ 6% (1)	2.67	£534
Revised Ground Rent		
Reviewed Rent	£2,500	
YP 50 years deferred 3 years @ 6%	13.23	£33,084
Reversion		
Vacant possession value	£1,250,000	
Less tenants improvements	£200,000	
	£1,050,000	
PV of £1 in 53 years @ 6%	0.046	£48,300
Current value of freehold		£81,918

Marriage Value (1)

Freehold VP excluding improvements		£1,050,000
Less		
Current value of freehold	£81,918	
Plus current value of lease		
excluding improvements	£725,000	
		£806,918
Marriage Value		£243,082
Freeholders share 50%		£121,541

Enfranchisement Price under section 9(1C) of the 1967 Act

Value of freehold	£81,918
+ Freeholders share of Marriage Value	£121,541
	£203,459
Enfranchisement Price (2) *Say*	£203,500
+ Any loss or reduction of value of other	
interests of loss of development value (3)	

Notes
(1) This approach follows that applied by the Lands Tribunal in *Re John Lyons Charity* (1996) LRA/4/96. See also the Lands Tribunal in *Trustees of the Eyre Estate* v *Saphir* (1999) LRA/18/47/1988

(2) See also the Lands Tribunal determination in *John Lyons Charity* v *Brett* (1998) (LRA/16/1997) and *Sharp* v *Cadogan Estates* (1998) (LRA/33 & 35/97).

(3)   These are the additional areas of compensation right specifically added to the landlord's entitlement in such cases under section 9(A) where the right arises under section 1A or 1B.

Study 12

Premises with redevelopment potential and the repossession rights under section 17

This examines the assessment of enfranchisement price where the house and the premises have development potential. The problem is how to take account of the landlord's rights to repossession for redevelopment under section 17 of the Act which are available where a tenant has taken advantage of the rights for a lower valued house to have a 50-year lease extension at a section 15 rent. The principles have not been widely tested before the Lands Tribunal or the courts, probably because of the fact that the gains available through marriage via enfranchisement are sufficient to ease successful negotiations. In the Leasehold Valuation Tribunal determination in *Booer* v *Clinton Devon Estates* (1989) HBLR LVT 108 the tribunal reflected the redevelopment right for six plots, but included in the enfranchisement price the full value of a reversion to a section 15 rent. It should be noted that for houses qualifying under section 1A and paying a price assessed under section 9(1C) the landlord specifically has to be compensated for any foregone development value under section 9A.

Facts

A large obsolete house on a 0.7 ha plot in a provincial city is held with a six-year unexpired term at a ground rent of £25 pa. The house had a rateable value of £195 at March 23 1965 and the tenant qualifies to enfranchise under the enfranchisement price assumptions in section 9(1) of the Act assuming a lease extension at a section 15 rent. The value of the plot with planning permission for the best feasible redevelopment would be £250,000 freehold with vacant possession. The value of the site restricted to use for one dwelling would be £55,000 and the entirety value of the house improved would be £120,000, both figures being freehold with vacant possession. The lessors are willing to sell the freehold interest without restrictions on redevelopment on the title and have accepted the lessee's notice of enfranchisement. The lessee requires a valuation of his current interest as he is now proposing to sell to a developer.

Valuation of the lessee's interest for sale (1) taking account of the landlord's redevelopment rights under section 17(2)

Cost of enfranchisement			
Term	£ pa	£	£
Rent payable	25		
YP 6 years at 7%	4.8		
			120
Reversion to full redevelopment value		250,000	
Less: section 17 and Schedule 2 compensation. The value to the lessee of a 50-year lease at a section 15 modern ground rent. (2)			
Annual value of house, Say (3)	12,000		
Less: section 15 rent based on 7% of £55,000	3,850		
Notional annual profit rent	8,150		
YP for 50 years at 10% and 4% (tax 40%)	9.0		
Schedule 2 compensation in 6 years' time (3)		73,350	
Value of reversion less compensation		176,650	
PV £1 in 6 years at 10% (4)		0.56	
Present value of lessor's reversionary rights under the enfranchisement hypothesis			98,924
Total enfranchisement price (5)			99,044
Add landlord's and tenant's enfranchisement fees and costs (6) Say			2,000
Total enfranchisement costs (7)			101,044
Say			101,000
Full development value of the site (as given)			250,000
Less: costs of enfranchisement			101,000
Value of lessee's interest for sale (8)		(9)	£149,000

Notes

(1) A lessee may assign his lease with the enfranchisement rights as soon as a valid notice of enfranchisement has been served: see section 5.

(2) Section 17 reserves to the landlord of a dwelling, where the lessee has taken an extended lease under the Act, the right to repossession for redevelopment. Here it is assumed in assessing enfranchisement price that the tenant has an extended lease; it would appear logical to incorporate in the valuation the notional right of the landlord to repossession. This right is available not earlier than one year prior to the original term date and the landlord must compensate the tenant under Schedule 2 for the value of a 50-year lease at a section 15 modern ground rent.

(3) The basis for and approach to these figures is open to argument and negotiation.

(4) The deferment allows for the assumption that the lessor cannot obtain possession until the end of the term, when he will then have to pay the lessee compensation under Schedule 2.

(5) If the rateable value were over £500 (£1,000 in London), so that the enfranchisement price fell to be assessed under the *Norfolk v Cambridge* and *Lloyd-Jones'* approaches (*supra*), the marriage value component might add further to the enfranchisement price and reduce the value of the lessee's interest, even though the assumptions under section 9(1A) preclude the notional right to repossession under section 17 and assume a reversion to a registered rent rather than a modern ground rent.

(6) The tenant is responsible for landlord's reasonable legal and valuation costs incurred in connection with the enfranchisement: see section 9(4).

(7) No cases from the Lands Tribunal or the courts on the points covered in this study have been reported. The Lands Tribunal briefly considered a related section 17 matter in *Cottingham-Mundy* v *Dover Borough Council* (1971) 220 EG 817, while a Leasehold Valuation Tribunal did consider the matter in *Booer* v *Clinton Devon Estate* (1989) HBLR LVT 108.

(8) If the lessee assigned his interest before agreement as to the enfranchisement price, negotiations might subsequently delay development. Therefore the full redevelopment value of the site might be deferred for a suitable period to allow for agreement.

(9) A valuation in this case of the lessee's interest but ignoring the section 17 redevelopment rights gives a very different answer.

The total enfranchisement costs on a cleared site approach
under section 9 (1) would be lower. It can be seen that this
section 17 approach is very much a landlord's argument.

Study 13

Flats: Individual right to acquire a new lease under the Leasehold Reform, Housing and Urban Development Act 1993

Section 39 of the Leasehold Reform Housing and Urban
Development Act 1993 grants, to qualifying tenants on long leases
at low rents who have occupied a flat as their principal residence
for the last three years or three out of the last 10 years, the right to
surrender and take a new lease for 90 additional years at a
peppercorn rent. The immediate landlord may not always be the
"competent landlord" see section 40 of the 1993 Act. A premium is
payable to the landlord, or to each landlord where there is more
than one, assessed under Schedule 13 to the 1993 Act and the new
lease is on the terms of the existing lease with modifications as
appropriate under the provision of section 57 of the 1993 Act. While
a further right to a new lease may be exercised, none of the
statutory security of tenure provisions will apply, (section 59). The
tenant has to pay the various and reasonable costs of the landlord
under section 50 and the landlord retains the right to apply to the
court for possession for redevelopment either during the final 12
months of the original term or during the last five years of the new
lease. In such circumstances the landlord would have to pay
compensation for the value of the tenant's interest under Schedule
14 of the 1993 Act.

The premium payable in these circumstances is the total of the
reduction in the value of the landlord's interest and the landlord's
share of marriage value as set out in Schedule 13 to the 1993 Act.
Key valuation assumptions include an open market value without
the tenant seeking to buy, excluding the right to the new lease and
any tenant's improvements.

Facts

A third-floor flat in a well located London mansion block with four
bedrooms, three living rooms and a net internal area of 250 m²
(gross internal area 272 m²). The flat was held on a lease from the
freeholder for 27 years with seven years expired at a fixed ground

rent of £150 pa. The tenant has been accepted to qualify to have the new lease following notice under section 42 of the 1993 Act and it is agreed that the landlord should receive a premium for the value of a lease for a further 90 years at a peppercorn rent. The vacant possession value of the new lease on the flat is agreed at £700,000, or about £2,574 per m2 of the gross internal area, with an uplift of 5% if it was freehold. There is a well run service charge arrangement for the block.

Valuation of the Premium Payable by the tenant in accordance with Schedule 13 to the 1993 Act

(a) *Diminution of the landlord's interest*
   Current Interest (1)
   Existing lease 7 years unexpired

Ground rent	£150 pa		
YP 7 years @ 6½%	5.49		£824

Reversion to capital value

Lease for 90 years agreed value		£700,000 (2)	
Add for freehold interest say 5%		£35,000	
Say		£735,000	

But less for Landlord and Tenant Act 1954 rights/risk of not obtaining possession

Say 40% leaving		£441,000	
PV of £1 in 7 years @ 6½%		0.6435	£283,786
			£284,610

Deduct

Value of freeholder's future interest - Capital value Say in 90 years (as above)	£441,000		
× PV of £1 in 97 years @ 6½%	0.0022		* £981
Diminution in landlord's value			£285,591

(b) *Marriage Value*

Tenant's interest under the new lease	£700,000		
Landlord's reversionary interest	£981	£700,981	
Less Tenant's existing leasehold interest agreed at say	£200,000		

Landlord's existing interest	£284,610	£484,610
Gain on marriage		£216,371
Landlord's share (3) (4) Say 60%		0.60 £129,823
Total premium payable for extended		
lease of flat plus fees & costs of landlord		£415,414
Say		£417,250

Notes

(1) See *Cadogan Estates Ltd* v *McGirk* (1998) (LRA/6/1997) where a similar approach was determined by the Lands Tribunal and a range of comparables were considered with the issue of whether the lease value should be discounted at the end of the existing term to reflect the loss of statutory protection.

(2) In *Goldstein* v *Conley* (1998) (LRA 20/1996 and LRA/5/1998) the Lands Tribunal made a further deduction of 40% from the freehold unencumbered value to reflect the potential impact of the Landlord and Tenant Act 1954 rights on this element of the computation. A 72½% marriage value share as applied in *Shahgholi* (below) was also applied in this case.

(3) In *Cadogan Estates Ltd* v *Shahgholi* (1998) (LRA/26 1996/LRA/51/1997) the Lands Tribunal accepted compelling evidence that the landlord's share of marriage value would have been struck at 72½% not at the convention of 50%, in a reference concerning a long leasehold flat in a six-storey period house.

(4) See also *Re Grosvenor Estate Belgravia* (1999) LRA/30/1998 which concerned, *inter alia*, the quantification of the loss in value of the freeholder's interest in the whole building caused by a tenant's new lease right. A 6% discount rate was applied and a 30% adjustment for the statutory tenancy right.

Study 14

Flats; Collective Enfranchisement under the Leasehold Reform, Housing and Urban Development Act 1993

Legal background

These provisions, sections 1 to 38, enable qualifying tenants to buy the freehold and intermediate leasehold interests in a block of flats through a nominee purchaser. The right extends to appurtenant property, including garages, outhouses and gardens and applies to self contained buildings or parts of buildings where there are two or more flats owned by qualifying tenants and at least two thirds of the total number of flats in the block are owned by qualifying tenants. Self-contained buildings are those structurally detached or vertically divided and capable of separate development. Excluded are buildings where more than 10% of the gross internal floor areas, excluding common parts, are non residential or the building has no more than four units and has a resident landlord.

Qualifying tenants are those of flats on long leases at low rents, with those on business leases and tenants of charitable housing trusts specifically excluded. The flats must have been the principal residence of those tenants wishing to enfranchise for the last 12 months, or three years out of the last 10 years. Long lease include ones over 21 years in length, perpetually renewable ones and continuation tenancies under Part I of the 1954 Act. Low rents are those which in the initial year were nil or less than two thirds of the rateable value (leases pre 1 April 1963), or less than two thirds of the rateable value for leases granted between April 1 1963 and April 1 1990, or rents of less than £250 pa (£1,000 pa in Greater London).

The processes for a qualifying group of tenants to combine and create the nominee purchaser and progress collective enfranchisement are complex and the reader is referred to sections 11 to 30 of the 1993 Act. The provisions relating to determination of price and costs of enfranchisement are in sections 32 and 33 and Schedule 6 to the Act and the owner of any interest will have a lien for any other outstanding debts following collective enfranchisement. The costs of the process are in general born by the nominee purchaser, effectively all the enfranchising tenants.

The price for the freehold is the open market value of the freeholder's interest, the freeholder's share of marriage value and compensation for other losses resulting from enfranchisement. The valuation assumptions are that the title is subject to the existing leases and that the leaseholders have no right to buy collectively or to acquire a new lease. Tenant's improvements are excluded and there are anti avoidance provisions. Marriage value is that amount achievable in the open market by a willing seller, or 50% whichever is the greater. In addition compensation for other loss or damage to the freeholder arising from the acquisition is to be paid to include:

(i)   Reductions in the value of other interests of the freeholder.
(ii) Loss of development value, even if that could have been reduced, for example by the freeholder taking a leaseback.

Facts

The nominee purchaser, established by qualifying tenants, wishes to acquire the freehold of a block of 20 identical self-contained flats outside London all held with 70 years unexpired of 99-year leases all at equal £100 pa ground rents, totalling £2,000 pa fixed for the balance of the leases. The lessees reimburse the lessor for the cost of insurances and services are effectively operated by the lessees through the nominee purchaser company and the freeholder has no ongoing service or management responsibilities. 16 of the 20 leaseholders are participating tenants. At the date of the nominees' notice the open market values of each existing 70-year leasehold interest in a flat is agreed to be £50,000. The landlord's share of marriage value is agreed to be 55% and the freehold reversionary interest in the whole block currently to be worth £26,000.

Collective Enfranchisement Price for the Nominee Company (1)

Value of existing leases of participating tenants		
16 × £50,000	£800,000	
Value of Virtual Freehold		
Say total value of leases plus 5% (2)	£840,000	
Add for elimination of ground rent		
£1,600 pa × 7 YP (3)	£11,200	£851,200
Less		
(i) Value of existing leases	£800,000	
(ii) Value of existing freehold	£26,000	£826,000
Marriage Value		£25,200
Freeholder's share 55% (4)		£13,860
Collective Enfranchisement		
Price		
(i) Value of existing freehold		£26,000
(ii) Share of Marriage Value		£13,860
		£39,860

Plus Any reduction in the value of the interests in property held by the freeholder arising from the acquisition and any loss in the development value.

Plus The freeholder's fees and costs.

Notes

(1) The nominee purchaser can subsequently grant new or extended leases to the flat occupiers on terms to be agreed between themselves. The funding and organisation of the nominee purchaser would have to be established, together with the various occupying tenants' joint agreement to underwrite the costs, whether abortive or otherwise prior to commencing.

(2) See *Maryland Estates Ltd* v *Abbature Flat Management Company Ltd* (1998) LRA/5 1997 where an adjustment of 5% was determined by the Lands Tribunal; also *Maryland Estates Ltd* v *63 Perham Road Ltd* [1997] 2 EGLR 98.

(3) See *Maryland Estates* (*supra*).

(4) The marriage value amount under Schedule 6 to the 1993 Act is a minimum of 50% or such greater sums as can be proved with relevant evidence: see *Cadogan Estates Ltd* v *Shahgholi* (1998) (*supra*) where 72½% was determined.

Study 15

Flats; Collective Enfranchisement under the Leasehold Reform, Housing and Urban Development Act 1993: Shorter Lease Terms Unexpired

A freehold purpose built 7-storey block of 13 flats in London is occupied by tenants all on 99-year leases with 20 years unexpired at the agreed valuation date. Nine of the leases are held by qualifying tenants who wish to collectively acquire the freehold through a nominee company. The ground leases are at fixed rents of £50 pa each and the vacant possession values of each current leasehold flat are agreed at £250,000 on a new 75-year lease but subject to the current unexpired lease only £165,000. There will be an agreed loss in value to the freeholder of about £100,000 to an adjacent development site, the access to which will be lost by the sale of the block.

Value of Freeholder's Interest in accordance with Schedule 6 to the 1993 Act (1)

Participating tenants' leases		
Ground rents × 9	£450	
YP for 20 years @ 8% (2)	9.818	4,418
Reversion in 20 years		
9 flats worth £250,000 each	£2,250,000	

Less risk of LTA 1954 rights		
Say 40% (3) reduction	£1,350,000	
PV of £1 in 20 years @ 8%	0.214	£289,640
		£294,058

Non Participating Tenants' Leases		
Ground Rents × 4	£200	
YP for 20 years @ 8%	9.818	£1,964

Reversion in 20 years		
4 flats worth £250,000	£1,000,000	
Less risk of LTA 1954 rights		
Say 40% (3)	£600,000	
PV of £1 in 20 years @ 8%	0.214	£128,728
		£130,692

Total Freehold Interest (£294,058 + £130,692)		£424,751

Marriage Value

New Nominee Freehold Value

(i) Nine new long 125 year leases at nominal rents in participating flats.		
9 × £250,000	£2,250,000	
(ii) Four existing non-participating flats. Existing freehold values of ground rents & reversions	£130,692	£2,380,692

Current Position

(a) Existing freehold interest in participating flats	£294,058	
(b) Existing leasehold interests of participating flats.		
9 × £165,000	£1,485,000	£1,779,058
Gain on Marriage		£601,634
Freeholder's share of marriage value say 60%		£360,980

Premium Payable

(i) Value of freehold interest pre-enfranchisement		£424,751
(ii) Freeholder's share of marriage value		£360,980

(iii) Loss in value of freeholder's adjacent development plot agreed at say reflecting cost of alternative access	£100,000
	£885,731
Total premium payable (i) + (ii) + (iii)	
Say	£886,000

Plus the freeholder's related fees and costs (4) (Cost of process £98,415 per flat for the 9 participants (5)).

Notes
(1) See *Becker Properties Ltd* v *Garden Court NW8 Property Co Ltd* (1997) LRA/24/1996 in which the Lands Tribunal explores "hope value" of premiums from non participating tenants and yield rates appropriate in such circumstances.
(2) The discount rates will need to be suitable to reflect current precedents and arguments. In *Becker Properties Ltd* (*supra*) discount rates of 9% were accepted but for longer lease terms albeit with rising ground rents.
(3) This allowance is for the risk of the tenant remaining in occupation as a protected tenant under the 1954 Act: see *Goldstein* v *Conley* (1998) (*supra*).
(4) See also *Kemp and Siegfried* v *Myers* (1998) LRA/39/1997.
(5) The participants as freeholders would have the potential to extract premiums from the non participants for lease extensions as the term dates for their leases draw closer.

Study 16

Rights of occupiers of buildings in two or more flats to acquire their landlord's interest under the provisions of Part 1 of the Landlord and Tenant Act 1987 as amended by the Housing Act 1996

The Landlord and Tenant Act 1987, as amended, provides, *inter alia*, the right of tenants to have first refusal on a relevant disposal of any premises containing two or more flats, where those flats are held by qualifying tenants, and the number of flats held by such tenants exceeds 50% of the total number of flats in the premises. The rights of such tenants are confined to privately owned blocks and to exercise these rights the following conditions must be satisfied:

(i)   That the tenant occupies a flat under a tenancy which is not a protected shorthold, a business tenancy under the Landlord and Tenant Act 1954 Part II, a tenancy terminable on cessation of employment or an assured tenancy.

(ii) The landlord must be proposing a relevant disposal of an estate or interest.

(iii) The premises must contain two or more flats held by qualifying tenants.

(iv) The number of such flats must exceed 50% of the total number of flats in the premises.

(v)   The premises may consist of the whole or part of a building.

There are a number of exemptions from the provisions of the 1987 Act which would not, for example, apply if any part of the premises were intended to be occupied other than for residential purposes and those parts exceeded 50% of the internal floor area. Resident landlords are excluded from the operation of the 1987 Act as are a number of exempt landlords including local authorities, housing associations and urban development corporations.

Facts: Shop with three flats above built 80 years ago

The freehold of the whole premises is held by a private property company. The shop is let separately from the flats on a five-year lease at a recently fixed rack-rent of £5,500 pa net. Two of the flats are occupied by tenants on long leases each with 27 years to run at £10 pa net with a separate reviewable service charge. One flat is let on a tenancy at a registered rent of £1,350 pa including a reviewable service charge.

The internal floor areas are as follows:

Shop	153 m²
Common parts	32.5 m²
Each flat	57.6 m²

The landlord is proposing to offer the freehold interest for sale.

Possible effects of the 1987 Act

The three flats, each apparently occupied by a qualified tenant, exceed the two required and 50% of the flats in the premises appear to be within the Act. The non-residential part appears to be insufficient to disqualify the premises as the internal floor area of the shop is only 47% of the whole after having disregarded the common parts.

$$\frac{153 \text{ m}^2}{(172 \text{ m}^2 + 153 \text{ m}^2)} = 0.47$$

If one of the flats was let as offices then the premises would not be subject to the 1987 Act as more than 50% of the internal floorarea would be occupied for other than residential purposes. The landlord is not exempt or resident and therefore before selling on the open market should serve "offer notices" under section 5 of the Act, as amended, on the three qualifying tenants of the flats. This notice will need to define the estate to be sold, the premises, the consideration required and give the tenants the statutory time to respond. Two out of the three would need to vote to accept the offer for them to be able to proceed and they would then need to arrange to organise and fund the purchase, If the tenants were not to proceed with an "acceptance notice" in the prescribed period the landlord may, during the following 12 months dispose of the "protected interests" for a price not less than and on other terms corresponding to those offered to the tenants. The landlord might offer the shop, as well as the flats, to the qualifying tenants of the flats, who might have to fund a pure investment property purchase of the shop. Alternatively, the landlord could choose to offer the tenants the reversionary interest in only the flats on long leasehold, although this might conflict with the good management of the shop and its future investment performance. A landlord's proposed sale might be delayed by the revised 1987 Act procedures introduced by Schedule 6 of the 1996 Act even if the tenants did not in the event proceed with the purchase. If the freehold of the premises were owned by a local authority or new town development corporation, however, it could be disposed of immediately.

Procedure and tactics

The cost of purchasing the shop could deter the flat tenants from accepting the offer or from negotiating an agreement. If the landlord failed to follow the offer notice procedure and sold the interest to a new landlord, the qualifying tenants could, if the majority agree, serve a "purchase notice" on the new landlord. If, following the "purchase notice", the parties fail to agree terms a leasehold valuation tribunal could be required to determine the estate or interest to be disposed of and other terms under section

12. A tenant may argue before a tribunal that the shop part should not be part of the interest to be disposed of.

Possible approach to the valuation of the landlord's interest under the 1987 Act
This is on the basis that the landlord is disposing of its interests in the whole premises.

	£ pa	£	£
Shop part			
Rent income net	5,500		
Less: management	300		
Net income	5,200		
YP in perp at 8%	12.5	65,000	65,000
Two flats occupied on long leases			
Term			
Rents received 2 × £10		20	
YP 27 years at 10% (1)		9.24	
		185	
Reversion to two flats to be			
sold on new 99-year leases			
2 × £45,000	90,000		
Less: factor to reflect the chance			
of tenants opting to continue			
under Part I of the Landlord			
and Tenant Act 1954,			
Say 20% (2)	18,000		
	72,000		
× PV of £1 in 27 years at 12 % (1)	0.047	3,384	3,569
One flat let on a registered rent			
Rent received		1,350	
Less: outgoings			
services	200		
repairs	150		
insurance	100		
management	105		
		555	
Net income		795	
YP in perp at 14%			
Say		7	5,565

Plus an amount to reflect the possible sale to the sitting tenant of a long lease or the sale of the flat with vacant possession (2)		
Say	6,000	11,565
Total value of the landlord's interest (3)		80,134
Say (4) (5)		£80,000

Notes
(1) The discount rates to be adopted would need to be drawn from justifiable comparable market evidence and no strong reliance should be placed on those used in this example.
(2) The factors and amounts to be allowed for to reflect the possible sale of flats with vacant possession will depend on the age profile, the financial and family circumstances and the intentions of the sitting tenants as these factors will influence the market's view of such opprotunities.
(3) The consideration figure to be inserted on any offer notices may need to be higher than this in order to provide a reasonable margin for balanced negotiation. The key elements are the shop value and the less-clear marriage value element. In *Twinsection Ltd* v *Jones* (1997) LRA 29/LRA 31/1997 the Lands Tribunal in dealing with a purchase by qualifying tenants considers the rights of tenants over gardens and amenity land and changes since the tenant's purchase notice.
(4) Given the high proportion of value attributable to the shop element it is unlikely that the leasehold occupiers of the flats would wish to, or be able to, fund the purchase of the landlord's interest and they probably would not serve an "acceptance" notice allowing the landlord to sell his interest on the open market.
(5) See *Davis* v *Stone* [1992] 2 EGLR 222 a determination of the purchase price under these provisions by a Leasehold Valuation Tribunal which also concerned the terms for the conveyance.

Further reading

Aldridge, T M *Leasehold Law*, Oyez Longman.

Aldridge, T M *Rent Control and Leasehold Enfranchisement*, Oyez.

Barnes, D M W *The Leasehold Reform Act 1967*, Butterworths.

Hague, N T *Leasehold Enfrancbisement*, Sweet & Maxwell (Third Ed) 1999

Hubbard, C C and D W Williams, *Handbook of Leasehold Reform*, Sweet & Maxwell (1988)

Wellings, V G Woodfall *Landlord and Tenant*, Sweet & Maxwell.

MacGilp J and G Fox, RICS/Institute of Housing, *Leasehold Reform, Housing and Urban Development Act 1993 – A guide to Part I and Part II*, 1994.

Chapter 4

The Landlord and Tenant Acts
(as they apply to business tenancies)

The Landlord and Tenant Act 1954 Part II (as amended by the Law
of Property Act 1969) together with the Landlord and Tenant Act
1927 gives security of tenure to tenants of business premises and a
right to possible compensation at the end of their tenancies. This
security is afforded by:

(a) automatic continuance of the tenancy notwithstanding expiry
     of the term at common law (section 24);
(b) compelling a landlord who desires possession to establish one
     or more of the grounds listed in section 30;
(c) giving the tenant a right to apply for a new tenancy by a
     section 26 request or on a counternotice to a section 25 notice
     to terminate.

Conditions for security of tenure to apply

Before these rights accrue, the tenancy must be one to which the
1954 Act applies, such conditions being outlined in section 23(1),
which provides

> ... applies to any tenancy where the property comprised in the tenancy
> is or includes premises which are occupied by the tenant and are so
> occupied for the purposes of a business carried on by him or for those
> and other purposes.

In practice, therefore, first there must be a tenancy, so that licences
are excluded. In *Street* v *Mountford* [1985] 1 EGLR 128 the House of
Lords held that where residential accommodation is granted for a
term at a rent with exclusive possession the grant is a tenancy.
While this decision relates to occupation of residential
accommodation, it is submitted that it is just as applicable to
business premises. In *Onyx (UK) Ltd* v *Beard* [1996] EGCS 55 the
court held that the absence of a provision for the payment of rent in
the agreement pointed to the fact that the arrangement was not of

a commercial character and raised the question of whether it might be a licence.

Second, the premises must be occupied by the tenant for the purposes of a business. Usually, this will not give rise to any problems, as occupation by an agent, for instance a manager, will suffice, as in *Cafeteria (Keighley) Ltd* v *Harrison* (1956) 168 EG 668. It has been suggested, in *Bagettes* v *GP Estates Ltd* (1956) 167 EG 249, that a tenant who occupies premises for the sole purpose of subletting parts of the building is outside the scope of the 1954 Act, so that the right to renew does not enure in favour of a tenant who has totally sublet: see *Narcissi* v *Wolfe* [1960] Ch 10. In *Graysim Holdings Ltd* v *P&O Property Holdings Ltd* [1996] 1 EGLR 109 the House of Lords held that as a matter of principle a tenant under a business tenancy cannot sublet part of the property to a business subtenant and at the same time continue to occupy a "holding" comprising all the property contained in the demise to him. The definition of "holding" in section 23(3) excludes property not occupied by the tenant indicating that two persons (other than when acting jointly) cannot be in occupation for the purposes of the 1954 Act at the same time. In *Linden* v *Department of Health and Social Security* [1986] 1 EGLR 108 flats which were occupied by persons employed by a district health authority which, in turn, was exercising the functions on behalf of the Secretary of State were held to be occupied for the purposes of a government department. An interesting decision was reached in *Cristina* v *Seear* [1985] 2 EGLR 128, where the business premises were occupied by a company of which the tenants owned all the shares. The tenants contended that the company was a mere vehicle or alter ego through which the business was carried on by them. The Court of Appeal held they were not entitled to apply for a new tenancy.

Finally, "business" is defined in section 23(2) as including "a trade, profession or employment and includes any activity carried on by a body of persons, whether corporate or unincorporate". This has been held to include the activities of a tennis club and the storage of goods in a lock-up garage: see *Bell* v *Alfred Franks & Bartlett Co Ltd* [1980] 1 EGLR 56. (See also *Groveside Properties Ltd* v *Westminster Medical School* [1983] 2 EGLR 68.) The term will obviously include such premises as shops and offices. However, it should be noted that if a tenant carries on a business in breach of a prohibition in the lease, the protection of the 1954 Act does not apply to the premises unless the landlord has consented to, or acquiesced in, the breach: section 23(4).

Excluded tenancies

If the tenancy comes within section 43 of the 1954 Act it will not be subject to the security of tenure provisions. In general, the Act does not apply to:

(i)   agricultural holdings but may apply to a field used for horse-riding lessons – see *Wetherall* v *Smith* [1980] 2 EGLR 6;

(ii) mining leases;

(iii) service tenancies;

(iv) short leases which, with certain exceptions, are for terms not exceeding six months;

(v)   extended leases as there is no right to renew an extended tenancy granted under section 16(1) of the Leasehold Reform Act 1967;

(vi) tenancies at will – see *Manfield & Sons* v *Botchin* [1970] 2 QB 612;

(vii) tenancies granted by exempt bodies: sections 57–60 of the 1954 Act. Following the introduction of the Landlord and Tenant (Licensed Premises) Act 1990 the position of licensed premises with regard to 1954 Act protection is as follows: the effect of section 1(1) of that Act is that section 43(1)(*d*) of the 1954 Act ceases to have effect in relation to any tenancy entered into on or after July 11 1989 otherwise than in pursuance of a contract made before that date.

If the tenancy in question was entered into before July 11 1989 and continues in existence until July 11 1992 it will be protected by the 1954 Act thereafter. In such a case a notice served under section 24(3)(*b*) does not have effect: section 1(2) of the 1990 Act.

Section 1(3) of the 1990 Act provides that in the circumstances envisaged by section 1(2) the landlord or the tenant can serve certain notices as if section 43(1)(*d*) had already ceased to have effect, namely:

(i)   a section 25 notice specifying as the date of termination July 11 1992 or any later date;

(ii) a section 26 notice requesting a new tenancy beginning not earlier than July 11 1992;

(iii) a section 27 notice stating that the tenant does not desire his tenancy to be continued.

Automatic continuance under section 24

Under section 24 a tenancy to which the Act applies cannot come to
an end save in the manner provided by the Act, so that if the
procedures laid down in sections 25 and 26 are not strictly adhered
to the tenancy continues in force indefinitely. Section 24 provides
that a tenancy to which the Act applies shall not come to an end
unless terminated in accordance with the provisions of the Act and
the tenant under such a tenancy may apply to the court for a new
tenancy:

(a) if the landlord has given notice under section 25 to terminate
     the tenancy or
(b) if the tenant has made a request for a new tenancy under
     section 26.

This section also expressly reserves several common law methods
of terminating a business tenancy. For instance, a tenant can
surrender his lease provided that the instrument of his surrender
was not executed, nor any agreement to surrender concluded,
before the tenant had been in occupation for one month. The right
to forfeit the tenancy or any superior tenancy is also provided for
as well as the ability of the tenant to serve a notice to quit. In *Esselte
AB* v *Pearl Assurance plc* [1997] 1 EGLR 73 the Court of Appeal held,
inter alia, that since sections 23 to 27 of the 1954 Act are all drafted
in the present tense it was not possible to construe the provisions as
being applicable to a tenancy to which the 1954 Act had applied in
the past but had ceased to apply by virtue of the tenant not being
in occupation of the premises. In that case a notice under section 27
was not required. In *Aireps Ltd* v *City of Bradford Metropolitan
Council* [1985] 2 EGLR 143 the 1954 Act did not apply as the
premises for which a new tenancy was sought no longer existed.

Termination of business tenancy

The renewal or termination procedure may be commenced either
by the landlord serving a section 25 notice to terminate or by the
tenant serving a section 26 request for a new tenancy. A schematic
diagram of the operation of the 1954 Act is contained in Diagram 1,
p139 opposite.

Diagram 1.
Landlord and Tenant Act 1954 Pt II: Procedure for Termination-Renewal.

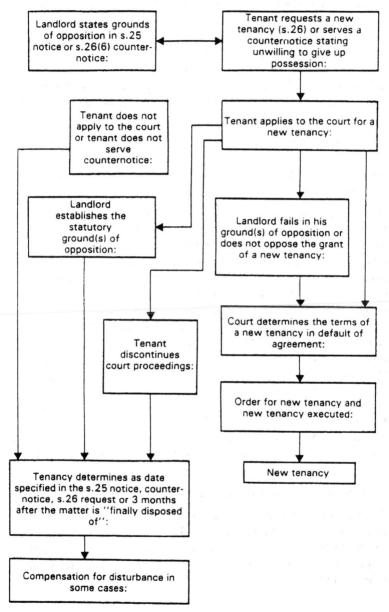

Note: At any of these stages (before a court order) the parties may agree a renewal by agreement.

(a) Section 25 notice

Such a notice must be in the prescribed form and must be given by the "competent" landlord: see Landlord and Tenant Act 1954 Pt II (Notices) Regulations 1983 (SI 1983 No 133) as amended by the Landlord and Tenant Act 1954 Pt II (Notices) (Amendment) Regulations 1989 (SI 1989 No 1548). The "competent" landlord is the first superior landlord having an interest that will not come to an end within 14 months by effluxion of time (section 44). Such a notice must state the date on which the current tenancy will come to an end and such a date must not be earlier than the date on which the tenancy would expire either by notice to quit (periodic tenancy) or by effluxion of time (fixed-term tenancy).

The section 25 notice must be given not more than 12 months and not less than six months before the termination date specified in the notice (section 25(2)). In computing the date of service of the section 25 notice or a tenant's counternotice the general rule is that only one day of the notice period is excluded and not both the date of service and the date of expiry; *Hogg Bullimore & Co* v *Co-operative Insurance Society Ltd* (1984) 50 P&CR 105. Further, the notice must require the tenant, within two months after the giving of the notice, to notify the landlord in writing whether at the date of termination he will be willing to give up possession. Failure to do so may not necessarily invalidate the notice; *Baglarbasi* v *Deedmethod Ltd* [1991] 2 EGLR 71. The notice must also contain a statement as to whether the landlord would oppose an application to the court for the grant of a new tenancy and, if so, on which of the grounds in section 30 he would do so.

If a 21-year lease contained a break clause giving the landlord an option to terminate the lease every seven years, what would be the effect of section 25 on the provisions in the lease? In *Weinbergs Weatherproofs Ltd* v *Radcliffe Paper Mill Co Ltd* [1957] 3 All ER 663, it was stated that the service of an ordinary break clause notice would not satisfy section 25, but may enable a section 25 notice to be served thereafter, while in *Scholl Manufacturing Co Ltd* v *Clifton (Slim-line) Ltd* [1966] 3 All ER 16 it was held that a single notice may, in certain circumstances, operate both the break clause and satisfy section 25.

It should be noted that the provisions of section 25(3) and (4) (the "not earlier than" provisions) do not apply where the tenancy has already expired or been terminated at common law and is continuing under section 24; *Lewis* v *MTC (Cars) Ltd* (1975) 29

P&CR 495. In such an example, the notice under section 25 can be given at any time to expire not more than 12 nor less than six months later than the date on which it is given.

(b) Section 26 request

For a tenant's section 26 request to be valid, the tenant's current tenancy must have been granted for either a term of years certain exceeding one year or a term of years certain and thereafter from year to year. However, a request cannot be made after the landlord has served a section 25 notice. Such a request gives the tenant the right to apply for a new tenancy within two months (sections 26(6) and 30(1)). Should the landlord fail to reply, he will lose his right to oppose a new tenancy.

Application to the court

Where a section 25 notice or a section 26 request has been served, the tenant can apply to the court for a new tenancy. Sections 24(1) and 29(1) provide that where the application is consequent upon a section 25 notice the tenant must have notified the landlord in writing within two months that he was unwilling to give up possession. Such application to the court must be made not less than two months and not more than four months after the section 25 or section 26 requirements have been complied with (section 29(3)). In computing the two and four months' time-limit for the tenant's application to the court for a new tenancy the corresponding date rule should be applied: see *EJ Riley Investments Ltd* v *Eurostile Holdings Ltd* [1985] 2 EGLR 124.

It is of the utmost importance to note that recourse to the court to determine the terms of the new tenancy is to be made only where there is disagreement. Section 28 provides that where the landlord and the tenant agree to the grant of a further tenancy of the holding, the current tenancy will continue until that date but will not be a tenancy to which the 1954 Act applies. To satisfy section 28 the agreement must be in writing and there must be agreement on all the material terms: see *Derby & Co Ltd* v *ITC Pension Trust Ltd* [1978] 1 EGLR 38. In *RJ Stratton Ltd* v *Wallis Tomlin & Co Ltd* [1986] 1 EGLR 104 an "agreement" for the purposes of section 28 was held to be a binding contractual arrangement enforceable by the parties at law. Application to the court within the strict time-limits is essential to protect the tenant's interest.

Landlord's grounds of opposition

Whether the landlord states his opposition in his own section 25 notice or in his counternotice to the tenant's section 26 request he will have to confine himself to the seven grounds outlined in section 30(1), paras (*a*) to (*g*), which are as follows:

(a) *Failure to repair.* In order to be able to rely on this ground the landlord must prove that the state of repair of the holding is such that the tenant should not be granted a new tenancy. There is a discretion in the court as to the degree of disrepair: see *Lyons* v *Central Commercial Properties Ltd* [1958] 1 WLR 869. In *Eichner* v *Midland Bank Executor & Trustee Co Ltd* (1970) 216 EG 169 the court was of the opinion that it was entitled to consider the whole of the tenant's conduct in relation to his obligations and was not limited to the landlord's grounds.

(b) *Persistent delay in paying rent.* It is clear from *Horowitz* v *Ferrand* [1956] CLY 4843 that the arrears of rent do not have to be sustained, but if the reason for the delay no longer applies, the landlord may not succeed under this ground: see also *Hopcutt* v *Carver* (1969) 209 EG 1069. In *Hurstfell Ltd* v *Leicester Square Property Co Ltd* [1988] 2 EGLR 105 the Court of Appeal was of the opinion that the crucial question of whether the arrears were likely to recur was one for the judge at first instance.

(c) *Breaches of other obligations.* This ground includes substantial breaches of the tenant's obligations under the current tenancy or any other reason connected with the tenant's use or management of the holding. In grounds (*a*), (*b*) and (*c*) the whole of the tenant's conduct may be considered: see *Eichner* (*supra*).

(d) *Alternative accommodation offered.* Where the landlord offers alternative accommodation to the tenant such accommodation must be provided or secured by the landlord. The terms of the alternative accommodation must be reasonable having regard to the current tenancy and be suitable for the tenant's requirements, enabling him to preserve goodwill.

(e) *Better return if let or sold as larger unit.* Where the competent landlord is a superior landlord (and not the immediate one) and the tenancy in question is a subtenancy the landlord may oppose a new tenancy on the ground that by letting all the premises contained in the head lease together more rent could be obtained.

(f) *Intention to demolish or reconstruct.* This ground will apply where the landlord establishes that on the termination of the current tenancy he intends to demolish or reconstruct the premises comprised in the holding (or a substantial part of it) or to carry out substantial work of construction that he could not reasonably so do without obtaining possession. In *Betty's Cafés Ltd* v *Phillips Furnishing Stores Ltd* [1959] AC 20, it was held that the landlord must have this intention at the date of the proceedings and he will fail if major obstacles still lie ahead of him, eg no planning permission obtained. Further, in *Fisher* v *Taylors Furnishing Stores* [1956] 2 QB 78, it was stated that the motive for effecting the work was immaterial.

The type or amount of work needed to satisfy ground (*f*) is a question of fact and degree. In *Housleys Ltd* v *Bloomer-Holt Ltd* [1966] 2 All ER 966, the court held that demolishing a garage and wall was sufficient. Demolition without reconstruction is also sufficient. In *Botterill* v *Bedfordshire County Council* [1985] 1 EGLR 82 the Court of Appeal held that infilling by removal of topsoil, depositing waste and replacing the topsoil was not reconstruction within section 30(1)(*f*). Under section 31A the tenant may still be able to obtain a new tenancy if he agrees to the inclusion in the new tenancy of terms giving the landlord sufficient access to undertake the work or that he is willing to accept a tenancy of a smaller part of the holding. In *Mularczyk* v *Azralnove Investments Ltd* [1985] 2 EGLR 141 the Court of Appeal rejected the contention that the works proposed could be undertaken under the provisions of section 31A(1)(*a*) because the landlord could not reasonably carry out the works without obtaining possession of the holding and without interfering for a substantial time or to a substantial extent with the business user of the land. Further, the tenant may be able to show that the current tenancy contains an access clause and the work falls within the terms of that clause: see *Heath* v *Drown* [1973] AC 498.

(g) *Intention to occupy himself.* The landlord can establish this ground where he proves that on the termination of the current tenancy he intends to occupy the holding for the purposes or partly for the purposes of a business to be carried on by him there or as his residence. In *Re Crowhurst Park, Sims-Hilditch* v *Simmons* (1973) 28 P&CR 14 it was held that the requirement of para (*g*) was satisfied where the landlord intended to use the premises for partnership purposes. In *Westminster City Council* v *British Waterways Board* [1984] 2 EGLR 109, it was held that a tenant council could not protect its occupation by intimating its refusal of planning permission to the landlord water board.

Some problems occur in practice regarding the intention to occupy where the landlord concerned is a company forming a group of companies. In such a case, section 42(3) provides that the intention is to be construed as including intended occupation by any member of the group for the purposes of a business to be carried on by that member. Finally, it should be noted that the court must be satisfied that the landlord has a sufficient intention and it will enquire into the matter as under para (*f*): see *Gregson* v *Cyril Lord Ltd* (1962) 184 EG 789. In *Europark (Midlands) Ltd* v *Town Centre Securities plc* [1985] 1 EGLR 88 the Court of Appeal accepted the following as showing a firm and settled intention on the part of the landlords:

(i)   minutes of board meetings;
(ii)   evidence of quotations received from the supplier of equipment: and
(iii) an affidavit from the landlords' property director.

Terms of a new tenancy granted by the court

Sections 32 to 35 provide guidance to the court when it is called to determine any of the terms of a tenancy. Given a successful application by the tenants, the court is bound to order the grant of a new tenancy (section 29). As a general rule, the following limits will apply in so far as the parties have failed to agree.

Property to be comprised in new tenancy (section 32)

As a general rule, the new tenancy must comprise the premises as they stood at the date of the order, except that where the landlord has sought to oppose a new tenancy on ground (*f*) the court may grant a tenancy of part of the holding under section 31A.

Duration of a new tenancy (section 33)

The length of the new tenancy shall be as agreed between the landlord and tenant or, if decided by the court, it shall not exceed 14 years. It is clear that the duration of the old lease is a relevant factor as in *Betty's Cafés* (*supra*) where a 14-year term was ordered but was reduced to a five-year term by the Court of Appeal, so that the court will rarely grant a term longer than the original lease. Where the landlord fails to establish any of the grounds (*d*) to (*g*) but persuades the court that he is likely to be able to satisfy the grounds in the near future, the court may order a short-term lease: see *Upsons Ltd* v *E Robins Ltd* [1956] 1 QB 131.

In view of the comments concerning the interrelationship of section 25 and a break clause it is significant to note that the court can insert a break clause into a new tenancy: see *McCombie* v *Grand Junction Co Ltd* (1962) 182 EG 369. An interesting example of the exercise of this power took place in *Adams* v *Green* [1978] 2 EGLR 46, where the landlord owned a block of 12 shops in one of which the tenant asked for a new lease. The landlord wished to have a break clause inserted which would enable him to determine the lease after giving notice. The Court of Appeal granted a 14-year lease with a break clause included. In *CBS (United Kingdom) Ltd* v *London Scottish Properties Ltd* [1985] 2 EGLR 125 the court stated that while it was perfectly fair and proper for the landlord to seek to maximise the value of his investment, the court has to decide what is reasonable in the circumstances and the matter ought to be decided with fairness and justice. As the tenants were in the process of moving to a new location the court granted a 12-month term, while the landlord had argued for a 14-year term. (Contrast with the decision in *Charles Follett Ltd* v *Cabtell Investments Ltd* [1986] 2 EGLR 76.)

Rent under new tenancy (section 34)

Such rent is to be determined by the court having regard to the terms of the tenancy (other than those relating to rent) at which the holding might reasonably be expected to be let in the open market by a willing lessor, there being disregarded:

(a) Any effect on rent of the fact that the tenant has or his predecessors in title have been in occupation of the holding,

(b) Any goodwill attached to the holding by reason of the carrying on thereat of the business of the tenant (whether by him or by a predecessor of his in that business),

(c) Any effect on rent of an improvement to which this paragraph applies,

(d) In the case of a holding comprising licensed premises, any addition to its value attributable to the licence if it appears to the court that having regard to the terms of the current tenancy and any other relevant circumstances the benefit of the licence belongs to the tenant.

The working of section 34 needs some explanation as it touches on other areas of landlord and tenant law concerning, for example, rent review. It is clear that the terms of the lease may have a direct bearing on the rent. The point was emphasised in *Charles Clements (London) Ltd v Rank City Wall Ltd* [1978] 1 EGLR 47, where the court rejected an attempt by the landlord, as a means of raising the rent, to force on the tenant a relaxation of a covenant limiting user which would have been of no benefit to the tenant. But what "tenant" must the court have in mind? Section 34 directs the court to the "willing lessor". The answer to the meaning of this phrase was given in *FR Evans (Leeds) Ltd v English Electric Co Ltd* [1978] 1 EGLR 93 where Donaldson J was of the opinion that "the willing lessee is an abstraction – a hypothetical person . . . He will take account of similar factors but he too will be unaffected by liquidity problems, governmental or other pressures". In *Dennis & Robinson Ltd v Kiossos Establishment* [1987] 1 EGLR 133 the Court of Appeal was of the opinion that under such an assumption the strength of the market is a matter for the valuer. On the other hand, it is suggested that the open market rent may be affected by the profitability of the tenant's particular business. Whatever the position on profitability, the amount of the previous rent is not a factor for determination of the new rent. Further, section 34(3) allows the insertion of a rent review clause and in *WH Smith & Son Ltd v Bath City Council* (1984) (unreported) local factors affecting the pattern of the rent review were held to prevail over other factors. In recent years the question of the inclusion of a rent review clause in the new lease at renewal has been considered by the courts on numerous occasions. Diagram 2 (opposite, page 147) contains examples of such cases for illustrative purposes.

Other terms of the new tenancy (section 35)

It is in determining the new rent and other terms of the new tenancy that the courts have faced problems in recent years. This problem was highlighted in *O'May v City of London Real Property Co Ltd* [1982] 1 EGLR 76, where the lease was of office premises for a term of five years. Under the lease the landlord was responsible for

Diagram 2: Lease length and rent review provisions at renewal

CASE	Length of current lease	Length of renewed lease current lease	Rent review provisions in renewed lease	Rent review provisions in	Pattern of review	Court
Stylo Shoes Ltd v Manchester Royal Exchange Ltd (1967) 204 EG 803	8 years	14 years	None	Upwards or downwards	At end of first seven years	High Court
Janes (Gowns) Ltd v Harlow Development Corporation [1980] 1 EGLR 52	21 years	20 years	None	Upwards or downwards	Five yearly intervals	High Court
W H Smith & Son v Bath City Council (1984) (unreported)	21 years		None	Upwards or downwards	Four yearly	Bath County Court
Charles Follett Ltd v Cabtell Investment Co Ltd [1986] 2 EGLR 76	14 years	10 years	Upwards only	Upwards only	At the end of the fifth year	High Court
Boots the Chemist Ltd v Pinkland Ltd [1992] 2 EGLR 98	21 years (along with other leases)	14 years	None	Upwards or downwards	1997 and 2002	Wood Green County Court
Blythewood Plant Hire Ltd v Spiers Ltd (in Receivership) [1992] 2 EGLR 103	5 years, yearly tenancy thereafter	10 years	None	Upwards only		Wood Green County Court
Abdul Amarjee v Barrowfen Properties Ltd [1993] 30 EG 98	Tenant from year to year	14 years	None	Upwards or downwards	Five yearly intervals	Wood Green County Court
Graham v Hodgson (1993) (unreported)	12 years	12 years	Complex – tied to state pension	Upwards or downwards	Three yearly intervals	St Austell County Court
Forbuoys plc v Newport BC [1994] 1 EGLR 138	21 years	9 years	None	Upwards or downwards	Three yearly intervals	Newport (Gwent) County Court

maintenance and repairs. When the lease expired, the landlord agreed to offer the tenants a new lease but wished to change it into a "clear lease" by which responsibility for maintenance and repairs was to be on the tenants. An additional service charge was to be levied in return for which the tenants would be compensated by a reduction in the proposed new rent. The tenants applied to the court for a new tenancy. At first instance, Goulding J enunciated four tests in deciding whether a particular landlord's proposals were justified:

1. Has the party demanding a variation of the terms of the current tenancy shown a reason for doing so?
2. If the party demanding the change is successful, will the party resisting it in principle, be adequately compensated by the consequential adjustment of open market rent under section 34?
3. Will the proposed change materially impair the tenant's security in carrying on his business or profession?
4. Taking all relevant matters into account, is the proposal, in the court's opinion, fair and reasonable as between the parties?

The House of Lords agreed with the first three tests but stated that the fourth test of discretion should be applied in each of the preceding stages. It thus seems that, to justify a change in any of the other terms in a lease, the landlord will have to satisfy the tests laid down in the *O'May* case: see also *Gold* v *Brighton Corporation* [1956] 1 WLR 1291.

Contracting out of the 1954 Act

Agreements which attempt to exclude the tenant's right to renew his tenancy are void unless they are authorised by the court (section 38(1)). Under section 38(4) the court may, on the joint application of both parties, authorise an agreement excluding the right to renew a tenancy to which the 1954 Act applies. In *Essexcrest Ltd* v *Evenlex Ltd* [1988] 1 EGLR 69 the Court of Appeal held that the application for (and grant of) the order to contract out must be made before an unequivocal execution of the lease. In *Cardiothoracic Institute* v *Shrewdcrest Ltd* [1986] 2 EGLR 57 there were successive extensions of a tenancy agreement negotiated subject to a condition that the extension should be the subject of a tenancy agreement approved under section 38(4). In such circumstances it was held that it was clearly intended by the parties that until the approval of the county court was obtained there was no legally binding tenancy agreement between them. In *Nicholls* v *Kinsey* [1994] 1 EGLR 131 the Court of Appeal held that section 38(4) provides that an agreement excluding the operation of sections 24 to 28 is only valid if

(i)   there is a prospective tenancy for a term of years certain; and
(ii) the court authorises an agreement excluding the relevant provisions of the Act in relation to that tenancy; and
(iii) the agreement is made in pursuance of such authorisation.

Following *Alnatt London Properties Ltd* v *Newton* [1983] 1 EGLR 73, "offer to surrender back" clauses infringe section 38.

Rent reviews and business tenancies

The problem of the implementation of a rent review clause may occur in a business tenancy at two stages, namely, at any time during the currency of the lease and/or on renewal. In most cases, the rent review or reviews will take place during the currency of the lease and will require careful consideration. Draftsmen now recognise that there are several essential matters that must be stipulated in such a clause, namely:

1.   Stipulations as to time in the service of notices.
2.   Interval of rent review.
3.   Formula and machinery for determining the new rent.
4.   Provisions in event of disagreement.

1. Service of notices

It is frequently the case that the rent review procedure is activated by the service of a "trigger" notice which commences the rent review process. In most cases, the notice will be served by the landlord on the tenant and the main problem that has arisen in this area is whether a failure to keep strictly to the timetable set in the rent review clause will result in the landlord losing its rights to a reviewed rent. In *United Scientific Holdings Ltd* v *Burnley Borough Council* [1977] 2 EGLR 61, the House of Lords laid down the general rule that time was not of the essence in the service of notices in a rent review process but stated that there could be exceptions to this general rule. It is possible to make time of the essence by stating expressly in the lease that it should be so, but time can also be made of the essence where there is an interrelationship between the rent review clause and some other clause in the lease. Such was the case in *Al Saloom* v *Shirley James Travel Service Ltd* [1981] 2 EGLR 96, where an underlease contained both a break clause and a rent review clause. The last date on which the landlord could serve the rent review notice was the same as that on which the tenant could give notice exercising his option to determine the underlease. In these circumstances, it was held that the presence of the break clause has the effect of making time of the essence. Where the lease provides for a period between service of the rent review notice and exercising the break clause, so as to allow the tenant to determine whether he wishes to continue in occupation, the service of the rent review notice will be of the essence: see *Coventry City Council* v *J Hepworth & Son Ltd* (1982) 265 EG 608. In *United Scientific* (*supra*) it was suggested that extreme delay in the service of a trigger notice may have a prejudicial effect on the landlord's cause. The question of what constitutes unreasonable delay has arisen on several occasions. In *H West & Son Ltd* v *Brech* [1982] 1 EGLR 113 a delay of 18 months was not sufficient to affect the landlord's rights: see also *Accuba* v *Allied Shoe Repairs* [1975] 1 WLR 1559, where also 18 months' delay was insufficient and contrast with *Telegraph Properties (Securities) Ltd* v *Courtaulds Ltd* [1981] 1 EGLR 104, where a six-year delay was fatal to the landlord's action. In *Amherst* v *James Walker (Goldsmith & Silversmith) Ltd* (1983) 267 EG 163 Oliver LJ commented:

> But I know of no ground for saying that mere delay, however lengthy, destroys the contractual right. It may put the other party in a position, where, by taking the proper steps, he may become entitled to treat himself as discharged from his obligation: but that does not occur automatically and from the mere passage of time ..."

In *United Scientific* (*supra*) the House of Lords recognised that a contra-indication may make time of the essence for the service of a rent review trigger-notice. What amounts to a contra-indication is the subject of dispute. In *Henry Smith's Charity Trustees* v *AWADA Trading & Promotion Services Ltd* [1984] 1 EGLR 116 the clause contained an elaborate time schedule and provided that the rent stated by the landlord be deemed to be the rent if the tenant's counternotice was not served in time. The Court of Appeal held that where the parties had not only set out a timetable but had provided what was to happen in the absence of strict compliance with that timetable, the general rule was rebutted. The Court of Appeal reached a different conclusion in the interpretation of a "deeming" provision in *Mecca Leisure Ltd* v *Renown Investments*

(Holdings) Ltd [1984] 2 EGLR 137. In *Greenhaven Securities Ltd* v *Compton* [1985] 2 EGLR 117 the rent review clause provided that if the parties had not, within a 15-month time-limit, agreed on an arbitrator or made an application for the appointment of an arbitrator the new rent should be a sum equal to the old rent. Goulding J distinguished the decision in *Mecca* and held that the default provision constituted a contra-indication. The opposite view was taken in *Taylor Woodrow Property Co Ltd* v *Lonrho Textiles Ltd* [1985] 2 EGLR 120, where the court noted that in *Henry Smith's* the deeming provisions were two-way while in *Mecca* they were one-way. In *Taylor Woodrow* the deeming provision was one-way, so that time was not of the essence in the service of the counter-notice.

2. Interval of rent review

It is clearly important for the interval of rent review to be expressly stated so that, for example, on a 21-year lease the rent review clause may become operative in the seventh and 14th years. Where it is not so, the rent review clause runs the risk of being inoperable. In *Brown* v *Gould* [1972] Ch 53 the option for a new lease was for a term of 21 years "at a rent to be fixed, having regard to the market value of the premises at the time of exercising the option". The court held that if no machinery was stated for working out the formula, the court will determine the matter itself. A liberal approach to the construction of an option to purchase was adopted in *Sudbrook Trading Estate Ltd* v *Eggleton* [1983] 1 EGLR 47.

3. Formula and machinery

The formula and machinery has, of necessity, a direct relationship with the interval of rent review. Valuers are faced with enough problems on rent review without the clause adding to those problems by failing to define the rent on review. Such was the case in *Beer* v *Bowden* [1981] 1 All ER 1071, where the clause provided only that the rent should be the fair market rent for the premises. In the particular circumstances of the case the Court of Appeal stated that the rent should be the fair market rent for the premises. See also *Thomas Bates & Son Ltd* v *Wyndham's (Lingerie) Ltd* [1981] 1 EGLR 91, where a term was implied that the rent was to be that which is "reasonable as between the parties". By way of contrast, in *King* v *King* [1980] 2 EGLR 36 the court refused to look at the defective rent review clause from a reasonableness viewpoint.

Some novel provisions on rent are being considered by draftsmen seeking to increase the landlord's benefit as in *Bovis Group Pension Fund Ltd* v *GC Flooring & Furnishing Ltd* (1982) 266 EG 1005, where the clause provided for the new rent to be assessed by reference to the rent that could be obtained if the premises were let for office purposes and the court stated that it was to be assumed that the building had planning permission for office use notwithstanding that no such permission had, in fact, been granted. Similarly, in *Pugh* v *Smiths Industries Ltd* [1982] 2 EGLR 120 it was held that the rent review should be on a literal construction of the lease where the formula provided that the presence of the review clause should be disregarded in calculating the new rent: see also *Lister Locks Ltd* v *TEI Pension Trust Ltd* [1982] 2 EGLR 124. The converse situation applied in *Grea Real Property Investments Ltd* v *Williams* (1979) 250 EG 651, where it was decided that the effect of improvements on a rent review of premises in shell-form only, had to be disregarded. If no such improvements disregard clause is present, the tenant will have to pay increased rent on his own improvements (contrast with the situation on a lease renewed under section 34 of the 1954 Act). If a strict user clause is present in the lease, this may also have a dampening effect on the new rental level as in *Plinth Property Investments Ltd* v *Mott Hay & Anderson* [1979] 1 EGLR 17: see also *Law Land Co Ltd* v *Consumers' Association Ltd* [1980] 2 EGLR 109. The factors contained in the hypothetical lease for the purpose of the determination of the revised rent may pose problems for the valuer. In *National Westminster Bank plc* v *Arthur Young McClelland Moores & Co* [1985] 2 EGLR 13 the court held that in that particular lease the fair market rent had to be ascertained on the assumption that there was no rent revision clause contained in the hypothetical terms which the arbitrator had to apply. A similar conclusion was reached (at first instance) in *Equity & Law Life Assurance Society plc* v *Bodfield Ltd* [1985] 2 EGLR 118. Different conclusions from these two cases were reached in *Datastream International Ltd* v *Oakeep Ltd* [1986] 1 EGLR 98 and *MFI Properties Ltd* v *BICC Group Pension Trust Ltd* [1986] 1 EGLR 115. In *British Gas Corporation* v *Universities Superannuation Scheme Ltd* [1986] 1 EGLR 120 Browne-Wilkinson V-C said that the correct approach in these circumstances was as follows:

(a)   words in a rent exclusion provision which require all provisions as to rent to be disregarded produce a result so manifestly contrary to commercial common sense that they cannot be given literal effect;

(b)   other clear words which require the rent review provision (as opposed to all provisions as to rent) to be disregarded, must be given effect to, however wayward the result; and

(c)   subject to (b), in the absence of special circumstances it is proper to give effect to the underlying commercial purpose of a rent review clause and to construe the words so as to give effect to that purpose by requiring future rent reviews to be taken into account in fixing the open market rent under the hypothetical letting.

These were stressed as being only "guidelines" by the Court of Appeal in the *Equity & Law Life* case [1987] 1 EGLR 124.

The interpretation of the *British Gas* guidelines was considered in *Co-operative Wholesale Society* v *National Westminster Bank plc* [1995] 1 EGLR 97 in relation to the question of whether deductions should be made to "headline" rents to reflect concessionary or rent-free periods other than for fitting out when determining the open market rent after the expiry of the rent-free period. The Court of Appeal held that a clause which excludes the assumption that an hypothetical tenant would have the expense of moving in and fitting out is more in accordance with the presumption in favour of reality than one which does not, under the guidelines enunciated in *British Gas Corporation* v *Universities Superannuation Scheme Ltd* [1986] 1 EGLR 120. On the other hand, a clause which deems the market rent to be the headline rent obtainable after a rent-free period granted simply to disguise the fall in the rental value of the property is not in accordance with the basic purpose of a rent review clause. Such a clause enables the landlord to obtain an increase in rent without any rise in property values or fall in the value of the money by reason of changes in the way the market is choosing to structure the financial packaging of the deal. Therefore, in the absence of unambiguous language, a court should not be ready to construe a rent review clause as having this effect.

4. Disagreement

A well-drafted rent review clause should always provide for a procedure in the event of disagreement between landlord and tenant on the new rental level. It should be made clear whether reference to an arbitrator or independent expert is desired. In most cases, if the parties cannot agree on whom should be appointed, the President of the Royal Institution of Chartered Surveyors is the person most frequently requested to appoint someone.

Compensation

Section 37 of the 1954 Act has been amended by section 149 of and Schedule 7 to the Local Government and Housing Act 1989. In addition, the Landlord and Tenant Act 1954 (Appropriate Multiplier) Order 1990 (SI 1990 No 363) provides for new multipliers. The interrelationship of the amended section 37 and the 1990 order provides for the following possibilities:

(a) Where the date of service of the landlord's section 25 notice or section 26(6) counternotice is on or before March 31 1990, the appropriate multiplier is three or, in the case where the conditions of section 37(3) of the 1954 Act are satisfied, the multiplier is six.

(b) Where the date of service of the landlord's section 25 notice or section 26(6) counternotice is on or after April 1 1990, the appropriate multiplier is one or, in the case where the conditions of section 37(3) of the 1954 Act are satisfied, the multiplier is two.

(c) Where the transitional provisions contained in Schedule 7 apply, March 31 1990 is deemed to be the date for the determination of the rateable value and not the date (after April 1 1990) of the landlord's section 25 notice or section 26(6) counternotice. The transitional provisions apply if the following conditions are met:

   (i)   the tenancy concerned was entered into before April 1 1990 or was entered into on or after that date in pursuance of a contract made before that date, and

   (ii) the landlord's notice under section 25 or, as the case may be, section 26(6) is given before April 1 2000, and

   (iii) within the period referred to in section 29(3) for the making of an application under section 24(1), the tenant gives notice to the landlord that he wants the special basis of compensation provided for by this paragraph, and

   (iv) there must be a rateable value shown on the valuation list as at March 31 1990.

   Where the conditions are met, the appropriate multiplier is eight times the rateable value on March 31 1990.

(d) Where the transitional provisions apply, compensation amounting to 16 times the rateable value on March 31 1990 is payable (section 37(3) of the 1954 Act) if:

(i)   during the whole of the 14 years immediately preceding the termination of the current tenancy, the premises being or comprised in the holding have been occupied for the purposes of a business carried on by the occupier or for those and other purposes; or

(ii) during those 14 years there was a change in the occupier of the premises and the person who was the occupier immediately after the change was the successor to the business carried on by the person who was the occupier immediately before the change.

Compensation for improvements

The Landlord and Tenant Act 1927 as amended by the Landlord and Tenant Act 1954 Part III provides that on quitting a holding at the termination of the lease a business tenant (section 17) may in certain cases claim compensation from the landlord for improvements carried out by the tenant or his predecessor in title (section 1).

In order to qualify for compensation, the tenant must follow the statutory procedure strictly as outlined in Diagram 2 on p147. Prior notice of the proposed work must be served on the landlord giving an opportunity for him to object, or to elect to undertake the work in consideration of a reasonable increase of rent, or such rent as the court determines (section 3(1)). If the landlord objects, and no agreement is reached, the tenant may apply to the court for a certificate that the work is a proper improvement. This will be granted if the work adds to the letting value of the holding, is reasonable to its character, and does not reduce the value of any other nearby property belonging to the landlord. Where no notice of objection is served, or a court certificate is obtained, the tenant may proceed with the work. A further certificate of completion must be obtained from the landlord or the court (section 3(6)). Any increased rent under a renewed lease should not include the value of improvements carried out by the tenant or his predecessors in the business within the previous 21 years (section 34, Landlord and Tenant Act 1954, as amended).

The tenant's claim for compensation must be made within the appropriate time-limits (section 47). The amount is the lesser of the net addition to the value of the holding resulting directly from the work or the reasonable cost of carrying out the work at the termination date less an allowance for obsolescence (section 1(1)).

As the basis of the compensation is the value of the improvements to the landlord, the amount may be reduced if the landlord intends to alter or change the use of the premises and no compensation, is payable if the premises are to be demolished (section 1(2)). If disputed, the amount may be determined by the court (section 1(3)).

Chapter 5

Industrial and Distribution Properties

Introduction

As a valuation field, industry contains not only the greatest range of sizes compared with shops, office and other commercial premises, but also by far the greatest number of different types of use. At one extreme the valuer will find a small workshop, possibly on an upper floor of an old building where only one or two people work, perhaps in the rag trade with a sewing machine. At the other extreme are the car assembly plants sometimes occupying many square kilometres. The valuer may have to consider properties that range from the fully automated plants incorporating a great deal of sophisticated computerised equipment to ancient workshops where the processes have not changed for centuries.

Before embarking on the valuation of any property the valuer should first familiarise himself with the current edition of the *RICS Appraisal & Valuation Manual* ("The Red Book"). The section "Purposes of Valuation and their Bases" will be a determining factor in the choice of valuation base selected. However valuers should be conscious that the bases of valuation are kept under constant review by the profession and to a certain extent are continuing to evolve. The Red Book requires the valuer to agree the appropriate valuation basis with his client and/or his professional advisers unless it is already specified by law or prescribed in a legal agreement. While the valuer should have a working knowledge of the uses of such bases as Market Value, Estimated Realisation Price and Estimated Restricted Realisation Price, the most familiar bases to the property and other professions are Open Market Value, Existing Use Value and Depreciated Replacement Cost, although the usage of the latter is relatively rare. The valuer of industrial property should also be familiar with the definitions of Open Market Value of Plant and Machinery and Value of Plant and Machinery to the Business and ensure that the basis of valuation adopted for the land and buildings is compatible with that for any plant and machinery, particularly where the two elements are to be valued by different valuers.

Factories and warehouses (be they vacant, partially vacant, or let and income producing) which are to be either bought or sold will usually be valued on an Open Market Value basis. Estimated Realisation Price or Estimated Restricted Realisation Price bases, sometimes required by lenders, have generally not found favour with valuers who have found it difficult in practice to distinguish Estimated Realisation Price from Open Market Value.

Valuations of property for company accounts or other financial statements which have regard to the land and buildings as investments or as part of a going concern will usually be on either an Open Market Value basis or an Existing Use Value basis. For specialist buildings, however, a Depreciated Replacement Cost value may be more appropriate. The Depreciated Replacement Cost basis will certainly be appropriate where plant and machinery forms an integral part of the building, such as a cement works.

Types of industries

While not necessarily directly germane to the mechanics of valuation, an understanding of the economic categorisation of industry will help in the valuer's appreciation of the context in which an industry has evolved and sometimes point to the determinants of that industry's location. The four basic categories, which should not be confused with Planning Use Classes are:

(i) **Primary Industries**

These are concerned with extracting material direct from the land or the sea and do not involve any processing or fabrication of a finished product. Mining and quarrying come into this category.

(ii) **Secondary Industries**

This includes the bulk of manufacturing industries from food and drink to chemical and mechanical engineering, printing and publishing.

(iii) **Tertiary Industries**

These provide services which are basically oriented towards the retail market. Such industries as construction, transport and communications are included in this group.

(iv) **Quaternary Industries**

These include research establishments and those concerned with the provision of information or expertise.

Influences on location

A wealth of economic and demographic information is now available from both local authorities and private research organisations. Examination of such information will help the valuer understand what has influenced the location of industry historically and the factors influencing current location decisions.

Clearly, primary industries can only be located at the source of raw materials. On the other hand, the locational requirements of secondary industries are much more complex. Some are mainly located with reference to the market which they serve, such as bakers, while others are strongly tied to raw materials, such as brickworks. The majority are, of course, intermediate between these extremes. Traditionally, service industries were almost always located where those services were required – at the market. However, improving communications have reduced the necessity for this. Again, warehouses serving the traditional distribution industry were located to serve local or sub-regional markets while in the late 20th Century locations strategically close to the national motorway network (to complement the highly sophisticated stock control systems of both the retail and manufacturing sectors) were favoured. The advent of e-commerce may switch preferences to more local distribution nodes.

Quaternary industries are very often more market oriented and businesses may be located to serve the needs of just one large customer or to be near the source of a specialist skills base such as on one of the science parks associated with a university.

Given an understanding of the general locational needs of different types of industry, what factors influence the choice of any particular location? For established businesses, the location of the factory will often to be due to historic reasons which bear little relation to the current market circumstances. However, in deciding to relocate a factory or warehouse, a business is likely to consider at least some of the following:

- The availability of labour, be it skilled or unskilled and, increasingly, its cost. The cost and availability of housing will be part of the equation.

- The availability of land and buildings. For smaller businesses the prior availability of suitable advance factories and warehouses or of second-hand stock will be highly influential in helping the relocation to be achieved more quickly than would be possible if land had to be acquired and a bespoke building constructed. For larger operations, for which a purpose built plant is the only likely option, the availability of land will be a major determinant. Given the competition from alternative uses, particularly housing, the availability of large sites is relatively limited.

- The quality of communications in terms of effective access for the workforce, access to raw materials, access to markets for the finished product, and access to ports and airports. Proximity to a motorway junction remains a major influence, but other factors such as drive times from the docks, for those who are import/export oriented, or proximity to airports for those with overseas business and interests will all play their part.

- For some industries the provision of services will be fundamental. Typically this requirement will be for water, but may include sewage disposal facilities, gas or electricity.

- Environmental and legislative considerations will also have an influence. A business where there is a risk, for instance, of chemical seepage may seek sites with advantageous geological conditions and avoid sites close to watercourses, while businesses depending heavily on road transport will consider the current law prescribing the hours that a truck driver may drive between rests.

From a valuation perspective, a recognition of these influences is important, particularly when considering generally comparable properties in, say, neighbouring towns.

Government grants and tax breaks

Encouraging industry and influencing its location has long been an instrument of economic policy and at any one time Government (in various guises) may offer both grants and tax concessions. Grants can take the form of direct cash incentives to assist with the construction of buildings or cash incentives to train staff. These incentives could be obtained from the European Union, Central Government or Local Government. While the availability of grants varies from time to time, they have proved to be an effective means by which Government has influenced the establishment and location of industry. Enterprise Zones, with attractive planning and rating regimes along with tax benefits for investors, have been designated in the past to promote the case for specific, usually run-down, areas.

The valuer should seek information as to the availability of these grants and designations and be conscious of the distortions that they may have on the market. Equally the valuer should be fully familiar with current rules regarding industrial building and capital allowances.

Regulatory Regime

A steady stream of legislation dealing with environmental and health and safety issues imposes on the valuer an obligation to keep abreast of rules that apply both to sites and to new and older buildings. Attention to the issues surrounding the Environmental Protection Act 1990 and the Environment Act 1995, and familiarity with the Factory Acts, are important as is a knowledge of planning legislation. It would be unrealistic, however, to expect any but the most experienced of valuers to be familiar with the regulations specific to each industry and a valuer may need to agree limitations with the client and qualify the valuation accordingly.

In preparing a valuation, the valuer should be cognisant of the effect that general health and safety rules will have on the maintenance of the building and of the impact of Construction, Design and Management (CDM) regulations on costs, particularly when compared to other buildings being relied upon as comparables.

Building Specification

The industrial property market is dominated by the general purpose industrial building which has evolved as a consensus building suitable as an envelope for a wide range of manufacturing and distribution purposes. As industry has had to react to markets that change ever more rapidly, owner/occupied buildings have become less common and companies have come to rely on the provision of standard buildings built as "advance" units by development corporations, institutions and property companies. The characteristics of buildings for industry and distribution converged in the 1970s and early 1980s so that the buildings were identical regardless of the use to which they might be put under planning controls. This has resulted in the ubiquitous institutional "shed".

When considering comparables, the valuer should be alert to the differences between properties, the most important of which are:

(i) **The Site**

How accessible is the site? Are there obstacles *en route* from the main arterial roads? Low bridges, for instance. Is the access to the site suitable for large vehicles? Is the road adopted? Is it properly lit? To what use is neighbouring land put? If neighbouring land is residential, are there any restrictions on working hours or noise emissions?

(ii) **Site Cover**

The footprint of the building (being the area of the ground floor of the building measured on a gross external basis in accordance with the Code of Measuring Practice of the RICS and ISVA) divided by the area of the site gives the site cover. (Plot ratio is the ratio of the gross external area of all floors to the site area and is used in town planning.) While the utility of the land not built upon is of more interest to the occupier than the site cover itself, the valuer should be aware of current institutional thinking as to an optimum site cover ratio. At present a site cover of between 35 and 40% would be deemed optimum. Care must be taken where there is a high office content at first-floor level as the site cover ratio may mask an over-development. A site cover of over 50% is likely to result in congestion.

(iii) **Eaves Height**

In recent years the institutional "standard" for eaves height has risen along with increased floor loading requirements. Whilst 5 metres was an acceptable standard in the mid-1970s, a height of 6 to 7 metres is now required for even smaller buildings, rising to 12 metres or more for large distribution facilities. The Code of Measuring Practice has a diagram distinguishing between external eaves height, internal eaves height and clear internal height. The difference between the latter two could be half a metre so valuers should take care to compare like with like.

(iv) **Floor Loading**

Many occupiers could manage with much less than the institutional requirement for a floor loading of between 30 and 40 kN/m^2. The design "rule of thumb" is 2.5 kN/m^2 per metre of eaves height. The institutional rationale for their apparent over specification is the need for their buildings to be able to accommodate the possible requirements of as wide a range of potential occupiers as possible without the need for those occupiers to strengthen the floors for specific uses.

(v) **Office Content**

Offices with a separate entrance and their own toilets are usually provided with the standard institutional shed. While there is a debate as to the optimum office content, 10% of the total area of the factory is the accepted norm, considerably less would not be unusual for very large properties.

(vi) **Car and Lorry Parking**

Providing the site cover ratio is low enough, adequate car and truck parking facilities should be easily accommodated on the site. Care should be taken where the site cover is high to ensure that trucks can manoeuvre safely up to the loading doors. Manoeuvring and parking areas for trucks should be in concrete.

(vii) **Fit-Out**

Advance units are usually fitted out to the lowest standard that the developer believes he can get away with. This will usually include fitting out the offices with carpets, central heating and suspended ceilings, a roller shutter or sliding loading door to the factory premises and some sort of landscaping to the perimeter of the site. A higher specification such as electrically operated loading doors, heating and lighting to the factory and perhaps dock levellers may also be included.

(viii) **Shape**

Modern general purpose buildings tend to be rectangular with the length rarely more than two and a half times the width. A very long and narrow building may, however, have more attractions to certain occupiers. The optimum office arrangement is usually considered to be a pod to the front or side of the building. The clearer the interior is of internal columns, the better.

(ix) **Size**

Where comparing evidence of rents per sq metre (or still, usually, rents per sq foot!), the valuer should consider carefully the size of the building from which the evidence comes.

Whether a bespoke building can be valued by reference to evidence from transactions involving institutional sheds will depend upon the extent that the specification is at variance with the institutional norm. Some buildings will be so specialist that they do not bear comparison. These buildings may, however, have a value to occupiers in the same industry somewhat in excess of the value that a Depreciated Replacement Cost Valuation would suggest. This is particularly true for the new technology industries for which the immediate availability of a modern fully equipped plant with, for instance, fully tested clean rooms would often outweigh the advantages of a purpose built facility for which there could be several months' or years' delay.

Specialist buildings will very often incorporate more plant than the typical advance unit. In preparing the valuation, the valuer should consider the attraction of any unutilised capital allowances to purchasers with a high tax capacity.

Plant and machinery

As indicated above, not only are there an enormous number of different industrial processes, but these are often tied up with detailed locational requirements, such as the need for pure water or a particular quality in the labour force. The valuation of such a range of operations requires highly specialised skills and a knowledge that goes beyond the expertise needed to value land and buildings. A plant and machinery valuer may not be familiar with every industrial process, especially if it is of an unusual or specialised type but experience will give the valuer an understanding of how to assimilate the nature of a specific plant. By enquiry and research, the valuer will be able to establish inherent obsolescence, which is an important factor in such valuations. Almost always, much of the associated plant and equipment in a factory is common to many others. This applies, for example, very much to the chemical industry, where a complex looking process may appear to represent an insoluble conundrum. On detailed examination, however, this will break down into a series of pipes, pumps, condenser tanks, filtration plants and so on, leaving only a few unfamiliar items which need to be researched.

Clearly, a plant valuer must have the ability to recognise what he sees in all the major branches of industry. He must have at his fingertips records giving the present day cost of all the plant he is likely to come across, and know where to obtain such information about unusual items.

The valuation of plant and machinery is an increasingly specialised field (and one where the valuer usually practising in the land and buildings sector should be wary. None the less, the property valuer should be familiar with plant which normally forms a component of industrial buildings. Such plant might include heating installations, air conditioning and ventilation. Items such as crane gantries, which would usually pass with the building on sale or letting but which are really fitted in order to help with a specific process, may need special consideration. The valuer should ensure that the appropriate adjustments have been made for the plant when considering the comparables. Some plant will add value while some will reduce it. Particular care should be taken when pipes and ducts are to be left *in situ*. The cost of cleaning chemical and gaseous contaminants can be very high.

Valuations for Financial Statements

The valuer is now regularly required to update the asset side of the balance sheet. This enables the correct present day values to be substituted for the existing values, which may well have been calculated on the basis of historic cost.

The Red Book sets out the bases to be used in most circumstances. Open Market Value and Existing Use Value are generally the appropriate bases for non-specialist property. Such properties that meet, or approximate to, standard institutional buildings can usually be valued by the conventional investment method or by direct comparison with prices paid per square metre, where such information can be obtained. When utilising the investment method of capitalising passing and estimated future income streams be aware that investment yields are usually quoted as net of a notional purchaser's acquisition costs (stamp duty, legal and surveyors' fees and VAT).

Valuers will be familiar with the five assumptions for Open Market Value:

> . . . the best price at which the sale of an interest in property would have completed unconditionally for cash consideration on the date of valuation, assuming:
> (a) a willing seller;
> (b) that, prior to the date of valuation, there had been a reasonable period (having regard to the nature of the property and the state of

the market) for the proper marketing of the interest, for the agreement of the price and terms and for the completion of the sale:

(c) that the state of the market, level of values and other circumstances were, on any earlier assumed date of exchange of contracts, the same as on the date of valuation:

(d) that no account is taken of any additional bid by a prospective purchaser with a special interest: and

(e) that both parties to the transaction had acted knowledgeably, prudently and without compulsion.

Existing Use Value is the appropriate basis for properties ". . . occupied for the purposes of the undertaking." Its definition includes the five assumptions made for Open Market Value but with the additional assumptions that:

(f) the property can be used for the foreseeable future only for the existing use; and

(g) that vacant possession is provided on completion of all parts of the property occupied by the business.

The Red Book also comments that, "In most cases, where a property is fully developed for its most beneficial use, it is expected that Existing Use Value will equal Open Market Value with vacant possession." In other words, the result of an Existing Use Valuation should be to establish, by reference to market transactions and/or analysis, the price a business would have to pay to replace a property if it were deprived of it. It would be logical, if not universal practice, to include the costs of acquisition. The Red Book is not prescriptive on this point.

> The Depreciated Replacement Cost basis of valuation is used for the valuation of specialised buildings. It is a method of using net current replacement costs to arrive at the value to the undertaking in occupation of the property as existing at the valuation date, where it is not practicable to ascertain Existing Use Value.
> (The Red Book).

Depreciation is the measure of wearing out, consumption, or other loss of value of a fixed asset, whether arising from use, effluxion of time or obsolescence through technology or market changes. It follows that the valuer must, for this basis, separate the land from the buildings in his calculations. Very often the valuer will find that it is just due to the buildings being so special that the Existing Use Value basis is precluded and that the land can be valued by reference to comparables. Problems may, however, occur where there is a *sui generis* planning use. In such circumstances the Red Book directs that, "the approach should be to value the land with the benefit of an assumed planning permission for a use, or a range of uses, prevailing in the vicinity of the actual site."

To establish the future economic life of buildings, depreciation should be considered under two heads:

(i) *Economic obsolescence*. This is the valuer's measurement of the physical obsolescence of the premises, taking into account age, present condition and future costs in use, particularly maintenance and repair.

(ii) *Functional obsolescence*. This involves a consideration of the suitability of the premises for their present use and their likely efficiency if continued in that use by the undertaking, or in some other use if such were reasonably conceivable.

The basic methodology requires the valuer to establish the cost of replacing the existing buildings, including fees and fit out, (the gross current replacement cost) and to then depreciate that figure to reflect the value attributable to the remaining economic working life of the building. The valuer should also take into account environmental and planning policies when considering the appropriate depreciation factor. Where there have been technological changes, it may be appropriate to consider a modern substitute building of smaller size but equivalent capacity

The efficacy of the Depreciated Replacement Cost basis is dependant upon the business having adequate potential profitability with regard to "the value of the total assets employed and the nature of the operation". (The Red Book).

No case studies have been included in this chapter. Reference should be made to those in chapter 16, Asset Valuations.

© JK Hayward 2000

Chapter 6
Offices

Office buildings are invariably valued by the investment method, which is the determination of the present value of the right to receive, and liability (if any) to pay, a future sum or a series of future sums by discounting them at a compound rate determined by direct comparison with yields obtained on other investments, ie it is the determination of the value of an interest by the capitalisation of rental income.

The attention of all surveyors carrying out any valuation is drawn to the *Appraisal and Valuation Manual* issued by the Royal Institution of Chartered Surveyors; often referred to as the "Red Book".

The methodical assembly of all the information required is of paramount importance to ensure that no relevant factors are overlooked. How this is done is illustrated in this chapter by considering the valuation of various interests in an office building. The following are introductory matters.

The freeholder (A) built on the site an office building which was completed and let in 1952 to a tenant whose lease expired in 1982. He then re-let the building to tenant (B) for a term of 75 years. B has subsequently underleased the building to various occupying tenants. A valuation is required as at 2000 of the following interests in the building:

the freehold;
the head leasehold;
the freehold in possession, ie the value to the freeholder if he was entitled to receive the rack-rents receivable by the head lessee.

The building is assumed to be in a first-class location in the West End of London. The accommodation comprises a basement used as storage and offices on the ground and six upper floors. The floors were leased in open plan form and these areas formed the basis for the apportionment of service charges (see later). Tenants have installed their own internal partitioning. The building is centrally heated by two gas-fired boilers which were renewed two years ago and this system is in good working order. The upper floors are served by three lifts, all of which have been renewed within the last two years. The sixth floor of the building was constructed three years ago by J Ltd, the occupational tenants, at their own expense.

The first essential in valuing any building is to read the various leases and to prepare a synopsis of the relevant details. The importance of this cannot be expressed strongly enough.

The Leases

The headlease

The freehold is owned by A, who has leased to B on the following terms:

Term:	75 years from 1982.
Initial rent:	£40,000 pa exclusive of all outgoings.
Rent reviews:	At the end of the 25th (2007) and 50th (2032) years.
Basis of rent review:	£40,000 pa plus 5/8th of the annual rack-rental value (Note 1) of the building at the date of review in excess of £64,000 pa subject to such figure being not less than the rent being paid at the date of review. The rent on review is to be assessed having regard to the terms of the head lease and on the assumption that the building is vacant and available for letting in accordance with market practice at the date of review (Note 2).
Insurance:	The tenant is required to insure against all risks at full replacement cost to include fees and for three years' loss of rent.
Repairs:	The tenant is required to carry out all repairs and decorations internally and externally and to maintain all plant and equipment (eg lifts and central-heating installations) in sound working order.

Notes

(1) It is customary to estimate the rental value at the date of review on the basis of rental values current at the date of valuation and this will need to be undertaken in order to arrive at the amount of the head rent on review.

(2) On this basis it must be assumed that the building is let to its
     best advantage. It could be assumed to be let as a whole or by
     floors or by parts of floors. Some head leases put restrictions
     on how a building may be let, ie no more than one tenant per
     floor or not more than a given number of tenants in the
     building as a whole. It is also to be assumed that if current
     market practice is to let on the basis of five-year rent reviews
     the tenant will let on that basis.

The underleases

The synopsis of the underleases is best prepared in schedule form
as shown in Schedule I. Although it may be purported that all
sublettings are in accordance with a standard form of underlease,
all leases should be perused, as variations from the standard are
frequently to be found. As the building has been let to a number of
tenants, their superior landlord B will endeavour to recover from
them a proportion of the cost of the outgoings he has to bear. These
will be recovered by service charges but the liability for payment by
individual occupiers may not be identical in all cases. The method
of apportioning service charges should also be noted. It is usual in
the case of office buildings to base the apportionment on the ratio
of the floor area occupied by a particular tenant to the total area of
the building and therefore it is important to note the definition of
floor area (if any) in the underleases. Many leases which intend to
recover 100% of the cost of services provided when a building is
fully let, do not in reality do so. Where in respect of any tenancy the
landlord does not recover the whole of the apportioned service
charge a deduction from rent will have to be made in respect of the
amount of service charge which is irrecoverable. It is advisable to
record on this schedule the area of each subletting, the floor on
which it is situated, the specific use and the estimated rental value
on reversion.

Notes on Schedule I
(1) Companies C, D, E and F are all substantial public companies.
     G Ltd is a public property development company which has
     lately been in financial difficulty and its covenant strength is
     doubtful. H is a firm of apparently successful solicitors and the
     two senior partners are guarantors under the terms of the
     lease. J Ltd is a foreign-based oil exploration company and for

Schedule I
Schedule of underleases

Tenant (1)	Floor	Use	Floor area	Rent paid £ pa	Service charge	Rent review / end of lease	Estimated Reversionary value £ pa
C Ltd	Basement	Storage	371	1,000 (4)	No provision for recovery of the cost of heating or lifts	End of lease 2032 No rent reviews	18,500
D Ltd	Ground	Offices	492	2,500 (4)	No provision for recovery of the cost of heating or lifts	End of lease 2032 No rent reviews	157,440
E Ltd	First	Offices	492	67,000	Full recovery	Rent reviews 2001 and 2006 End of lease 2011	157,440
F Ltd	Second	Offices	492	76,000	Full recovery	Rent reviews 2002, 2007 and 2012 End of lease 2017	157,440
G Ltd	Third	Offices	492	172,200 (2)	Full recovery	Rent reviews 2003 and 2008 End of lease 2014	157,440
H & Co	Fourth	Offices	492	67,000	Full recovery	Rent review 2001 End of lease 2006	157,440
J Ltd	Fifth	Offices	492	67,000	Full recovery	Rent reviews 2001 and 2006 End of lease 2011	157,440
J Ltd	Sixth (3)	Offices	232	Nil (tenant's improvements)	Full recovery		22,504

the last two years it has disclosed substantial losses in its annual accounts.

(2) The lease to G Ltd is subject to upward-only rent reviews and the rent from this floor, which was negotiated six years ago, is now considered to be in excess of the current market rent by about 25%. There may therefore be no increase in rent on review in 2003.

(3) This floor was constructed by J Ltd at its own expense in 1996. There is no provision in the lease requiring them to construct this floor and while it is provided in the lease that on review improvements carried out at the expense of the tenant are to be ignored in assessing the rent, owing to an oversight the tenant has failed to serve the required notice on the landlord under section 3 (1) of the Landlord and Tenant Act 1927.

(4) The rents payable under these leases are fixed until the expiration of the terms in 2031 (31 years) and the tenants do not contribute towards the cost of lifts or heating. Owing to likely further inflation, the head lessee will receive a steadily declining income on this account.

Cost of services

The head lessee B is responsible for all outgoings and these need to be carefully analysed in order to calculate the landlord's residual liability in those cases where the apportioned cost is not fully recoverable from the underlessees. A detailed analysis of the service charge is shown in Schedule II. It will be noted that there is no provision for the recovery of any sum in respect of the renewal of plant or equipment. Where this is the case a deduction will need to be made at the end of the valuation to allow for obsolescence where renewal is imminent. The landlord has, however, renewed the boilers and lifts recently, so that no deduction is required in this case. The cost of services varies from building to building and where actual figures are not available estimates must be built up by pricing the individual services supplied.

Schedule II

Schedule of items included in the service charge and analysis of the actual cost per m² of providing these services

Item	Cost per m² £
Staff wages	6.90
Rent for caretaker's flat	1.10
Cleaning of common parts	2.25
Electricity	2.05
Central heating of building (1)	7.50
Heating maintenance contract (1)	2.35
Building and third party insurance	4.80
General maintenance	11.85
Lift maintenance (1)	1.60
Gardening	0.15
Managing agent's fee (2)	5.10
Auditor's fee	0.30
Total cost of the provision of services per m²	£45.95

Notes on Schedule II

(1) Cost of lifts and central heating: it is necessary to calculate these separately in respect of the basement and ground floor as they are not recoverable from the underlessees.

Central heating	£7.50
Heating maintenance contract	£2.35
Lift maintenance	£1.60
	£11.45

	Area in m²
Basement	371
Ground floor	492
	863 at £11.45 per m² = £9,881

(2) Managing agent's fees based on 12½% of the total cost of services and not on a percentage of rent.

Reversionary rental values

Where parts of the building are let at rents which do not reflect current rental values these will have to be estimated. Such rents on reversion should be calculated on open market rental values as at date of valuation and not on the estimated rental value at the date of review, on the assumption that there will be an increase owing to inflation. The best comparables for this purpose will be recent lettings or sublettings which have taken place in the building itself. As a check on their validity, recent lettings in comparable buildings in the vicinity should be examined. Schedule III illustrates the methods used in the devaluation of comparables. More comparables would be required in practice. From a consideration of the devalued comparables the valuer will determine his basic current market valuation rates per sq metre. This determination will be based not entirely on mathematical calculations but also on the valuer's judgment based on experience. In the example, he has decided on £50 per m^2 for the basement storage and £320 per m^2 for the office floors.

On any rental and service charge analysis the valuer should be careful not to mix metric and imperial measurements. For example, rents will often be given in imperial terms.

Notes on Schedule III

(1) F Ltd (see Schedule I) have retained 218 m^2 of office accommodation for their own use. In order to provide access to the office suite let to K Ltd and to its own accommodation, a common access passage has been partitioned off from the area originally leased by F Ltd. This has an area of 88m^2.

As the basis for valuation is the total internal floor area for each floor exclusive of lavatories, lift lobbies and staircases, the net rent per sq metre of the sublet portion needs adjustment to allow for a proportionate share of the internal corridor as follows:

Rent of subletting	£50,000	
$\dfrac{\text{Net area of subletting}}{}$ =	186m^2	= £268.82 per m^2

Adjustment:

Total net area of occupied parts

$\dfrac{\text{Total area of subletting}}{}$ × £268.82

$\dfrac{186m^2\ +\ 218m^2}{492}$ × £268.82 = £220.74 per m^2

(2) C Ltd (see Schedule I) have let the entire basement area at a

Schedule III
Schedule of underleases

Address	Tenant	Floor	Area in m²	Service charge	Date of letting	Rent paid £ pa	Adjustment	Adjusted rent per m² £ pa
The subject property	K Ltd (1)	Part 2nd	186 (1)	Full	Letting just agreed. Lease to 2012. Rent reviews 2002 and 2007	50,000 (268.82 per m²)	See Note (1)	220.74
The subject property	L Ltd (2)	Basement	371	No provision for the recovery of the cost of heating or lifts	Lease just agreed for 5 years	22,800 (61.50 per m²)	For service charge see Note 2	50.00
A similar property	M Ltd (3)	4th floor	550	Full	Agreed 1 year ago. Lease for 20 years to 2019. Rent reviews 5 yearly	193,875 (352.50 per m²)	See Note 3	320.00

rack-rent, but otherwise on the same terms on which they themselves hold it. This rent therefore reflects the fact that the tenant has no liability to reimburse the landlord with a proportion of the cost of heating, maintenance of heating or providing lifts. An adjustment to the rent will have to be made to take this into account. Rent £22,800 less £4,248 (irrecoverable cost of heating and lifts) = £18,552 (£50 per m^2).

(3) This floor has been partitioned by the landlord to provide individual offices. It is estimated that the partitioning adds approximately £32.50 per m^2 to the value of this accommodation. In the case of the subject property, no partitioning has been provided by the landlord and therefore this must be deducted for the purpose of comparison.

Inspection of the building

The collection of data may take some time but having put the work in hand an inspection of the building should now be made. It is not proposed to go into detail on matters to be noted on inspection but the following are the principal ones:

- Location – general and particular.
- General description, construction and condition of the building.
- Services: eg lifts, central heating and public services available.
- Degree of obsolescence of building and services with particular reference to renewals for which the head lessee may be liable.
- Floor areas should be checked against plans if available, otherwise a survey will be necessary. The measurement of any building should be in accordance with the Code of Measuring Practice issued by the Royal Institution of Chartered Surveyors for offices it is the net internal area that is required.

Enquiries of public authorities

Information should be obtained on town planning consents, any established use rights and Building Regulation consents. Local authority development plans should be inspected to see whether any developments are projected in the vicinity and the highway authority should be consulted on road proposals as either of these matters might affect the potential value of the property.

Assessment of the rental data (Schedule IV)

Having collected all the relevant data it is necessary to analyse and co-ordinate them. Since the valuation will be arrived at by capitalisation of net income, the reversionary rental values of all underleases will need to be calculated and thereby the increased income at each rent review is ascertained. This has been summarised in Schedule IV to give the total rent receivable by the head lessee at each rent review year based on current rental values. From these totals will be deducted the head rent and any irrecoverable service charges. The head rent falls to be reviewed in 2007 and is calculated as follows:

Calculation to determine rent to be paid on review under head lease rack-rental value of building

	Area in m²	Rent per m²	Rent £pa
Basement storage (2)	371	50	18,550
Ground/fifth-floor offices	2,952	320	944,640
Totals (1) (2)	3,323		£963,190
Rack-rental value	963,190		
Less: Head Rent	64,000		
	£899,190		

Revised head lease rent $= 5/8 \times £899{,}190 + £40{,}000 = £601{,}994$

Schedule IV

Notes

(1) The sixth floor is excluded from the calculation as this is a tenant's improvement which the lease provides is not to be taken into account for rent review purposes.

(2) The rent review is based on rack-rental value and not rent received.

A further deduction to arrive at net income has been made in respect of that part of the income arising from the rent of the third floor which is in excess of current rental value. This unsecured income will be valued separately.

For more information on rents see Chapter on "Lease Renewals and Rent Reviews of Commercial Properties".

Schedule IV

Schedule showing the present and future net income at current rental values receivable by the head lessee

	2000	2001	2002	2007
	1	2	3	4
	£ pa	£ pa	£ pa	£ pa
Basement (1)	1,000	1,000	1,000	1,000
Ground floor (1)	2,500	2,500	2,500	2,500
First floor	67,000	157,440	157,440	157,440
Second floor	76,000	76,000	157,440	157,440
Third floor	172,200	172,200	172,200	172,200
Fourth floor	67,000	157,440	157,440	157,440
Fifth floor	67,000	157,440	157,440	157,440
Sixth floor	–	–	–	–
Total rents receivable	452,700	724,020	805,460	805,460
Less: irrecoverable service charge	9,881	9,881	9,881	9,881
Total net rents receivable	442,819	714,139	795,579	795,579
Less: rent received for 3rd floor in excess of rental value	14,760	14,760	14,760	14,760
	428,059	699,379	780,819	780,819
Less: head lease rent	40,000	40,000	40,000	601,994
Net rents receivable(pa)	£388,059	£659,379	£740,819	£178,825

Note on Schedule IV

(1) The reversionary incomes in respect of the basement, ground and sixth floors have been ignored because they are so long deferred. In the case of the basement and ground floor the underlease still has 32 years unexpired while in the case of the sixth floor, despite the fact that no notice has been served under the Landlord and Tenant Act 1927, the tenant will still have the benefit of these improvements if a new lease is granted by virtue of the Law of Property Act 1969.

In order to comprehend this information it is often advantageous to illustrate the income pattern in graph form. From this it is easy to appreciate that the head lessee suffers a substantial reduction in income in 2007 when the head rent is reviewed.

Two factors affecting the capitalisation rate at which income is valued are the frequency of rent review and the quality of the covenant of the tenant. It is therefore advisable to analyse the rental data to illustrate the position. This has been done in Schedule V.

Schedule V

Analysis of existing and future rental income from the property and analysis of rent received by frequency of review

	£pa	%
Rent received at present subject to 5-year reviews	449,200	99.2
Rent fixed for 32 years	3,500	0.8
	£452,700 pa	100%

Analysis of rental value of property by frequency of rent reviews

	£pa	%
Open market rental value of parts subject to 5-year reviews	787,200	75.9
Open market rental value of part where rent fixed for 32 years	175,990	16.9
Open market rental value of tenants' improvements	74,240	7.2
	£1,037,450 pa	100%

Analysis of rent received from underlessees according to covenant value of tenants

	£pa	%
Major public companies	146,500	32.4
Public company in financial difficulties	172,200	38.0
Private partnership	67,000	14.8
Foreign-based company	67,000	14.8
	£452,700 pa	100%

The capitalisation rates to be applied will be ascertained by analysis of market transactions. It is important that transactions are analysed on the same methods as are adopted for the valuations. Office property is but one form of investment which is competing for investment funds and its performance must be measured against other investments in terms of current yield, rental and capital growth. Other sources of investment whose current

performance can readily be ascertained from the *Financial Times* are
fixed interest investments such as Government stock, debentures
and also equities.

Yields are constantly varying thus affecting the popularity of
particular types of investment, so that any valuer must keep
himself up to date with the investment market in general.

Valuation is, however, not an exact science: it is the valuer's
personal interpretation of the market in relation to the particular
property being valued which is important. It has been assumed in
the valuations which follow that the current yield obtainable on
prime freehold office investments is of the order of 7%. "Prime" is
not easy to define but in simple terms it means a well-built modern
building, fully let at current market rents, subject to five-year
upward-only rent reviews and to tenants whose covenant is blue
chip. Yields adopted for the valuation take into account the
variations in the property being valued from the prime freehold
investment. Adjustments to the yield will need to be made in
respect of age, location, quality of building and frequency of rent
reviews. Although it is customary to adopt values current at the
date of valuation in estimating reversionary values, the extent to
which a property is a hedge against inflation will depend on the
frequency of rent reviews. Corresponding yields for different
frequencies of rent review at varying growth rates are to be found
in *Donaldson's Investment Tables.*[1]

Valuation of the freehold interest subject to the head lease

	£ pa	£
Present income	40,000	
YP in perp at 10.5% (1)	9.5	380,000
Increase in rental income received in 2007		
(revised headlease rent £601,994 less £40,000	561,994	
YP in perp at 12.5% deferred 7 years (2)	3.5	1,966,979
Value of freehold subject to the head lease		2,346,979
Less: purchaser's costs 5.7625% (3)		127,845
		2,219,104
	Say	£2,220,000

---

[1] *Donaldson's Investment Tables* are available from Donaldsons, London SW1Y 6PE.

Notes

(1) This part of the income is fixed for the full term of the head lease, so it has been valued in perpetuity. Assuming that long-dated gilts are showing a return of 13%, this investment, though long dated, has the advantage that (a) there is an ultimate reversion to a central London freehold, (b) there is a substantial "marriage value" (see below), so justifying a yield of somewhat less than long-dated gilts.

(2) The estimated additional income receivable in seven years is fixed for 25 years. The reversion to full market value is so long deferred and so uncertain, with the building 100 years old at the end of the lease, that it has been ignored and the reversionary income receivable in 2002 valued in perpetuity. The increased income is also less secure than the basic ground rent because it is dependent on rental values in six years' time, hence the yield has been adjusted to 12.5%.

(3) Purchaser's costs in acquiring the interest being valued are deducted so as to give the actual rate of return to the investor based on his total capital outlay, which will include costs of acquisition. The costs comprise 4% stamp duty on purchase, 1% agent's fee and 0.5% legal fees plus VAT on agent's and legal fees. The percentage deducted needs to be altered as changes occur, particularly on stamp duty which at 2000 is at 4% for values above £500,000. Also note the costs of purchase relate to the end value. To arrive at the end value divide the gross value by 1.057625 to give value net of costs. Do not take 5.7625% of the gross figure as this will inflate costs.

An examination of Schedule IV and the graph will show that the income receivable by the head lessee falls into three parts. Owing to the substantial increase in head rent in 2007 only £178,825 (referred to below as "Minimum income") is secured for the full term of the head lease after deducting the non-recoverable service charge and the excess of rent payable over rental value for the third floor (Schedule 1 Note (2)). Of this income £3,500 pa is not reviewable until 2032 (32 years hence) but the balance is subject to five-year rent reviews. Nevertheless, the head rent payable on review in 2007 represents about 77% of estimated net rent receivable at that date after deducting the excess rent of the third floor and the non-recoverable service charge expenditure.

The second part of the income is that received only until the head lease rent review in 2007 (referred to in the valuation as "Excess

Current & Future Income Projection

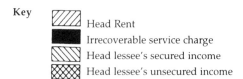

Key
- Head Rent
- Irrecoverable service charge
- Head lessee's secured income
- Head lessee's unsecured income

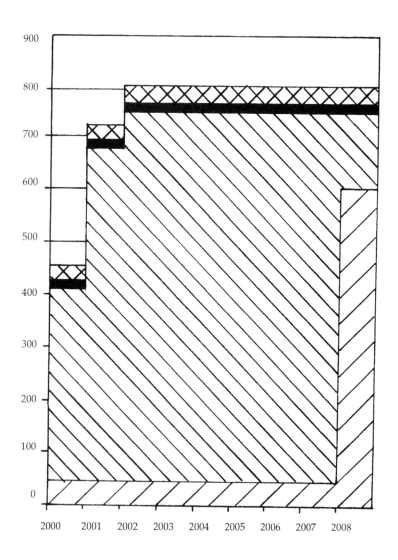

income received before review in 2007"). This income is receivable for only seven years and its capital value must therefore be amortised over the seven-year term. This type of income pattern is unattractive to an institutional investor who favours a steady income growth over a period, whereas in this case the income received from the property falls substantially in seven years' time. Consequently, while this income is reasonably well secured a relatively high rate of interest, namely 10%, is used.

Finally, there is the excess income over rental value in respect of the third floor payable by a tenant whose covenant is doubtful. This income has therefore been valued at a high rate (12%) and amortised over the remainder of the term using the taxed sinking fund rate.

Valuation of head lessee's interest

(a) *Minimum income* (see Schedule IV column 4)

			£ pa	£
(i) Years 2000/2007				
Net rent			178,825	
YP 7 years at 8.5% and 4%				
(tax at 40%)(1)			3.38	604,428
(ii) Years 2007/2057				
Net rent (2)			178,825	
YP 50 years at 9% and 4%				
(tax at 40%)		9.909		
PV £1 in 7 years at 9%		0.55	5.45	974,596
				£1,579,024

Notes

(1) For sinking funds tax at 40p in the £ is the usual rate adopted by valuers but some valuers use 30p in the £ to reflect rate of corporation tax. The rates of interest and tax are interdependent.

(2) Although the net income is the same as in the first seven years it is now top slice income subject to the payment of a head rent of about 77% of the total income.

(b) *Excess income received before review in 2007*

	£pa	£
Years 2000/2001		
Net rent (£388,059 − £178,825)	209,234	
YP 1 year at 10%	0.91	190,403

Year 2001

Net rent (£659,379 – £178,825) (1)		480,554	
YP 1 year at 10%	0.90		
PV £1 in 1 year at 10%	0.91	0.819	393,573

Years 2002/2007

Net rent (£740,819 – £178,825) (1)		561,994	
YP 6 years at 10%	4.35		
PY £1 in 2 years at 10%	0.83	3.61	2,028,798
			£2,612,774

"Pannell fraction" – which is the adjustment to allow for amortisation over seven years.

$$\frac{\text{YP 7 years at 10\% and 4\% (tax at 40\%)}}{\text{YP 7 years at 10\%}} \quad = \quad \frac{3.21}{} \quad = \quad 0.66$$

$$0.66 \times 2,612,774 \quad = \quad £1,724,431$$

An alternative approach would be by the use of a double sinking fund.

(c) *Rent received in excess of rental value for third floor (for alternative methods of valuing excess rents see end of chapter)*

	£pa	£
Rent	14,760	
YP 14 years at 15% and 4% (tax at 40%)	4.15	61,254

Total value of head lessee's interest	£
(a)	1,579,024
(b)	1,724,431
(c)	61,254
Total value	3,364,709
Less: purchaser's costs 4.7625%	183,328
	3,181,381
Say	£3,180,000

Note

All rents have been reviewed to current rental values by 2003 except for the basement and ground floor. As these are not reviewable for 32 years the value on reversion is negligible.

Valuation of freehold interest in possession
(ie subject to the occupation leases)
The figures of net income have been extracted from Schedule IV.

	£ pa	£
Current rental income 2000	428,059	
YP in perp at 7.5%	13.33	5,706,026
Reversion 2001	699,379	
Less: current rent	428,059	
Increase in rent	271,320	
YP in perp deferred 1 year at 7.5%	12.4	3,364,368
Reversion in 2002 to (1)	780,819	
Less: current rent 2001	699,379	
Increase	81,440	
YP in perp deferred 2 years at 7.5%	11.54	939,817
		10,010,211
Excess rental income received for third floor		
Rent	14,760	
YP 14 years at 15% and 4% (2) (tax at 40%)	4.15	61,254
		10,071,465
Less: purchaser's costs 5.7625%		548,747
		9,522,718
Say		£9,520,000

Notes
(1) The rental value of fifth-floor tenant's improvements has been ignored, since the tenant has the benefit of these improvements for another 12 years under his existing lease and for the term of any subsequent lease or 21 years whichever is longer (Law of Property Act 1969). Similarly the rental value of the basement and ground floor has been ignored as these areas are subject to leases without review for an unexpired term of 32 years.
(2) Although this is in the nature of short leasehold interest it is unlikely to be of interest to a gross fund, who above all want a secure income, and consequently it has been valued using a tax adjusted table.

The freehold subject to the head lease is not an attractive purchase for an institution because of the infrequency of the rent reviews. Equally, the head leasehold interest is unattractive because of the sharp fall in income in seven years' time. The valuation of the freehold subject to the occupation leases illustrates the considerable increase in value by amalgamating the freehold and head leasehold interests. The difference between the value of the freehold in possession and the value of combined freehold and leasehold interests valued separately is known as the marriage value.

	£	£
Value of freehold in possession		9,520,000
Value of freehold subject to head lease	2,220,000	
Value of head lessee's interest	3,180,000	5,400,000
Marriage value		£ 4,120,000

It would be usual that, at some stage, either the freeholder or the head lessee will endeavour to effect an amalgamation by buying out the other at a price equivalent to the value of his own interest together with somewhere around 50% of the marriage value, the final outcome depending on the negotiating skills of the parties.

Valuation of rents in excess of open market rents

During the 1990s it has not been unusual for valuers to come across situations where a property has a rental stream in excess of the open market rent, usually due to rental values falling.

To the landlord this means that the rental investment is not adequately secured by the property and the additional income above the open market rent is at risk and where continuation of this excess rent is dependent on the financial strength of the tenant. Should the tenant become bankrupt then the additional rent will no longer flow from the property, especially given the recent changes in the law on assignment of leases.

For many years the simple way this was dealt with was for valuers to apply a higher rate and capitalise over the remainder of the term using a sinking fund rate. This is the method adopted earlier in the chapter.

However, the method currently being adopted by the market is what is known as the short cut Discounted Cash Flow (DCF) approach.

In this method the valuer assesses the expected growth in rent from the property, and applies this to the open market rent (as opposed to the passing rent) to ascertain initially at which review or lease termination (whichever is the earlier) the current open market rent has grown to be equal or in excess of the passing rent. Having ascertained this review date, the passing rent is capitalised until the appropriate rent review date or lease renewal date at the top-sliced discounted rate. From the review date the anticipated rental income is capitalised at the appropriate yield of the property, discounted at the equated yield.

An example is as follows:

Office area	500m²
Passing rent £400/m², ie	£200,000 p.a.
Assessment open market rent £300/m	£150,000 p.a.
Over-rented element	£50,000 p.a.
Review pattern 5-yearly – lease 3 years in existence	
Assumed all risk yield	6%
Equated yield	10%
Expected growth is rent	5%

At first review in two years' time, assuming expected growth of 5% pa, open market rent will be £165,375 pa, ie still below passing rent.

At second review in seven years' time, again assuming rental growth of 5% pa, the open market rent will be £211,064, ie above passing rent. This means you adopt the equated yield until the review in seven years' time.

Valuation:

Passing rent		£200,000	
Y.P. 7 years at 10%		5.39	1,078,000
Estimated rent at review 7 years,			
Say		£211,000	
Y.P. in perpetuity @ 6% =	16.67		
PV £1 in 7 years @ 10% =	0.513	8.55	1,804,050
			2,882,050
Less costs at 4.7625%			131,018
Value			£2,751,032
Say			£2,750,000

It should be pointed out that in the DCF approach different growth rates could be utilised for each year to reflect the valuer's judgement on anticipated growth in rents for the property. Deciding on growth rates is not always easy. If in doubt it is recommended the valuer utilises growth rates in the economy forecast by the Treasury.

Chapter 7

Retail Properties

The valuation of retail properties for non-statutory purposes is affected to only a limited extent by legislation, the principal enactments being Part II of the Landlord and Tenant Act 1954 (as amended by the Law of Property Act 1969), Part I of the Landlord and Tenant Act 1927 and the Offices, Shops and Railway Premises Act 1963. Reference may also need to be made to the Use Classes Order 1987 and General Permitted Development Order 1995.

The wide variation in retail types, including the isolated "round the corner" shop, the departmental store of 40,000 m² occupying an island site in a premier shopping position in a major city, and the out of town superstore, probably provides the valuer with a greater test of his expertise than is generally encountered in the valuation of other types of commercial property.

With few exceptions, retail outlets, whether freehold or leasehold, owner-occupied or held for investment, are normally valued on an "investment" basis, ie by the capitalisation of rents and rental values, and it is the assessment of rental value to which the valuer has initially to apply his mind.

Rental Value

Situation

In arriving at retail rental values the location is of paramount importance. In an established town or city centre the pattern of trading will usually be readily ascertainable from inspection and from commercially produced street plans which are available for the major centres giving the names of all the traders. Retail valuers commonly identify the section of a street which commands the maximum rental value as the "100% position" and from analysis of rentals elsewhere in the street are able to describe the less valuable sections in percentage terms, ie the "90% position", "75% position" etc. The more general expressions of "prime", "secondary" or "tertiary" are also often used.

Where there are long, continuous parades of shops the points at which rental values change may be difficult to identify and may in fact be constantly moving. Where parades are interrupted by side roads or by non-retail frontages such "breaks" often become the divisions between sections of different rental values.

The level of rents achieved in the "100% position" will largely depend on the size and importance of the centre concerned and the rental value in the prime position in one town may be several times greater than the rental value for a similar-sized unit in the peak position in a smaller town even though the occupier of both units is the same multiple retail company.

In practice the identification of the rental value pattern has become increasingly difficult in recent years. There has been a great deal of development activity in the shopping field and in some cases existing shopping centres have been entirely redeveloped. In other cases new centres have been built adjacent to existing centres and have caused a radical distortion of the previous trading pattern. In the latter cases inspection alone may not be sufficient to identify those positions which have increased or decreased in value and, where evidence is lacking or is inconclusive, discussions with traders and such aids as pedestrian counts may be necessary. Established regional centres such as the Metro Centre at Gateshead, Meadowhall at Sheffield or Lakeside at Thurrock have undoubtedly affected levels of rents in competing towns, and new out of town schemes such as Cribbs at Bristol, and Bluewater in Kent continue to be monitored carefully for their effect on nearby established centres. Increasingly sophisticated research techniques are also being employed to estimate the catchment area likely to be served by a centre and the income groups and spending habits of the potential shoppers.

Known future developments may have to be taken into account. For instance a proposed multi-storey car park in a centre badly provided with parking facilities may have the effect of increasing the value of shops close to the car park, possibly at the expense of shops previously of higher value.

Identification of trends can be important. New Bond Street, W1, and Camden Passage, N1, are examples of London streets, which have changed in value dramatically in recent years because they have become centres for the sale of designer clothes and antiques respectively. Similarly the change to "speciality" shopping in South Molton Street, W1, produced a rapid escalation in rental value in the late 1980s. Even established shopping streets such as King's Road, Chelsea, can undergo changes which produce a greater-than-average variation in rental values over a short period. It should be borne in mind that changes in rental value brought about by special demand might be short term only which could quickly be reversed. Local, regional and national trends can vary considerably both in rate and in direction, and it is this irregular movement of rental values which makes it difficult to construct indices which are anything more than a general guide.

Tourism can materially affect the turnover of shops in central London and in historic towns in the regions. Changes in the economic climate in the country of origin of visitors, or in currency exchange rates, can have a substantial and sometimes very rapid effect on turnover, and consequently on rental value. In some cases the effect may be too short-lived to influence rental values; if prolonged, however, rents agreed in times of peak tourist booms may not be maintained subsequently. Oxford Street in Central London is a particularly good example of such movement and it is one of the most volatile locations in the UK, relying to a significant extent on tourist trade.

In a limited number of cases, development costs may be a factor in determining rental values. This could happen, for instance, where a new development takes place in an elderly shopping centre or where an entirely new centre is built. It should be emphasised, however, that a rent calculated to give an adequate return on the landlord's development costs will still have to stand the test of profitability in the eyes of potential tenants.

All potential tenants will first make an estimate of the turnover likely to be achieved in the retail unit in question, and from the "mark up" (ie the amount by which the sale price exceeds the cost) appropriate to their particular merchandise they will then calculate the rent they can afford as a residual of all other operating costs. The process is not wholly objective, and a wish to be represented in a particular location, or the prospect of above-average turnover growth, can be among factors which override purely actuarial considerations. Rents related directly to turnover are common in the USA, but there are still relatively few cases in the United Kingdom – the shopping centres at Eldon Square, Newcastle Upon Tyne, the Metro Centre, Gateshead and the Harlequin, Watford, being some of the major examples. Even these rents are usually linked to a minimum base figure which may be 80% of the open market rent. Elsewhere, owners and occupiers in the UK are concerned at the uncertainty of rent/income which a turnover rent introduces.

Size and Type

The traditional shopping street in the smaller town often started life as a series of terraces of houses which were progressively converted into shops, and this origin accounts for the limited size of many of the shops in the older centres. The tendency in recent years has been for tenants to demand larger units, and new developments will now contain a significant number of larger shops. It will be necessary for the valuer to know the level of demand for the various sizes of shops, otherwise the interpretation of rental evidence can be misleading. For instance, demand for small shops suitable for "niche" operators in a particular location may produce rental values which cannot be safely applied to larger shops, even by zoning or other discounting methods. Similarly, there may be a special demand for large units, particularly in major locations. The special demand might exist, however, for only a limited time. The kind of problem which often faces a valuer is the valuation of, say, an old-established store which has a very large ground floor in a good position in a multiple shopping street. If it is known that one or more of the major space users at present occupying badly planned premises in an inferior position are in the market for better premises, the rent which the valuer estimates might be received may be in excess of rents previously achieved. Conversely, if all the major space users are already locally represented in adequate premises, the valuer might decide that there would be no demand for the unit as a whole and arrive at his rental value by dividing the property into viable units, making allowance for the cost of conversion in the capital value.

Merchandise display is an important factor in retailing, and a major retail company can spend as much on finishing and fitting a unit as on the cost of the structure. Consequently, the internal design has to be taken into account in assessing rental value. Irregular shapes, low ceiling heights, badly sited staircases, differences in floor levels, intrusive stanchions and beams and such like can adversely affect value.

Parking restrictions and pedestrianisation make off-street loading facilities increasingly important. Lack of rear-loading facilities can make shops in heavily restricted areas unlettable for certain trades, eg supermarkets, where large and frequent deliveries are necessary.

Although it has been customary for corner shops with display windows on the return frontage to be regarded as having a greater value than non-corner units, the difference in value is largely a matter of judgment in the particular circumstances. In some cases any additional value could be completely offset by the cost to the tenant of maintaining display space, and subsequent interference to the internal layout. A corner unit at the junction between a primary and a good secondary shopping street would, however, normally be expected to command a higher rent than a similar unit without a return frontage, although the difference has reduced in recent years.

The majority of retailers require staff or stockroom facilities within their units either at the rear, in the basement or on an upper floor. The analysis of rents of lock-up shops, shops with basements and shops with basements and/or upper floors in a shopping centre will usually provide the valuer with a guide to the value of the ancillary accommodation, but experience is necessary when building up a composite value, eg of a shop with a basement and first floor, when rental evidence is not available of directly comparable properties. Attempts are sometimes made to value basements and upper floors on the basis of a proportion of the value of the ground floor. In practice, however, the market is usually too erratic for any reliable analysis of this nature to be made and it cannot be over-emphasised that this type of mathematical analysis plays virtually no part in the process by which a tenant arrives at his estimate of the rent he thinks he can afford to pay. The retail valuer should guard against the tendency to substitute theory for experience in cases where direct evidence is not available.

It is customary in the case of existing retail units to calculate floor areas on a net basis, ie excluding lavatories, staircases, lifts, landings, escalators, plant rooms etc. In new developments, however, where units are being let in "shell" form (ie. with no wall or ceiling finishings and where the tenant is to install lavatories etc.) it is not uncommon for letting areas to be quoted on a "gross internal" basis. The definitions of gross internal area and net internal area given in the Code of Measuring Practice issued by the Royal Institution of Chartered Surveyors should be followed in all cases.

Although zoning of ground floors plays an essential part in rating valuations and in the preparation of valuations for Lands Tribunal or court cases it should be borne in mind that in the open market the zoning system has to be adapted to suit the facts and should not be used on a pre-conceived basis. Ideally, the depth and number of zones should be arrived at by analysis of known rents, but the frequent absence of open market lettings has resulted in a tendency for the adoption of "standard" Zone A depths. Over-reliance on standard depths, particularly when used in conjunction with the analysis of secondary evidence, such as rent review settlements, can produce misleading results. Mathematical precision is not a feature of the valuation of retail properties for open market rental purposes and even between skilled valuers the margin of opinion may be surprisingly wide because of the imperfections in the market.

The conventional zoning basis is to divide the ground floor into sections of equal depth and to assume that each section or zone is worth 50% of the preceding zone, the zone on the street frontage (Zone A) having the 100% value. When dealing with deeper shops, however, the "halving back" process normally stops after four or five zones. To continue halving back can produce what may appear to be unrealistically low values for the rear zones, and ground-floor sales rates, even at the rear, should generally always exceed ancillary and storage rates elsewhere in the shop. With very deep shops, however, the reality might be that a Zone A value derived from the analysis of rents of smaller shops may be of little or no relevance and an alternative approach, such as a valuation on an "overall" basis, may be more appropriate.

Irregular-shaped shops where, for instance, the unit widens at the rear or extends sideways behind the front zone or zones present difficulties for the zoning process and to avoid arriving at an excessive rental value it may be necessary to regard areas at the side of the rear zones as if they were part of one of the deeper zones (eg a side extension to Zone B being regarded as if it were in Zone C).

Irregularities and disabilities are also frequently dealt with by application of "end" or "quantity" allowances, usually in the form of percentage discounts applied to the aggregate rental value. Where arbitrary allowances have to be made to arrive at realistic rental values, however, it is frequently an indication that over-reliance has been placed on the zoning system and that the search for comparable properties needs to be widened or other approaches adopted.

A frequently adopted variation to applying a progressively halved Zone A value is to calculate the area of each zone and then to apply the "halving back" process to the area of each zone, so that the area of Zone B is reduced by 50%, that of Zone C by a further 50%, *et seq*. The chosen Zone A value is then applied to the aggregate of the adjusted areas. The resulting total area is referred to as the area "in terms of Zone A", usually abbreviated to ITZA. The ITZA approach is useful for comparison purposes but in some cases can be misleading as it is possible for two shops to have identical ITZA areas despite having widely differing area distributions between the various zones. In practice it is usually advisable to state both the ITZA area and the actual zone areas.

It cannot be over-emphasised that zoning is only a tool and that the acid test is whether or not the rental value which has been arrived at is one which is achievable in the open market.

Study 1

A rental valuation is required for lease renewal purposes. The shop has a frontage of 6.5 m, a depth of 22.5 m and a net internal area of 133.1 m².

Evidence is available of the following current transactions:

Shop 1: Lease recently renewed at £30,000 pa. Similar position. Frontage 6.5 m, depth 15 m. Net internal area 90.5 m².

Shop 2: New letting at £44,000 pa. Better position. Frontage 7.2 m, depth 25.1 m. Net internal area 173.6 m².

Shop 3: New letting at £37,200 pa. Similar size and position but better property. Net internal area 129.2 m².

Shop 4: Lease renewed 6 months ago at £34,000 pa. Identical to the subject unit.

Analysis of the rents of the four comparables shows that by adopting two 7.5 m zones and a remainder and "halving back", a reasonably consistent Zone A figure can be deduced:

	Area in m²	£	£ pa
Shop 1: Ground floor:			
Zone A	46.3	439	20,326
Zone B	44.2	439/2	9,702
No remainder			–
	90.5		30,028
		Say	£30,000
Shop 2: Ground floor:			
Zone A	52.5	459	24,097
Zone B	52.5	459/2	12,049
Remainder	68.6	459/4	7,872
	173.6		44,018
		Say	£44,000

Shop 3: Ground floor:

Zone A	50.1	466	23,347
Zone B	39.8	466/2	9,273
Remainder	39.3	466/4	4,578
	129.2		37,198
		Say	£37,200

Shop 4: Ground floor:

Zone A	48.2	421	20,292
Zone B	45.1	421/2	9,494
Remainder	40.2	421/4	4,231
	133.5		34,017
		Say	£34,000

Shop 2 is in a better position than the subject property and should have a higher Zone A value. Shop 3 is a better shop and will also have a higher Zone A value. Shop 4 gives a good indication of the rental value six months ago but is now low. The analysis suggests a Zone A value of £445 per m^2 for the subject property and this is supported by Shop 1.

Valuation

	Area in m^2	£	£ pa
Ground floor:			
Zone A	47.2	445	21,004
Zone B	43.6	445/2	9,701
Remainder	42.3	445/4	4,706
	133.1		35,411
Say			£35,400

Notes

It has been assumed that the lease terms are similar in each case. In practice, adjustments may have to be made, eg for differing repairing liabilities to produce comparable rents. The analysis will also include any basements or other floors included in the leases.

Restaurants. In recent years there has been a significant increase in the number of restaurant outlets, although these are mainly located in secondary non-retail locations due to problems in obtaining planning consents in the main retail pitches, and the higher rental costs involved in these locations. Many traditional restaurants which were a feature of most High Streets have disappeared owing to their inability to produce profit margins comparable with those of other multiple retail traders. Where conventional restaurants do exist in retail locations they do not normally present valuation problems as their rental values will be dictated by retail rents in that location, assuming there is no restriction on general retail use. Sometimes the appearance of "fast food" restaurants, usually providing a limited or specialised range of food, has produced, for the better-located units, rental values considerably in excess of those obtainable from a normal retail use in the same location. The valuation of such properties and those in secondary locations normally requires a considerable degree of specialised knowledge on the part of the valuer.

Supermarkets, which are mainly associated with the grocery trade, have in recent years developed on separate lines from normal shops. The tendency has been for size requirements to increase, with great importance being placed on the proximity of car-parking facilities. In new centres supermarkets with a total area of at least 3,000 m^2 are not uncommon. The relationship between gross area and sales area varies between retailers but ancillary space (storage, preparation, etc) equivalent to about 25%–35% of the ground-floor sales area will usually be required. Initially, most supermarkets were planned on ground-floor level only. Second generation supermarkets often included first-floor sales areas, but the present trend is a reversal to ground-floor sales areas only. Early supermarkets were predominantly food based, but later developments contained an increased range of merchandise such as hardware, clothing and furniture. Changes in operational requirements tend to be relatively rapid and it is important that the valuer should be aware of current developments.

Normal rental valuation practice for purpose-built supermarkets is to apply an overall rate to the total gross internal area. It is obviously essential that the same approach should be used for analysis and synthesis.

Purpose-built supermarkets, particularly the larger ones, have become a specialised category of retail property as their size and location have made comparison with normal retail properties increasingly difficult. It is now generally accepted that, in the absence of evidence of rental values of similar sized units in the same location, comparison may have to be made with the rents of supermarkets in other locations.

Departmental and variety stores present particular problems, as their number is limited and there is rarely any direct evidence of comparable open market transactions available. The majority of the very large older stores are owner-occupied, the tenure being either freehold or long leasehold, and where stores do change hands it is more often than not as part of a transaction involving the sale of the owning company. Some rental evidence is found from sale and leaseback transactions and rack-rented stores are occasionally found in new schemes, although in the latter the tendency is for the store operator to ground-lease a site in the scheme. There is, however, a body of rent review evidence for variety stores in the range of 3,000 to 10,000 m² total area. Many of these reviews are based on previous rent review evidence in the region, a direct result of the scarcity of open market transactions. As a result it is a commonly held view that the better stores are under-rented, ie reviewed at less than their true open market value, while the reverse is true for poorer stores where, in reality, little demand may exist.

It should be appreciated that whereas the proprietor of a normal sized retail shop will usually stock merchandise which is both limited in range and chosen as far as possible to produce the maximum profit margin, a departmental or variety store will usually offer a very wide range of merchandise with varying profit margins and may even operate some sections at a loss in order to be able to offer a full service. As a general rule the smaller and faster moving lines are sold on the ground floor and the larger and slower moving lines, such as furniture, on the upper floors or in the basement. In practice, an analysis of turnover by departments can show that, in terms of profitability, the basement and the first floors of a multi-floor store are worth more in relation to the value of the ground floor than would be the case if a similar analysis were made of the turnover of a conventional shop. The analysis will also probably show, however, that the value of the floors above first-floor level, despite service by lifts and escalators, falls much more rapidly than would be indicated by the analysis of normal sized retail properties, and that fourth and fifth floors, if used for retailing, may have a minimal or even a nil value in profitability terms.

The extent to which profitability should be taken into account is a matter of experience but it is common for store operators to think of rent in terms either as a percentage of overall turnover or as an overall rate per m². In arriving at his final answer the valuer might have to use a combination of a conventional overall rate valuation, an estimation of rent as a proportion of turnover and an estimate of the alternative use value of the premises. The alternative use value may be of little consequence in the case of a modern, purpose-designed store but may produce the only viable valuation in the case of an old, out of position store. The alternative use value is sometimes referred to as the "break up" value where the only practical alternative is the division of the ground and other floors into smaller units with appropriate provision being made for the costs of conversion.

It is now common practice for department and variety stores to be valued on the basis of an overall rate applied to the total gross internal area. The valuer uses his skill to adjust this rate for size, location, number of floors and layout. A common topic of discussion is the size threshold over which a retail unit should be valued on an overall, rather than zoning basis. Much depends on the comparable evidence and size of ground floor, but generally stores over 2,000 m² are valued on an overall basis.

Care should be taken as to whether a store is to be valued on a fitted or shell basis. Valuing fittings (eg escalators, lifts and plant) can produce a significant increase in the rental value, often in the order of 10% of the total rent. Existing use valuations are generally carried out on a fitted basis, as are rent reviews following a sale and leaseback transaction. Many rent review clauses do, however, direct a shell assumption.

Study 2

A valuation is required for rent review purposes of a department store on the fringe of a good multiple shopping position. The store was purpose-built in 1925 but has been modernised and is adequately served by lifts and escalators. The lease provides that for the purpose of the rent reviews the rental value shall be that of a department store and that the store should be valued as fitted.

The store comprises ground floor, basement and four upper floors; the ground floor has a frontage of 30 m and a depth of 100 m. The gross internal area of the entire store is 12,000 m².

	m²	£	£ pa	
Total GIA	12,000	45.00	540,000	
Plus 10% for fittings				54,000
Total			594,000	
Say				£595,000 pa

Notes
(1) Evidence of similar sized stores suggests a range of £40.00 to £47.50 per m^2.
(2) Evidence of rent reviews of fitted stores suggests that 10% is the appropriate adjustment.

Superstores vary from isolated "green fields" stores unsupported by other shops to stores designed as the hub of new shopping centres. Wide variations in design can be found ranging from older buildings of warehouse-type construction to fully fitted purpose built stores on one floor only. Overall size is, however, likely to be in excess of 3,000 m^2 and extensive parking facilities are required (in the case of very large superstores it is not unusual for over 1,000 car parking spaces to be provided). Superstores are generally valued by applying an overall rate to the gross internal area and comparisons may need to be made on a regional or even a national basis.

Retail Warehouses are now an important sector of the retail market. They are generally found in out of town locations, either as solus units (usually DIY), as clusters of two or more units, or on fully fledged purpose built retail parks containing as many as 20 units varying in size from 500 to 10,000 m^2. They are valued on an overall rate basis applied to the total gross internal area. Such matters as parking, access, prominence, size, competition, loading facilities and design all play their part in determining the rental value. Open market evidence is plentiful and in a relatively short number of years this has become very much a sector for the specialist valuer. Most retail warehouses only have a "bulky goods" planning consent which restricts demand. Some have open retail consents which can substantially increase their value where demand exists from high street retailers. Fosse Park in Leicester and the Fort at Birmingham are good examples of the latter category.

Lease terms

Where a retail property is valued subject to a lease, the terms of the lease will have to be taken into account in the valuation. Landlords will almost always seek to pass the responsibility for all outgoings to the tenant, and first lettings of new properties will provide for the tenant to be responsible for rates, insurance and all repairs. In the case of entire properties, the tenant will usually be required to undertake all repairs; lettings of parts of properties can require the tenant to pay a proportion of the expenditure incurred by the landlord in repairing the property. Many leases of parts of older properties still provide for the tenant to be responsible for internal repairs only, and where such leases are renewed the provisions of the Landlord and Tenant Act 1954 make it difficult for the landlord to succeed in widening the repairing covenant (*O'May* v *City of London Real Property Co Ltd* [1982] 1 EGLR 76).

Leases of shopping centres frequently provide for the payment of an annually reviewable service charge by the tenant to cover all outgoings, including normal services, special services such as security staff and the maintenance of large areas of common parts such as service roads. The charge is usually related to floor areas but is sometimes linked to rateable values. In covered centres with air conditioning and heating, the service charges can be high and in some centres it is possible that the level of these charges may adversely affect rental value when the rents are due for review.

The frequency of rent reviews is a factor which has a major effect on the value of properties subject to existing leases. The norm for the past 15 to 20 years has been for rents to be reviewed at five-year intervals, previously review periods of 7, 14, 21 or even 35 years were common.

Rental valuations may now still have to be undertaken on properties which are the subject of leases which were granted in the early days of rent reviews with these longer intervals. The assessment of the rent to be paid at the review dates can cause difficulties as there is likely to be no evidence of new lettings with a review frequency longer than, say, five years and very little information as to settlements in similar cases. It is generally assumed that a tenant would pay a higher rent for a lease with a term of, say, 14 years without review than for a lease with reviews at five or seven yearly intervals. Because of the lack of evidence, calculations are sometimes made which are intended to show the rent which a tenant could afford to pay for the longer term and which are largely dependent on assumptions as to the likely rate of increase in rental values. It is debatable whether assumptions as to future rates of inflation or future rental values should ever be employed by valuers concerned with the assessment of open market values, but the biggest objection to such wholly hypothetical calculations is that it is very unlikely in practice that a retail tenant could be found who would be prepared to commit himself to a rent greatly in excess of the normal open market rental value in the hope of enjoying a substantial profit in the later part of the review period, particularly when he would be competing with other retailers on a reduced margin of profit in the early part of the period. However, a rule of thumb developed in the 1980s that 1% pa should be added to the rent for each year over a normal 5 yearly cycle. Therefore a 14 yearly cycle would attract a 9% increase. This percentage reduced as prospects for rental growth diminished in the early 1990s, but a recent return to rental growth has seen the level of increase back to 0.75% to 1% pa.

Leases which provide for payment of a rent based on a percentage of turnover are sometimes encountered, mainly in fairly modern shopping centres. The most common form of percentage rent clause provides for the payment of a base rent and/or the payment of an additional rent based on turnover. Where there is a base rent there is usually also a base turnover, eg the lease might stipulate a rent of £50,000 pa plus an additional rent of 7% of the turnover in excess of £1,000,000 pa. The base rent is often related to a percentage (often 80%) of the open market value of the unit, and may be subject to review at intervals. The additional rent is usually calculated annually on audited accounts. The valuer may be called upon to assess both the initial base rent and the rent at the review dates and this calls for a knowledge of the relative profitability of different trades as the percentage of turnover will vary from trade to trade. It may be necessary to vary the percentage on different parts of the turnover of a particular retailer, eg in the case of a confectioner and tobacconist the percentage of the turnover of cigarettes and tobacco, a large proportion of the price of which is duty, will be less than the percentage of turnover on sweets and chocolates. Considerable experience is necessary when valuing properties subject to leases which contain percentage rent clauses as it is not uncommon to find that the turnover of similar shops let to similar trades varies widely – this can happen, for instance, with ladieswear shops which tend to cater for particular age or fashion groups.

It sometimes happens that a shopping centre may contain shops let on normal rack rental leases and shops let on percentage leases, and there will undoubtedly be discrepancies in the relative rental levels. Experience is required on the part of the valuer in deciding which basis provides the true rental value.

User clauses in retail leases vary between fairly open users and those limited to a particular trade, and can have a considerable bearing on rental value. It can be argued that a limitation to a specific trade will always restrict the potential market and will, therefore, adversely affect rental value. In practice the effect will be less where there is likely to be sufficient demand from other operators in the specified trade to generate the full open market rental value. Where the specified trade is one which has gone "down market", or for which there is likely to be little demand if the shop was offered in the open market with the same restriction, the adverse effect can be considerable. In the case of shopping centres, however, it is sometimes found that a tenant mix policy, which restricts certain types of trade to specified units, provides an advantage as occupiers will be assured that direct competitors will not be allowed to trade elsewhere in the centre.

Study 3

The leases of three similar shops in the same parade are due for renewal. Rental valuations are required by the landlord. The position has improved considerably since the leases were granted 14 years ago and Marks & Spencer and WH Smith have opened new stores in the adjoining block. Originally there was a mixture of private traders and small multiples, but recent lettings and assignments have been to major multiple companies. Evidence suggests that the open market rental value of each of the three shops is £75,000 pa.

Shop 1: let to a chemist, with no user restriction in the lease.
Shop 2: let to a confectionary operator and the lease contains an absolute prohibition against change of use. Although the tenant operates three other shops and trades efficiently, the maximum rent which he could afford to pay for this branch is £50,000 pa and this is supported by audited accounts.
Shop 3: let to a mobile phone operator with a restriction against assigning or subletting to any trade already represented in the parade.

Valuations

Shop 1 is straightforward and a valuation of £75,000 pa can be reported.

Shop 2 presents the valuer with a problem. The tenant's trade is one which in recent years has been unable to compete with the profit margins achieved by other trades except in secondary positions. If the confectioner decides to seek a renewal of his lease he will argue that the restriction adversely affects rental value, and in the event of court proceedings will probably succeed in having the restriction continued in the new lease (*Charles Clements (London) Ltd* v *Rank City Wall Ltd* [1978] 1 EGLR 47). Evidence of the effect of such a restriction is unlikely to be available, in which case the valuer will probably advise that the landlord may have to be prepared to accept a rent lower than £75,000 pa (tenants' accounts are not, however, generally admissible as evidence of rental value – (*QH Barton Ltd* v *Long Acre Securities Ltd* (1981) 262 EG 877)).

Shop 3 is also not straightforward. If there are a number of trades not represented in the parade, the valuer may take the view that the restriction does not affect the rental value, and report a figure of £75,000 pa. If, however, most of the potential uses are already present, the valuer may believe some reduction is necessary, but not of the same magnitude as that of Shop 2.

In the case of new shopping centres in good locations where all the units are marketed simultaneously, there is generally little or no variation between the rents achieved for units of similar size and in similar positions, regardless of whether or not there are user restrictions in the leases. In such cases there are grounds for arguing that a similar result should be achieved when reviewing rents or renewing leases. Provided there is not an excess of the less profitable use, the semi-monopoly status mitigates the effect of the user restriction. In small local parades restrictions can enhance the value of individual units.

By their nature, retail properties are more likely to have alterations and improvements made to them than other types of commercial property. The retail valuer must always, therefore, take steps when dealing with properties subject to leases to identify alterations which have been made by the tenant or his predecessor which could rank as improvements for the purposes of section 34 of the Landlord and Tenant Act 1954. Where qualifying improvements have been made and the purpose of the valuation is to advise the landlord or the tenant as to the rental value for renewal of the lease, the property has in effect to be valued as if the improvement had not been carried out. Where an improvement was made during the currency of the expiring lease, and regardless of the length of that lease and of the date of the improvement, the effect on rent of the improvement is to be disregarded for the duration of the subsequent lease (the duration being at the discretion of the court with a maximum of 14 years). Where an improvement was made during a previous lease, and more than 21 years prior to the date of service of notice for a new tenancy, the tenant loses the right to have the improvement disregarded.

Study 4

A 35-year lease is about to end and the tenant requires a rental valuation prior to negotiating a new tenancy. The building comprises a shop with two upper floors. Present floor areas are: ground floor 97 m²; first floor 92 m²; second floor 78 m². Twenty-five years ago the tenant, at his own expense, extended the ground and first floors at the rear, the net additional area on each floor being 11 m². He also removed the original staircase between the ground and first floors and replaced it with a new staircase in the extension. The original staircase occupied an area of 5 m² on the ground floor and 3 m² on the first floor. Current rental values are: ground floor (valued overall) £250 per m²: first floor £45 per m²; second floor £30 per m².

Valuation

	Area in m²	£	£ pa
Second floor	78	30	2,340
First floor	78 (92 − (11 + 3))	45	3,510
Ground floor	81 (97 − (11 + 5))	250	20,250
			£26,100
Say			£26,000

Notes
(1) For the purpose of the study it has been assumed that the same rate per m² can be applied to the area of the original and of the extended ground floor; in practice different rates may be applicable.
(2) The tenant is entitled to the benefit of his improvement during the currency of the new tenancy up to a maximum term of 14 years.

It is sometimes suggested that the value in the unimproved state should be increased to reflect an element of potential, possibly by ascribing a notional ground rental value to the area occupied by the improvement, if it extends the built area. It is also sometimes permissible to value a property in its improved state, if it can be shown that any reasonable tenant would have carried out those improvements. For instance a tenant may, at his own cost, construct a new staircase to improve access to an ancillary floor, where previously access may have been difficult, or possibly non-existent. The valuer may consider valuing the property in its improved state, although the cost of the improvement must, of course, be taken into account. It is worth bearing in mind that this approach was generally rejected in the case of *Iceland Frozen Foods Plc* v *Starlight Investments Ltd* [1992] 1 EGLR 126, although this judgment has been much criticised.

It should be appreciated that the provisions of section 34 of the Landlord and Tenant Act 1954 do not apply to rent reviews. It is common practice for wording either identical or similar to that of section 34 to be incorporated in rent review clauses, but in the absence of such wording improvements carried out by the tenant need not be disregarded when assessing rental value for review purposes (*Ponsford* v *HMS Aerosols Ltd* [1978] 2 EGLR 81).

Where a rent review clause is silent on the subject of improvements an anomaly can arise in that the tenant may have to pay the full rental value for the property in its improved state at the rent review date(s), but on termination of the lease may be entitled to a new lease at a rent disregarding the improvements.

The question of whether or not a shopfront is an improvement sometimes arises. In the case of a shop in a multiple shopping position, it is customary for the major retailers to provide their own individual and distinctive shopfronts. The existence of a shopfront in these situations is, therefore, likely to be of little interest to potential tenants for they would immediately remove the shopfront and replace it with one of their own pattern, which would be unlikely to qualify as an improvement as it is particular to that trader. In a secondary shopping street, a modern shopfront which could be used by any of a number of private traders would, in most circumstances, add to the rental value of the shop and be properly regarded as an improvement.

Where a retail unit is let in shell form and the lease is not granted in consideration of the tenant completing the fitting out, the works, if extensive, may qualify as improvements. This can cause problems for the valuer at the end of the lease (or at the review date(s)); there could well be no evidence of the rental value of units let in shell form (*GREA Real Property Investments Ltd* v *Williams* [1979] 1 EGLR 121).

In recent years the interpretation of leases has been a major source of litigation. It is essential that the valuer should be aware of current court decisions, particularly in respect of repair, user, alienation, improvements and rent reviews.

Capital Value

The rate of interest at which the rents and rental values of retail units are to be capitalised can vary considerably over a relatively short period and will depend on the demand for retail investments at the time of the valuation. Where a property is let, the demand will be primarily from the investment market and the interest rate will be derived from experience and from the analysis of transactions involving similar properties. In the case of vacant units there may be competition from potential owner occupiers and the valuer will need to identify situations where the investment or the occupational market is in the ascendancy in any particular case.

The primary factors which decide the yields likely to be acceptable to investors are the location, age and condition of the property, the calibre of the tenant, the tenure and the terms of the lease, and the likely future rate of growth of rental values. The lowest yields occur when there is an optimum combination of these factors.

Location is important. The majority of investors favour retail properties situated in centres where there is an established and stable pattern of trading, and where a majority of the traders are well known retail companies. As location deteriorates, yields rise disproportionately and different classes of investors may be involved. Traditionally, the highest rents and lowest yields, and consequently the highest capital values, are to be found in the cities and the larger towns. Retail warehouse and superstore yields move on a more national basis, and do not necessarily reflect the strength or otherwise of the nearest town centre.

Maximum security of income is usually associated with the financial standing of the tenant. Units let to major retailers with large numbers of branches provide the highest level of demand and the lowest yields. Since the larger multiple retailers have branches in centres of widely differing size and importance, the particular combination has to be taken into account in assessing value.

A feature of the retail property market is the considerable variation in the type and age of the properties concerned. They can range from units in newly erected shopping centres to converted period buildings, and yields may have to be adjusted accordingly.

Although the repairing liability imposed on the tenant by the lease may make a lack of routine repairs of little concern to the landlord, the presence of more serious defects can adversely affect the likely yield.

Investors pay the highest prices for properties where the outgoings and liabilities of the landlord are at a minimum and where the rental income is at a maximum. The lowest yields are obtained from properties let for optimum terms (currently 15 to 25 years for most retail properties) at full open market rents and on modern leases which provide for rent reviews at frequent intervals and for the tenant to be responsible for all repairs and outgoings.

Chapter 8

Taxation

(*Throughout this chapter Finance Act is abbreviated to FA*)

Note: the following studies have been prepared in accordance with legislation in force at the date of the assumed transaction. Taxation law is subject to annual change. Readers should therefore check for any amendments in relevant legislation before using these studies for current transactions.

The main part of this chapter deals with the computation of tax on a sum or gain of a capital nature arising out of land. Such an amount will be treated for tax purposes as either capital or income and taxed in one or more of the following ways:

1. As a chargeable gain under capital gains tax or, in the case of a company, corporation tax on the chargeable gain;
2. As the value transferred under a chargeable capital transfer made on or within seven years of death and subject to inheritance tax;
3. As the profit or gain from a trade or adventure in the nature of trade and taxed as income under Case I of Schedule D;
4. As a gain of a capital nature arising out of the disposal or development of land and taxed as income under Case VI of Schedule D;
5. As a premium on a short lease and taxed as rental income under Schedule A.

The computation of the amount subject to tax involves both calculation and valuation. In some cases the taxable amount is found by calculation only, eg the profit from a trade in land; in others, the taxable amount can be found only by a combined process of calculation and valuation, eg the computation of the chargeable gain under capital gains tax. The main part of this chapter is concerned with the second group of computations and deals with capital gains tax (CGT) and inheritance tax (IHT). The object is to explain the valuations and computations necessary to produce the amount of gain or value assessable to tax. Calculations showing the amount of tax payable have been omitted.

The last part of this chapter (page 250 onwards) deals with Value Added Tax and property.

Open Market Value

The statutory provisions governing valuations for CGT and IHT are summarised under each of the taxes, but they have a common base, "open market value". This is defined in similar terms for both taxes and has received judicial interpretation which may be generally applied. It is summarised as follows.

A sale in the open market assumes a seller who is a free agent, not anxious to sell without reserve, but one who would sell only at the best possible price that is obtainable. It assumes a purchaser with a knowledge of the condition and situation of the property and all surrounding circumstances. The market is open and includes every possible purchaser, including a special purchaser who would bid more because the property has a special value to him. A special purchaser would not necessarily need to make only one bid more than the general price to secure the property but may be forced to bid higher by speculators hoping to buy the property and sell it on to him at a higher price. The price which a property is expected to fetch in the open market is found by reference to the expectations of properly qualified persons who have fully informed themselves of all relevant information about the property, its capabilities, the demand for it and the likely purchasers (*IRC* v *Clay* [1914] 3 KB 466; *Glass* v *IRC* [1915] SC 449; *Lynall* v *IRC* [1972] AC 680).

The open market is not a purely hypothetical market, exempt from restrictions imposed by law, but the actual market and any restrictions on either seller or purchaser may be taken into account (*Priestman Collieries* v *Northern District Valuation Board* [1950] 2 All ER 129). The property must be assumed to be capable of sale, even if it is subject to restrictions on sale or actually unassignable. However, the purchaser must take the property subject to any such restrictions and adjusts his price accordingly (*IRC* v *Crossman* [1937] AC 26; *Baird dec'd* v *IRC* [1991] SLT 9). The sale assumes a hypothetical seller and buyer of the property, but the valuation must reflect any actual third parties with interests in the property, eg on the valuation of a share in a joint tenancy the intentions of the actual landlord and the other actual joint tenant are relevant (*Walton dec'd* v *IRC* [1996] STC 68).

Market value must be assessed on the assumption that the

property is sold in its actual state of repair at the date of valuation, but that the vendor would arrange for the property to be sold in the most advantageous manner without incurring unreasonable expense of time and effort in making the arrangements. In the case of a landed estate this may involve dividing the property into easily established lots for sale if this would produce a higher price than selling the estate as a single unit (*Duke of Buccleuch* v *IRC* [1967] 1 AC 506; *Earl of Ellesmere* v *IRC* [1918] 2 KB 735). If two or more property interests are being valued for the same taxable event, it may be necessary to assume that separate interests are sold together if that would produce a higher price, eg where a deceased owned freehold farmland and a share in the partnership owning a farming tenancy over that land (*Gray* v *IRC* [1994] STC 360). Alternatively where a single asset only is the subject of valuation, it cannot be assumed to be aggregated with other assets owned by the taxpayer, eg if freehold property is let to a company controlled by the freeholder, the interest is to be valued as subject to that tenancy and not as if it were an unencumbered freehold (*Henderson* v *Karmel's Executors* [1984] STC 572). No allowance is given for hypothetical costs of sale which might be incurred by the hypothetical vendor (*Duke of Buccleuch* v *IRC* above).

A valuation should take into account any inherent possibilities or prospects attaching to the property at the date of valuation, eg hope value for future development or the possibility that a lease might be surrendered (*Raja Vyricherla Narayana Gajapatiraju* v *Revenue Divisional Officer, Vizagapatam* [1939] AC 302). Where the valuation is actually made after the valuation date, however, it may not take into account events which occurred subsequent to that date and which showed how those possibilities or prospects in fact turned out, eg the grant of planning permission or the surrender of the lease. The valuation must reflect those events as possibilities and not as the certainties they have become by the time the valuation is made (*Gaze* v *Holden* [1983] 1 EGLR 147).

For property held jointly, either as tenants in common or as beneficial joint tenants, where the object of the trust for sale is to enable the parties to have the proceeds of sale, the normal rule for valuing one of the joint interests has been to allow a reduction in value of 10% (*Cust* v *CIR* [1917] 91 EG 11). Where the trust, under which shares in property are held, was created for the purpose of continuing occupation by the joint owners, the Court would be unlikely to order a sale to benefit one of them. In that case the discount should be 15% (*Wight* v *IRC* [1982] 264 EG 935). In

Charkham v *IRC* [1997] Lands Tribunal 1997 (unreported), it was held that for undivided minority shares in investment property the hypothetical purchaser would discount the open market vacant possession value for the uncertainty of whether the Court would exercise its discretion to order a sale under a section 30 Law of Property Act 1925 application, and for the uncertainty of the liability for costs of the action, rather than valuing on a capitalised yield basis; the same discount should apply to different sized minority holdings in the same property; and that the discount to be applied should take into account the particular factors affecting each property at the valuation date, with the result that for one property the discount was determined at 15%, and for others 20% and 22.5%.

Capital Gains Tax

The FA 1965 introduced capital gains tax (CGT) in respect of chargeable gains accruing to a person on the disposal or assumed disposal of assets after April 6 1965. The statute law on CGT is now contained in the Taxation of Chargeable Gains Act 1992 (TCGA 1992) as amended by subsequent Finance Acts. CGT is payable by individuals and trustees. Chargeable gains less allowable losses in excess of the annual exemption limit are taxed on an individual as the top slice of the taxpayer's savings income. Trustees are taxed at "the rate applicable to trusts". Companies do not pay CGT, but a company's chargeable gains (calculated as for CGT) are included in the profits and charged to corporation tax.

From 1965 to the introduction of betterment levy in 1967, CGT stood alone, but between 1967 and 1970 it was restricted to a tax on increases in current use value. The abolition of betterment levy on July 22 1970 restored CGT to its former position as the sole tax on capital gains, but it again suffered restrictions in scope on two further occasions. First, on December 17 1973 when the so-called development gains tax (DGT) was introduced to tax as income that part of the chargeable gain termed a development gain. Second, on August 1 1976 when development land tax (DLT) generally replaced DGT while CGT continued to tax increases in current use value. DLT has been abolished for disposals after March 19 1985 and CGT is once again the sole tax on capital gains.

A gain of a capital nature arising out of the disposal of land, which is not specifically exempt, will not automatically suffer CGT. Two hurdles must first be overcome. First, the disposal must not be a trading disposal or an adventure in the nature of trade. If it is, then it will not be subject to CGT, but the profit or gain will be taxed as trading income under Case I of Schedule D. Second, the disposal must not fall within the anti-avoidance provisions of section 776 of the Income and Corporation Taxes Act 1988. This is a widely drawn section designed to tax as income under Case VI of Schedule D a gain of a capital nature arising out of the disposal of land which was acquired, held as stock or developed, with the objective of realising a gain.

CGT applies to the disposal or assumed disposal of land. The term "disposal" is not defined but includes a sale, assignment, gift, grant of a lease at a premium, grant of an option, receipt of a capital sum, settlement of land on trustees and the demolition or destruction of a building. The conveyance or transfer of an asset by way of security (eg a mortgage) is not a disposal for the purposes of CGT.

In general terms up to April 5 1998, a CGT computation comprises the deduction from the consideration for the disposal of certain expenditure, such as the cost of acquisition, improvements and costs of disposal, and then the further deduction of an allowance for inflation ("indexation") since March 31 1982 or the later date of acquisition. Where a property was owned on April 6 1965, only the gain after that date will be taxable. Where a property was owned on March 31 1982, the market value at that date can be substituted for the earlier acquisition cost to reduce the gain. In practice, therefore, the gain on the disposal of a property acquired before or after March 31 1982 will usually be restricted to the real gain since that date or the later date of acquisition. The studies which follow, however, show the alternative methods of calculating the chargeable gain, although in practice it will usually be obvious that rebasing of the gain to the March 31 1982 value will produce the lowest gain without the need for further calculations.

The FA 1998 changed the CGT regime for disposals by individuals, trustees, and personal representatives. (The CGT regime for disposals realised by companies is unaffected by FA 1998 and so the remarks in the rest of this paragraph do not apply to them.) For disposals occurring after April 5 1998 of assets acquired before April 1 1998 the allowance for inflation only goes up to April 1998. For assets acquired from April 1 1998 onwards no allowance is given for inflation. Instead for disposals after April 5 1998 the chargeable gain may be reduced by "taper relief". In appropriate circumstances this reduces the amount on which tax is charged in proportion to the number of whole years of ownership.

Qualifying business assets benefit from more favourable taper relief than other assets. The notes to Study 1 give an example of the effect of taper relief on a chargeable gain.

A CGT computation often requires the incorporation of values, depending on the circumstances of ownership and disposal. Various valuations may be required. Those most usually needed are market value at the date of disposal (eg where there is a gift or disposal otherwise than by way of a bargain at arm's length) (studies 2 and 7); market value at March 31 1982 for the purposes of rebasing the gain to that date or calculating the indexation allowance (studies 2–6); and market value at April 6 1965 (eg where land is sold at a price above current use value (Study 3) or material development has been carried out after December 17 1973 or an election has, or might be, made to calculate the gain by reference to market value at April 6 1965 (Study 2)). Other valuations that may be required include: the value of the interest retained following a part disposal (for incorporation in the part disposal fraction) (Study 4), current use value at the date of disposal (where it is not immediately apparent whether the disposal consideration included development value) (Study 3), the capital value of a rent-charge or income (where the consideration for the disposal comprised or included such an annual sum) (Study 6) and the notional premium for the gratuitous variation or waiver of lease terms.

The statutory definition of "market value" for CGT purposes is contained in section 272 TCGA 1992. It is the price which the property might reasonably be expected to fetch on a sale in the open market, no reduction being made for the assumption that the whole of the property holding has been placed on the market at one and the same time. Where a valuation is made on a disposal, the date of that disposal is the date of the contract and not the conveyance or, in the case of a conditional contract, the date when the condition is satisfied (section 28 TCGA 1992).

Where a property is subject to any interest or right by way of security it shall be assumed, both on acquisition and disposal, that the property is not subject to such security (section 26 (3) TCGA 1992). In certain circumstances, where a person making a disposal retains a contingent liability, no allowance for this is made in the valuation. If the liability subsequently becomes enforceable then a retrospective adjustment is made (section 49 TCGA 1992).

An appeal concerning a dispute as to the valuation of any land for the purposes of CGT is to be heard by the Lands Tribunal (section 46D Taxes Management Act 1970).

Study 1

This Study sets out the basic framework of a CGT computation where the property disposed of was acquired after March 31 1982.

In March 1984 A purchased the freehold interest in a shop for £200,000. The incidental costs of purchase were £7,500. In June 1985 he spent £10,000 on improvements and sold the property in March 1998 for £500,000. His incidental costs of sale were £12,500.

All references are to the TCGA 1992.

CGT computation (pre FA 1998 rules)

	£	£
Consideration for disposal		500,000
Less: Allowable expenditure:		
Consideration for acquisition plus incidental costs (section 38 (1) (a))	207,500	
Enhancement expenditure (section 38(1)(b))	10,000	
Relevant allowable expenditure	217,500	
Incidental costs of disposal (section 38(1)(c))	12,500	230,000
Unindexed gain (section 53(2)(a))		270,000

Less: Indexation allowance (sections 53, 54, and 55(2)) by reference to relevant allowable expenditure only:

Consideration for acquisition plus incidental costs £207,500 multiplied by

$$\frac{RD - RI}{RI} \text{ where}$$

RD = RPI for month in which disposal
occurs March 1998 =   160.80
RI = RPI for month of acquisition
March 1984 =   87.48

$$\frac{RD - RI}{RI} = \frac{160.80 - 87.48}{87.48} = \quad 0.838$$

£207,500 × 0.838   173,885

Enhancement expenditure £10,000 multiplied by

$$\frac{RD - RI}{RI} \text{ where}$$

RD = RPI for month in which disposal
        occurs =   160.80

RI = RPI for month in which
         expenditure occurred =   95.41

$\dfrac{RD - RI}{RI} = \dfrac{160.80 - 95.41}{95.41} =$ 0.685

£10,000 × 0.685   6,850   180,735
Chargeable gain   £89,265

FA 1998 changes

If the above disposal had occurred in June 2000, and the property qualified as a business asset through being used by A for the purposes of his trade throughout his period of ownership, the CGT computation would be modified as follows:

Unindexed gain (as above)   270,000

Less: Indexation allowance (restricted
to the period to April 1998 (section
122 FA 1998):
Consideration for acquisition plus
incidental costs £207,500 multiplied by

$$\dfrac{RD - RI}{RI} \quad \text{where}$$

RD = RPI for April 1998 =   162.60
RI = RPI for March 1984 =   87.48

$\dfrac{RD - RI}{RI} = \dfrac{162.60 - 87.48}{87.48} =$ 0.859

£207,500 × 0.859   178,243

Enhancement expenditure £10,000 multiplied by

$$\dfrac{RD - RI}{RI} \quad \text{where}$$

RD = RPI for April 1998 =   162.60
RI = RPI for month in which
          expenditure occurred =   95.41

$\dfrac{RD - RI}{RI} = \dfrac{162.60 - 95.41}{95.41} =$ 0.704

£10,000 × 0.704   7,040   185,283
Chargeable gain   84,717

Less : Tapering Relief – 1 qualifying holding year – 7.5% Note 1	6,354
Taxable Amount	£78,363

Notes

(1) To calculate the amount of taper relief available on the disposal of a business asset, the period from April 5 1998 or later date of acquisition is measured to establish the number of whole years of qualifying holding period. For disposals of business assets after April 5 2000 taper relief accrues at the rate of 12.5% after one whole qualifying holding year; after two such years 25%, after three years 50% and after four years the maximum taper relief is obtained of 75% (Section 66 FA 2000 modifying section 2A TCGA 1992 (5) & (8)).

The business asset was owned by A on April 5 1998 and so qualifies for two years of qualifying holding period for taper relief purposes.

(2) Had the asset not been used for business purposes by A, but instead it had been an investment, eg let to a third party, the asset would have accrued 5% taper relief by the date of disposal.

For non-business assets taper relief is given at the rate of 5% pa from the third whole qualifying holding year after April 5 1998. Such assets held on March 17 1998 are deemed to have a minimum of one qualifying holding year for taper relief (section 2A TCGA 1992 (5) & (8)). Accordingly for non-business assets the earliest date a disposal may benefit from taper relief is April 6 2000. The maximum taper relief for non-business assets is 40% achieved after 10 whole years of qualifying holding period.

Study 2

This study sets out the basic framework of a CGT computation where the property disposed of was acquired before April 6 1965.

A purchased the freehold interest in a commercial building in June 1957 for £8,000. His incidental costs of acquisition were £300. In June 1964 he spent £1,000 on improvements. On April 6 1965 the property was let on a full repairing and insuring lease, with five years unexpired, at a rent of £750 pa exclusive. The rack-rental value was £1,150 pa.

In December 1969 A spent £750 in legal costs successfully defending his title to the property. On March 31 1982 the property was let on a full repairing and insuring lease, with three years unexpired, at a rent of £1,500 pa exclusive. The rack-rental value was £2,500 pa.

In February 1997 A transferred the freehold to his son for a consideration of £20,000. The incidental costs of disposal were £600. At the date of disposal the property was let on a full repairing and insuring lease, with 12 years unexpired, at a rent of £5,000 pa exclusive with five-year rent reviews: the next review was in four year's time. The rack-rental value was £6,000 pa.

All references are to TCGA 1992.

This transaction is between connected persons (section 286 (2)): it is, therefore, treated as otherwise than by way of a bargain at arm's length and market value must be substituted for the actual consideration (sections 17 and 18).

(i) *Market value at disposal (section 272)*

	£ pa	£
Rent on lease	5,000	
YP 4 years at 7%	3.4	17,000
Rack-rental value	6,000	
YP perp deferred 4 years at 7%	10.9	65,400
		£82,400
Market Value at date of disposal: February 1997		
Say		£82,500

(ii) *Market value at March 31 1982 (sections 35, 55 and 272)*

	£ pa	£
Rent on lease	1,500	
YP 3 years at 7.5%	2.6	3,900
Rack-rental value	2,500	
YP perp deferred 3 years at 7.5%	10.7	26,830
		£30,730
Market Value at March 31 1982		
Say		£31,000

(iii) *Market Value at April 6 1965 (section 272 and Schedule 2 para 17)*

	£ pa	£
Rent on lease	750	
YP 5 years at 6%	4.2	3,150
Rack-rental value	1,150	
YP perp deferred 5 years at 6%	12.5	14,320
		£17,470
Market Value at April 6 1965		
Say		£17,500

CGT Computation
There are three different methods of calculating the gain on which tax is charged.

1. The calculation may be based on the original date and cost of acquisition, with the resulting gain being reduced by "time apportionment".
2. The taxpayer may elect that the calculation will be based on an assumed sale and repurchase of the property at open market value on April 6 1965, with no time apportionment.
3. The calculation may be based on the open market value on March 31 1982, again assuming sale and repurchase, and similarly with no time apportionment.

If the taxpayer has already made an election for all his assets to be rebased to March 31 1982 values, only the third method of calculation will be available (section 35(5) TCGA 1992).

If no such election has been made, the taxpayer first should compare the gains under (1) and (2) above and choose the lower gain. (An election for April 6 1965 valuation is irrevocable so that once made the option of calculating the gain under method (1) is no longer available).

Then he should compare the gain under (1) or (2), as the case may be, with the gain under (3) and choose whichever will give him the lower chargeable gain.

Method (1): Time Apportionment (Schedule 2 para 16) (Note 1)

	£	£
The first step is to calculate the overall gain.		
Deemed consideration for disposal (market value)		82,500
Less: Allowable expenditure:		
Consideration for acquisition plus incidental costs (section 38(1)(a))	8,300	
Enhancement expenditure (section 38(1)(b))	1,000	
Legal costs defending title (section 38(1)(b))	750	
Relevant allowable expenditure	10,050	
Incidental costs of disposal (section 38(1)(c)) (Note 2)	600	10,650
Unindexed gain (section 53 (2) (a))		71,850

Less: Indexation allowance (sections 53, 54, and 55(2)):

(a) by reference to relevant allowable expenditure multiplied by	10,050	

$$\frac{RD - RI}{RI} \quad \text{where}$$

RD = retail price index (RPI) for
month in which disposal occurs =   155.00
RI = RPI for March 1982 =   79.44

$\dfrac{RD - RI}{RI} = \dfrac{155.00 - 79.44}{79.44} =$	0.951	
	9,558	

Or

(b) by reference to market value at March 31 1982 multiplied by	31,000	

$$\frac{RD - RI}{RI} \quad \text{(above)}$$

	0.951	
	29,481	
(b) is higher		29,481
Overall gain		£42,369

Apportionment of overall gain
The overall gain is allocated proportionately to the items of allowable expenditure under section 38 (1) (a) and (b) (Schedule 2 para 16(4)), and then each allocated part of the gain is then subjected to "time apportionment".

		£
Consideration for acquisition plus incidental costs	"E 1"	
$\dfrac{8,300}{10,050} \times £42,369$		34,991
Enhancement expenditure	"E 2"	
$\dfrac{1,000}{10,050} \times £42,369$		4,216
Legal costs defending title	"E 3"	
$\dfrac{750}{10,050} \times £42,369$		3,162
Overall gain		£42,369

Chargeable gain
The proportion of each of the above items of expenditure relating
to the period after April 6 1965 is now calculated. The chargeable
gain is the sum of the above items each multiplied by the fraction

$$\frac{T}{P + T}$$

where P = the period from the date of acquisition, or date when the
expenditure was incurred or reflected in the property, to April 5
1965; and T = the period from April 6 1965 to the date of disposal
(Schedule 2 para 16) (Note 3).

	£
Consideration for acquisition plus incidental costs	
£34,991 "E 1" × $\frac{382}{94 + 382}$ months =	28,081
Enhancement expenditure	
£4,216 "E 2" × $\frac{382}{10 + 382}$ months =	4,108
Legal costs defending title	
£3,162 "E 3" × $\frac{382}{0 + 382}$ months =	3,162
Chargeable gain	£35,351

Assumed acquisition at April 6 1965 (Schedule 2 para 17)
A may elect to have his gain calculated on the assumption that he
sold and immediately reacquired his freehold interest at market
value on April 6 1965 (Schedule 2 para 17):

	£	£
Deemed consideration for disposal (market value)		82,500
Less: Allowable expenditure:		
Consideration for acquisition (market value at April 6 1965(section 38 (1) (a)) (Note 4)	17,500	
Legal costs defending title (section 38(1)(b))	750	
Relevant allowable expenditure	18,250	
Incidental costs of disposal (section 38(1)(c)) (Note 2)	600	18,850
Unindexed gain (section 53 (2) (a))		63,650

Less: Indexation allowance (sections 53, 54, and 55(2)):

(a) by reference to relevant allowable expenditure (above)	18,250
multiplied by $\dfrac{RD - RI}{RI}$ (above)	0.951
	17,356

Or

(b) by reference to market value at March 31 1982	31,000	
multiplied by $\dfrac{RD - RI}{RI}$ (above)	0.951	
	29,481	
(b) is higher		29,481
Chargeable gain		£34,169

This is smaller than the time apportioned gain and, provided A is confident that he can agree a 1965 value of about £17,500 with the district valuer, he could elect to use this alternative method of calculating the chargeable gain. This gain must, however, now be compared with the gain which would have arisen if A had sold and then reacquired the property at market value on March 31 1982. The smaller of these two gains will be the chargeable gain.

Assumed acquisition at 31 March 1982 (section35)

	£	£
Deemed consideration for disposal (market value)		82,500
Less: Allowable expenditure:		
Consideration for assumed acquisition (market value at March 31 1982) (section 38(1)(a)) (Note 4)	31,000	
Incidental costs of disposal (section 38(1)(c))	600	31,600
Unindexed gain (section 53(2)(a))		50,900
Less: Indexation allowance (sections 53, 54 and 55(2)):		
(a) by reference to relevant allowable expenditure (as above)	9,558	

Or

(b) by reference to market value at March 31 1982 (as above)	29,481	
(b) is higher		29,481
Chargeable gain		£21,419

This is smaller than the gain calculated on the basis of an assumed acquisition at April 6 1965 and will, therefore, be the chargeable gain.

Notes

(1) This is a disposal after April 5 1988 where the deemed consideration is not in excess of current use value and no material development has been carried out: therefore, any of the three methods of calculating the gain may be used (section 35 and Schedule 2 paras 16 and 17).

(2) Actual costs of disposal must be used (section 38 (1) (c)).

(3) The above calculations follow the rule in *Smith* v *Schofield* [1993] STC 268 in deducting the indexation allowance before apportioning the overall gain.

(4) On an assumed sale and reacquisition, it is not to be assumed that incidental costs were incurred (section 38 (4)).

Study 3

This study illustrates the CGT computation where land acquired before April 6 1965 is sold at a price in excess of current use value. It also illustrates roll-over relief on the disposal of business assets.

In 1964 A purchased the freehold interest in a site of 0.364 ha with planning permission for industrial development for £7,500. His incidental costs of acquisition were £300. He commenced building works in June 1965 and in the following December he completed the construction of a factory of 1,440 m^2 at a total cost of £65,000. He occupied the factory for the manufacture of ball bearings until February 1997 when he sold it for £500,000, incurring incidental costs of disposal of £15,000. He reinvested £480,000 in the purchase of another factory to carry on his business.

All references are to TCGA 1992.

(i) Current use value at disposal (section 272 and Schedule 2 para 10)
"Current use value" (CUV) is defined as market value calculated on the assumption that it is, and would continue to be, unlawful to carry out any material development of the land, other than

development authorised by planning permission and started before the valuation date. CUV cannot, therefore, contain any element of hope value or development value (*Watkins* v *Kidson (No 2)* [1979] 1 WLR 876; *Morgan* v *Gibson* [1989] STC 568).

"Material development" is the making of any change in the state, nature or use of land other than the minor works and uses listed in Schedule 2 para 13.

(a) Value as existing factory	£
Rental value, 1,440 m² at £17.50, Say	25,200
YP perp at 12%	8.33
	£209,916
Say	£210,000

Or

	£
(b) Site with assumed planning permission to demolish existing factory and build a new factory with a cubic content of up to 110% of existing factory (Schedule 2 para 13 (1) (b))	
0.364 ha at £625,000 per ha	227,500
Less: Cost of demolition	10,000
	£217,500

Both valuations are less than the sale price: the disposal was, therefore, at a price in excess of CUV (Schedule 2 para 9 (1) (b)).

(ii) *Market value at April 6 1965 (section 272 and Schedule 2 para 9(2))*
Site with planning permission for the construction of a factory of 1,440 m²:

	£
0.364 ha at £35,000 per ha	12,740
Say	£12,750

(iii) *Market Value at March 31 1982 (sections 35, 55 and 272)*

	£
Rental value, 1,440 m² at £16.00, Say	23,000
YP perp at 10%	10
	£230,000

CGT Computation

This disposal was at a price in excess of CUV: the chargeable gain must therefore be calculated by reference to either the market value at April 6 1965 or the cost of acquisition, whichever produces the smaller gain or loss. Time apportionment is not available because the sale was at a price in excess of CUV (Schedule 2 para 9 (2)). This gain must then be compared with the gain on an assumed acquisition at market value at March 31 1982. The smaller of these two gains will be the chargeable gain.

By reference to market value at April 6 1965 (Schedule 2 para 9 (2))

	£	£
Consideration for disposal		500,000
Less: Allowable expenditure:		
Consideration for assumed acquisition (market value at April 6 1965 (section 38(1)(a))	12,750	
Enhancement expenditure (section 38(1)(b))	65,000	
Relevant allowable expenditure	77,750	
Incidental costs of disposal (section 38(1)(c))	15,000	92,750
Unindexed gain (section 53 (2) (a))		407,250

Less: Indexation allowance (sections 53, 54 and 55(2))
(a) by reference to relevant allowable expenditure (above)   77,750
multiplied by

$$\frac{RD - RI}{RI} \quad \text{where}$$

RD = RPI for month in which disposal occurs =   155.00
RI = RPI for March 1982 =   79.44

$$\frac{RD - RI}{RI} = \frac{155.00 - 79.44}{79.44} =$$ 0.951
73,940

Or

(b) by reference to market value at
March 31 1982   230,000
multiplied by $\dfrac{RD - RI}{RI}$ (as above)   0.951
218,730

(b) is higher		218,730
Chargeable gain (subject to relief – see below)		£188,520

By reference to cost of acquisition (Schedule 2 para 9 (4))

	£	£
Consideration for disposal		500,000
Less: Allowable expenditure:		
Consideration for acquisition plus incidental costs (section 38 (1) (a))	7,800	
Enhancement expenditure (section 38(1)(b))	65,000	
Relevant allowable expenditure	72,800	
Incidental costs of disposal (section 38(1)(c))	15,000	87,800
Unindexed gain (section 53(2)(a))		412,200

Less: Indexation allowance (sections 53, 54 and 55(2))

(a) by reference to relevant allowable expenditure (above) 72,800
multiplied by

$$\frac{RD - RI}{RI} \quad \text{(as above)} \qquad \underline{0.951}$$

$$\underline{69,233}$$

Or

(b) by reference to market value at March 31 1982 230,000
multiplied by

$$\frac{RD - RI}{RI} \quad \text{(as above)} \qquad \underline{0.951}$$

$$\underline{218,730}$$

(b) is higher	218,730
Chargeable gain (subject to relief – see below)	£193,470

The chargeable gain calculated by reference to market value at April 6 1965 (£188,520) is smaller but must now be compared with the gain calculated on an assumed acquisition at market value at March 31 1982.

Assumed acquisition at market value at 31 March 1982 (section 35)

	£	£
Consideration for disposal		500,000
Less: Allowable expenditure:		
Consideration for assumed acquisition (market value at March 31 1982 (section 38(1)(a))	230,000	

Incidental costs of disposal (section 38(1)(c))	15,000	245,000
Unindexed gain (section 53 (2) (a))		255,000
Less: Indexation allowance (sections 53, 54 and 55(2)):		
(a) by reference to relevant allowable expenditure (as above)	69,233	
Or		
(b) by reference to market value at March 31 1982 (as above)	218,730	
(b) is higher		218,730
Chargeable gain		£36,270

This chargeable gain is smaller than the gain calculated by reference to market value at April 6 1965 (£188,520) and will be used.

Roll-over relief (sections 152–158)
A disposed of a qualifying asset (the factory) used exclusively for the purposes of his trade, and spent most of the proceeds on the purchase of new premises for the continuance of his business. He may therefore, claim to defer his CGT liability by deducting part of his chargeable gain from the acquisition cost of the replacement factory, provided the amount reinvested (£480,000) exceeds the indexed base cost of the asset sold (£245,000 + £218,730 = £463,730).

	£	£
Chargeable gain eligible for relief		36,270
Less: deduction for amount of sale proceeds not reinvested:		
Chargeable gain	36,270	
Less: amount not reinvested (£500,000 – £480,000)	20,000	
Rolled-over gain		16,270
Chargeable gain		£20,000

The acquisition cost of A's replacement factory will be reduced to £463,730 (ie £480,000 purchase price less £16,270 rolled-over gain). The chargeable gain not rolled-over, £20,000, is subject to CGT.

Notes

(1) Where part only of the overall gain is a chargeable gain (eg where the gain is time apportioned) and part only of the proceeds of sale is reinvested, the following adjustments are made. First, the part of the proceeds not reinvested is reduced to the proportion that the chargeable gain bears to the overall gain when calculating the amount of the chargeable gain which cannot be deferred (section 152(2)). Second, the balance of the chargeable gain is then deducted from the acquisition cost of the replacement property (section 153).

(2) The roll-over relief available to A does not alter the acquisition cost to the purchaser of A's factory, nor the sale proceeds to the vendor of the new factory purchased by A (section 152 (1) and 153 (1)).

Study 4

This study illustrates the CGT computation where there is a part disposal of a freehold interest by the grant of a lease not exceeding 50 years at a premium. It also illustrates the relationship between CGT and income tax Schedule A. (Schedule 8 para 5(1) TCGA 1992 and Section 34(1) ICTA 1988).

G purchased the freehold interest in a commercial building in 1966 for £15,000. His incidental costs of acquisition were £300. In 1970 he spent £1,000 on improvements. He subsequently let the whole property on full repairing and insuring terms at a rack-rent. On March 31 1982 this lease had three and three quarter years unexpired at a rent of £2,000 pa exclusive. The rack-rental value was £2,250 pa. At the expiry of the lease it was renewed at a rack rent for a further 11 years after which the tenant vacated.

In January 1997 G granted a new full repairing and insuring lease for 15 years at a fixed rent of £3,250 pa exclusive and a premium of £80,000. The rack-rental value was £12,500 pa. His incidental costs of this part disposal were £2,500.

All references are to TCGA 1992 except where indicated otherwise.

(i) Market value at March 31 1982 (sections 35, 55 and 272)

	£ pa	£
Rent on lease	2,000	
YP 3.75 years at 8%	3.1	6,200

Rack-rental value	2,250	
YP perp deferred 3¾ years at 8%	9.4	21,150
		£27,350
Say		£27,500

(i) Market value of the freehold reversion after the part disposal (section42(2)(b) and Schedule 8 para 2 (2))

	£ pa	£
Rent on lease	3,250	
YP 15 years at 15% (Note 1)	5.8	18,850
Rack-rental value	12,500	
YP perp deferred 15 years at 7½%	4.5	56,250
		£75,100
Say		£75,000

This amount is B in the part disposal fraction $\dfrac{A}{A + B}$ (see below).

CGT computation

Apportionment of premium (Schedule 8 para 5 (1) and section 34 Income and Corporation Taxes Act 1988)
The premium on a lease granted for not more than 50 years is subject to both CGT and income tax Schedule A. The part of the premium subject to CGT is calculated as follows:

	£	£
Total Premium		80,000
Less: Part taxed as income under Schedule A:		
Total premium	80,000	
Deduct: 1/50 th for each complete year of the term other than the first: 14/50 × £80,000	22,400	57,600
Part of premium treated as consideration for CGT part disposal		£22,400

Chargeable gain calculation
The grant of a lease at a premium is a part disposal of the landlord's interest (Schedule 8 para 2 (1)).

Disregarding rebasing March 31 1982 (section 35)

	£	£
Consideration for part disposal (premium, as above)		22,400

Less: Allowable expenditure:

	£	£
Consideration for acquisition plus incidental costs (section 38(1)(a))	15,300	
Enhancement expenditure (section 38(1)(b))	1,000	
	16,300	

Apportioned to the part disposal by the fraction (section 42)

$$\frac{A}{A + B} \quad \text{where}$$

A = consideration for the part disposal (Note 2)
  = £22,400 in numerator but
    £80,000 in denominator (Schedule 8 para 5(1))
B = market value of property
    Undisposed of = £75,000

		£
$\dfrac{A}{A + B} = \dfrac{22,400}{80,000 + 75,000} =$		0.145

	£	£
Relevant allowable expenditure	2,363	
Incidental costs of disposal (section 38(1)(c))	2,500	4,863
Unindexed gain (section 53(2)(a))		17,537

Less: Indexation allowance (sections 53, 54 and 55(2)):

(a) by reference to relevant allowable expenditure   2,363
multiplied by

$$\frac{RD - RI}{RI} \quad \text{where}$$

RD = RPI for month in which
       disposal occurs =   154.40
RI = RPI for March 1982 =   79.44

$\dfrac{RD - RI}{RI} = \dfrac{154.4 - 79.44}{79.44} =$		0.944
		2,230

Or

(b) by reference to market value at March 31 1982   27,500
apportioned to the part disposal by the fraction

$$\frac{A}{A + B} \quad \text{(as above)} \qquad \frac{0.145}{3,987}$$

multiplied by indexation allowance

$$\frac{RD - RI}{RI} \quad \text{(as above)} \qquad 0.944$$

(b) is higher   3,763   3,763
Chargeable gain   £13,774

Assumed acquisition at 31 March 1982 (section35)

	£	£
Consideration for disposal (ie premium as above)		22,400

Less: Allowable expenditure:
Consideration for assumed acquisition (market
value at March 31 1982 (section 38(1)(a)),
apportioned to the part disposal by fraction
(section 42 and Schedule 8 para 5 (1))   27,500

$$\frac{A}{A + B} \quad \text{(as above)} \qquad \frac{0.145}{3,987}$$

Incidental costs of disposal (section 38(1)(c))   2,500   6,487
Unindexed gain (section 53(2)(a))   15,913

Less: Indexation allowance (sections 53, 54
and 55(2)):

(a) by reference to relevant allowable
expenditure (as above)   2,230
Or
(b) by reference to market value at
March 31 1982 (as above)   3,763
(b) is higher   3,763
Chargeable gain   £12,150

This chargeable gain is smaller than the gain calculated by
disregarding rebasing to March 31 1982 (£13,774) and will be used.
The part of the premium not taxed under CGT (£57,600) is taxed as
income under Schedule A.

Note

(1) High capitalisation rate used to reflect the fact that a fixed rent for 15 years would be unattractive to a prospective purchaser.

(2) The part disposal fraction $\dfrac{A}{A + B}$ in section 42 is modified in the case of a part disposal of property by the grant of a lease not exceeding 50 years for a premium. In such a case, part of the premium is taxed as income under Schedule A (section 34 Income and Corporation Taxes Act 1988). The numerator "A" is the premium less the amount taxable as income, whereas in the denominator "A" is the full premium received (Schedule 8 para 5(1) TCGA 1992).

Study 5

This study illustrates the CGT liability arising on the disposal by assignment of a short lease for a premium.

In December 1977 J took an assignment of the headlease of a shop which had 45 years unexpired at a fixed rent of £100 pa exclusive. He paid the assignor £20,000 and incurred incidental costs of £1,000. J spent £15,000 on improvements to the property in March 1980. He sublet the property on a full repairing and insuring lease for 20 years from March 25 1982 at a rent of £9,000 pa exclusive with rent reviews every five years.

In January 1997 A disposed of his headlease for £250,000. His incidental costs were £7,500.

All references are to TCGA 1992.

(i) Market Value at March 31 1982 (sections 35, 55 and 272)

	£ pa
Rent receivable	9,000
Less: rent payable	100
Profit rent	8,900
YP 40¾ years at 9% and 4% (taxed at 40%)	9.4
	83,660
Say	£83,500

CGT computation
Disregarding rebasing March 31 1982 (section 35)

	£	£
Consideration for disposal		250,000

Less: Allowable expenditure: (Note 1)
Consideration for acquisition plus incidental
costs £21,000 (section 38(1)(a))

Reduced by: Amount written off (Schedule 8
para 1 (3) & (4)):

$$£21,000 \times \frac{P(1) - P(3)}{P(3)} \text{ where}$$

P(1) = % derived from table for
      duration of lease at acquisition
      (45 years) =   98.059
P(3) = % derived from table for
      duration of lease at disposal
      (25 years 9 months) =   82.147

Amount written off:

$$\frac{98.059 - 82.147}{98.059} = \qquad 0.162$$

£21,000 × 0.162 = £3,402

Reduced amount = £21,000 − £3,402 =   17,598

Enhancement expenditure £15,000
(section 38(1)(b))

Reduced by: Amount written off
(Schedule 8 para 1(3) & (4)):

$$£15,000 \times \frac{P(2) - P(3)}{P(3)} \text{ where}$$

P(2) = % derived from table for duration
      of lease when enhancement
      expenditure first reflected in the
      lease (42 years 9 months) =   96.979
P(3) = % derived from table for duration
      of lease at disposal (25 years
      9 months) =   82.147

Amount written off:

$$\frac{96.979 - 82.147}{96.979} = 0.153$$

£15,000 × 0.153 = £2,295

Reduced amount = £15,000 − £2,295 =	12,705	
Relevant allowable expenditure	30,303	
Incidental costs of disposal (section 38(1)(c))	7,500	37,803
Unindexed gain (section 53(2)(a))		212,197

Less: Indexation allowance (sections 53, 54 and 55(2)):

(a) by reference to relevant allowable expenditure (Note 2)   30,303
multiplied by

$$\frac{RD - RI}{RI} \quad \text{where}$$

RD = RPI for month in which
     disposal occurs =   155.40
RI = RPI for March 1982 =   79.44

$\dfrac{RD - RI}{RI} = \dfrac{155.4 - 79.44}{79.44} =$	0.956	
	28,970	

Or

(b) by reference to market value at
March 31 1982 £83,500 (Note 2)

Reduced by: Amount written off (Schedule 8
para 1(3) & (4) and section 53(2)(b) and (3)):

$$£83,500 \times \frac{P(1) - P(3)}{P(3)} \quad \text{where}$$

P(1) = % derived from table for
       duration of lease at deemed
       acquisition at March 31 1982
       (40 years 9 months) =   95.895
P(3) = % derived from table for
       duration of lease at disposal
       (25 years 9 months) =   82.147

Amount written off:

$$\frac{95.895 - 82.147}{95.895} = 0.143$$

£83,500 × 0.143 = £11,940

Reduced amount = £83,500 − £11,940 =	71,560

Multiplied by

$\dfrac{RD - RI}{RI}$ (as above)		0.956	
		68,411	
(b) is higher			68,411
Chargeable gain			£143,786

Assumed acquisition at 31 March 1982 (section35)

	£	£
Consideration for disposal		250,000

Less: Allowable expenditure:
Consideration for assumed acquisition (market
value at March 31 1982 (section 38 (1) (a)), £83,500

Reduced by: Amount written off (Schedule 8 para
1(3) & (4) and section 53(2)(b) and (3)):

$£83,500 \times \dfrac{P(1) - P(3)}{P(3)}$ (as above)

$£83,500 \times 0.143 = £11,940$

Reduced amount = £83,500 − £11,940 =	71,560	
Incidental costs of disposal (section 38(1)(c))	7,500	79,060
Unindexed gain (section 53(2)(a))		170,940

Less: Indexation allowance (sections 53, 54
and 55(2)):
(a) by reference to relevant allowable expenditure
(as above)

	28,970	

Or

(b) by reference to market value at		
March 31 1982 (as above)	68,411	
(b) is higher		68,411
Chargeable gain		£102,529

This chargeable gain is smaller than the gain calculated by disregarding
rebasing March 31 1982 (£143,786) and will be used.

Note
(1) A lease is treated as a wasting asset when the unexpired term
does not exceed 50 years at the relevant time, ie at acquisition or
disposal (Schedule 8 para 1(1)). The allowable expenditure in
respect of such a lease is written off over the duration on a
reducing basis in accordance with the table in Schedule 8 para 1.

(2)   Where the acquisition value (actual or deemed), or enhancement expenditure, used in computing the unindexed gain, is amended by any enactment (eg the short lease curved line restriction of allowable expenditure in Schedule 8), indexation is calculated on the amended amount (section 53 (3)).

Study 6

This study illustrates the CGT computation where there is a part disposal of a leasehold interest by the grant of a sublease at a premium out of a headlease with less than 50 years unexpired.

In May 1988 F granted to H Ltd a full repairing and insuring lease of a shop for 35 years at a fixed rent of £100 pa exclusive and a premium of £40,000. H Ltd's incidental costs were £1,250.

In January 2000 H Ltd sublet the shop to T on a full repairing and insuring lease for 10 years at a fixed rent of £250 pa exclusive and a premium of £100,000. H Ltd's incidental costs were £2,500.

The CGT liability for H Ltd is to be calculated.

All references are to TCGA 1992 except where indicated otherwise.

Notional premium on sublease to T if rent under sublease (£250 pa) had been the same as rent under headlease (£100 pa) (Schedule 8 para 4 (2) (b))

	£ pa	£
Premium under sublease		100,000
Add: additional premium if rent under sublease same as rent under headlease (ie capital value of profit rent):		
Rent received from sublease	250	
Less: Rent paid under headlease	100	
Net Profit Rent	150	
YP 10 years at 12% and 4% (taxed at 30%)	4.2	630
		100,630
Say		£100,650

CGT computation

	£	£
Consideration for part disposal (premium)		100,000

Less: part of premium taxed under Schedule A (Schedule 8 para 5 (1) and section 34 Income and Corporation Taxes Act 1988):

Premium	100,000	

Deduct: 1/50th for each year of the term
other than the first: $9/50 \times £100,000$ 18,000

82,000

Less: allowance for premium paid on grant of headlease, being the amount chargeable to income tax on F for granting the headlease (section 37 Income and Corporation Taxes Act 1988): (Note 1)

$£40,000 - (34/50\text{ths} \times £40,000) = £12,800$

Multiplied by

$$\frac{\text{duration of sublease to T at grant}}{\text{duration of headlease to H at grant}} = \frac{10}{35}$$

Allowance $10/35 \times £12,800$	3,657	
Part of premium chargeable to Schedule A income tax		78,343
Part of premium treated as consideration for CGT part disposal		21,657

Less: Allowable expenditure:
Consideration for acquisition plus incidental costs (section 38(1)(a)) 41,250
Apportioned to part disposal by fraction (Schedule 8 para 4(1) and (2) (b)): (Note 2)

$$\frac{P(1) - P(3)}{P(2)} \quad \text{where}$$

$P(1)$ = % derived from table for duration
  of headlease at grant of sublease
  (23 years 4 months) = 78.557
$P(2)$ = % derived from table for term of
  headlease at grant of that lease
  (35 years) = 91.981
$P(3)$ = % derived from table for duration
  of headlease at termination of
  sublease(13 years 4 months) = 57.102

$\dfrac{P(1) - P(3)}{P(2)} = \dfrac{78.557 - 57.102}{91.981}$		0.233
		9.611

Less: for reduction in premium on sublease
due to increased rent over headrent
(Schedule 3 para 4 (2) (b)):

$$\frac{\text{Premium paid on sublease}}{\text{Notional Premium on sublease}} = \frac{100,000}{100,650} = \qquad 0.99$$

Relevant allowable expenditure	9,514	
Incidental costs of part disposal (section 38(1)(c))	2,500	12,014
Unindexed gain (section 53(2)(a))		9,643

Less: Indexation allowance (sections 53, 54
and 55(2)):
by reference to relevant allowable expenditure only   9,514
multiplied by

$$\frac{\text{RD} - \text{RI}}{\text{RI}} \quad \text{where (Note 3)}$$

RD = RPI for month in which disposal occurs =	166.10	
RI = RPI for May 1988 =	106.20	

$$\frac{\text{RD} - \text{RI}}{\text{RI}} = \frac{166.10 - 106.20}{106.20} = \qquad 0.564 \qquad 5,366$$

Chargeable gain   £4,277

As H Ltd is a company the chargeable gain is subject to corporation tax
(Section 1(2) TCGA 1992). Part of the premium (£78,343) is taxed as income
under Schedule A in the company's corporation tax computation.

Notes
(1) Part of the premium paid by H Ltd to F (ie £3,657 out of
£40,000) on the grant of the headlease is set-off against H Ltd's
corporation tax Schedule A liability on the premium of
£100,000 received on the grant of the sublease to T (section 37
Income and Corporation Taxes Act 1988).

For the period while H Ltd remained in occupation and
used the shop for the purposes of its trade, it will also have
been able to claim a deduction annually in computing its
corporation tax liability on its trading profits under Schedule
D Case I of $\frac{1}{35}$ of the amount chargeable to income tax on F
(section 87 Income and Corporation Taxes Act 1988). There is a
similar relief for a deduction in a Schedule A business
computation for property sublet (section 37 (4) ibid).

(2) The normal part disposal fraction (see Study 4) does not apply where the part disposal is the grant of a sublease out of a lease with less than 50 years unexpired (Schedule 8 para 4 (1)).

(3) As H Ltd is a company it continues to benefit from indexation after April 1998, but it does not qualify for taper relief (Sections 1(2) and 53 (1A) TCGA 1992).

Inheritance Tax

The FA 1975 abolished estate duty and introduced a new tax, now known as Inheritance Tax (IHT). The legislation for IHT was consolidated in the Inheritance Tax Act 1984 which has been amended by subsequent Finance Acts. (References to legislation in this part of this chapter are to the IHT Act 1984 unless indicated otherwise).

Subject to certain exceptions, the tax applies to gifts and other gratuitous transfers of property (actual or deemed) made during a person's lifetime after March 26 1974 and the deemed transfer on death after March 12 1975. Following FA 1986 IHT has largely been restricted in scope to transfers of property within seven years of death. Lifetime gifts made more than seven years before the donor's death are (with certain exceptions) now outside IHT.

IHT is based on the value transferred by a "transfer of value". This is broadly a disposition whereby the transferor's estate is reduced in value, unless it is excluded because it is not intended to confer a gratuitous benefit and it is an arm's length transaction between unconnected persons, or is a disposition such as might be expected to be so made.

Thus a gift, a transfer at undervalue, a deliberate omission to exercise a right, or the creation of a settlement may be transfers of value. Permitting another the use of assets for less than full consideration may result in a transfer of value. On death, a person is assumed to have made a transfer of value immediately before his death of all his remaining estate.

FA 1986 introduced the "potentially exempt transfer". This is a transfer of property (eg a gift) made between individuals, or into an accumulation and maintenance trust, or into a trust for the disabled, or the disposal or termination of a beneficial interest in possession in settled property under certain circumstances. If the transferor lives for at least seven years after the date of transfer it will be exempt from IHT. Thus, IHT is largely restricted to the transfer of property on or within seven years of death. If a potentially exempt transfer becomes chargeable because the transferor does not survive seven years, it is necessary to take into account any earlier transfers made in the seven years before the chargeable transfer to establish the tax rate applying to that chargeable transfer. (Bringing such earlier transfers into account does not of itself make them taxable). Lifetime gifts are liable to CGT, subject to certain deferral and other reliefs in particular cases.

Certain lifetime gifts do not qualify as potentially exempt from IHT (eg a transfer into a discretionary settlement, where no beneficiary has a right to income as it arises) and are liable to IHT at half the rate applicable on death. In such cases the charge will be increased to the full rate applicable on death if the transferor does not survive seven years after the transfer.

IHT and its forerunners has been the sole tax on death since March 31 1971 when CGT ceased to apply to disposals on death.

IHT is chargeable on the loss to the transferor by his transfers of value, as they occur throughout each seven-year period on a cumulative basis, provided that the transfer is not an exempt transfer, eg a gift made more than seven years before the donor's death. On death the deceased is deemed to have made a last transfer of value of all his remaining property. His death may also trigger a tax liability on potentially exempt transfers made within the previous seven years. There is now only one table of tax rates for IHT which is applicable to the transfer on death and potentially exempt transfers which become taxable by falling within seven years of death. The rate of tax payable on a potentially exempt transfer caught by the transferor's death is subject to tapering relief if the transfer was made between three and seven years before death. (This "tapering" relief for IHT should not be confused with the "tapering" relief for CGT introduced by FA 1998.)

Questions of valuation arise whenever a transfer of value is made or deemed to have been made. On such a transfer the valuation required is the loss to the transferor's estate. This loss will usually be the value of the property transferred, plus the IHT payable thereon by the transferor, and therefore a valuation of the transferred property only will be required.

In some cases, however, the loss to the transferor will exceed the value of the property transferred owing to the consequent depreciation in value of the transferor's retained estate and any "related property" (see below). Here, valuations will be required of the whole of the transferor's estate both before and after the disposal. Indeed, a strict application of the valuation provisions of IHT requires that *every* transfer of value needs valuations of the *whole* of the transferor's estate both before and after the transfer. This poses innumerable valuation problems, particularly where the estate is large or varied: questions of "lotting" and yields before and after the transfer could be fertile ground for argument. In practice, and particularly where the property transferred forms a clearly separate unit, it seems to be readily agreed that the value of the property transferred will also represent the loss to the transferor's estate for IHT purposes.

"Related property" is property comprised in the estate of the transferor's spouse for IHT purposes, or property which is, or has been in the preceding five years, held by a charitable trust, political party, housing association, or qualifying public body as a result of an exempt transfer by the transferor after April 15 1976. If the value of any property in the transferor's estate is less than it would be if it were the appropriate portion of the aggregate of that and the related property, then the value shall be taken to be the appropriate portion of that aggregate (section 161). Related property can take various forms, such as jointly held property, successive interests in property, parts of a set, and property with close physical proximity. For example, if A owned a block of land with poor access with a value of say £20,000, and his wife owned an adjoining block on its own worth £10,000, but which gave good access to both blocks so that together they would be worth £70,000, then on a transfer by A the related property provisions would apply. The value would be the "appropriate portion" of £70,000. This is calculated on the proportionate values of the separate assets valued as such, ie 20,000 / (20,000 + 10,000). The loss to his estate from the transfer would be £46,667. Thus, in related property situations, it is necessary to value the separate assets as such, and the aggregate value.

A person with a beneficial interest in possession in settled property is treated as owning the property in which the interest subsists (section 49). This may have an effect on the value of both the settled property and other property forming part of the person's estate for IHT purposes. However if the interest in possession comes to an end during the beneficiary's lifetime, the transfer of value is computed on the value of the trust's assets in isolation (section 52). A beneficial interest in settled property is one where the beneficiary has the present right to present enjoyment of income as it arises without the need for any decision by the trustees (*Pearson* v *IRC* [1980] 2 All ER 479, HL).

The basic IHT valuation provision is the definition of "market value" in section 160 IHT Act 1984. The value of property for IHT purposes is defined as the price which the property might reasonably be expected to fetch if sold in the open market; no reduction in price is to be assumed on the ground that the whole property is to be placed on the market at one and the same time.

In determining the value of a person's estate his liabilities are to be taken into account, provided they were imposed by law (eg rates) or incurred for consideration (eg a mortgage) (section 5(3) and (5)), but a liability in respect of which there is a right of reimbursement is taken into account only to the extent that reimbursement cannot reasonably be expected to be obtained (section 162 (1)). A liability is valued at the time of the chargeable event and, if it is a future liability, it may be discounted (section 162 (2)). A liability which is an encumbrance on a particular property shall, so far as possible, be taken to reduce the value of that property (section 162 (4)).

Where the property comprises a residential property acquired under the right to buy provisions contained in the Housing Act 1980, and therefore subject to a liability to repay the discount on a sale within five years, the market value of the property will be the amount which would be received on a hypothetical sale, subject to the obligation which would fall on the hypothetical purchaser to repay the discount in the event of a disposal by him within the specified period; but on the assumption that the hypothetical sale did not itself give rise to the obligation to repay the discount (*Alexander* v*IRC* [1991] STC 11).

Most importantly, the transferor's IHT liability on the chargeable transfer under consideration, but not generally his other tax liabilities thereon, is to be taken into account when valuing his estate immediately after the transfer (section 5 (4)). Thus the net value transferred must be grossed up by the amount of IHT in order to arrive at the value transferred. This amount then takes its place in the aggregate of chargeable transfers during each seven-year period. This grossing up process does not apply where the tax is paid by the transferee or on the deemed transfer on death.

Where under a contract the right to dispose of any property has been excluded or restricted then, on the next transfer, that exclusion etc shall be ignored unless consideration was given for it. However allowance will be made for the part of the value transferred attributable to the exclusion or restriction, if the contract imposing the exclusion etc was itself a chargeable transfer (section 163).

Special rules apply to the deemed transfer on death. For example, any change in the value of all or part of the deceased's estate due to his death shall be taken into account in the valuation as if it had occurred before death (section 171). Reasonable funeral expenses may be deducted in computing the value of the deceased's estate (section 172).

On an appeal to the Special Commissioners or the High Court against a determination by the Board of Inland Revenue concerning an IHT matter, any question as to the valuation of land in the United Kingdom shall be referred to the Lands Tribunal (section 222).

Special reliefs apply in respect of agricultural property, business property and woodlands, necessitating special valuation rules.

Agricultural property relief for IHT

Agricultural property is defined in section 115(2), (4) and (5) to be agricultural land in the United Kingdom, the Channel Islands or Isle of Man, together with cottages, farm buildings and farm houses as are of a character appropriate to the property; and includes woodlands, and buildings used for the intensive rearing of livestock or fish, provided their occupation is ancillary to the occupation of agricultural property. The breeding and rearing of horses on a stud farm qualifies as agriculture. From April 6 1995 land and buildings used for short rotation coppice will qualify as agricultural property (FA 1995 section 154).

The relief is given on the "agricultural value" of the property. This is taken to be the value of the property as if it were subject to a perpetual covenant prohibiting its use otherwise than as agricultural property (section 115 (3)).

In valuing agricultural property for IHT, farm cottages occupied by farm workers are to be valued ignoring any value attributable to their suitability for occupation by anyone else (section 169). A cottage occupied by a retired farm worker may benefit from this provision subject to certain conditions (Inland Revenue Concession F16).

In the case of a working farmer occupying agricultural land for which there is some additional non agricultural value, eg hope value for development, the excess value will not be covered by agricultural relief but may instead be covered by "business relief", see below.

100% relief is given for transfers of qualifying agricultural property after March 9 1992 where immediately before the transfer the transferor had the right to obtain vacant possession within the next 12 months (section 116(1) and (2)), or where he had held the property since before March 9 1981 (subject to conditions specified in section 116 (3)). 100% relief is also due if the land is let on a tenancy commencing on or after September 1 1995 when the Agricultural Tenancies Act 1995 came into force, and also in relation to occasions of charge on or after November 26 1996 where farmland and buildings are dedicated in a habitat scheme under one of the various Habitat Regulations (section 124 C). In other cases the relief is 50%. (For transfers before March 10 1992 the rates were 50% and 30% respectively).

Inland Revenue Concession F17 permits the transfer of tenanted farmland to be regarded as qualifying for agricultural relief on the vacant possession basis if it carries the right to vacant possession within 24 months of the transfer (eg a *Gladstone* v *Bower* [1960] 3 All ER 353 23 month tenancy granted before September 1 1995), or is valued at an amount broadly equivalent to vacant possession notwithstanding the terms of the tenancy (eg related property).

To qualify for agricultural relief the property must have been occupied by the transferor for agricultural purposes for a minimum of two years to the date of transfer, or owned by the transferor for a minimum of seven years to the date of transfer and occupied (by anyone) for agricultural purposes during that period (section 117). For the relief to apply to a potentially exempt transfer which becomes chargeable on the transferor's death within seven years of the transfer, the transferee must have owned the property continuously from the date of transfer to the date of the transferor's death and it must remain qualifying agricultural property (section 124A). There are provisions for allowing the disposal and replacement of agricultural property without loss of relief (sections 118 and 124B).

Business property relief for IHT

Business property relief applies where relevant business property is transferred. Essentially this is the transfer of a business (ie by sole trader) or of an interest in a business (eg a partnership share); or of shares in a company not listed on a recognised stock exchange; or of shares so listed which give the transferor control of the company. For these cases the relief has been 100% for transfers since March 10 1992 (section 105 (1)) (except that for unquoted shares the transferor needed to control 25% of the voting power prior to April 6 1996, otherwise the rate was 50%).

This relief also applies to land and buildings owned by the transferor which are used for business purposes by a partnership of which he is a member or by a company which he controls. For these cases the rate has been 50% since March 10 1992 (section 105 (1) (d)).

The business property must be owned by the transferor for a minimum of two years prior to the transfer (section 106). There are provisions for allowing the disposal and replacement of business property without loss of relief (section 107).

"Business" is not defined other than by excluding activities wholly or mainly involved in dealing in land or shares, or investment holding (section 105 (3)). Thus claims for business property relief were denied for letting industrial property units (*Martin* v *CIR* [1995] STC 5 (SCD 02)), for letting furnished flats (*Burkinyoung* v *CIR* [1995] STC 2 (SCD 03)), and for long term letting of caravan units (*Hall & Hall* v *CIR* [1997] STC (SCD 126)). The business of property development (as opposed to property investment) does qualify for the relief.

For a transfer of business property within seven years of death to benefit from business property relief it is necessary for the transferee to continue owning the property throughout the period from the transfer to the death of the transferor. Also the property must be used for a qualifying business property relief purpose by the transferee (section 113A). Disposal and replacement with other qualifying business property is permitted (section 113 B)

Woodlands regime for IHT

There is a special regime for woodlands in section 125. If woodlands relief is to be claimed the deceased must have owned the land for the five years prior to his death. On death a claim can be made to leave out of account the value of the timber in computing the deemed transfer of the deceased's estate on death. However, woodlands which qualify for agricultural relief as being ancillary to an agricultural holding do not qualify for woodlands relief but automatically receive agricultural property relief. Commercially managed woodlands which are not part of an agricultural holding qualify for business property relief. Since both these reliefs are more favourable than woodlands relief, a claim under section 125 is rare.

If a claim has been made but the timber is disposed of before a subsequent death, then the "net" sale proceeds, or on a gift the "net value" of the timber at that time, is charged to IHT as the top slice of the deceased's estate at the rates of IHT in force at the time of the disposal or gift. If the deceased could have claimed business property relief instead of woodlands relief on death, the value charged to IHT is reduced by 50%. "Net proceeds" or "net value" is defined as the proceeds or value after deducting such expenses of disposing of the timber, and the costs of replanting within three years of the disposal, which are not allowable as deductions for income tax (sections 127(1) and 130(1)(b)).

Study 7

This study illustrates the IHT computation where a potentially exempt transfer (a lifetime gift) becomes subject to tax by the death of the donor within seven years of the transfer. It also deals with agricultural relief, hold-over relief for a gift of business assets and the relationship between IHT and CGT on gifts.

In September 1984 A's father died leaving him the freehold interest in a farm let on an Agricultural Holdings Act tenancy to a tenant in his late 30s. The probate value was agreed at £190,000. The farm continued to be let, and on March 25 1994 A made a gift to his son S of his freehold interest in the farm, subject to the tenancy. A's legal costs for the gift were £2,500. The gift was valued at £240,000. A's only previous transfer was a gift to his daughter of £152,000 cash on January 1 1990. A died on 30 April 1997, when the value of the tenanted farm was £270,000.

All references are to IHT Act 1984 except where indicated otherwise.

IHT computation
(i)   The gift on March 25 1994 is a potentially exempt transfer and accordingly is assumed to be exempt from IHT (section 3A (5)).
(ii) On the death of A within seven years of the gift made on March 25 1994, the gift is established as a chargeable transfer (section 3A (4)). IHT is charged at the rate applicable on death on the charging scale in force at the date of death (Schedule 2 para 1A).

	£	£
Transfer of value (Notes 1 & 2)		240,000
Less: Agricultural relief – 50% (section 116(1) & (2)) (Note 3)		120,000
		120,000
Annual exemptions 1993/94	3,000	
1992/93 (Note 4)	3,000	6,000
Chargeable transfer		114,000
IHT payable		
Cumulative chargeable transfers		
January 1 1990 Gift of cash (Note 5)	152,000	
Less: Annual exemptions 1989/90	3,000	
1988/89	3,000	146,000
March 25 1994 Chargeable transfer as above		114,000
		260,000
Deduct: Zero rate band (Note 6)		215,000
		£45,000
IHT at 40% on £45,000	18,000	
Less: Tapering relief – reduction to 80% of full rate (Note 7)	3,600	
IHT payable (Note 8)		£14,400

Notes

(1) The value transferred by a lifetime transfer is the reduction in the transferor's estate resulting from the disposition (section 3(1)). Relief may be claimed where the market value of property comprised in the transfer has fallen between the date of transfer and the transferor's death (section 131(2)). The relief is given by deducting from the value of the lifetime transfer the reduction in market value, and charging that reduced amount to IHT at the date of death. The relief does not alter the deceased's total of cumulative transfers for the purposes of computing IHT on other transfers, eg the deemed transfer on death of the remainder of his estate.

(2) The incidental costs of the transfer are left out of account if borne by the transferor; if borne by the transferee they reduce the value transferred (section 164).

(3) Agricultural property relief is at 50% as the transferor's interest in the property did not carry the right to vacant possession within 12 months (section 116(2)). Agricultural property relief applies in this case as the transferor had owned

the property for more than seven years, and the transferee owned the property continuously from the date of gift to the date of the transferor's death and it continued to be occupied for agricultural purposes.

(4) The annual exemption £3,000 can be carried forward for one fiscal year only, if not previously used. The value transferred is reduced by agricultural property relief before the annual exemption is utilised.

(5) The gift on January 1 1990 is more than seven years before the transferor's death and thus is itself exempt from IHT. However it is still taken into account in computing the cumulative total of transfers for calculating the IHT on any other chargeable transfer made within seven years after the gift and within seven years of the death.

(6) The table of tax rates used is the lower of that applicable at the date of the transfer and at death (Schedule 2 para 1A).

(7) As the gift of the farm occurred more than three but less than four years before the death, the rate of tax is subject to tapering relief reducing it to 80% of the full rate. Tapering increases as the period from the transfer to death lengthens: 4 to 5 years – 60%, 5 to 6 years – 40%, 6 to 7 years – 20% (section 7(4)).

(8) The transferee is liable to pay the IHT on death (section 204 (8)), and consequently there is no grossing up. The IHT paid by him may be deducted in computing any future chargeable gain on disposal of the property, provided a hold-over claim was made in respect of the gift by the donor and him under TCGA 1992 section 165 (1) (TCGA 1992 section 165 (10) and (11)). See note 4 below.

CGT computation on the gift in 1994

	£	£
Deemed consideration for disposal, market value (Notes 1 & 2)		240,000
Less: Allowable expenditure: Consideration for acquisition, ie probate value (Note 2)	190,000	
Relevant allowable expenditure	190,000	
Incidental costs of disposal (section 38(1)(c))	2,500	192,500
Unindexed gain (section 53(2)(a))		47,500

Less: Indexation allowance (sections 53, 54 and 55(2)) by reference to relevant allowable expenditure only:
Relevant allowable expenditure £190,000 multiplied by

$$\frac{RD - RI}{RI} \quad \text{where}$$

RD = RPI for month of disposal,
$3/94$ = 142.50
RI = RPI for month of acquisition
$9/84$ = 90.11

$\frac{RD - RI}{RI} = \frac{142.50 - 90.11}{90.11} = 0.581$

£190,000 × 0.581	110,390
Loss – no CGT to pay (Notes 3, 4 & 5)	£(62,890)

Notes

All references are to TCGA 1992 except where indicated.

(1) This transaction is between connected persons (section 286 (2)): it is therefore treated as otherwise than by way of a bargain at arm's length, and market value (section 272) must be substituted for the actual consideration (sections 17 and 18).

(2) Where the value of an asset has been ascertained for the purposes of charging IHT on death, that value is taken as being the market value at that date for the purposes of CGT (section 274).

(3) Following FA 1994 section 93, indexation cannot create an allowable loss for CGT for disposals on or after November 30 1993. In this case the disposal is to a connected person and so if there were an allowable loss its use would be restricted to gains on future disposals to the same person (section 18 (3)).

(4) If there had been a chargeable gain on the gift, it could have been held-over by a claim being made by the donor and donee since the property qualified for agricultural property relief for the purposes of IHT. In that case the donor would be treated as making a no gain no loss disposal for CGT, and the donee's acquisition would be treated as being at market value less the held-over chargeable gain (section 165 (1) and (4) as extended by Schedule 7 para 1).

(5)   Had a CGT liability arisen it would have been the transferor's liability. Any CGT paid by the transferor is ignored for IHT purposes in computing the value transferred by the transfer of value (section 5 (4) IHTA 1984).

Value Added Tax and Property

Overview of VAT Liability

For VAT purposes property is divided into three broad categories and the VAT treatment varies for each category:

1.   qualifying buildings: dwellings, relevant residential buildings and relevant charitable buildings;
2.   commercial buildings and civil engineering works; and
3.   bare land.

Qualifying buildings: dwellings, relevant residential buildings and charitable buildings

Qualifying residential buildings are dwellings (VAT Act 1994, Schedule 8, Gp. 5, Nt. 2) and "relevant residential buildings" which are communal residential buildings such as homes for the elderly, hospices, children's homes (VAT Act 1994, Schedule 8, Gp.5, Nt (4)).

Qualifying charitable buildings are those used for a "relevant charitable purpose", ie charitable non-business purposes or as a village hall or similar in providing social or recreational facilities for a local community (VAT Act 1994, Schedule 8, Gp.5, Nt. (6)) .

Construction (goods and services)

The construction of qualifying buildings is zero-rated. However, the zero-rating for relevant residential and relevant charitable buildings is restricted to supplies to the person who intends to use the building for the relevant purpose and subject to that person issuing a certificate to the supplier confirming the use (VAT Act 1994, Schedule 8, Gp.5, Items 2 and 4, and Nt. 12).

Zero-rating also covers the construction of a self contained annex to an existing relevant charitable building provided that the annex is capable of functioning independently, and the main access to the annex is not through the existing building or vice versa (VAT Act 1994, Schedule 8, Gp.5, Nt. (17)) .

Sub-contractor's services are zero-rated only for the construction of new dwellings.

Supplies of goods and services in the course of conversion of a non-residential building into a dwelling or relevant residential building are standard-rated (except when supplied to a registered housing association).

Where a building is enlarged or extended and that enlargement or extension creates an additional new dwelling, the separate disposal of which is not prevented by any planning or similar consent, the construction goods and services are zero-rated.

Special rules apply for "protected buildings" (see below).

All repair and maintenance is standard-rated. Apart from the exceptions mentioned above, standard-rating applies to any work done to an existing building, including any alteration, extension, reconstruction, enlargement or annexation of an existing building.

Sale

The first grant of a major interest (freehold or lease exceeding 21 years) by the person constructing a qualifying building is zero-rated (VAT Act 1994, Schedule 8, Gp.5, Item 1). For relevant residential and relevant charitable buildings, zero-rating applies only if the recipient provides a certificate to the grantor confirming the use of the building (VAT Act 1994, Schedule 8, Gp.5, Nt.12).

A person converting a non-residential building to create a new dwelling(s) or a building for relevant residential purposes can zero-rate the first grant of a major interest in the building (VAT Act 1994, Schedule 8, Gp.5, Item 1). Zero-rating for conversion to a relevant residential building applies only if the recipient provides a certificate to the grantor confirming the use of the building (VAT Act 1994, Schedule 8, Gp.5, Nt.12).

Study 8 Conversion of non-residential building to residential

	£	VAT
Purchase of old redundant hotel	170,000	nil
Conversion: 2 Retail shops 4 Flats }	120,000	21,000
Professional fees	30,000	5,250

Shops & flats to be let

Notes

(a) Intended short letting will be exempt, with no VAT recovery on expenses;

(b) Election to waive exemption on property will make letting of shops standard-rated to facilitate VAT recovery on expenses relating to the shops. Election will not be effective for flats;

(c) Consider zero-rated grant of major interest in flats to separate entity (which will then short let) to enable VAT on expenses relating to flats to be reclaimed.

The sale of a dwelling, relevant residential or relevant charitable buildings which does not fall within the scope of zero-rating is exempt from VAT.

Letting

All letting (other than a lease exceeding 21 years by the person constructing) of dwellings, relevant residential or relevant charitable buildings is exempt from VAT.

Protected buildings

"Protected buildings" are broadly listed buildings and scheduled monuments which are also qualifying buildings (VAT Act 1994, Schedule 8, Gp.6, Nt.1).

"Approved alterations" to protected buildings are zero-rated, but repairs and maintenance are standard rated. In most cases an "approved alteration" means an alteration to a protected building for which listed building consent is both needed, and has been obtained from the appropriate local planning authority prior to commencement of the work. Special rules apply to churches, buildings on Crown or Duchy land and scheduled monuments (VAT Act 1994, Schedule 8, Gp.6, It.2 &3 and Nt.6).

The grant of a major interest by a person substantially reconstructing a protected building is zero-rated (VAT Act 1994, Schedule 8, Gp.6, It.1).

Change of use

VAT must be accounted for if, within 10 years of completion of a relevant residential or relevant charitable building which has been zero-rated by certificate, the whole, or part of the building is put to a non-qualifying use (VAT Act 1994, Schedule 10, para. 1). Where there is continued occupation but a change of use, a deemed standard-rated supply arises. Any sale or letting for a non-qualifying use is standard-rated.

Study 9 Change of use New Building for Residential Use

1.4.91   Zero-rated construction (by certificate)

1.4.97   Sale of freehold OR
          Letting (all or part of building)   } standard-rated
          for use as HOTEL

Notes
Sale or letting for non-qualifying use is within 10 years of obtaining zero-rating for construction of the building therefore the supplies are standard-rated.

Holiday accommodation
Special rules apply to holiday accommodation (VAT Act 1994, Schedule 9, Gp.1, Nts.11 &13) and (VAT Act 1994, Schedule 8, Gp. 5, Nt. 13).

Commercial buildings and civil engineering works

Commercial buildings and civil engineering works do not qualify for zero-rating. These include offices, retail premises, industrial buildings, such as factories and warehouses, roads, drains and airfields.

Construction
The construction of commercial buildings and civil engineering works is standard-rated.

Freehold sale
The VAT treatment of the freehold sale of a commercial building or civil engineering work depends on whether the building/work is new or old. For VAT purposes "new" means less than three years from the date that the building/work is completed. All freehold sales which take place within the three-year period are standard-rated. Freehold sales of buildings/works over three years old are exempt from VAT (VAT Act 1994, Schedule 9, Gp.1 It.1). However, an election to waive exemption can be made (see below).

Letting

All letting (including the sale of long leasehold) of commercial buildings and civil engineering works is exempt from VAT (VAT Act 1994, Schedule 9, Gp.1 It.1). However, an election to waive exemption can be made (see below).

Bare land

The freehold sale and all letting of bare land is exempt from VAT (VAT Act 1994, Schedule 9, Gp.1 It.1), subject to certain specific exceptions. However, an election to waive exemption can be made (see 2.). The exceptions to the general exemption for land which are taxed at the standard-rate are specified in VAT Act 1994, Schedule 9, Gp. 1 (a) to (n) and include:

(a) the grant of any interest, right or licence to take game or fish, other than the freehold sale of land which includes the sporting or fishing rights;
(b) the provision of pitches for tents or of camping facilities, including caravan pitches on seasonal sites;
(c) the granting of facilities for parking a vehicle;
(d) the grant of a separate right to fell and remove standing timber.

Mixed supplies can arise when freehold land is sold including new civil engineering work which does not cover the whole of the land. The consideration should be apportioned on a fair view and reasonable basis, eg cost.

Election to waive exemption

The election to waive exemption (option to tax) was introduced in 1989 for land, civil engineering works and commercial buildings (VAT Act 1994, Schedule 10, para. 2) . To be effective all elections must be notified to Customs in writing within 30 days of the election being made. If previous exempt supplies of the property have been made it is usually necessary to obtain Customs' permission to elect. As a general rule, once an election is made and notified to Customs, supplies of the land or property become taxable thus allowing VAT incurred on expenses relating to the supplies to be reclaimed as input tax.

Study 10 Option to tax – Freehold sale

	With Option		*Without* Option
	Cost	VAT	Cost (VAT inclusive)
	£M	£	£m
Building purchase	2.0	350,000	2.350
Refurbishment costs	1.0	175,000	1.175
	3.0	525,000	3.525
Sale of Freehold	(3.5)	612,500	(3.5)
Profit/(Loss)	0.5		(0.025)

Special rules apply for elections made by members of a VAT group. Broadly an election by one group member binds the other members of the group.

The gift of a property on which an election has been exercised is a standard-rated supply. An election can only be revoked, subject to certain conditions (including obtaining Customs' permission), within three months of being made or after 20 years.

An election cannot apply for certain supplies eg land sold to a housing association or an individual for construction of a dwelling and property intended for use as a dwelling, or solely for relevant residential or relevant charitable purposes. If the vendor/landlord of a mixed use building has elected to waive exemption, the election applies only to the commercial element. For example, in the case of a shop with a flat above, only the shop element will be standard-rated, the election will not apply to the selling price/rent attributable to the flat.

An anti-avoidance measure was introduced effective from March 19 1997 to restrict the option to tax in certain circumstances where the property will be used other than for taxable purposes (VAT Act 1994, Schedule10, para. 2(3AA) and 3A). The disapplication of the option affects freehold sales, leases, assignments, surrenders and any other commercial property supplies to which the option to tax could apply.

Under the provisions an option for taxation is disapplied in respect of a grant if, *at the time of the grant*

(a) the buildings or land fall within the scope of the capital goods scheme in the hands of the grantor; and

(b) it is the intention or expectation of the grantor, or a person responsible for financing the purchase or development of the land, that the land will become "exempt land".

Study 11 Option to tax – disapplication

(1) Developer constructs Doctor's surgery £500,000 plus VAT
(2) Doctors pay £100,000 premium on agreement to lease
(3) Developer lets building to Doctors

Notes
(a) Building is Capital Goods Scheme asset of developer
(b) Doctors are providing finance and occupying for exempt use
(c) Option to tax would be disapplied
(d) Take account of non-recovery of VAT on the developers costs when setting the rental value.

Study 12 Option to tax – supplies not affected by disapplication

(1) *Pension Fund*
    – purchase/construction
    – let to fully taxable connected party for own occupation
(2) *Property Investment Co.*
    – purchase/construction
    – let to unconnected third party for exempt use (provided third party is not involved with finance of the development).

Whether or not to opt to tax

The main factors which affect the decision on whether to opt to tax are the amount of VAT on expenses which will be lost if an option to tax is not made and the VAT status of the recipient of the supply. For example, if the tenant of a building is to be a business in the VAT exempt sector eg a bank, building society, insurance company, educational institution, any VAT charged on the rent will be an additional cost to the tenant who will have a partial or total restriction on recovery of the VAT. In some situations it may be possible to reach agreement for the landlord not to opt to tax in return for an increased rent value to compensate him for the VAT on expenses which he is unable to recover. The same principle applies for freehold sales. The landlord needs to be certain that the short term advantage of opting to tax does not create long term problems eg when he wishes to sell the freehold of the property. It should also be remembered that opting to tax will involve compliance costs associated with VAT registration.

The wording of an existing lease may affect the landlord's

decision on whether to opt to tax. If an existing lease is silent on VAT the landlord can add VAT to the rent or service charges payable, unless the lease specifically prevents it (VAT Act 1994, section 89). If the lease contains a provision preventing the landlord charging VAT in addition to the rent and service charges, he may still opt to tax, but will have to account for VAT out of the actual rent and service charges payable.

The following are some examples of tenants that will have a restriction on the right to reclaim VAT on rent and service charges:

- Financial institutions eg banks, building societies, finance houses;
- Financial agents e.g. stock brokers, mortgage brokers;
- Insurance companies, agents and brokers;
- Universities and private schools;
- Pension funds;
- Betting shops;
- Charities;
- Private hospitals;
- Non-profit making sports clubs;
- Undertakers.

© 2000, Hilary Allan AIIT as to the VAT section,
© 2000, C P Freeman FCA FTII as to the remainder of the chapter.

Chapter 9

Compulsory Purchase

The following abbreviations are used in this chapter:

LCA 1845	Lands Clauses (Consolidation) Acts 1845
LCA 1961	Land Compensation Act 1961
CPA 1965	Compulsory Purchase Act 1965
LCA 1973	Land Compensation Act 1973
ALA 1981	Acquisition of Land Act 1981
T&CPA 1990	Town and Country Planning Act 1990
P&CA 1991	Planning and Compensation Act 1991
TCGA 1992	Taxation of Chargeable Gains Act 1992

The Lands Clauses (Consolidation) Acts 1845 were a code designed to regulate the procedure, compensation and any other disputed matters likely to arise between owners and acquiring authorities. The provisions of the Acts have since been incorporated in enabling powers of compulsory purchase although certain sections may be specifically excluded by the terms of enabling Acts. The Acts and their interpretation by the courts form the basis of the existing law of compensation on compulsory purchase.

The ALA 1981 lays down a detailed procedure for the making and confirmation of a compulsory purchase order applicable to acquisitions by local authorities and the Minister of Transport and applies the 1845 Acts, except for certain sections thought to be outdated. Thus section 92 relating to acquisition of part only is modified to include a "material detriment" provision. The CPA 1965 consolidates the 1845 Acts thus amended.

The statutory basis of compensation under the 1845 Acts was the value to the owner, but the Acquisition of Land (Assessment of Compensation) Act 1919 provided modifications to that basis. The "six rules" of that Act were re-enacted in section 5 of the LCA 1961 and the Act made applicable to all cases of compulsory acquisition. The LCA 1973 dealt with many deficiencies in the then compensation provisions while the P&CA 1991 similarly dealt with later compensation problems that had come to light providing, *inter alia*, for additional compensation for residential owner-occupiers in the form of substantially increased home-loss payments, for advance payments and the payment of interest. Thus most acquisitions will now involve the rules of the 1965 Act, the procedure of the 1981 Act and the compensation provisions of the 1961, 1973 and 1991 Acts.

Notice to treat

Its effect. Right to withdraw a notice to treat. Abandonment of powers. Valuation effect of notice to treat.

Section 4 of the CPA 1965 provides that the powers of compulsory purchase shall not be exercised after three years from the date the compulsory purchase order became operative. It was held in *Co-operative Insurance Society* v *Hastings Borough Council* [1993] 2 EGLR 19 that a notice under section 3 of the Compulsory Purchase (Vesting Declarations) Act 1981 did not amount to exercise of powers. The High Court declined to follow *Westminster City Council* v *Quereshi* [1991] 1 EGLR 256. When notice to treat has been received, the "Special Act" must always be considered. There may be special provisions as to taking part only or the right to acquire easements or special compensation provisions.

Following the service of a notice to treat the owner may be forced to sell and the acquiring authority to purchase the land in question after the statutory time-limit of six weeks for withdrawal of notice to treat on receipt of a claim (section 31(1) LCA 1961). In the absence of a valid claim, a notice to treat may be withdrawn within six weeks after the decision of the Lands Tribunal (section 31(2)).

Compensation for any loss or expenses occasioned by the giving and withdrawal of the notice is payable under section 31 (3) but this will be limited if withdrawn under section 31(2) and the claim was made late. It was held in *R* v *Northumbrian Water Ltd, ex parte Able UK Ltd* [1995] EGCS 194, that the water company had an unqualified right to withdraw a notice to treat under section 31(1) within six weeks of a claim even though it had entered into possession of the land. This does not apply under section 31(2). In *Williams* v *Blaenau Gwent Borough Council* (1994) 40 EG 139 the Lands Tribunal decided that they had jurisdiction to award compensation. The authority had decided not to proceed with the acquisition and the owners had accepted a withdrawal of the notice to treat upon payment of the proper compensation. An earlier county court judgment for professional fees did not affect the claimant's rights.

Section 67 P&CA 1991 inserts a new section 2(A) in the CPA 1965 and provides that a notice to treat shall cease to have effect after three years unless compensation has been agreed, awarded, paid, or paid into court or possession has been taken or the question of compensation referred to the Lands Tribunal. The parties may agree, however, to extend the period. Compensation is payable for any loss or expenses occasioned by the serving of the notice to treat and its ceasing to have effect, section 2(C) including interest 2(E).

Notice to treat does not fix the date for valuation and the owner should continue to insure, maintain and protect the property from vandalism: see *Lewars* v *Greater London Council* (1981) 259 EG 500 and *Blackadder* v *Grampian Regional Council* [1992] 2 EGLR 207. Disturbance claims can in some circumstances now relate to losses before notice to treat: see *Prasad* v *Wolverhampton Borough Council* (1983) 265 EG 1073, CA.

Compensation can be claimed only in respect of such interests in the land as existed at the date of the notice to treat and the acquiring authority's total burden cannot be increased by the creation of fresh interests after that date if done with the object of increasing the compensation – *Mercer* v *Liverpool, St Helens & South Lancashire Railway* [1904] AC 461. Thus, the grant of a lease to a limited company to give them a compensatable interest and a claim for disturbance may well have to be disregarded. There have been many cases dealing with such occupational difficulties: see eg *Smith, Stone & Knight Ltd* v *Birmingham Corporation* [1939] 4 All ER 116 where it was held that a subsidiary company without a lease operated as the servant of a parent company owning the premises. In a 1976 Court of Appeal decision, *DHN Food Distributors Ltd* v *Tower Hamlets Borough Council* [1976] 2 EGLR 7, CA, three companies, having no real separate identity, were treated as one and were able to claim compensation accordingly. However, the House of Lords in *Woolfson* v *Strathclyde Regional Council* [1978] 2 EGLR 19 concluded that the strict view should prevail and the "corporate veil" was not to be pierced.

In this connection it should be noted that sections 37 and 38 of the LCA 1973 provide for disturbance payments for persons without compensatable interests comprising removal expenses and, if a trade or business, a disturbance claim while section 47 LCA 1973 allows for the right of a business tenant to apply for a new tenancy under Part II Landlord and Tenant Act 1954 to be considered where the land is held on a business tenancy.

The rule against increasing the burden to an acquiring authority also applies to events before notice to treat as the ALA 1981 directs the tribunal to disregard the effect of any work done or interest created if it is satisfied that it was not reasonably necessary and was done only with the object of increasing compensation.

Date of valuation on compulsory acquisition

Following *West Midland Baptist (Trust) Association (Inc)* v *Birmingham Corporation* (1968) 19 P&CR 9, CA, the date of the valuation is :

For *Rule (2) cases*. The earlier of the date on which the valuation is made (ie the date of the negotiations) or the date on which the acquiring authority took possession. *Courage Ltd* v *Kingswood District Council* [1978] 2 EGLR 170 decided that a council had taken possession after service of a notice of entry under section 11(1) of the CPA 1965. They had carried out certain site works which increased the value although a key to the entrance gate was not handed over by the claimants until seven months later. See also *C&J Seymour (Investments) Ltd* v *Lewes District Council* [1992] 1 EGLR 237 where the date of valuation was held to be the last day of the tribunal hearing following the Court of Appeal decision in *W&S (Long Eaton) Ltd* v *Derbyshire County Council* [1975] 2 EGLR 19.

For *Rule 5 cases*. The earliest of the date on which the valuation is made, the date on which the work on the new premises could reasonably have been expected to start, and the date of possession.

The *Bwllfa* principle is important in considering what weight, if any, may be attached to evidence or events after the date of valuation: see *Bwllfa & Merthyr Dare Steam Collieries (1891) Ltd* v *Pontypridd Waterworks Co* [1903] AC 426. As Lord Macnaghten said in that case:

> Why should the arbitrator listen to conjecture on a matter which has become an accomplished fact? With the light before him, why should he shut his eyes and grope in the dark?

See also *Bolton Metropolitan Borough Council* v *Waterworth* [1982] JPL 33, CA.

Time barring of claims

It was held in *Hillingdon London Borough Council* v *ARC Ltd* [1998] EGCS 95, CA, that a claim for compensation must be referred to the Lands Tribunal within six years of date of entry on the land unless by reason of its conduct the authority was not entitled to rely on section 9 of the Limitation Act, 1980. Also in *Co-operative Wholesale Society* v *Chester le Street District Council* [1996] 46 EG 158, where it was held that, based on the conduct of the parties, a reference to the Tribunal made after a six-year period of limitation in section 10(3) of the Compulsory Purchase (Visiting Declarations) Act, 1981 was not time barred.

Compensation where depreciation in value is occasioned by the scheme but no land is taken

Compensation under section 10 CPA 1965

Prior to the LCA 1973 the only legislation dealing with this was section 10 CPA 1965. The effect of this section was clarified by the four rules in *Metropolitan Board of Works* v *McCarthy* (1874) LR 7 HL 243. To obtain compensation depreciation in the value of land has to be caused by the lawful execution of the works which would have been actionable but for the statutory powers to proceed with the works. In *Argyle Motors (Birkenhead) Ltd* v *Birkenhead Corporation* [1974] 1 All ER 201, HL, it was confirmed that section 10 does not count loss of profit. *Wilson's Brewery Ltd* v *West Yorkshire Metropolitan County Council* [1977] 2 EGLR 175 was a claim for loss of support for a gable end wall by the demolition of an adjoining property. The Tribunal decided that a mutual right of support had not been abandoned, nor were any rights subsequently obtained "clam" (that is acquired secretly). A claim under section 10 was therefore proper. A case involving the over-riding of a restrictive covenant was *Wrotham Park Settled Estates* v *Hartsmere Borough Council* [1993] EGCS 56. The Court of Appeal held that it was not inappropriate for compensation to be assessed by reference to the diminution in value of the benefited property consequent on the carrying out of the authorised works and not a sum representing a price for relaxing the covenant.

Section 10 can apply to reduction in the letting value during the execution of works but it would be almost impossible for damage caused by noise, dust or ventilation to satisfy all of the rules: *Wildtree Hotels Ltd* v *Harrow London Borough Council* (2000) 31 EG 85 HL.

Study 1

Note: *to save repetition no mention is made of surveyors' fees and legal costs and interest on the agreed consideration, where appropriate, in this and all subsequent studies in this chapter.*

The land A B C D has been compulsorily acquired and a primary school is being built close to Blackacre. A restrictive covenant in favour of Blackacre limits the use of the site to agricultural use. A vehicular right of way to Greenacre crosses its northern part and will be extinguished, as the land is essential for inclusion in the primary school site.

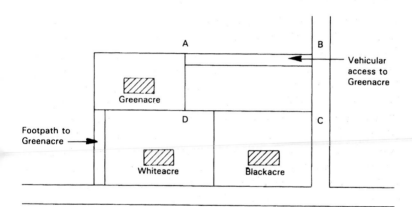

Claim for freeholder of Greenacre under section 10 CPA 1965 for loss of right of access

	£
Before value: value with both vehicular and footpath access	120,000
Less: after value: footpath access adjoining Whiteacre only	110,000
Claim for depreciation	£10,000

Claim for freeholder of Blackacre under section 10 of the CPA 1965 for overriding a restrictive covenant.

	£
Before value: open market value with the benefit of the restrictive covenant	150,000
Less: after value: value with primary school adjoining	130,000
Claim for depreciation	£20,000

In each case, interest on compensation is payable (Section 63(1) LCA, 1973).

The owner of Whiteacre has no claim under section 10 of the CPA 1965 as he does not fall within the *McCarthy* rules.

Compensation under Part 1 of the LCA 1973

Part 1 of the LCA 1973 greatly increased the scope for compensation for injurious affection where no land was taken by providing that, in addition to any rights under section 10 of the CPA 1965, compensation is payable for depreciation by the use of certain public works, particularly highways, but only for depreciation caused by the physical factors specified in section 1 (2) – noise, vibration, smell, fumes, smoke, artificial lighting and the discharge on to the land of any solid or liquid substance. Public works are defined as any highway, aerodrome or other works on land provided or used in the exercise of statutory powers. Section 9 deals with alterations to public works, both carriageway alterations (section 9(5)) and runway aprons (section 9(6)) but mere intensification of use of public works does not qualify.

Section 1(6) prescribes that compensation under Part 1 does not apply unless immunity from nuisance in respect of the use other than a highway is conferred by the relevant statute. *Vickers* v *Dover District Council* [1993] 1 EGLR 193 concerned the provision of a car park. It was held that the Tribunal had no power to award compensation as section 32 of the Road Traffic Regulation Act, 1984 does not either expressly or by implication, confer immunity from actions for nuisance. Also *Marsh* v *Powys County Council* [1997] 33 EG 100 where a similar decision was reached concerning the use of a primary school.

Section 2 sets out interests which qualify for compensation; an "owner" or owner-occupier in the case of residential property, an owner-occupier of a farm or an owner-occupier of other premises with an annual value of £24,600 or below. Section 171 of the T&CP 1990 defines annual value as the rateable value if the property is non-domestic and no part is exempt from rating. Special rules apply to hereditaments which include domestic or exempt property or which are entirely exempt from rates. An "owner" includes a leaseholder with not less than three years unexpired at the date of the notice of claim. The interest must have been acquired before the relevant date which is the date on which the works were first used (see section 11 for the rules for inherited interests.)

Section 3 as amended by section 112 of the Local Government, Planning and Land Act, 1980 deals with the submission of claims which must be made within the six years provided by the Limitation Act, 1980.

Section 4(1) specifies that compensation will be based on prices current on the first day of the claim period, which is 12 months after the works began to be used. Account is to be taken of the use of the works at that date and of any intensification that could then have reasonably been expected.

Section 5 deals with the assumptions as to planning permission on the relevant land. Planning permission is not to be assumed for other than Schedule 3, T&CP 1990 rights. Thus depreciation to development value is not covered. Existing planning permissions that have not been implemented are also to be disregarded.

Section 6 provides for a reduction of compensation for any betterment conferred by the works, including to contiguous and adjacent land, and any works undertaken by the authority to mitigate damage. So, for example, in *Hillard* v *Gwent County Council* [1992] 2 EGLR 204 the Tribunal took depreciation at 6% of the unaffected value, but reduced the award to 4% to reflect 2% betterment.

Shepherd v *Lancashire County Council* (1976) 33 P&CR 296 concerned a council tip: the Tribunal found that the property had been depreciated by the works but this was due to the proximity of the tip and not the use and the claim therefore failed. The Tribunal accepted, however, that depreciation need not be permanent to be compensatable.

It was held by the Lands Tribunal in *Blower* v *Suffolk County Council* [1994] 2 EGLR 204 that depreciation caused by the view of distant street lighting from the house was compensatable.

In *Hallows* v *Welsh Office* [1995] 21 EG 126 the Tribunal said that the depreciation in value may be calculated in many ways such as by a percentage of the pre-works market value (ie unaffected value), by taking part of the total reduction in value caused by the works as due to the physical factors or even as a spot figure based on the valuer's experience of the locality. The Tribunal may also be able to assess the depreciation from agreed settlements.

Wakely v *London Fire & Civil Defence Authority* [1996] 50 EG 98 included a claim in respect of depreciation from the use of a new fire station on the freehold interest in a house let on a regulated tenancy, the award being 6.25% of the unaffected value subject to the tenancy, while *Clwyd Alyn Housing Association Ltd* v *Welsh Office*

[1996] 17 EG 189 was concerned with a claim for the owner of sheltered housing let on secure tenancies. It was held that the evidence of settlements for owner occupied houses could not be applied to sheltered housing and the claim failed through lack of evidence of any depreciating effect of a by-pass on rents.

Study 2

"Fairview" is a house on the edge of a town which had open country at the rear. Some 2 ha of this land immediately behind the house has been purchased by the local council from a farmer for a waste depot. There are two cranes in use for some 10 hours a day and the site is also a disused car disposal point and is also used for sorting glass and waste paper. About 50 vehicles a day use the road alongside "Fairview" to obtain access to the rear. The value of the house has been depreciated by the loss of privacy and view and also by the operations in the area, particularly noise.

Compensation is limited to the "physical factors" in section 1(2) and will not cover loss of view or privacy. A before and after approach can be adopted as below:

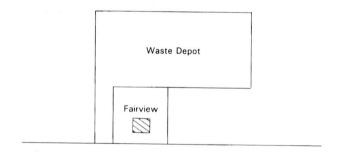

Before value	£
Open market value assuming that the works are in existence but without the "physical factors" (to confine compensation due to use). This is obviously a valuation that cannot be substantiated by comparables.	120,000

Less after value
Open market value allowing for the
"physical factors" and any betterment   100,000

£20,000 plus interest as
allowed in section 18(1)

Claim
Alternatively
Say 12½% of unaffected (pre-scheme) value of £160,000 = £20,000
OR 33⅓% of total reduction in value of £60,000 = £20,000

Had the land been taken for a modern sewage works capable of operating without noise or smell but nevertheless the market value of the house had been substantially reduced because of public prejudice against its close proximity, a claim under Part 1 of the LCA 1973 would fail as the depreciation would not be caused by the physical factors specified in section 1(2).

If any part of Fairview had been taken, even a small part of the garden to improve access, then compensation would not have been limited as under Part 1 of the LCA 1973.

Compensation where part only of the land is taken

The first consideration where notice to treat is served in respect of part of a property is whether section 8(1) of the CPA 1965 or section 53 of the LCA 1973 applies and whether they should be invoked by serving a counter-notice to take the whole property. Section 8(1) applies if

(a) the notice to treat is in respect of part of a house, building or manufactory or of part of a park or garden belonging to a house and
(b) the part cannot be taken without material detriment to the house, building or manufactory or, in the case of a park or garden, without materially affecting the amenity or convenience of the house.

Ravenscroft Properties Ltd v *Hillingdon London Borough Council* (1968) 206 EG 1255 clarifies the tests for material detriment. This concerned the taking of the rear access, garage and garden of Old Bank House, High Street, Uxbridge, said to be the cradle of Barclay's Bank. The Tribunal concluded that the taking of the rear portion clearly transformed the property into something quite different from what it was before, the change being wholly for the bad.

In considering the material detriment, section 58 of the LCA 1973 applies; regard shall be had not only to severance but to the use of the whole or the works.

Section 8 also applies to blight notices (section 166 T&CPA 1990). Thus in *Hurley* v *Cheshire County Council* (1976) 31 P&CR 433, the Tribunal held that while the inconvenience to the house through having part of the garden taken away could be minimised by landscaping, the amenity would be seriously affected by the road construction.

The Lands Tribunal decided in *Glasshouse Properties Ltd* v *Department of Transport* [1994] 1 EGLR 207 that a notice seeking to operate section 8 had to be served before the date of entry into possession.

Section 8(2) CPA 1965 applies to land not in a town or built upon and deals with circumstances under which an owner can require the authority to acquire severed land of less than half an acre. Section 8(3) contains provisions whereby the acquiring authority may require the owner to sell severed land if the owner requires certain accommodation works. Section 53 of the LCA 1973 applies where notice to treat is served in respect of part of a farm where the remainder cannot be farmed efficiently either by itself nor with other "relevant land".

If it has been agreed that part only will be acquired, the first step is to consider whether accommodation works will be carried out by the acquiring authority. Although the authority does not have to carry out accommodation works, it invariably does so because:

(i) it will reduce the amount of compensation payable as the works will be reflected in the after value of the retained land and

(ii) the works can almost certainly be carried out by the same contractors who carry out the remainder of the work.

The Lands Tribunal cannot order accommodation works to be carried out by the acquiring authority but it may take their effect into account. The cost of the works should also be agreed if possible because the cost of such works is normally added to the remainder of the compensation payable for the purpose of calculating the fee to be paid to the claimant's surveyor. It may also be relevant to the unconditional offer ("sealed offer") procedure, the purpose of which is to protect a party's position as to costs of any reference to the Tribunal. In *Stanford Marsh Ltd* v *Secretary of State for the Environment* [1997] 17 EG 170, a sealed offer was accepted 11 days after its receipt (immediately after the start of a hearing) instead of the permitted five working days. It was held that the claimants were entitled to their costs up to the permitted time but must bear their own and the acquiring authorities costs after then.

In *Stedman* v *Braintree District Council* [1990] EGCS 72 the claimant spent £22,844 on a wall when part of his land, a private roadway, was taken, without discussing the works with the acquiring authority. He than sought to recover this sum but the Lands Tribunal pointed out that there is no specific right to recover the cost of accommodation works. Furthermore building the wall was not the only solution. He was awarded £5,128, being the reasonable cost of fencing and gates.

Statutory basis of compensation where part only is taken

Section 7 of the CPA 1965. If part only is taken, the owner is entitled not only to the value of the land to be purchased but also to compensation for damage sustained as a result of severance, or injurously affecting other land held with it. *Cowper Essex* v *Acton Local Board* [1889] 14 App Cas 153 clarifies the meaning of "held with". It is enough to show that the land taken and the land alleged to be injuriously affected are so near together or so situated in relation to each other that the possession and control of each gave an enhanced value to the whole. Section 44 of the LCA 1973 provides that compensation for injurious affection is to be assessed in relation to depreciation arising from the whole of the works and not just that arising from the land previously held with the land retained. Compensation is restricted to injury likely to arise from the proper exercise of statutory powers. Thus injuries caused by unauthorised acts give rise not to compensation but to an action for damages, or possibly an injunction.

A case having wide implications is *Norman* v *Department of Transport* [1996] 24 EG 150 where the Tribunal held that highway subsoil in that case was part of a cottage adjoining the highway, occupation of which constituted occupation of the whole. A blight notice was therefore valid. In certain circumstances, therefore, if a road improvement is to be made which involves the subsoil , the adjoining owner may be brought within the compensation provisions of section 7.

Section 7 of the LCA 1961 applies the principle of set-off where the person entitled to the relevant interest is also entitled in the same capacity to an interest in other contiguous or adjacent land which increases in value owing to the "scheme" as defined in Schedule 1 to the LCA 1961. Thus, in *John* v *Rhymney Valley Water Board* (1964) 192 EG 309 the acquisition of land for a reservoir enhanced the value of retained housing land. Section 261 Highways Act 1980 provides special set-off provisions in the case of acquisitions of land for highway purposes and is not limited to contiguous or adjacent land.

Betterment of the retained land can exceed the value of the land taken, resulting in only nominal compensation. This applied in *Grosvenor Motor Co Ltd* v *Chester Corporation* (1963) 188 EG 177 and *Cotswold Trailer Parks* v *Secretary of State for the Environment* (1972) 27 P&CR 219 and also in *Leicester City Council* v *Leicestershire County Council* [1995] 32 EG 74 where the Tribunal said that whether or not it was possible to set off benefit to retained land depended on the relevant statutory provisions. Section 8(1) of the LCA 1961 applies where such contiguous or adjacent land is subsequently compulsorily acquired. The price will be inclusive of the value of the betterment. Similarly section 8(2) provides that if injurious affection had previously been paid, the price will reflect the depreciated value. Two cases illustrating severance and injurious affection in the case of agricultural properties are *Cuthbert* v *Secretary of State for the Environment* (1979) 252 EG 1023, at pp 1115 and 1176 and *Wilson* v *Minister of Transport* [1980] 1 EGLR 162.

Study 3

A is the freeholder of a large stone house standing in grounds of about 1½ ha situated on the eastern outskirts of an industrial town. Permission to develop the whole for housing purposes has been refused on several occasions over the past 10 years on the grounds that any further extension of the town in the direction proposed would be inappropriate. The planning position, however, has recently been changed as a by-pass on the eastern side of the town has been approved which will entail the acquisition of an area of your client's land 15 m wide for the whole of its width of 80 m. The proposed road will be elevated along this section and will be, at its nearest, some 85 m from the house. In these changed circumstances planning permission would be forthcoming for residential development on the remainder of the land.

The following preliminary matters have to be considered: section 8 of the CPA 1965 and section 58 of the LCA 1973. A decision must first be reached as to whether to accept a notice to treat in respect

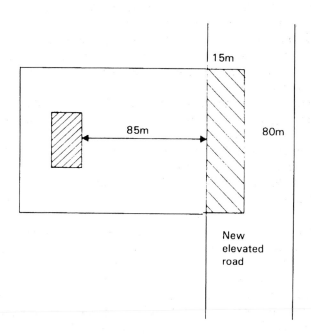

of part only or whether to serve a counternotice to take the whole. If served, a counternotice would probably be upheld. However, on the assumption that the owner wishes to remain and is prepared to sell part only, the first step must be to agree the accommodation works which will be carried out by the acquiring authority. When agreed, a claim for compensation can then be prepared. This will comprise:

> Value of land taken *plus* severance and injurious affection to land retained: (section 7 of the CPA 1965 and section 44 of the LCA 1973) *less* set-off for betterment of land retained (section 7 of the LCA 1961).

In addition a disturbance claim could be made if consistent with the basis of claim; that is if the land taken is acquired on the basis of existing use value. A claim for injurious affection must also be consistent with the basis of claim for land taken. For example, if part of a farm is taken and the basis of compensation is development value for housing, injurious affection to the part of the farm retained due to the housing use on the land taken would not be proper.

A "before" less an "after" value gives a composite total for compensation but this must be broken down into its constituent parts to comply with section 4 of the LCA 1961 dealing with submission of claims with possible effect on costs. This before and after approach cannot be used in every case, however, and it was criticised by the Lands Tribunal in the circumstances of *Abbey Homesteads Group Ltd* v *Secretary of State for Transport* [1982] 2 EGLR 198 in favour of valuing the land taken and severance and injurious affection to the retained land separately. This concerned compensation for a strip of land for a road through land having residential development potential and meant that the land taken had to be valued as a strip only, as although severance depreciation in the claimant's retained land is payable this does not apply to severance depreciation in the value of the land taken from him: see also *Hoveringham Gravels Ltd* v *Chiltern District Council* (1977) 243 EG 911, CA and *English Property Corporation plc* v *Kingston-upon-Thames Royal London Borough Council* [1998] EGCS 35.

However; adopting the before and after approach:

Assessment of compensation

	£	£
Before value		
Value of house and 1½ ha garden land prior to the scheme for the ring road; no reasonable prospect of planning permission		250,000
After value		
Either		
Value of house with garden of 1,200 m² *less* in extent having regard to the construction and use of the motorway and also to any works of mitigation carried out by the acquiring authority under Part II of the LCA 1973 and to agreed accommodation works, Say,	220,000	
OR		
1.38 ha housing land at, Say, £600,000 per ha, *less* cost of demolition, Say	820,000	

The after value therefore, is £820,000 as it will clearly be profitable to develop the retained land after the scheme. The betterment due to the scheme exceeds the value of the land taken and compensation will be nominal, say £5.

Notes
(1) Where land is taken, there can be no claim under Part I of the LCA 1973 (section 8(2)).
(2) A certificate of appropriate alternative development under section 17 of the LCA 1961 as amended by section 65 of the P&CA 1991 could have been applied for. A certificate specifying no alternative use apart from the acquiring authority's scheme was assumed for the purpose of the above. If a certificate specifying residential development had been obtained, the implications of the *Abbey Homesteads* decision would have been relevant (value as a strip of land) but clearly betterment results in nominal compensation.

Acquisition of part of a property subject to a lease – apportionment of the rent payable

If a part only of a property held on lease is taken, it will first be necessary to apportion the lease rent as to a rent for the part taken and the part retained. Only the latter will continue to be payable under the lease. The Lands Tribunal have power to apportion in the absence of agreement. Compensation to both freeholder and leaseholder for land taken may then be arrived at on the basis of "before" and "after" values.

Study 4

A shop has a net frontage of 6 m and a depth of 20 m. There is also a private forecourt 3 m deep and two floors of living accommodation set back above the shop. It is held on lease having five years unexpired at a net rent of £3,400 pa. The private forecourt and the front 3 m of the shop are to be acquired for road widening but the upper part will not be affected. Accommodation works to reinstate the shop front have been agreed. Comparables would indicate a present net full rental value of £200 per m^2 for a zone A of 6 m deep on the basis of two 6 m zones and a remainder and "halving back".

Suggested apportionment of rent payable:

Estimated full rental value "before"

		£
Zone A 6 m × 6 m × £200		7,200
Zone B 6 m × 6 m × £100		3,600
Zone C 6 m × 8 m × £50		2,400
Forecourt, Say,		800
Upper part		2,000
Full net rental value		£16,000

Estimated full rental value "after"
(effectively loss of zone C and forecourt)

		£
Zone A 6 m × 6 m × £200		7,200
Zone B 6 m × 6 m × £100		3,600
Zone C 6 m × 5 m × £50		1,500
Upper part		2,000
Full net rental value		£14,300

Apportionment of rent of £3,400 pa

		£
Part retained $\dfrac{14,300}{16,000} \times 3,400$	Say	3,040
Part taken $\dfrac{1,700}{16,000} \times 3,400$	Say	360
		£3,400

Assessment of compensation
Freeholder's claim

Before value	£	£	£
Net income	3,400		
YP 5 years 7%	4.1		
		13,940	
Reversion to full net rental value	16,000		
YP perp deferred 5 years 8%	8.5		
		136,000	
			149,940
Less: After value	3,040		
YP 5 years 7%	4.1		
		12,464	
Reversion to	14,300		
YP perp deferred 5 years 8%	8.5		
		121,550	
			134,014
Claim for freeholder for land taken			£15,926

Leaseholder's claim

	£	£
Before		
Net full rental value	16,000	
Less: rent payable	3,400	
Profit rent	12,600	
YP 5 years 9% and 4% (tax at 40%)	2.5	
		31,500
Less: After value		
Net full rental value	14,300	
Less: rent payable	3,040	
Profit rent	11,260	
YP 5 years 9% and 4% (tax at 40%)	2.5	
		28,150
Claim for leaseholder for loss of profit rent		£3,350

Compensation for the lessee for disturbance for, at the minimum, temporary disruption to trade during the carrying out of the accommodation works will also be claimable. Further, claims for severance and injurious affection for both parties might be justifiable under section 7 CPA 1965, reflected in lower "after" valuations.

Compensation for land taken

Section 5 of the LCA 1961 contains the "six rules" which relate to the six subsections. The following notes refer to detailed points on each of these subsections.

Rule 1 "No allowance shall be made on account of the acquisition being compulsory".

Nevertheless Home Loss Payments (sections 29 to 33 LCA 1973) and Farm Loss Payments (sections 34 to 36 LCA 1973) may be payable (for details see later individual paragraphs).

Rule 2 "The value of land shall, subject as hereinafter provided, be taken to be the amount which the land if sold in the open market by a willing seller might be expected to realise".

This rule was clarified in *IRC* v *Clay* [1914] 3 KB 466 and excludes any special value to the owner which is not reflected in market value but the actual owner may be considered a bidder in the market. It allows for potentialities and "hope" value in so far as they would be reflected in market value: In *Ali* v *Southwark London Borough Council* (1977) 246 EG 663 the fact that the tenants were close friends of the claimant landlord and there was an understanding that they would vacate if he wished to reoccupy was decided by the tribunal to be of no relevance and compensation was awarded on an investment value approach.

Compensation would nevertheless be based on vacant possession value if the occupier was a licensee – *John David* v *Lewisham London Borough Council* (1977) 243 EG 608. Section 50 of the LCA 1973 prescribes that no abatement in vacant possession value may be made on account of the rehousing of an owner-occupier. In the case of *Sullivan* v *Broxtowe Borough Council* (1986) 26 RVR 243, LT, it was held that the claimant was entitled to take into account the possibility of obtaining a home improvement grant in assessing his claim for compensation.

In *Stokes* v *Cambridge Corporation* (1961) 180 EG 839, followed in *Challinor* v *Stone Rural District Council* (1972) 27 P&CR 244 and *Haron Development Co Ltd* v *West Sussex County Council* (1974) 230 EG 515, an allowance of one-third of the development value was made for the cost of acquiring land to gain access to developable backland without existing access, if it was the key while in *Crown House Developments* v *Chester City Council* [1997] 09 EG 155, 30% of the development value was taken as the ransom value of the subject land which was clearly the most suitable access.

The deduction of one-third is not sacrosanct, as in the circumstances of *B&H Oberman* v *Arun District Council*, unreported, (LT REF/3/1976) 50% was deducted as the cost of securing access in the open market.

The *Stokes* principle was considered in *Wards Construction (Medway) Ltd* v *Barclays Bank plc* (1994) 68 P&CR 391. The Court of Appeal said that the *Stokes* principle was one of valuation, not of law, and although the Tribunal's valuation in the case was very high, there were no grounds on which it could be interfered with. The Tribunal's basic approach was to take a ransom value of 15% less one-third for the risk of an alternative scheme (reported under *Batchelor* v *Kent County Council* [1992] 1 EGLR 217.

Rule 3 Schedule 15 para 1 P&CA 1991 made a very important amendment to rule (3) which now reads:

> The special suitability or adaptability of the land for any purpose shall not be taken into account if that purpose is a purpose to which it could be applied only in pursuance of statutory powers or for which there is no market apart from the requirements of any authority possessing compulsory purchase powers.

Prior to this amendment there were two limbs to rule (3), one of which was not normally contentious and concerned special value to the acquiring authority. This first limb still remains and was an issue in *Hertfordshire County Council* v *Ozanne* [1989] 2 EGLR 18, which concerned the compulsory acquisition of land for a highway improvement which would facilitate the development for residential purposes of a large area in a different ownership after a statutory closing order not on the reference land. On eventual appeal to the House of Lords [1991] 1 EGLR 34 it was held that the rule applied only if the statutory powers relating to the purpose for which powers were sought related to the use of the land being acquired. Statutory powers conferred on the Secretary of State to order the stopping up of a highway on land that was not part of the land being acquired could not form the basis of the application of the rule to the land to be acquired from the claimant.

The second limb was designed to cover the circumstances in *IRC* v *Clay* [1914] 3 KB 466 and to exclude special purchaser value in other cases. It is this limb that has now been repealed after many cases which substantially reduced its effect.

In *Batchelor* v *Kent County Council* [1990] 1 EGLR 32, the Court of Appeal pointed out that rule (3) could apply only if the land had a "special suitability or adaptability". The Tribunal had found that although the most suitable access from the south was on the order land, this would not have been the only possible access and there were other options. It may therefore have been most suitable but "most suitable" did not correspond with "specially suitable" and the prefatory words of the rule were not satisfied (see also *Wards Construction (Medway) Ltd* v *Barclays Bank plc* (1994) 68 P&CR 391, a further reference of the same case where Barclays Bank plc was substituted in place of Mr Batchelor.

Rule 4 "Where the value of land is increased by reason of the use thereof or of any premises thereon in a manner which could be restrained by any court, or is contrary to law, or is detrimental to the health of the occupants of the premises or to the public health, the amount of that increase shall not be taken into account".

Section 4 P&CA 1991 deals with time-limits on enforcement action and introduced a new section 171B to the T&CPA 1990. Where there has been a breach of planning control consisting in the carrying out without planning permission of building, engineering or mining operations or similarly where there has been a change of use of any building to use as a single dwelling-house, no enforcement action may be taken after four years from completion of the operations or, in the latter case, of the change of use. In the case of any other breach of planning control, the period is 10 years from the date of the breach.

Section 10 P&CA 1991 also substituted new sections 191 to 194 of the T&PA 1990 replacing established use certificates with certificates of lawful use or development.

The possibility of planning consent being assumed under sections 14 to 17 of the LCA 1961 must also be borne in mind.

Hughes v *Doncaster Metropolitan Borough Council* [1991] 1 EGLR 31, HL, concerned land used without planning permission but, immune from enforcement procedure. A point at issue on a disturbance claim was whether the value of the land was increased by a use "contrary to law" and falling within rule (4). The House held that although, in practice, the value of the land and disturbance were separately assessed the courts had consistently adhered to the principle that the value of the land taken and compensation for disturbance were inseparable parts of a single whole in that together they made up "the value of the land" to the owner.

However, in the light of the "well-known principle that a statute should not be held to take away private rights of property without compensation unless the intention to so do is expressed in clear and unambiguous terms" (*Colonial Sugar Refining Co Ltd* v *Melbourne Harbour Trust Commissioners* [1927] AC 343) and in the light of the provisions of the T&CP 1971 (now the 1990 Act) it seemed impossible to treat an established use under that Act as being "contrary to law" within the meaning of rule (4).

Section 10 (2) of the P&CA 1991 now provides a new section 191(2) T&CPA 1990 stating that uses and operations are lawful if no enforcement action may be taken.

Rule 5 "Where land is, and but for the compulsory acquisition would continue to be, devoted to a purpose of such a nature that there is no general demand or market for the land for that purpose, the compensation may, if the Lands Tribunal is satisfied that reinstatement in some other case is *bona fide* intended, be assessed on the basis of equivalent reinstatement".

Sparks v *Leeds City Council* [1977] 2 EGLR 163 laid down four essentials before rule (5) can apply, the burden of proof being on the claimant. The premises were a social club. These are:

(1)   That the land is devoted to a purpose and would continue to be so devoted but for the compulsory acquisition.

(2) The purpose is one for which there is no general demand or market.

(3) There is a *bona fide* intention to reinstate and

(4) If these conditions are satisfied the tribunal's discretion should be exercised in the claimant's favour.

In *Zoar Independent Church Trustees* v *Rochester Corporation* (1974) 230 EG 1889 rule (5) was awarded even though the congregation fell to only three or four by the date of entry. Rule (5) has been held to apply to a theatre in Kidderminster, *Nonentities Society (Trustees)* v *Kidderminster Borough Council* (1970) 215 EG 385 although the Tribunal thought the position might have been different had the theatre been in London. *Harrison & Hetherington Ltd* v *Cumbria County Council* [1985] 2 EGLR 37 concerned a livestock market at Botchergate, Carlisle. It was held by the House of Lords that a latent demand was not a general demand within the meaning of rule (5) and only manifested itself in the rare event of market premises being offered for sale. Rule (5) was therefore applicable.

Wilkinson v *Middlesborough Borough Council* [1982] 1 EGLR 23, CA, concerned the acquisition of premises occupied by a veterinary surgeons' multi-principal practice. The Court of Appeal held that rule (5) did not apply as the claimants had not discharged the burden of proof of establishing that there was no general demand or market for property for carrying on veterinary surgeons' practices.

The date for valuation is the earliest of the date of valuation, the date of possession or the date when redevelopment might reasonably have commenced: *West Midland Baptist (Trust) Association (Inc)* v *Birmingham Corporation* [1968] 2 QB 188, CA. A further point in this case is that no deduction should be made for the age of the building, as if allowance was made, the whole object of rule (5) would be defeated and extinction of charities might result from a rule which was intended to enable them to survive. However in *Roman Catholic Diocese of Hexham & Newcastle* v *Sunderland CBC* (1964) 186 EG 369, the cost of essential repairs to the existing hall and fees were deducted from the compensation payable.

Compensation under rule (5) could alternatively cover the cost of acquiring another building and converting it. If the rule (2) basis exceeds the rule (5) figure, the owner will, of course claim the former.

In *Festinog Railway Co* v *Central Electricity Generating Board* (1962) 13 P&CR 248, CA, it was held that the Tribunal was not bound to assess compensation on a rule (5) basis, though satisfied that the conditions specified in the rule were met. The relation between the cost of reinstatement and the value of the undertaking was a relevant consideration for the Tribunal to take into account.

Rule 6 "The provisions of rule (2) shall not affect the assessment of compensation for disturbance or any other matter not directly based on the value of the land".

Disturbance claims must always be consistent with the basis of value under rule (2): see *Horn* v *Sunderland Corporation* [1941] 2 KB 26. Thus an owner will compare compensation on an existing use basis together with a disturbance claim, with compensation based on redevelopment value only and will choose that which is the more favourable to him. His tax position in each case must be borne in mind and the possibility of "rollover" relief from capital gains tax.

Leasehold Reform Act 1967

If a valid notice to enfranchise is served before a notice to treat, the notice to enfranchise will "cease to have effect" but the compulsory purchase compensation will have regard to it (section 5(6)(*b*)). If, however, the notice to treat precedes a notice to enfranchise, section 5(6)(*a*) states that the latter notice will be of no effect. For the effect on valuations: see *Boaks* v *Greater London Council* (1978) 18 RVR 17. Section 5(6)(*a*) will be of particular importance where a lease has less than three years unexpired on the notional open market sale date, so that a prospective purchaser can never qualify to enfranchise. The value of the leasehold interest will therefore be considerably less, reflecting Landlord and Tenant 1954 Part I rights only and the freehold interest correspondingly more. *Sharif* v *Birmingham City Council* (1978) 249 EG 147 concerned a leasehold interest with the benefit of a valid notice to enfranchise; the award was on the basis of the freehold value less the enfranchisement price and surveyor's and legal fees. No end deduction for contingency costs and time and trouble was made.

Assumptions as to planning permission

Certain planning assumptions may be made in ascertaining the value of the relevant interest and the owner will choose that which is most valuable:

A specific case may result in a number of such assumptions. Thus in *Co-operative Retail Services Ltd* v *Wycombe District Council* [1989] 2 EGLR 211 the existing use of a site was a coal yard; the town map showed it allocated for wholesale warehouse development, the acquiring authority required it for bus station development and there was an outstanding permission for high-tech purposes. Furthermore, on the evidence it was clear that planning permission for office development would readily be granted, which was the basis adopted.

Section 14(2) LCA 1961 – any planning permission actually in force at the date of notice to treat.

Section 14(3) permits the payment of "hope" value if the evidence is that the market would allow it, although any development value would be appropriately discounted.

Para 15 of Schedule 15 P&CA 1991 substitutes a new provision section 14(3A) LCA 1961. In considering planning assumptions under section 16 LCA 1961 regard shall be had to any contrary opinion to development expressed in a section 17 certificate. In this connection section 65(2) P&CA 1991 modified section 17 so that either a positive or negative certificate must include "but would not be granted for any other development".

Section 15(1) – permission such as would permit development in accordance with the proposals of the acquiring authority (although rule (3) might apply to the valuation). There are no qualifying rules as to its likelihood in a "no-scheme" world and permission is to be assumed: see *Myers* v *Milton Keynes Development Corporation* [1974] 1 WLR 696.

Section 15(3) and (4) – permission is to be assumed for development in both Part I and Part II Schedule 3 T&CPA 1990, subject to Schedule 10 to that Act, unless compensation under Part II has become payable. Such compensation has been repealed by Schedule 19 P&CA 1991. As a point of valuation interest, however, in *Richmond Gateways Ltd* v *Richmond upon Thames London Borough Council* [1989] 2 EGLR 182, CA the court upheld a decision of the Lands Tribunal in respect of a claim for refusal to Part II Schedule 3 rights and held that, in a residual valuation to determine the value with planning permission for the erection of a penthouse on the roof of a block of flats, it was in order not to deduct for developers' risk and profit as there was no necessity to assume a sale to a purchaser who will himself want to resell at a substantial profit to himself, as soon as he has built the hypothetical structure. There is nothing in the Acts to require the benefit to be attributed, as a result of a valuation exercise, to some hypothetical purchaser.

Any development specified in the Town and Country Planning (General Permitted Development) Order 1995, SI 1995 No 418 must be taken into account as also the Town and Country Planning (Use Classes) Order 1987 SI 1987 No 764.

Under section 16(2) permission is to be assumed for development for which planning permission might reasonably be expected to be granted for the zoned use in the development plan but in considering that question and also whether any other section 16 assumptions are applicable, regard shall be had to any contrary opinion expressed in a section 17 certificate (now section 3(A) LCA 1961 inserted by para 15(2) Schedule 15 P&CA 1991).

The Court of Appeal in *Provincial Properties (London) Ltd* v *Caterham and Warlingham Urban District Council* [1972] 1 QB 453 decided that the test of reasonableness could entail a planning permission being completely disregarded for the zoned use if that permission would not, in fact, have been granted.

Schedule 2 to the T&CP 1990 deals with "old development plans" and their continued relevance unless there is conflict with the proposals in a local plan, structure plan or the Greater London Development Plan. In particular para 5 of Part III deals with development plans for compensation purposes where there is no local plan in force in a district. The owner may consider the most valuable of the planning assumptions derived from the structure plan or the Greater London Development Plan so far as applicable and also the old development plan. The "old development plans" cease to have any effect on the adoption of a local plan.

Section 16(4) and (5). In action areas (areas selected for the commencement of comprehensive treatment under section 36(4) T&CP 1990) the planned range of users in the area has to be considered and then there must be determined that which might have been appropriate had there been no such defined area and the development plan had not contained any proposals for the area. It must further be assumed that no development or redevelopment had taken place in the area in accordance with the plan and none of the land was proposed to be acquired for public purposes. The intention is to ascertain what consents would have been likely in the no-scheme world.

Section 15(5) states that regard should be had to any certificate of appropriate alternative development and any conditions specified therein. Section 65 of the P&CA 1991 now provides that an application for a Section 17 certificate of appropriate alternative development may be made by either party in respect of any land in which an interest is to be acquired.

Section 17(5) deals with conditions in such certificates.

Circular 48/59 stated that the certificate system should be worked out on a broad common-sense basis; a certificate was not a planning permission but to be used in ascertaining the fair market value of the land.

Section 65(2) P&CA 1991 also substitutes new paras (a) and (b) in section 17(4) LCA 1961 dealing with information to be included in all certificates and section 65(3) adds a new subsection (9A) to section 17, providing for expenses in connection with the issue of a certificate and for appeals determined in the applicants' favour to be recoverable.

The date of the proposal to acquire as defined by section 22(2) LCA 1961 is the date at which all relevant facts are to be considered, and in determining such a section 17 application it had to be assumed that the proposal had been cancelled on the relevant date and then ordinary planning principles should be applied to the existing circumstances. See *Secretary of State for the Environment* v *Fletcher Estates* TLR February 17, 2000 HL and *Jelson Ltd* v *Ministry of Housing and Local Government* [1970] 1 QB 243.

A certificate may sometimes be granted perhaps for land in a green belt for "institutional use in large grounds". For valuation problems in these and similar circumstances see *Lamb's Executors* v *Cheshire County Council* (1970) 217 EG 607.

Section 64 P&CA 1991 adds additional subsections (5) to (8) to section 14 LCA 1961 dealing with special planning assumptions for the construction of highway schemes, including alteration or improvement. If a planning determination for the purposes of assessing compensation is to be made either under section 16 or 17 it is to be assumed that if the subject land was not to be acquired, no highway would be constructed to meet the same, or substantially the same, need.

Valuation rules in assessing market value

Under section 9 LCA 1961 any depreciation in value of the land acquired by reason of designation or other indication of intention to acquire compulsorily is to be disregarded, eg comparables of sales within the "scheme" may well be ignored: see *Jelson Ltd* v *Blaby District Council* (1977) 243 EG 47, CA, where reduction in value stemmed from past "indication" of likely acquisition and also *Trocette Property Co Ltd* v *Greater London Council* (1974) 231 EG 1031 and *London & Provincial Poster Group Ltd* v *Oldham Metropolitan Borough Council* [1991] 1 EGLR 214. In *Thornton* v *Wakefield Metropolitan District Council* [1991] 2 EGLR 215 it was held that a notification of likelihood of acquisition in a planning permission to convert a former theatre fell within section 9 but the claimant remained in full control and was not precluded in law from implementing his scheme until the deemed notice to treat.

The *Pointe Gourde* principle and section 6 of and Schedule 1 LCA 1961 cause particular difficulties. In *Pointe Gourde Quarrying & Transport Co Ltd* v *Sub-Intendent of Crown Lands* [1947] AC 565 it was stated that:

> It is well settled that compensation on compulsory purchase of land cannot include an increase in value which is entirely due to the scheme underlying the acquisition.

This also applies to diminution in value: see *Salop County Council* v *Craddock* (1969) 213 EG 633, CA, and *St Pier Ltd* v *Lambeth London Borough Council* [1976] 1 EGLR 189. "Scheme" for the purposes of *Pointe Gourde* may differ from the scheme as defined in Schedule 1 LCA 1961: see *Bird* v *Wakefield Metropolitan District Council* [1978] 2 EGLR 16.

In *Batchelor* v *Kent County Council* [1990] 1 EGLR 32, CA, the court decided that if a premium value is "entirely" due to the scheme underlying the acquisition then it must be disregarded under the *Pointe Gourde* principle; if it was pre-existent to the acquisition it must be regarded. To ignore a pre-existent value would be to expropriate it without compensation and it would contravene the fundamental principle of equivalence – see *Horn* v *Sunderland Corporation* [1941] 2 KB 26. The Tribunal had identified the scheme as the construction of a roundabout and associated road works. The Lands Tribunal's remitted decision is reported at [1992] 03 EG 127.

In a further reference of this case to the Court of Appeal, *Wards Construction (Medway) Ltd* v *Barclays Bank* (1994) 68 P&CR 391 the court decided that the Lands Tribunal had applied the *Pointe Gourde* principle correctly and no error of law could be detected.

Hertfordshire County Council v *Ozanne* [1989] 2 EGLR 18, CA, concerned the compulsory acquisition of land for a highway improvement in connection with a large development area. The award was remitted to the Tribunal as the Tribunal should have identified the "scheme" for the purposes of the *Pointe Gourde* principle and had not done so. It was not possible to say, as the Tribunal had done, that in the imagined no-acquisition world the land had a value as a ransom strip unless one identified the scheme. The remitted decision is reported in *Ozanne* v *Hertfordshire County*

Council [1992] 2 EGLR 201. *Abbey Homesteads (Developments) Ltd* v *Northamptonshire County Council* [1991] 1 EGLR 224 was concerned with lands subject to a section 52 agreement reserving it for school purposes. The Court of Appeal had decided [1986] 1 EGLR 24 that the covenant in the agreement was a restrictive convenant running with the land and intended to be permanent and declared that compensation for the land was to be determined on the basis that it was affected by the covenant. On remit, however, the Tribunal decided that the imposition of the restrictive convenant was part of the scheme underlying the acquisition and also that the agreement itself constituted an indication within section 9. On appeal against this decision (1992) JPL 1133, the Court of Appeal stated that neither *Pointe Gourde* nor section 9 operated to remove the restriction nor require it to be disregarded. However, in the "no-scheme world" there would still have been a need for a school to be provided on an alternative site, in which case the section 52 agreement on the subject land could be discharged.

In *JA Pye (Oxford) Ltd* v *Kingswood Borough Coucil* [1998] 2 EGLR 159, the Court of Appeal held that the Lands Tribunal had not made a perverse finding as to the "scheme" but had overlooked two clauses in a section 52 agreement. The award was remitted to the Tribunal.

Section 6 and Schedule 1 LCA 1961 provides that any increase or decrease in the value of land taken is to be ignored in so far as it is due to development or prospective development under the scheme except in so far as it would have occurred in the absence of the scheme.

Schedule 1 sets out four cases:

Case 1. Normal CPO cases

Case 2. Comprehensive development areas and action areas

Case 3. New towns

Case 4. Town Development Act 1952 cases

The Local Government, Planning and Land Act 1980 added a further case:

Case 4A. Urban development areas (but the remaining corporations in England were wound up by the Urban Development Corporations in England (Area and Constitution) Order, 1998, SI 1998 No 769).

In each case the schedule sets out the development which is to be taken as part of the scheme, the effect on the value of which is to be disregarded, except in so far as it would have taken place but for the scheme.

In *Davy* v *Leeds Corporation* (1964) 16 P&CR 244, the question was whether the effects of the demolition of other property in the area had to be excluded from the valuation by virtue of section 6. Clearance is specifically development for the purposes of section 6 (section 6 (3) (b)) and as there was no prospect of the land being cleared other than by the corporation, the claimant could only claim the value of his sites cleared for development but with the surroundings still encumbered with buildings.

Section 6 and Schedule 1 have not superseded the *Pointe Gourde* principle. They act both independently and in conjunction with each other: *Viscount Camrose* v *Basingstoke Corporation* (1966) 64 LGR 337, CA. In this case, the "relevant land" exclusion referred to in Schedule 1 comprised a substantial part of the scheme and the claimant argued that betterment from this land could therefore be included. The *Pointe Gourde* principle was invoked to defeat this contention.

In *Myers* v *Milton Keynes Development Corporation* (1974) 230 EG 1275, land was taken for a new town providing for it to be developed for residential purposes in 10 years' time. The effect on value of the new town scheme had to be disregarded when compensation was assessed, but deferred planning permission for houses in accordance with section 15 LCA 1961 had to be taken into account. Neither section 6 nor *Pointe Gourde* require any assumed permission to be disregarded, only value due to the scheme which gives rise to it (but for the new town scheme, permission would not have been forthcoming). It was also pointed out that section 15(1) contains no qualifying words as in section 16(2)(b) and consent for the acquiring authority's proposals must therefore be assumed.

In *Halliwell* v *Skelmersdale Development Corporation* (1965) 16 P&CR 305, both valuers substituted for what had actually happened what they thought would have taken place in the "no-scheme world". It was said that this was wrong and that the valuation rule was *rebus sic stantibus*. What has to be ignored is the effect on value, not the development itself. This rather tenuous difference illustrates the legal and valuation difficulties involved in discounting betterment.

Study 5

To illustrate alternative planning assumptions
A freehold site (0.15 ha) of a factory built in 1890 but demolished

after a fire eight years ago is to be acquired for housing purposes, for which purpose it is allocated in the development plan at a density of 200 persons per ha. It has been occupied for the past five years by a car breaker. The present rent is £6,000 pa (exclusive) on a yearly tenancy, the rent having been fixed two years ago. The present full net rental value is £8,000 pa. Temporary planning permission for this use expires in three years' time. The rateable value is £5,000.

Claim for freeholder

The following matters have to be considered:

Section 5 (2) LCA 1961 – open market value

Section 14 (3) LCA 1961 – it must not be assumed that planning consent would automatically be refused for uses other than industrial or residential which, in this case, are to be assumed under section 15(3) and (4) and section 16(1). If, therefore, further permission might be obtained for car breaking but for the compulsory acquisition, this may be taken into account.

A quick estimate would show, however, that this would produce a lower capital value.

Section 15(3) and (4) LCA 1961. Assume Part I, Schedule 3 to T&CP 1990 (subject to Schedule 10) – the right to rebuild factory plus 10% as the building demolished was an "original" building.

Section 16 (1). Residential development at a density of 200 persons per ha.

Valuations based on the following planning assumptions are allowable, the claimant being able to adopt whichever gives the highest value:

(a) Car-breaking income for three years with reversion to the higher of either industrial or residential value (no Landlord and Tenant Act 1954 compensation to car breaker as he would serve a tenant's notice to quit if he could not use the site)

or

(b) Car-breaking income for, say, one year to allow time for possession after a landlord's notice to quit with similar reversions but less compensation to tenant under the Landlord and Tenant Act 1954 (Appropriate Multiplier) Order 1990 (SI 1990 No 363) of once or twice the RV, depending on the 14-year rule.

Valuations required:
(a) Site of 0.15 ha with planning consent to reinstate the
     original building plus 10% cube subject to 10% gross
     floor space addition not being exceeded. A residual
     valuation might well be required, as comparables
     would allow for expected permitted plot ratio and
     more economic design and layout for modern
     factories, Say, value for section 15(3) and (4) purposes   £250,000

(b) Site of 0.15 ha with planning consent for housing
     purposes at a density of 200 persons per ha. Gross
     site area, Say, 0.18 ha = 36 persons, Say, 36 rooms
     at £6,000 per room (based on comparables)   £216,000

Industrial site value is higher and therefore housing development
can be ignored for valuation purposes. The question is whether
possession would be sought as soon as possible for industrial
purposes or whether it would be more profitable to enjoy the car-
breaking income for the full remaining three years and then secure
possession.

*Value if possession sought for redevelopment by service of a notice to quit
at the first opportunity*

	£	£
	6,000	
YP at 1 year at 11% (allow one year to secure possession)	0.9	
		5,400
Reversion to industrial site value	250,000	
Less: compensation to tenant, say 1 × RV	5,000	
	245,000	
PV 1 year at 8%	0.93	
		227,850
(unsecured ground rent rate taken although cost of capital rate could be argued)		233,250
Say		235,000

Value if car-breaking income allowed to continue for full 3 years

	£	£	£
	6,000		
YP 1 year at 11%	0.9	5,400	

Reversion to full rental value for car-breaking		8,000			
YP 2 years at 13%	1.67				
PV 1 year at 13%	0.88	1.47			
			11,760		
Reversion to site value		250,000			
PV £1 3 years at 8%		0.79	197,500		
			214,660		
Say				215,000	

The compensation to the freeholder is therefore £235,000. The yearly tenant is not entitled to receive a notice to treat. If notice of entry is served and possession taken, his compensation will be assessed under section 20 of the CPA 1965 having regard to section 47 LCA 1973. From the above figures, it appears that, in the absence of the scheme, the landlord could have been expected to secure possession as soon as possible for redevelopment. This will limit compensation for disturbance under section 20.

Study 6

Purchase notice and planning assumptions in an Action Area

Your client owns the unencumbered freehold interest in the remains of a factory built in 1932, which was destroyed by fire last year. The factory had a net usable floor space of 1,400 m², the site area being 800 m².

The site is shown on the action area map for a future primary school. For this reason planning permission for the rebuilding of the factory was refused and the refusal upheld on appeal, although the proposal for the school cannot be implemented for many years. What action can your client now take and what are the financial consequences?

The following preliminary matters have to be considered:

Section 137 T&CP 1990.

The conditions for the service of a purchase notice after refusal of planning permission are satisfied as the land has become incapable of reasonably beneficial use in its existing state as clarified by section 138; that is no account shall be taken of any unauthorised prospective use of the land as defined in section 138 (2). This means that the value after the planning refusal can only be compared with Schedule 3 value not taking into account any other possible use of the land. The purchase notice must relate to the whole of the land the subject of the refusal or conditional grant of planning permission. *Cook* v *Winchester City Council* [1995] 07 EG 129 and the claimant must own the whole of the land *Smart & Courtney Dale Ltd* v *Dover Rural District Council* (1972) 23 P&CR 408.

As the factory was built before July 1 1948 valuable Schedule 3 rights exist to rebuild and clearly there is no reasonably beneficial use.

Section 139(3) T&CP 1990 prescribes that the date of a deemed notice to treat is that date on which the authority (or another authority) have served a "response notice" indicating that they are willing to accept the purchase notice, while section 143 provides that if the Secretary of State confirms a purchase notice, the deemed notice to treat will be such date as he may direct.

Basis for valuation
Section 5 (2) LCA 1961. Open market value. Date of valuation. *West Midland Baptist (Trust) Association (Inc)* v *Birmingham Corporation* (1968) 19 P&CR 9, CA

Planning assumptions. Section 15 (3) LCA 1961. Schedule 3 subject to Schedule 10 T&CPA 1990. The right to rebuild plus 10% cubic content subject to gross floor space not being exceeded by more than 10%.

Section 16(4) LCA 1961. As the land is in an action area one has to consider the planned range of users and decide which would have been appropriate in the hypothetical circumstances envisaged by the section. Assume residential as the area was primarily residential in the pre-scheme world.

The section 17 certificate procedure is also now applicable in action areas: section 65 (1) P&CA 1991 – assume residential use.

Summary
What is required is the higher of the value for either industrial or residential purposes disregarding in accordance with section 6 and Schedule 1 LCA 1961 any increase or decrease in value owing to development or potential development in the action area except in so far as development would have taken place in the "no-scheme" world.

Valuation
Factory site value Value of a factory site of 800 m² with assumed

planning permission to re-erect the previous building + 10% (if considered economic), a building of 1,540 m². The value may be enhanced by the existence of usable remains or depreciated by the necessity to remove them. This would be reflected in reduced or increased costs in a residual valuation. Say £120,000 as it stands.

Comparable industrial land transactions must be treated with caution as the plot ratio and other planning requirements may be substantially different.

Residential site value
Net site area 800 m², including half the surrounding roads; Say, 900 m². 900 m² at a density of 200 persons per ha.

Accommodation for 18 persons
occupancy ratio, say 1:1
18 rooms

	£
18 rooms at £5,000 per room (Assuming this is the most convenient unit for comparison in this instance)	90,000
Less: demolition, Say	10,000
	£80,000

Compensation therefore is £120,000 on industrial site value basis.

Study 7

Equivalent reinstatement basis
A non-conformist chapel with a small congregation of about 30 is some 200 years old, in poor repair, and is sited in a shopping street. The adjoining shops are old and the area is allocated for residential purposes but with shopping frontage to the shopping street in the approved local plan.

The chapel completely covers a site of 12 m frontage and 20 m depth and is to be compulsorily acquired. The chapel site with adjoining properties is to be used for shopping, each shop to have one floor for storage above. Notice to treat was served two years ago. Some one and a half years ago a site in a residential area was suggested for reinstatement but the trustees have only recently decided to go ahead with its purchase at a cost of £80,000. The planning authority have made a condition in the grant of planning permission for the new chapel that a car park be provided.

Basis of valuation
Either section 5(2) LCA 1961. Open market value of building (or site value less demolition costs) having regard to planning assumptions:

Section 14(3). Consent to use the building for storage purposes probably available in the absence of compulsory purchase.

Section 15(1) and section 16 (2) shopping site value less cost of demolition.

Or section 5(5) LCA 1961. Cost of equivalent reinstatement. Date of valuation – earlier of date when valuation being made or date when redevelopment could reasonably have commenced. This could be a matter of dispute owing to trustee's delay in arriving at a decision. As to the car park, as there was not one with the original chapel, unless the planning authority was insisting on the provision, the cost of providing it could not be recovered.

Valuation
(a) Site value for shopping disregarding any increase in value due to the "scheme". A residual valuation may be required. For a good example of a residual valuation, see *St Pier Ltd* v *Lambeth Borough Council* [1976] 1 EGLR 189 where a higher plot ratio in the "no-scheme" world was the basis taken by the Lands Tribunal.

The Tribunal's views on the use of residual valuations in compulsory purchase references are well set out in *Clinker & Ash* v *Southern Gas Board* (1967) 203 EG 735 where it was said that the discipline of open market conditions achieves a balance so that the residual method can be a precision valuation instrument in valuing development sites. This discipline is absent in an arbitration case as there is in effect, a captive purchaser and a captive vendor. In *First Garden City Ltd* v *Letchworth Garden City Corporation* (1960) 200 EG 123, the Tribunal said; "once valuers are let loose upon residual valuations however honest the valuers and reasoning their arguments, they can prove almost anything".

	£
Say, however, site value	150,000
Less: cost of demolition	3,000
	£147,000

(b)	Storage income from building 200 m² at £30 per m²	6,000
	YP perp, Say	9
		£54,000

(c)	Rule (5) Compensation		£
	(a)	Purchase of alternative site	
		Purchase price, legal costs and other disbursements of conveyance, surveyors' fees and interest on these items	90,000
	(b)	Cost of new equivalent building	
		Cost, architects' and quantity surveyors' fees and interest on these items	150,000
	(c)	Site works, costs, fees and interest	4,000
	(d)	Removals and other disturbance items	4,000
	(e)	Cost of providing car park on the site	15,000
			£263,000

The rule (5) basis will therefore be adopted and compensation will be, say £265,000 subject to a possible deduction for repairs to the existing hall and fees.

Study 8

Acquisition of back land and houses to provide access

The land shown hatched on the plan is allocated for educational use on the development plan and a compulsory purchase order has been made to build a primary school. The order comprises:

Reference no 1 Freehold back land, previously allotments, and at present unused with a four metre wide access to Woodlands Road.

Reference no 2 A freehold house subject to a regulated tenant and worth £25,000 as an investment in the open market.

Reference no 3 A freehold owner-occupied house worth £80,000 in the open market with vacant possession.

The houses indicated on the plan are all detached, having frontages of 8m, those marked X on the plan being owner-occupied. The remainder are let to statutory tenants.

Compensation for reference no 1.

A certificate of appropriate alternative development should be applied for under section 17 of the LCA 1961.

Assume that the certificate specifies residential use by the erection of 40 three-roomed flats subject to an access of a minimum of 16 m to Woodlands Road being provided.

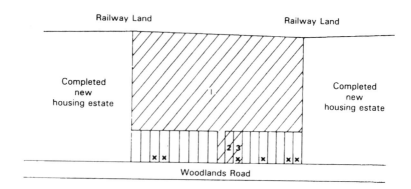

	£
40 flats at £20,000 per flat plot value	800,000

Less: cost of procuring an access in the open market. No one here possesses the "key" to the development as any two houses would suffice, particularly the CPO houses as they adjoin the entrance. It is assumed the freehold interest in reference no 2 could be purchased for £30,000 and reference no 3 for £90,000. It is further assumed that the tenant in the former would vacate for £10,000, a total of £135,000, including £5,000 for demolition and clearance costs. Say, however, £160,000 to allow a reasonable margin for risk.

	£
	160,000
Open market value of the back land	£640,000

NB: if one owner possessed the "key" to the development, the approach in *Stokes* v *Cambridge Corporation* (1961) 180 EG 839 would be adopted.

Compensation for reference no 2 Section 5, rule (3) does not now operate to exclude the special value of references nos 2 and 3 to provide access to the back land. (Schedule 15 para 1 P&CA 1991).

Freeholder
Freehold interest including value to special purchaser £30,000 (no home less payment as investor).

Tenant
Possible home loss payment together with a disturbance claim under section 37 of the LCA 1973.

Compensation for reference no 3.

Freeholder

Freehold house with vacant possession including value to special purchaser. Section 50 of the LCA 1973 precludes an abatement in compensation if the owner is rehoused by the acquiring authority. Claim, therefore, £90,000 plus 10% home loss payment under sections 29 and 30 LCA 1973 as amended by sections 68 and 69 P&CA 1991 = £99,000 plus possible disturbance claim.

It might be argued that a disturbance claim under section 5(6) of the LCA 1961 was inconsistent as some development value has been included in the purchase price.

Section 39 of the LCA 1973 is concerned with the duty to rehouse residential occupiers on compulsory purchase where suitable alternative residential accommodation on reasonable terms is not otherwise available. Trespassers and blight cases are excluded.

Study 9

Acquisition for new town purposes

An owner-occupied farm of 150 ha is in an area recently designated as part of the site of a new town. It is on the edge of a village about 4 km from an existing motorway. The village has a railway station on the main line to London. When acquired, 135 ha of the farm will be used for housing, 12 ha for a school and 3 ha for road purposes.

Prior to the new town proposal, the land was allocated for agricultural purposes but owing to pressure of demand some land in the village was annually being released for housing.

The following preliminary matters have to be considered.

Date of valuation
West Midland Baptist (Trust) Association (Inc) v *Birmingham Corporation* (1968) 19 P&CR 9, CA.

Statutory basis of compensation
Section 5(2) LCA 1961. Open market value.

Planning assumptions
Section 14 (3) LCA 1961. If the market would have paid more than agricultural value, then that may be taken into account: it would be a matter of evidence.

Section 15(1) LCA 1961. Acquiring authority's use (housing on 135 ha in this case).

Section 16(2) LCA 1961. Plan allocated use, as for the intended use in this case.

Section 15(5) LCA 1961. Section 17 certificate of appropriate alternative development on the 15 ha not allocated for housing – likely to be residential also if not required for school or road purposes.

Effect of section 6 and Schedule 1 LCA 1961
Case 3. Disregard any increase in value of the land taken due to development or prospective development of the new town which would not have taken place but for the new town scheme.

Summary
It is necessary to value 150 ha with planning permission for housing in the "no-scheme" world reflecting the rate of natural growth without any artificial influx due to the scheme, remembering that planning permission without demand does not create value: *Viscount Camrose* v *Basingstoke Corporation* (1966) 64 LGR 337, CA.

The basic problem here is to assess what the demand for this land with planning permission would have been in the "no-scheme" world. Suppose that it is agreed that there would have been a demand for 5 ha pa for five years but that thereafter any further demand would have been unlikely. If residential value, based on comparables, is £300,000 ha then the total sale price of the housing land over the next five years will be £7.5m. This must be deferred for the average time for receipts – 2½ years.

	£	£
Total proceeds from sale of housing land	7.5 m	
PV £1 2½ years at 12%, Say	0.75	
(12% reflects the risk of the estimated demand		5,625,000
not being achieved.)		
(An alternative approach giving the same result		
would be to capitalise the yearly income. £1.5m		
by YP 5 years at 12%.)		
Add: Agricultural income from the 25 ha also for		
an average period of 2½ years.		
25 ha at, Say, £200 (net) per ha pa	5,000	
YP 2½ years at 5%	2.29	11,450

Balance of 125 ha
125 ha with planning permission but with the
remote possibility of demand – Say £20,000 ha.   2,500,000
                                                                  ─────────
                                                                  8,136,450
Say   £8.1 million
                                                                  ─────────

NB: If it is unlikely that there would be a demand for any of the land, albeit with planning consent, in the "no-scheme" world, then only a small amount in respect of hope value over agricultural value would be justified.

Land subject to agricultural tenancies

In valuing either the freehold interest subject to a tenancy or compensation to tenants of agricultural holdings dispossessed by an acquiring authority, the issue of whether possession could be obtained under the Agricultural Holdings Act on the grounds that planning permission has been given for a non-agricultural purpose is important: see case B, Schedule 3 to the Agricultural Holdings Act 1986. The acquiring authority's scheme is normally such a ground and in *Rugby Joint Water Board* v *Foottit* (1972) 222 EG 815 and *Minister of Transport* v *Pettitt* (1968) 209 EG 349, it was held that the interests had to be valued allowing for possession which increased the freeholder's compensation at the expense of the tenant. Section 48 of the LCA 1973 now provides, in effect, that possession shall be considered only if it would have been obtainable apart from the authority's scheme. It is necessary to determine how long the occupancy might have continued but for the compulsory purchase. The Lands Tribunal's approach to the uncertainties involved in valuing, in effect, a life expectancy of profit was illustrated in *Wakerley* v *St Edmundsbury Borough Council* (1977) 241 EG 921. The rights to succession on death or on retirement of the tenant are now covered by Part IV of the Agricultural Holdings Act 1986 which consolidated and amended certain enactments relating to agricultural holdings. Schedule 3, cases A to H list the circumstances in which the consent of the Agricultural Land Tribunal to the operation of notices to quit is not required.

After September 1 1995 new tenancies cannot be granted under the Agricultural Holdings Act 1986, other than succession tenancies. The parties are free to fix the terms of the tenancy and an open market rent applies on review, unless agreed otherwise. The Agricultural Tenancies Act, 1995 also provides that to obtain possession, for a tenancy which is either from year to year or a fixed term of greater that two years, the notice requirement is 12 to 24 months expiring on the term date. For terms of two years or less, no notice is required.

Compensation when permission for additional development is granted after acquisition

Similar provisions, later repealed, were originally in Part IV LCA 1961 and have been revived with amendments by section 66 and Schedule 14 P&CA 1991. This now provides that where within 10 years of completion of the acquisition a planning decision is made granting permission for the carrying out of additional development of any of the land, compensation is to be reassessed on the assumption that permission for such additional development was available at the date of the original valuation and any additional compensation paid to the claimant with interest. The new section 23(3) LCA 1961 excludes from these provisions, *inter alia*, acquisitions by urban development corporations and by highway authorities in connection with urban development areas and acquisitions made under the New Towns Act 1981. The provisions apply to sales completed after Schedule 14 came into force.

Disturbance

There is no statutory provision which expressly gives a right to disturbance on dispossession of the owner and whatever compensation is payable to an owner in respect of disturbance is an element in assessing the value of the land to him, not a distinct and independent head of compensation. The two elements were inseparable parts of a single whole in that together they made up the value of the land to the owner which was the only compensation which the 1845 code awarded to him. *Hughes* v *Doncaster Metropolitan Borough Council* [1991] 1 EGLR 31, HL.

The 1919 Act (now section 5 (2) LCA 1961) altered the basis of valuation for the land taken to open market value, but section 5(6) specifically leaves unaffected the right which the owner would previously have had to a disturbance claim.

The general rule is that any loss suffered by a dispossessed owner which flows from acquisition is compensatable provided that:

(a) It is not too remote. *Bennett* v *Northampton Borough Council* (1989) 29 RVR 114 concerned, *inter alia*, a claim for partial extinguishment of a business. A complex claim was made to substantiate a loss over a 12-year period but the Lands Tribunal held that the use of the Retail Price Index was too sophisticated and the loss was too remote.

(b) It is the natural and reasonable consequence of dispossession: *Harvey* v *Crawley Development Corporation* (1957) 55 LGR 104, CA.

An owner cannot obtain more than his losses and the principle of *Horn* v *Sunderland Corporation* [1941] 2 KB 26 must always be considered. A disturbance claim must be consistent with the basis of valuation of the land taken and thus an owner cannot claim redevelopment value for land taken plus disturbance to an existing business.

Prasad v *Wolverhampton Borough Council* [1983] 1 EGLR 10, CA, concerned a person who, when threatened with inevitable displacement from land because of compulsory acquisition acted reasonably in moving to other accommodation before notice to treat. It was held that he was displaced in consequence of the acquisition of the land and entitled to a disturbance payment. This Court of Appeal decision followed earlier Scottish decisions, *Park Automobile Co Ltd* v *Strathclyde District Council* [1983] 1 EGLR 195 (LTS) and *Aberdeen City District Council* v *Sim* [1982] 2 EGLR 22.

Emslie & Simpson Ltd v *Aberdeen City District Council* [1994] 1 EGLR 33 concerned loss of profits preceding acquisition caused by the general blighting of the area due to the scheme. The Court of Sessions held that while in certain cases a loss of profits prior to dispossession might be recoverable such as where it is due to steps taken to anticipate it, a loss due to the effect of blight on trading generally cannot be said to have been caused by the dispossession as it affected all traders whether their land was being taken or not.

The matter was further considered in *Director of Buildings and Lands (Hong Kong)* v *Shun Fung Ironworks Ltd* [1995] 1 EGLR 19. The Judicial Committee of the Privy Council decided that compensation was payable for losses occurring before resumption of land by the Crown provided that they arose in anticipation of resumption and because of the threat thereof and were not too remote or losses which a reasonable person would have avoided.

There is also a duty to mitigate losses as far as possible; a claimant must take all reasonable steps to minimise his losses so that his claim does not exceed the natural and reasonable consequences of dispossession. Disputes on this often relate to whether other premises should have been taken. Thus in the following cases the claimant did not take other premises and the acquiring authorities fought total extinguishment claims.

W C Jones & Co v *Edmonton Corporation* (1957) 8 P&CR 86
A lease of premises of which the freehold was owned by the acquiring authority was reasonably rejected by the claimant.

Rowley v *Southampton Corporation* (1959) 10 P&CR 172
One of a number of cases where the rent of the alternative premises was higher than the claimant felt he could pay. His refusal was taken to be justified.

Bede Distributors Ltd v *Newcastle upon Tyne Corporation* (1973) 26 P&CR 298
The claimants were held entitled to reject other suitable alternative accommodation owing to their financial position.

Freeman, Hardy & Willis Ltd v *Bradford Corporation* (1967) 203 EG 1099 also *Ind Coope (London)* v *Enfield London Borough Council* [1978] 1 EGLR 163
Similar cases where other premises were available but the claimants made decisions not to move. In the latter case the claimants said that the projected trade at the new premises did not justify the rent and the net profits were likely to be reduced. The decision was held not to be unreasonable and one that they were, in their discretion, entitled to make.

Knott Mill Carpets v *Stretford Borough Council* (1973) 26 P&CR 129, LT
The carpet company rejected relocation in a precinct as they needed competition. They were held justified to do so as they were unlikely to increase turnover sufficiently to pay higher rent and maintain profit.

Bailey v *Derby Corporation* [1965] 1 All ER 443
The claimant, a builder, did not take alternative premises as he was too ill to move. The Court of Appeal supported the Tribunal and

held that there was sufficient evidence to show that the ill-health was not due to the acquisition but was coincidental. It was therefore "an extraneous and independent matter which must be put on one side". This decision seems particularly hard and appears to conflict with the "loss to the owner" rule.

Hall v *Horsham District Council* [1977] 1 EGLR 152
The claimant's failure to proceed with the relocation, coupled with his general conduct, was fatal to the claim for total extinguishment.

Section 46 of the LCA 1973 deals with persons who have attained the age of 60, who are displaced from trade or business premises and who may now elect for a total extinguishment basis subject to giving certain undertakings prescribed in section 46(3). The Court of Appeal held in *Sheffield Development Corporation* v *Glossop Sectional Buildings Ltd* [1994] 2 EGLR 29 that section 46 was satisfied where the claimant was trading when a compulsory purchase order was made and when a deemed notice to treat came into effect although he vacated 20 months before the acquiring authority took possession. He was "required to give up possession" as specified by section 46.

A disturbance claim should be for the actual loss suffered by the owner. Every foreseeable loss or expense which is likely to result from dispossession should be claimed, provided that it is the direct, natural and reasonable consequence of being dispossessed. Once there is a binding contract to purchase, it is too late to claim further items, even though they would have been properly payable.

In *Palatine Graphic Arts Co Ltd* v *Liverpool City Council* [1986] 1 EGLR 19, CA, the claimant qualified for a 22% regional development grant in respect of certain disturbance items. The Court of Appeal held that the grant need not be deducted to arrive at the claimant's loss as it was not related to compulsory purchase.

Basis of compensation for disturbance

The acquisition may lead to the total extinguishment of a business, partial permanent loss of profits or only temporary disruption to trade, but what has to be considered is the loss to the owner. Thus where a business depends upon the personality of the owner reflected in higher profits, he will be entitled to claim on the basis of loss to himself. In *Sceneout Ltd* v *Central Manchester District Council* [1995] 34 EG 75, a total extinguishment case, loss to the owner of a laundry business exceeded market value and was the basis of the award.

Total extinguishment claims

The main items of claim will be for the value of the land taken and a claim for total loss of profits. In the case of industrial properties there may also be a claim for loss on forced sale of plant and machinery. It will be necessary for the owner to dispose of such items and his claim will be as follows:

Open market value of the plant and machinery
Less: Forced sale value (Possibly only scrap value)
Equals: Loss on forced sale (Payable by the acquiring authority).

In *Shevlin* v *Trafford Park Development Corporation* [1998] 08 EG 161, it was held that "value to the business" in para H 13.2 of the RICS *Appraisal of Valuation Manual* (the "Red Book") is consistent with value to the owner for compensation purposes, although in *Shevlin* there was a special extra value to the claimant to be allowed for.

Tamplins Brewery Ltd v *Brighton County Borough Council* (1971) 22 P&CR 746 was a removal case and dealt with the proper method of ascertaining loss where new plant and machinery were installed in alternative premises even though the old plant and machinery had a useful life. The relocation scheme adopted, with the concurrence and assistance of the corporation, was the best and cheapest way of mitigating the claimant's loss. It was held that the loss could not be measured by the open market value of the existing plant and machinery and that the cost of the new plant and machinery should be taken as the first step less the scrap value of the old and the value of the new plant after 10 years (the estimated life of the old plant) less also capitalised savings on running costs.

Loss on forced sale of fixtures and fittings is normally arrived at by taking a percentage of their value in situ and loss on forced sale of stock, by taking a percentage of its cost.

No list of items claimable can possibly be comprehensive, as any reasonable loss or expense is compensatable. Thus in *Widden & Co Ltd* v *Kensington and Chelsea Royal Borough Council* (1970) 213 EG 1442 the following were properly claimable:

- Redundancy payments and interest on a loan raised to pay them.
- Cost of cancellation of specific contracts.

- Holiday pay due on the termination of employment.
- Wages to staff to clear up after close-down.
- Advertising expenses in abortive search for new premises.
- A percentage of "bad debts" as these would be more difficult to recover once the business was wound up.

The Judicial Committee of the Privy Council in *Director of Buildings and Lands (Hong Kong)* v *Shun Fung Ironworks Ltd* (1995) 19 EG 147 1 EGLR 19 were unable to accept the Crown's submission that a claimant could never be entitled to compensation on a relocation basis that would exceed the amount of compensation payable on an extinguishment basis. It all depended on how a reasonable businessman, using his own money, would behave in the circumstances. In this case, however, a claim based on relocation should fail as the Lands Tribunal were entitled to conclude that the business planned for the alternative site was not the same business as that on the acquired site.

Removal cases

As any reasonable loss suffered by the owners is compensatable, no comprehensive list of items under this heading can be made but the following may well be involved:

Removal costs

Removal costs including the cost of dismantling, adaptation and reinstalling of plant and machinery or of fixtures and fittings and loss on forced sale of those items which cannot be taken. A claim may extend to extra temporary supervision after reinstallation of machinery. Minor items would include notifications of new address, reprinting stationery and telephone removal.

In *Succamore* v *Newham London Borough Council* [1978] 1 EGLR 161 removal 31 miles away was held to be reasonable and properly compensatable although other premises were available within a shorter distance.

Duplicated expenditure

These claims arise from the necessity to operate two premises for an overlap period. A claim would extend to double rent, rates and additional wages, heating, lighting and telephone. If the new premises are not rented, but purchased, then loss of interest on capital would be included. The claim should run until the date of completion of purchase or, if earlier, the date the acquiring authority take possession, as they will pay interest on the purchase price from that date and the claimant should have no financial liability in respect of the property after the acquiring authority has taken possession.

Repairs and cost of adapting new premises

The purchase price or lease terms of new premises should have had regard to disrepair, and thus expenditure on repairs would not be recoverable. As far as adaptions are concerned, the expenditure would be recoverable in so far as it does not increase the value of the new premises: see *M&B Precision Engineers Ltd* v *Ealing London Borough Council* (1972) 225 EG 1186 and *Smith* v *Birmingham City Council* (1974) 14 RVR 511.

Cost of new premises and increased overheads

A higher price for alternative accommodation will not normally be compensatable as the owner will be presumed to have obtained value for money, but in *Metropolitan and District Railway Co* v *Burrow* November 22, 1884 The Times, HL, and *Mogridge (WJ) (Bristol 1937) Ltd* v *Bristol Corporation* (1956) 168 EG 714 circumstances arose where increased overheads were allowed. The decisions in *Greenberg* v *Grimsby Corporation* (1961) 177 EG 647 concerning a new shop and *J Bibby & Sons Ltd* v *Merseyside County Council* [1979] 2 EGLR 14 dealing with new and larger replacement offices, applied the principle that greater efficiency of the new premises should be set off against increased operating costs.

In *Service Welding* v *Tyne & Wear County Council* [1979] 1 EGLR 36, CA, bank interest and loan charges to build an alternative factory were held not to be compensatable as the claimants had obtained "value for money".

Additional distribution costs were capitalised at 7 YP rather than a normal lower YP applicable for the valuation of goodwill in *West Suffolk County Council* v *W Rought Ltd* (1955) 5 P&CR 215.

Reasonable fees incurred in obtaining new premises

Reasonable fees incurred in obtaining new premises of a

comparable kind were allowed in *Harvey* v *Crawley Development Corporation* (1957) 55 LGR 104, CA. A claim may extend to reasonable abortive expenditure in seeking alternative accommodation, the costs of terminating a mortgage and taking out a new one, travelling expenses and the loss of earnings during time reasonably spent in searching. However, Denning LJ in the *Harvey* case said of an investor:

> if he chooses to buy another house as an investment, he would not get the solicitor's costs on the purchase. These costs would be the result of his own choice of investment and not the result of the compulsory purchase.

Schedule 15, para 2 P&CA 1991 provides, that where the interest of a person not in occupation is acquired, charges and expenses of re-investment in other land incurred within one year after date of entry are claimable. Interest on a bridging loan for a new property would be a proper item and also financial losses involved in having to terminate a low-interest fixed-rate mortgage and taking a new one at a higher rate. Moving might necessitate the purchase of new school uniforms, which would be payable. However claims must be reasonable. Thus in *WJ Mogridge (Bristol 1937) Ltd* v *Bristol Corporation* (1956) 168 EG 714 costs of drawings and preparation of quantities were disallowed, being unnecessary to arrive at a decision whether to build.

Goodwill

On compulsory purchase any goodwill is not purchased by the acquiring authority but the owner is compensated for what he will lose by being displaced. Thus the issue is always to decide how much, if any, of the existing trade will be transferred. If none, then the minimum compensation will be the market value of the goodwill as, but for the acquisition, such a sum could have been obtained as part of the consideration on sale of the business. "Loss to the owner" might conceivably be a higher figure as, for example, in the case a small one-man business.

The value of goodwill is usually measured by calculating an average adjusted net profit and applying a YP. The last three years' average profits from the actual accounts are usually considered but any year which is affected by the scheme should be disregarded, eg if profits have been reduced by nearby clearance. A higher YP than normal will be adopted if the profits show a rising trend as past profits are only a guide as to what future profits might have been. Conversely, falling profits justify a lower YP than would be normal.

The following adjustments to the profit from the accounts must be considered.

Adjustment for rent

If a claimant is the freeholder the full rental value must be deducted. If he is a leaseholder, the profit rent must be deducted (arrived at by adding back the rent paid and deducting the full rental value). It was decided in *Widden & Co Ltd* v *Kensington and Chelsea Royal Borough Council* (1970) 213 EG 1442 that an increase in rent to full rental value could not merely be passed on by way of increased charges and the profit rent had to be deducted.

Deduction for interest on capital employed in the business (other than land and buildings)

This will comprise plant and machinery, stock, fixtures and fittings, and vehicles etc and also cash in hand and in a bank current account.

This was traditionally taken at 5% but see *Reed Employment Ltd* v *London Transport Executive* [1978] 1 EGLR 166 where 10% was taken although the claimants stated that their financial arrangements were such that the money was borrowed interest free: see also *RC Handley Ltd* v *Greenwich London Borough* (1970) 214 EG 213 where 10% was taken and *Bostock Chater & Sons* v *Chelmsford Corporation* (1973) 26 P&CR 321 where 8½% was adopted.

Proprietor's remuneration in the case of a one-man business

Following *Perezic* v *Bristol Corporation* (1955) 165 EG 595, it is now clear that no deduction need be made for the value of an owner's own services in such cases.

Deductions might have to be made for the value of a wife's services in the business if nothing or insufficient appears in the accounts, but this was not done in *Zarraga* v *Newcastle upon Tyne Corporation* (1968) 205 EG 1163.

As far as companies are concerned, a reasonable sum should appear for the value of the work of each director: see *Shulman (Tailors)* v *Greater London Council* (1966) 17 P&CR 244. If, however, the company is virtually one man, his fees might be dealt with in the same way as proprietor's remuneration in the case of a one-man business: see *Lewis's Executors and the Palladium Cinema (Brighton)* v *Brighton Corporation* (1956) 6 P&CR 318.

Deduction from profits of a branch of a proportion of head office expenses

It is necessary to deduct from the branch profits the saving of head office expenses (if any) due to the closure: see *Reed Employment Ltd* v *London Transport Executive* [1978] 1 EGLR 166.

Compensation for loss of goodwill

In the case of total extinguishment, the minimum that a claimant will lose will be the market value of goodwill.

A YP which should be based on market evidence is applied to the adjusted net profit as previously determined. However, the Lands Tribunal in *W Clibbett Ltd* v *Avon County Council* [1976] 1 EGLR 171 drew attention to the lack of market transactions adduced as evidence and claims are often made on the basis of previous tribunal awards.

It must be emphasised that compensation is for goodwill that will not be transferred. A milkman with a shop and a milkround might well retain the latter on moving even if goodwill attached to the shop is lost.

In *F Neubert Ltd* v *Greater London Council* [1969] RVR 263, the conventional compulsory purchase "analytical" approach of valuing the land taken and goodwill separately was challenged and on the evidence the Tribunal preferred the "lock, stock and barrel" approach of the claimant's business transfer agent. This was that the market would pay 2¼ YP of the adjusted net profit, the latter figure being arrived at without any deduction for director's services. This approach does not seem to have been followed in later cases.

In *Widden & Co Ltd* v *Kensington and Chelsea Royal Borough Council* (1970) 213 EG 1442, 3½ YP was awarded in respect of a business of car repairs, spraying and sales and in *Shevlin* v *Trafford Park Development Corporation* [1998] 08 EG 161, 2½ YP for total extinguishment of an industrial heat treatment plant.

RC Handley Ltd v *Greenwich London Borough* (1970) 214 EG 213 concerned a substantial and well-established car business. The Tribunal expressed the view that 1 to 1½ YP was only appropriate to small businesses in clearance areas. Compensation was awarded on the basis of 3 YP being the correct figure had the business been totally extinguished.

W Clibbett Ltd v Avon County Council [1976] 1 EGLR 171 (an engineering business) was a case where the Tribunal adopted a "robust" approach similar to that used by the courts in assessing general damages as there was a lack of any evidence other than the *ipse dixit* of each of the valuers. This approach was followed in *Tragett* v *Surrey Heath Borough Council* [1976] 1 EGLR 175 concerning a sports shop. *Roy* v *Westminster City Council* [1976] 1 EGLR 172 concerned a national health service doctor's practice which legally was not a marketable asset. Compensation was nevertheless awarded on the basis of the loss to the owner. 3 YP was awarded for loss of goodwill of a long-established family café business in *Viazzani* v *Afan Borough Council* unreported (1976) (Lands Tribunal REF/150–1/1975) and 3¼ YP for total extinguishment of a hairdressers business at Waterloo Station, *Klein* v *London Underground Ltd* [1996] 1 EGLR 249. In *Ind Coope (London)* v *Enfield London Borough Council* [1978] 1 EGLR 163, 4½ YP was awarded for brewer's loss of wholesale profits in respect of an off-licence subject to a tied tenancy. The Tribunal felt that tied trade had more security than from a free house or than any other retail business but also decided that an off-licence should be treated no differently from any other shop.

Payne v *Kent County Council* [1986] 2 EGLR 218 was concerned with the acquisition of a petrol filling station ripe for demolition and rebuilding, the date of valuation being December 1982. The Tribunal took one-half of the net profit per gallon (7.1p), say 3.5p, as the rental element and applied this to a throughput of 1.5 million gallons giving £52,500 pa, capitalised at 6% to give say £875,000. £50,000 was added to reflect room for a small catering unit and £175,000 deducted as the cost of redevelopment. The award was £750,000.

Home loss payments

Sections 29 and 30 LCA 1973 (as amended by sections 68 and 69 P&CA 1991). The amended section 29 provides that a person is now entitled to a home loss payment if the following conditions are satisfied, throughout the period of one year (previously five years), ending with the date of displacement:

(a) He has been in occupation of the dwelling, or a substantial part of it, as his only or main residence, and

(b) He has been in such occupation by virtue of a qualifying interest which includes statutory tenancies under the Rent Acts (section 29(4)).

The amended section 30 provides that, if the interest of a person who so qualifies is that of an "owner" as defined in section 7 ALA 1981 the amount of the home loss payment is 10% of the market value of the owner's interest in the dwelling, subject to a maximum of £15,000 and a minimum of £1,500. If the qualifying interest is other than that of an "owner" the home loss payment is £1,500.

Discretionary payments may also be made if (a) and (b) above are satisfied but the occupation has been for less than one year (section 29(2)). Also section 29(5) LCA 1973 has been repealed so that payments now apply to blight notice cases (section 68(2) P&CA 1991).

Farm loss payments

Section 34 LCA 1973 applies to owners in occupation displaced in consequence of compulsory acquisition and who, not more than three years after displacement, begin to farm another agricultural unit.

Schedule 15 para 6 P&CA 1991 extends the entitlement to payments to claimants displaced from part only of their holding (of a minimum area of not less than 0.5 ha), yearly tenants and claimants who give up possession at any time after the making or confirmation of the compulsory purchase order but before being required to do so by the acquiring authority. Section 34(6) of the LCA 1973 is repealed so that blight notice cases become eligible.

Section 35 prescribes that the amount of a farm loss payment is based on the average annual profit from the acquired land after having made allowance for the full rental value and any profit which falls within compensation for disturbance.

Tax position of compensation for temporary loss of profits

Following *West Suffolk County Council* v *W Rought Ltd* (1955) 5 P&CR 215 tax became deductible on compensation for temporary loss of profits as it was held that such compensation would be tax free to the claimant. This was subsequently extended to certain expenses which were admissible deductions from profits for tax purposes. This position held until reversed by *Bostock Chater & Sons v Chelmsford Corporation* (1973) 26 P&CR 321. It was explained that since the Finance Act 1965, capital gains tax was levied on the basis that the consideration for disposal included any compensation for temporary loss of profits. As such, compensation was therefore taxable in the hands of the claimant and it was inappropriate to make any tax deduction from the compensation otherwise payable.

Under sections 52(4) and 245 TCGA 1992, an inspector of taxes is empowered to apportion the compensation as between capital and revenue, the effect being to free the compensation for temporary loss of profits of its capital nature and enable it to be treated as a trading receipt. An Inland Revenue statement of practice (8/79) endorses this treatment and deals similarly with loss on trading stock, removal expenses and interest. There is thus no case now for the deduction of tax in respect of temporary loss of profits or removal expenses: see also *Hobbs (Quarries) Ltd v Somerset County Council* (1975) 234 EG 829 which concerned compensation resulting from a discontinuance order.

Penine Raceway Ltd v *Kirklees Metropolitan Council* [1989] 1 EGLR 30, CA, concerned compensation for the refusal of a planning application after an article 4 direction. The court stated that tax should only be deducted if it was clear beyond doubt that the claimants would not be charged tax on compensation which would have been taxed in the normal course of events.

Short tenancies – section 20 CPA 1965

The right of a lessee to apply for a new lease of business premises under Part II, Landlord and Tenant Act 1954, is taken into account and reflected in the YP for loss of goodwill. However, yearly tenants or less are not entitled to a notice to treat and may be dealt with either by purchase of the superior interest and service of a notice to quit or by notice of entry and the taking of possession when compensation is assessed under section 20 CPA 1965. Section 47(2) LCA 1973 provides that compensation under section 20 will now reflect any right to continue in occupation.

Under P&CA 1991 Schedule 15 para 4 tenants may now claim for losses due to severance of the tenanted land from other land which they occupy or own. Section 20 (2) CPA 1965 related only to land held on the same tenancy.

Persons without compensatable interests

Sections 37 and 38 LCA 1973 as amended by Schedule 13 para 39 Housing Act 1974 apply to persons without compensatable interests displaced in consequence of compulsory acquisition provided they are in lawful possession. Disturbance payments must now be made to such persons assessed under section 38, comprising expenses in removing and, in addition, in the case of a business, disturbance having regard to the period the business might have continued. Section 37(4) restricts compensation to either section 38 compensation or Landlord and Tenant Act 1954 compensation, but not both.

In *Wrexham Maelor Borough Council* v *MacDougall* [1993] EGCS 68, the Court of Appeal held that MacDougall, who was the lessee, was in possession for the purposes of section 37 and a company, which earned his living had exclusive occupation from him. The company were therefore in lawful occupation under section 37.

Compensation paid to a tenant under section 37 Landlord and Tenant Act 1954 is not liable to capital gains tax: *Drummond (Inspector of Taxes)* v *Austin Brown* [1985] Ch 52, nor is compensation for disturbance under section 34 (1) of the Agricultural Holdings Act 1948: *Davis (Inspector of Taxes)* v *Powell* [1977] 1 WLR 258. In neither case did the sum payable derive from the asset.

Study 10

Factory in an action area under section 36 (4) of the T&CP 1990
An action area contains predominantly old warehouse and industrial buildings interspersed with two-storey Victorian cottages. A CPO has been made in respect of the whole area under section 226 T&CP 1990, the planned range of users in the action area map being shops, public buildings (including a library), offices, warehouses and flats.

You are acting both for the freeholder and the lessee of an old factory of 400 m^2 net sited within the area and allocated for use as part of a library site. The factory is subject to a lease now having eight years unexpired at a rent of £5,500 pa without review on tenant's internal repairing terms. The current net full rental value is £15,000 pa. Notices to treat were served on the owners of both interests two years ago and possession of the leasehold interest was taken six months ago after the tenant had moved to a new factory nearer the outskirts of the town, the only reasonable alternative.

That factory has an area of 300m² net and the rent is £16,000 pa on full repairing and insuring terms.

The following matters have to be considered:

Section 5(2) LCA 1961. Open market value.

Section 15(1) LCA 1961. Development such as would permit the acquiring authority's proposals but development value approach precludes a disturbance claim. *Horn* v *Sunderland* [1941] 2 KB 26.

Section 15(3). Existing use value and a disturbance claim.

Section 16(4). Consider which of the planned range of users would have been appropriate in the circumstances of section 16(5). Probably warehouse use.

Section 17 certificate procedure also now applies in action areas (section 65(1) P&CA 1991). Likely to be warehouse or industrial use.

Section 6(1). Disregard any increase or decrease in value in the circumstances of the second case in the first column of Schedule 1 as is attributable to development in column 2, subject to the qualifications of section 6(1)(*a*) and (*b*).

Conclusion

Valuation on the basis of the existing factory including Schedule 3 rights together with a proper disturbance claim will be the most profitable approach. The value in the "no-scheme" world must be taken ignoring any increase in value of the factory due to redevelopment in the area, as but for the "scheme" it is considered no major redevelopment would have taken place.

Freeholder's interest

Date of valuation = date when valuation being made.

	£	£	£
Rent reserved		5,500	
Less: external and structural repairs, Say,	1,500		
Insurance (based on reinstatement costs)	250		
Management (5% of rent passing)	275	2,025	
Net income		3,475	
YP 8 years at 11%		5.15	17,896
Reversion to full rental value		15,000	
YP perp deferred 8 years at 13% (freehold full rental value rate)		2.89	43,350
			61,246
Say			£61,250

Leaseholder's interest

Date of valuation = date of possession, six months ago but assume values have not materially changed since then.

	£
FRV net	15,000
Add: External and structural repairs, management and insurance (as in freehold valuation)	2,025
Full rental value (internal repairing lease)	17,025
Less: Rent payable	5,500
Profit rent	11,525
YP 8½ years at 15% and 4% (Tax at 40%)	3.12
Compensation for land taken	35,958
Say	£36,000

Disturbance claim

Basis of claim – the actual loss suffered by the leaseholder.

(It is assumed that the additional rent at the new premises is not compensatable in view of increased business efficiency.) Removals of plant and machinery, fixtures and fittings and stock and loss on forced sale where appropriate. Insurances against loss during the move. Duplicated expenditure, rent, rates, heating and lighting etc from the date of renting the new accommodation until the date of possession by the acquiring authority. *Crawley* costs, legal and surveyor's fees in renting the new premises and director's time and expenses in the search.

Capitalised yearly increased distribution costs.

Permanent loss of profits if the level of production will be lower in the new premises as a result of the reduced floor area.

Temporary loss of profits as a result of disruption of the business.

Any other proper disturbance items.

Study 11

Leasehold disturbance claim involving adjustment of accounts

A small corner back street shop with a residential upper part on two floors is to be acquired compulsorily for public open space purposes. The property has been occupied for 10 years by the lessee, who lives in the upper part and carries on a small grocery business from the shop. The rent payable for the whole is £1,500 pa exclusive on an internal repairing lease having 10 years unexpired without review. The full rental value of the property is £6,000 pa net of which £5,000 is attributable to the shop.

The accounts show that the outgoings for the upper part have been deducted as well as the rent of £1,500 for the whole premises.

Other relevant details are as follows:

The net profit after deducting mortgage interest of £400, repairs £360, and business rates of £920 is £22,000. The figures for repairs relate to the whole structure and rates to the business part only No deduction appears for the owner's or his wife's services. She works about 10 hours a week in the shop and he is virtually full time. There are no suitable alternative shop premises available and the acquiring authority agree that this is a total extinguishment case.

Prepare a claim for compensation for the lessee.

Leasehold claim on existing use value plus disturbance basis section 5(2) and (6) LCA 1961

	£
Land taken	
Full rental value of whole premises (net)	6,000
Add: Landlord's outgoings (external repairs, insurance and management), Say	600
Full rental value, internal repairing lease terms	6,600
Less: Rent payable under lease on the same basis	1,500
Profit rent	5,100
YP 10 years 12% and 4% (Tax at 40%)	3.86
	£19,686
Value of land taken, Say	£19,750

Disturbance claim

The accounts include outgoings for the residential upper part. The profit for the shop is isolated by adding back the rent paid and internal repairs (both relating to the whole property) and deducting figures applicable to the shop only.

	£	£
Unadjusted net profit		22,000
Add back		
Mortgage interest	400	
Rent payable	1,500	
Internal repairs	360	
		2,260
Profit if no deductions for rent or repairs		24,260

Less: Full rental value of the shop on internal repairing terms: Say	5,500	
Add: tenant's internal repairs on shop	300	5,800
		18,460
Less: interest on tenant's capital 5% of £10,000		500
		17,960
No deduction made for proprietor's remuneration but, say, £50 per week for wife (however see *Zarraga v Newcastle upon Tyne Corporation* (1968) 205 EG 1163)		2,600
Adjusted net profit		15,360
YP based on comparables (back street shop), Say		2
Compensation for total loss of goodwill		30,720
Loss on forced sale of tenant's fixtures and fittings, Say 80% of £3,000		2,400
Loss on forced sale of stock, Say 50% of £3,000		1,500
Abortive costs of seeking alternative premises, Say		400
Home loss payment for residential upper part. The lessee qualifies under the substituted section 29 (2) LCA 1973 and possesses an "owner's" interest under the substituted section 30. A 10% home loss payment based on the apportioned value of the lease of the residential upper part would only be about £330 (10% of approx. $\frac{1}{6}$th of £19,750). Therefore the minimum of £1,500 applies.		1,500
"*Crawley*" costs and removals for the upper part, Say,		300
Total disturbance claim		36,820
Value of land taken		19,750
		56,570
Total claim, Say		£57,000

Study 12

A total extinguishment claim in a blight notice case

You have been asked to advise a client, a grocer. He is now 70 years old and wishes to retire but his efforts to sell his business have been frustrated as the property is allocated as a future library site, a proposal which is to be implemented in about 10 year's time. His shop and upper part are in a poor trading position. He has a lease having nine years unexpired at £1,000 pa without review on a tenant's internal repairing lease. The rateable value of the business part is £4,700 and the present full net rental value of the whole is £5,350 pa.

He has been trading as a grocer for the past 10 years and his net profits for the last three years were:

£15,500
£17,000
£16,200 (most recent)

Both he and his wife work in the business but no figure for remuneration appears in the accounts for either of them. The wages of a part-time help and delivery-boy have been deducted in the accounts.

The following preliminary matters have to be considered:

The blight notice provisions – sections 149 to 171 T&CP 1990

These provisions enable specified owner-occupiers affected by planning proposals to require purchase of their interests.

If part only is affected by a scheme, the notice must be served in respect of the whole property (section 150(3)), but if the authority serve a counter notice that they require part only (section 151(4)(c)), the Lands Tribunal can decide whether the blight notice will apply to the whole or part (section 153(6)). In *Smith* v *Kent County Council* [1995] 44 EG 141 a counter notice was not upheld when a small part of a front garden was required for road widening.

The categories of blighted land are to be found in Schedule 13 to the Act. The prescribed ceiling annual value of the hereditaments where blight notices may be served by owner-occupiers under Section 149(3)(a) is £24,600 (SI 2000 no 539). Annual value is defined in section 171 of the 1990 Act and in the case of a property which is wholly non-domestic and no part of which is exempt, it is the rateable value at the date of service of the blight notice. For hereditaments which include domestic property or property which is exempt from rates, "annual value" is the sum of the rateable value of the non-domestic part and the "appropriate value" of the remainder ascertained in accordance with section 171(2) and (3).

Blight notices in respect of residential properties may be served by resident owner-occupiers, regardless of value and such notices may also be served by owner-occupiers of agricultural units or part thereof. Section 150 sets out the requirements that must be met before a valid blight notice may be served. The P&CA 1991 Schedule 15, para 13 removes the former requirement to have made reasonable endeavours to sell the interest before serving a blight notice in cases where a compulsory purchase order or a special enactment authorising compulsory purchase is in force but notice to treat has not been served.

The date for determining whether the land is blighted is the date of the counternotice. *Carrell* v *London Underground Ltd* [1996] 12 EG 129, and *Sinclair* v *Secretary of State for Transport* [1997] 34 EG 92, a change of plan after that date being irrelevant *Entwistle Pearson (Manchester) Ltd* v *Chorley Borough Council* (1993) 66 P&CR 277.

Basis of claim

A certificate of appropriate alternative development could be applied for, but a claim on existing use value and disturbance basis is likely to be the most favourable.

Section 46 of the LCA 1973. As the person carrying on the trade or business has attained the age of 60, compensation for disturbance may, at the option of the owner, be assessed on a total extinguishment basis subject to his giving certain undertakings.

Section 29(5) of the LCA 1973. The provision that no home loss payment may be made where a blight notice is served has been repealed by section 68(2) of the P&CA 1991.

Open market value of leasehold interest

	£	£	£
Full rental value of whole premises, net		5,350	
Add: Landlord's external repairs, insurance and management		550	
Full rental value (internal repairing lease basis)		5,900	
Less: Rent payable on the same basis		1,000	
Profit rent		4,900	
YP 9 years at 10% and 4% (Tax at 40%)		3.88	
Value of land taken			19,012

Disturbance claim for shop part

	£
Average unadjusted net profit	16,233
Assuming that the rent of £1,000 pa for the whole but repairs for the shop alone have been allowed in the accounts, it is necessary to make only an adjustment for rental value	
Add back: rent paid	1,000
	17,233
Deduct: full rental value of shop only on tenant's internal repairing lease, Say	4,200
Profit after making a deduction for the full rental of the shop	13,033

Less: Interest on tenant's capital 5% of £10,000	500		
Wife's wages: Say £80 per week (but see *Zarraga* v *Newcastle upon Tyne Corp* (1968) 19 P&CR 609)	4,160	4,660	
Adjusted net profit		8,373	
YP for total extinguishment based on comparables for this type of business		3	25,119
Loss on forced sale of fixtures and fittings, Say 80% of £2,500			2,000
Loss on forced sale of stock, Say 30% of £3,000 (mainly tinned food which could be disposed of easily)			900
Claim for land taken and disturbance claim on a total extinguishment basis (but excluding items below)			47,031

Disturbance claim for the residential upper part

Crawley costs based on actual and reasonable costs incurred.

Removal costs (including telephone) and expenses incidental thereto similarly assessed. Depreciation of fixtures and fittings due to removal or loss on forced sale, as appropriate.

The minimum home loss payment of £1,500 will clearly exceed 10% of the value of the lease of the residential upper part. Home loss payment is therefore the minimum of £1,500.

Note If the claimant's wife had been an employee he would have had to make her redundant and a redundancy payment would have been a valid item of disturbance claim.

Unfit housing

Schedule 9 to the Local Government and Housing Act 1989 made important changes to the Housing Act 1985 and the law relating to unfit housing and compensation. In particular para 83 of Schedule 9 substituted a new section 604 into the Housing Act 1985 containing a revised fitness for habitation standard.

The site value compensation provisions in sections 585 to 595 of the Housing Act 1985 ceased to have effect (para 76). Consequently section 10 of and Schedule 2 LCA 1961 dealing with ceiling values, owner-occupied supplements and well-maintained payments were no longer necessary and were repealed by Schedule 12. The normal open market value basis applies and this will reflect any lack of fitness and amenities and disrepair.

Section 584A provides for compensation for properties subject to closing or demolition orders based upon the diminution in the value of the owner's interest as a result.

Right to advance payment of compensation

Section 52 LCA 1973, as amended by section 63 P&CA 1991, provides that if an acquiring authority have taken possession they must, if requested, make an advance payment on account of compensation payable, of 90% of the agreed amount, or, if not yet agreed, of the compensation estimated by the authority.

An additional section 52 A provides for the advance payment to include accrued interest from the date of entry. To avoid double counting section 52A(9) provides that in such a case, the right to interest under section 11(1) CPA 1965 does not apply.

Interest on compensation moneys

Section 80 P&CA 1991 provides for the payment of interest on compensation at the rate prescribed under regulations made under section 32 LCA 1961 in cases listed in Part 1 of Schedule 18, which lists circumstances where existing legislation did not provide for interest, eg section 31 LCA 1961 dealing with compensation for withdrawal of notice to treat. Interest is payable from the date specified in the schedule until payment of compensation.

Part 2 of the schedule lists provisions that already provided for interest, *inter alia*:

Section 11 (1) CPA 1965 (entry on land)

Section 18 LCA 1973 (claims under Part 1 of that Act)

Section 63 (1) LCA 1973 ("*McCarthy*" claims where no land taken)

Section 11(1) CPA 1965 was at issue in *Manchester City Council* v *Halstead* [1997] RVR 266, CA, a cost of equivalent reinstatement case. It was held that although the local authority had paid for the replacement land and made stage payments for the building, interest was nevertheless payable as prescribed by section 11(1).

In *Mallich* v *Liverpool City Council* (1999) CA 33 EG 77 a claim for loss of investment income as a result of a very low advance payment was rejected as the claimant had been compensated on a total extinguishment basis and any delay in full payment could only be compensated by way of interest at the prescribed rate.

Tax on compulsory purchase compensation

If land is disposed of under a compulsory purchase procedure, or in the face of a threat thereof, the total compensation can be apportioned for tax purposes between its constituent parts (section 52(4) TCGA, 1992). Compensation for loss of profits or on stock and reimbursement of removal expenses are assessable under Case I or II of Schedule D. (Statement of practice SP8/79). Compensation for the land itself will be subject to capital gains tax and any amount apportioned as injurious affection or severance compensation is treated as a part disposal of the land retained for capital gains tax: see *Stoke-on-Trent Council* v *Wood Mitchell & Co Ltd* [1978] 248 EGLR 21.

Under the TCGA 1992, sections 247 and 248 a landowner who realises a capital gain on the disposal of land to an authority having compulsory purchase powers may be able to claim "roll-over" relief if he applies the proceeds of the disposal in the acquisition of "new-land" which meets certain conditions. He must, however, not have advertised the land for sale or taken any other steps to make known to the authority, or others, that he was willing to sell the land.

Further reading

Encyclopaedia of Compensation – Sweet and Maxwell.
Compulsory Purchase and Compensation (5th ed) – Barry Denyer-Green, Estates Gazette.

Chapter 10

Compensation for Planning Restrictions

This chapter has been shortened since the last edition to make space in the book. Readers who require a more detailed examination of planning compensation will find one in the author's *Handbook of Land Compensation*[1].

All references in this chapter are to the Town and Country Planning Act 1990, as amended, unless otherwise stated. The following abbreviations are used:

1991 Act	Planning and Compensation Act 1991
TP Act	Town and Country Planning Act 1990
GPDO	The Town & Country Planning (General Permitted Development) Order (presently 1995), as amended.
HS Act	Planning (Hazardous Substances) Act 1990
HSA	Hazardous Substances Authority
HS Consent	Hazardous Substances Consent
LB Act	Planning (Listed Buildings and Conservation Areas) Act 1990
LB Consent	Listed Buildings Consent
LCA	Land Compensation Act 1961
LPA	Local Planning Authority
SM Consent	Scheduled Monument Consent
T & CP	Town and Country Planning
The Tribunal	The Lands Tribunal

General Principles

There is no general right to compensation for adverse planning decisions: the compensation code is statutory and claims must meet the statutory requirements if they are to succeed. The events which may give rise to a claim, those who may be entitled to claim, the type of loss, the structure of the claim and the method of claiming compensation are all prescribed, either by the statutes or by regulations. The rules are different for the different sources of compensation.

---

[1] Published by Sweet and Maxwell.

As a general principle, it is likely that a claim will be possible if
an existing right is cut back or interfered with, but not if a consent
is refused from the outset. For example, there is no compensation
because planning permission is refused, but there is if it is revoked
or modified.

Eligible claimants are sometimes defined as those having an
interest in the land or being interested in it; in other cases the net is
cast more widely to include anyone who has suffered a direct,
consequential, loss.

The measure of compensation varies between the types of claim. It
may be any direct loss or the depreciation in the value of an interest
in the land. The former category embraces the latter. The loss may be
calculated by reference to land values or, on occasion, by reference to
a loss of profits; the claim may include abortive expenditure and
additional running costs. In some instances, notably discontinuance,
a full disturbance claim may arise. In most cases, professional fees
incurred in the preparation of the claim are recoverable.

Claims must be correctly submitted, usually within six months,
and are governed by regulations[2].

There are some instances where there is no prescribed form for
the claim, but where an Act states that particulars must be given it
is necessary to specify the amount sought. It can prove fatal to leave
the matter imprecise. "An amount to be agreed above £50" was
insufficient to satisfy the requirement of Part I of the Land
Compensation Act 1973: *Fennessy* v *London City Airport* [1995] 2
EGLR 167. This may be contrasted with the position under section
186 of the TP Act (stop notices), where there is no prescribed form
– see the text following study 15.

Not all activities require the approval of the local planning
authority. Section 55 defines development and lists several things
which do not constitute development at all. The GPDO gives
planning permission for a lot. The reader is referred to the order for
details.

Compensatable events

A claim for compensation may arise as a result of an adverse
decision with regard to any of the following. Disputes go to the
Lands Tribunal.

---

[2] T&CP General Regs 1992; Planning (LB & Conservation Areas) Regs 1990; and
others.

- An order revoking or modifying a planning consent.
- The refusal, or conditional grant, of consent for permitted development.
- An order requiring the discontinuance of a lawful use (whether an established use or one having planning consent).
- Certain refusals in connection with stop notices.
- The closure of a highway to vehicles.
- The imposition of restrictions on, or the revocation of, certain minerals consents.
- The revocation or modification of listed building consent.
- The imposition of a building preservation notice if the building is not subsequently listed.
- Consequential upon the preservation of ancient monuments.
- For additional protection of SSSIs, restrictions on agriculture and some drainage works.
- The revocation or modification of hazardous substances consent.
- The attachment of a blight notice.
- The attachment of purchase notice and/or a listed building or conservation area purchase notice.

Valuation principles

There are some principles of general application. The first concerns the definition of Schedule 3 value, formerly existing use value, which is a frequent ingredient in the after value used when calculating depreciation.

Schedule 3 development and the concept of existing use value

Any development will either fit within the descriptions set out in Schedule 3 or it will not. The distinction between these two groups has been central to the concept of existing use value ever since the Act of 1947, and still is. Third Schedule development formed part of the existing use value. Its descendant, Schedule 3, is a valuation schedule. It does not give planning consent but lists development for which consent may be assumed when valuing for compensation.

Development lying beyond the boundaries of the Third Schedule used to be called "new development", but the 1991 Act wiped the term "new development" from the legislation while preserving the distinction between Schedule 3 development and the rest.

The scope of Schedule 3 has also been amended. Part I contains two paragraphs. The first deals with rebuilding (plus an increase in the gross volume of 10%). This is subject to the additional restrictions imposed by Schedule 10, of which more later. The second paragraph deals with the conversion of any building used as a single dwellinghouse into two or more separate dwellinghouses and is not subject to Schedule 10.

Part II was repealed by the 1991 Act.

The term "existing use value" has been replaced by "Schedule 3 value".

Existing use value replaced by Schedule 3 value

The concept of an existing use value which includes certain notional development rights remains. It is the measure of those rights which is altered, and it is this change which is reflected by the change of name.

Schedule 3 value is of significance in compulsory purchase, in purchase notice cases and as the basis for calculating the depreciated value of an interest in some planning compensation cases. It is the value of the interest assuming consent for any development described in Part I of the Schedule, which is to say:

Para 1: Rebuilding subject to a maximum increase of 10% in the gross volume and to the restrictions of Schedule 10. The latter imposes a 10% limit on any increase in the gross floorspace, and no floorspace increase is permitted for replacement buildings.

Para 2: The use as two or more dwellinghouses of a building used as a single dwellinghouse.

So far as rebuilding is concerned, the replacement must be a replacement and not a wholly new creature bearing little resemblance to what went before, although new construction methods and some internal rearrangement – such as the omission of internal walls – could perhaps be claimed as falling within Part I.

The rebuilding work may incorporate an enlargement of not more than 10% in the gross volume, except for dwellings, where the maximum is the greater of 1, 750 cu ft or 10% of the volume.

Effect of Schedule 10 on Schedule 3 value

An "original building" can be regarded as the first building on the site, regardless of whether it was erected before or after July 1 1948. If and when it is rebuilt (after that date), it is the original which remains original, for the purposes of Schedules 3 and 10, and any increases in the volume or gross floorspace are measured by reference to that building and not to the replacement. It is important, therefore, to keep records of the original, as its ghost will survive into the future[3].

When an original building is first replaced, regardless of whether it was erected before or after July 1 1948, planning permission for an increase of not more than 10% in the gross volume and gross floorspace may be assumed for valuation purposes[4].

When a post-July 1 1948 replacement building has itself to be replaced or altered, Schedule 10 limits the amount of gross floorspace which may be used for any purpose in the new or altered building to the amount which was last used for that purpose in the replacement building immediately before it was replaced. This blocks second helpings of enlargement in terms of floorspace[5]. Schedule 3, paragraphs 1 and 13, block any increase in volume for such a second replacement.

Volume and gross floorspace are measured externally. Increases in the floorspace are measured by reference to the areas used in the original building. Where floorspace is used for more than one purpose, it is apportioned rateably[6].

Schedule 3 value and permitted development value

It is important not to confuse permitted development value with the classes of permitted development contained within Schedule 2 to the GPDO, and equally important to avoid confusion between these classes and the development described in Schedule 3.

There are times when the measure of compensation is the difference between permitted development value and Schedule 3 value – for example if the Secretary of State gives a direction as to the availability of some other consent in a purchase notice case. If this consent is less valuable than the Schedule 3 value, compensation payable for the difference and the purchase notice does not take effect.

---

[3] Schedule 3, para 13, and Schedule 10, para 5.
[4] Schedule 3, paras 1 and 10, with Schedule 10, para 1.
[5] Schedule 10, paras 2 and 5.
[6] Schedule 3, para 10 and Schedule 10, para 4.

Schedule 3 value is the value of an interest with the benefit of permission for the development specified in Schedule 3, Part I, subject to the restriction of Schedule 10 so far as rebuilding is concerned.

Permitted development value (PDV) is the value of an interest on the basis that planning consent will be given for development specified by the Secretary of State, but for nothing else, and that no other consents will be forthcoming.

Both Schedule 3 value and permitted development value are defined in section 144.

Heads of claim – the assessment of loss or damage

General principles

The statute will specify which, if any, of the six valuation rules of section 5 of the LCA are to be applied "modified as necessary". Commonly, one is required to value for sale in the open market, assuming willing parties. Where lost development value or lost developer's profit is at issue, one is not obliged to hypothesise a sale to a speculator: the claimant is in the market and may carry out the development: *Richmond Gateways Ltd* v *Richmond upon Thames London Borough* [1989] 2 EGLR 182, CA.

Claimants are duty-bound to minimise their losses. A compensation case is not an invitation to raid the resources of the compensating authority. Losses incurred in a genuine attempt to reduce losses are recoverable: *Pennine Raceway Ltd* v *Kirklees Metropolitan Borough Council* (1984) LT Ref/45/1980 (unreported), but see [1989] 1 EGLR 30.

Valuation evidence should be based on the analysis of market comparables. Where evidence is derived from the subject property, care must be taken to ensure that the transactions were genuine and at arm's length. The residual method is suspect unless the calculations were prepared for purposes other than the claim: they are much more credible if prepared in advance of the adverse planning decision – as part of the pre-purchase appraisal, for example.

Great care must be taken to avoid double counting – for instance, a claim for lost development value or profit must reflect the development costs without which that value could not be realised. This can require particular clarity of thought when claiming for abortive expenditure where planning permission has been modified.

The costs of compliance with other legislation must be included in the development costs – one is attempting to shadow what would have happened had there been no interference by the planning authority: *Church Cottage Investments Ltd* v *Hillingdon London Borough* [1990] 1 EGLR 205, CA.

Each case must be considered and valued according to its own facts. It is both dangerous and wrong to rely on earlier Tribunal decisions as precedent other than as giving an indication of possible methods and of how the law should be applied. It is invariably a mistake to take percentage deductions for risk, or other valuation data, from one case to another. Each case is unique so far as its facts and valuation evidence are concerned.

Depreciation in the value of the interest

The before-and-after approach is recommended. It is far more reliable to value the entire holding rather than just that part affected by the adverse decision, as this helps one to avoid mistakes such as double counting or the inadvertent omission of consequential losses such as severance.

The before value is that obtainable on the open market assuming that there had been no adverse planning decision. In a revocation case, for example, one assumes consent for the planning permission in hand prior to the making of the order.

The after value is the value subject to the adverse decision, with the benefit of any planning consents in hand, but the statutory provisions must be referred to in each case to discover which planning assumptions are to be made. In revocation, for example, one assumes consent for Schedule 3 development (as defined above).

The before and after approach has been criticised in connection with the calculation of injurious affection and severance, but was approved by the House of Lords in *Duke of Buccleuch* v *Metropolitan Board of Works* (1872) LR 5 HL 418.

Other direct losses

Where recoverable, these are what they say and may include disturbance. The loss must be a consequence of the adverse decision: see *Loromah Estates Ltd* v *Haringey London Borough Council* (1978) 248 EG 877, LT.

Depreciation and disturbance in the same claim

In *Horn* v *Sunderland Corporation* [1941] 2 KB 26, a compulsory purchase case, it was held that one may claim so as to receive the greater of development value on its own or existing use value plus disturbance. One cannot have both development value and disturbance.

In places, the planning compensation rules use the word "or" – sometimes in ways which imply "and/or". One must not forget the Golden Rule: the purpose of compensation is to put the claimant – so far as money can do it – in the position of not having been interfered with.

In planning cases, one must distinguish carefully between appreciation and depreciation in the value of an interest. In revocation, the original planning consent appreciated the land value, the revocation then depreciated it. If the claim is based on the recovery of the development value which came and went, one should not also claim for disturbance as "other losses" – *Horn* would prevent it.

Discontinuance, however, presents other variants and a more difficult picture. The order may require either a total cessation or merely a diminution of the established use. If the order results in a decrease in the land value, in the loss of business profits and in various expenses, clearly one will be entitled to claim for everything, including disturbance. However, an alternative planning consent may be forthcoming, either as part of the order itself or subsequently (following the service of a purchase notice) and its value must be brought into the reckoning.

Where appropriate, similar considerations may apply to other categories of compensation claim.

Loss of developer's profit

This can be an alternative to a claim for loss of development value, typically in a revocation case – one may opt for either, but not both. It is not to be confused with the loss of business profits or damage to goodwill.

Where a claim for loss of the developer's profit is made, one must recognise that such profits would have been speculative upon the successful completion and sale of the project which has been frustrated, and a deduction must be made for the risk and for the length of time it would take to secure the profit. The amount of any deduction for risk will vary according to the facts of the case; the better prepared the scheme, and the more advanced the work, the lower the risk. Another factor will be the amount retained (if any) against contingencies over and above that identified as being available for risk and profit. Thus, a carefully specified scheme for which tenders have been sought will contain less inherent risk than one which is only at an early stage of detailed preparatory work, while one for which a tender has been accepted will require even less.

Cases in which loss of profit was discussed or allowed include: *Excel (Markets) Ltd* v *Gravesend Borough Council* (1968) 207 EG 1061 & 1175, LT, loss of developer's profit following revocation of consent; *Hobbs (Quarries) Ltd* v *Somerset County Council* (1975) 234 EG 829, LT, damage caused by loss of profit from working the minerals rather than to the value of the minerals; *Burlin* v *Manchester City Council* (1976) 238 EG 891, LT – despite the appearance of a claim for lost profit, the claim was really for lost interest on the compensation, and that part was disallowed; *Cawoods Aggregates (South Eastern) Ltd* v *Southwark London Borough Council* (1982) 264 EG 1087, LT, damage to business following revocation – both depreciation in the existing use value of the land and loss of trading profit were allowed; *Richmond Gateways Ltd* v *Richmond upon Thames London Borough Council* [1989] 2 EGLR 182, no deduction for developer's profit need be made, since the claimant was a developer; *Pennine Raceway Ltd* v *Kirklees Metropolitan Borough Council* (1981) 258 EG 174, LT; (1983) 263 EG 721, CA: (i) licensee may claim CA (1982)[7], (ii) loss of projected trading profit for each of five years allowed, plus other matters including losses on an abortive attempt to mitigate loss by opening elsewhere[8], (iii) compensation not to be reduced for tax[9].

Interest on the claim

The 1991 Act[10] sets out the events from which interest will run. These are not restricted to planning compensation cases. The Tribunal has power to award interest. Rule 9 of the Lands Tribunal

---

[7] [1982] QB 382, 263 EG 721.

[8] [1989] 1 EGLR 30, CA.

[9] LT 1988 (1984 reissued consequential upon CA ruling re tax); CA [1989] 1 EGLR 30.

[10] Section 80 and Sched 18.

(Amendment) Rules 1997 governs the position and applies section 49 of the Arbitration Act 1996 to all proceedings.

Losses arising on other land

Provided the loss is a direct result of the adverse planning decision, a claim may be considered. A reduction in business profits is one example: see *Cawoods Aggregates (South Eastern) Ltd* v *Southwark London Borough Council* (1982) 1 EGLR 30, LT. Depreciation in the value of other land would be much harder to establish, although a revocation might have an adverse effect on the value of contiguous land. The key issues would be the physical relationship of the sites, the interest of the claimant in each of them, and the precise wording of the statute. The rules against remoteness would have to be met.

Professional fees

Valuer's and solicitor's fees are usually allowable. Note the singular!

Tax on compensation

Following *Pennine Raceway Ltd* v *Kirklees Metropolitan Borough Council* [1989] 1 EGLR 30, CA, it is now clearly established that compensating authorities must not reduce the compensation for tax just because they do not think the claimant will actually be taxed: tax is a matter between the Revenue and the claimant. Compensation may be so reduced only where it is clear "beyond peradventure" that the Revenue will not be levying a tax on the compensation and that the claimant could not have made the money ordinarily without paying tax, otherwise compensation is to be paid gross of tax and the tax accounted for to the Inland Revenue by the claimant. The question of tax is ignored in all the examples which follow.

Some compensating authorities deduct tax from interest payments unless the claimant produces a tax exemption certificate.

If the interference with the claimant's rights causes the loss of an exemption from tax, as happened in *Colley* v *Canterbury City Council* (1990) LT Ref/27/1990, it would be necessary to gross up the compensation to leave the claimant in the same position after paying the tax on the compensation as would have been enjoyed had the compensation been received without such liability for tax. In *Colley*, the revocation of consent for rebuilding a house which had already been demolished resulted in a capital gains tax charge incurred after the sale of what would otherwise have been the claimant's main residence.

In *Loromah Estates Ltd* v *Haringey London Borough Council* (1978) 2 EGLR 202 among other things, the claimant sought extra compensation to reflect a liability to pay development land tax. This liability arose for the first time in the period between the confirmation of the revocation order (May 1973) and the payment of the compensation (1978) – the appointed day for the Development Land Tax Act 1976 being August 1 1976. If the compensation had been paid before August 1976 DLT would not have been payable. This part of the claim was dismissed on the ground that the liability for DLT was a consequence of the passing of the DLT Act and not a consequence of the revocation order – in other words, the loss was too remote. It is more important to note that the local planning authority paid compensation in accordance with the valuation and not that sum minus the new tax liability.

Refusal of consent for development lying within Schedule 3 Part I

This does not rank directly for compensation, but section 137 allows the owner to serve a purchase notice.

Study 1

Purchase Notice following refusal of Schedule 3 development
In this case the land is rendered incapable of reasonably beneficial use.

A small freehold factory, built in 1930 and occupied by the freeholder, has been burned down. The local planning authority has refused consent for rebuilding as the site lies on the line of a proposed by-pass. The owner's purchase notice, served on the district council under section 137, has been accepted by the highway authority under section 139. The purchase price is to be assessed.

The site measures 18 m frontage by 45 m deep; the building had been of 600 m^2 gross floorspace and 2,200 m^3 gross volume. Industrial land is worth £900,000 per ha.

Site value assuming planning permission for a factory of 2,420 m^3: (1)
Schedule 3 value 810 m^2 at £90 per m^2 industrial site value (2)

Say	£73,000

Purchase price equals value of site: £73,000

Notes

(1) When considering whether land has become incapable of reasonably beneficial use, one must disregard any unauthorised prospective use. This is defined in section 138 (2) as anything involving the carrying out of development except that specified in Schedule 3, paragraph 1 (subject also to Schedule 10) and paragraph 2, but see the section about purchase notices at the end of this chapter.

(2) The land value should be found from the analysis of sales of comparable sites, reduced to some appropriate unit of comparison, and taking into account differences between sites. Only if such evidence is scant should a residual approach be used, working from the value of the completed development. It must be remembered that any valuations of comparable sites will have taken account of the plot ratios applicable to those sites, whereas all that can be claimed for the subject land is the right to rebuild the old factory plus 10%. The effect of this restriction should be reflected in the valuation.

(3) The purchase notice having been accepted, the valuation is as for compulsory purchase. See Chapter 9.

(4) The loss of the building itself ought to be an insurance matter.

(5) Paragraph 1 of Schedule 3 deals with an owner's "right" to rebuild, which is not the same as putting up a different type of structure by way of improvement. A distinction must be made between "original" and "non-original" buildings. An "original" building is the first building put up on the site. A "replacement" building is a replacement structure put up after July 1 1948.

   If rebuilding an "original" building (whether it is a pre-July 1948 structure or one first erected after that date), consent for the replacement to include an enlargement of not more than 10% gross volume and 10% gross floorspace may be assumed.

   Should it ever come about that the replacement building has itself to be replaced, it is the measurements of the old original building that remain relevant[11]. One has only one 10% increase no matter how often the building is rebuilt.

(6) The compulsory purchase rules apply to valuations for purchase notices – one is not restricted to Schedule 3 as the only source of planning assumptions. However, the reality of

---

[11] Schedule 3, para 13 and Schedule 10, para 5.

the situation where a purchase notice has had to be used may well be that Schedule 3 is the only (or best) planning assumption available. When considering whether or not the land has been rendered incapable of beneficial use, the planning assumptions are restricted to Schedule 3 development subject to the limitations of Schedule 10[12].

Study 2

Purchase Notice and Permitted Development Value (PDV)
This illustrates the calculation to be made when some worthwhile planning consent is available. In these circumstances the owner retains the site and is compensated for its loss in value.

The same factory as in study 1, but in this case placed in a residential area. Planning permission for its reconstruction has been refused on the ground that a factory is a non-conforming and unacceptable user.

The owner's purchase notice has been sent to the Secretary of State under section 139 and he has directed[13] that planning permission for three houses shall be granted if applied for.

Industrial land is worth £900,000 per ha and residential building plots are worth £20,000 each.

The compensation to be paid would be assessed as follows:

	£
Schedule 3 Value: (1)	
Industrial site, 810 m^2 at £90 per m^2 site value.	
Assume consent for factory of 2,420 m^2	72,900
Less: Permitted Development Value (PDV) (2)	
3 residential building plots at £20,000	60,000
Compensation (3), under section 144(2)	£12,900

Notes

(1) The Schedule 3 value is calculated by reference to an assumed consent for Schedule 3 development, paragraph 1 (rebuilding), subject to the restrictions of Schedule 10. The factory being an original building, one has the right to assume permission for a 10% increase in the volume and gross floorspace.

---

[12] Section 138.
[13] Under section 141(3).

(2) The PDV is calculated on the assumption that no consent would be given other than in accordance with the Secretary of State's direction: see section 144(6).

(3) The claimant need not be the freeholder. "Owner" is defined in as the person entitled to receive the rack-rent[14].

Study 3

Schedule 3: Replacement building

This shows the position where a pre-1948 structure which has already been replaced once after July 1948 has again to be rebuilt.

In this case the 1930 factory was of 2,200 m³ (600 m²) gross, but it was replaced in 1950 with a similar structure of 2,420 m³ (660 m²) designed to support overhead gantry-cranes. Planning policy has since been reversed and the area is now zoned for residential use. Two years ago the replacement factory was burned down and planning consent for rebuilding was refused. The subsequent history is the same as in study 2.

The site is of 810 m², but there is no comparable evidence to determine industrial site value. Residential site value is £36,000 per plot and permission for two houses would be granted.

The compensation would be assessed as follows:

Schedule 3 value: (1)

Gross development value (GDV)
660 m² gross, say 630 m² at £100 pa per m²
(full rental value)

	£ pa	£	£
Net income	63,000		
YP perp at 9%	11.11		420,000
Less: costs of development:			
Construction 660 m² at £575 per m²,			
Say		379,500	
Site clearance, Say		16,200	
		363,300	
Professional fees at 12%		43,596	
		406,896	

---

[14] Section 138.

Interest on £203,448, 1 year at 12%	24,414	
Legal and agents' fees at 3% of GDV	21,000	
Developer's profit at 15% of GDV (4 & 5)	105,000	
		332,353
		142,620
PV £1 in 1 year at 12% (6)		0.8929
Schedule 3 value, industrial site value, Say		127,345
Less: Permitted Development Value (PDV)		
PDV, 2 buildings plots at £36,000		72,000
Compensation equal to depreciation		£55,345

Notes

(1) Schedule 3 value and PDV are defined in section 144(6). Schedule 3 value takes account of development within Schedule 3, subject to the provisions of paragraph 13. This makes the 1930 building the original building, not the 1950 replacement, and so bars any attempt to increase the volume to 2,662 m³ (with a corresponding increase in the floorspace).

(2) If the factory had been erected for the first time after July 1 1948, it would still be permitted to assume a 10% increase when calculating Schedule 3 value because section 144(6) makes no distinction between buildings first erected before or after July 1 1948.

(3) Valuations which rely on the residual method are not readily accepted by the Lands Tribunal. The lack of hard market evidence of site value in this study is artificial, but see Note (5) below.

(4) An owner who could show that, if given permission to rebuild, the work would be put in hand for owner-occupation, could well avoid the item. See Note (5).

(5) The question of whether a deduction for developer's profit need be made will be answered by the facts of the case. Some valuers may argue that it ought always to be included as a cost when the average open market site value is being sought. Others may prefer to say (where they can justify it) that the owner would rebuild without reference to a speculative developer and accordingly need not deduct the latter's allowance for risk and profit. The omission of this deduction would, of course, increase the estimated depreciation, which is the measure of the compensation in these cases. In *Richmond Gateways Ltd* v *Richmond upon Thames London Borough Council*

[1989] 2 EGLR 182, CA, a Part II case, the claimant property company successfully maintained that it would have carried out the development itself, and the compensating authority lost its argument for the inclusion of developer's profit as a cost. Much will depend on the facts of the case. A further consideration will be the measure of the compensation. In some instances, the measure is the loss suffered by the claimant and one must then take the claimant as he or she is found. In other cases, as here, the compensation is the depreciation in the value of the interest, and in such cases an open market approach would be applicable. What would the market have paid for the claimant's interest on each of two sets of planning assumptions? The difference is the compensation. The general market would have regard to the need to allow for risk and profit, but under the six rules of compulsory purchase valuation (which apply by virtue of section 117: see section 144(5)), the claimant is deemed to be in the market as a prospective purchaser and may therefore be able to refute the argument as to retentions for risk and profit, particularly as here where the rebuilding would be an insurance matter in which neither retention would feature.

(6)   This represents an allowance for the interest on the purchase of the land. Strictly, the other costs of purchase (stamp duty and legal fees) have also to be deducted.

Refusal of development not coming within Schedule 3

There is no right to compensation of what used to be known as New Development, but if, following a planning refusal or the conditional grant of permission, land is found to be incapable of reasonably beneficial use, a purchase notice may be served. This topic is discussed at the end of the chapter.

Rather than confirm a purchase notice the Secretary of State may instead indicate the availability of some alternative consent. If this happens, and the alternative development is less valuable than the Schedule 3 value of the land, compensation for the difference becomes payable.

Study 4

Alternative consent following a Purchase Notice
Land is rendered incapable of reasonably beneficial use, but the

Secretary of State promises an alternative consent instead of confirming the purchase notice.

The law is contained in sections 137–144.

No 7 is an old, dilapidated, detached cottage adjoining some shops but protruding beyond the rear line of the pavement. The owner has repeatedly tried to sell it in its present condition but has been unable to do so because the council is threatening to make a demolition order.

No 10, a comparable cottage, was renovated by its owner at a cost of £65,000 and later sold for £125,000, but it was never in so serious a state of disrepair as No 7. Further down the road a vacant building plot with consent for a small house fetched £22,000 at a recent auction.

Following the refusal of planning consent for material changes to the external appearance of No 7, the owner served a purchase notice on the district council under section 137 which the council refused to accept. The Secretary of State has directed that consent be given for a modern house set back to the building line. The amount of compensation is to be assessed.

Compensation would be assessed as follows:

	£	£
Schedule 3 value: (1)		
Value with the right to rebuild and/or extend the cottage by up to the greater of 1/10th its cubic content or 1,750 cu ft (2)		132,500
Less: costs of renovation and extension (3)		81,000
Schedule 3 value		51,500
Less: Permitted Development Value: (1)		
Single vacant building plot, value based on comparable	22,000	
Less: cost of demolition and clearance net of salvage	4,500	
Permitted development value		17,500
Compensation equals depreciation		£34,000

Notes

(1) See section 144(6) for the definitions of Schedule 3 value and PDV.

(2) Value based on the sale price of No 10 with an allowance for the increase in size allowed to No 7.

(3) In practice these should be accurately costed, and detailed.

(4) The owner must claim his compensation from the local planning authority: see section 144(2).
(5) Had the Secretary of State confirmed the purchase notice then the ordinary compulsory purchase valuation rules would apply and the owner would receive £51,500 for his property: see Chapter 9.

Compensation for the revocation or modification of a planning consent

Sections 97–100 deal with the compensation. Sections 137–148 have relevance if, as a result of the authority's section 97 order, a purchase notice is served.

Where consent derives from an order such as the GPDO, but a direction requires that specific consent be obtained, compensation is available under section 107 if that consent is refused or is granted subject to conditions.

Section 117 applies to cases coming within Part IV. This attaches the six valuation rules of section 5 of the LCA 1961, "modified as necessary", and deals with cases where an interest is mortgaged. Part IV encompasses revocation and discontinuance (including mineral consents) and the apportionment and repayment of that compensation.

Schedule 18 to the 1991 Act provides for the payment of interest from the date of the section 97 order for revocation and modification cases and from the date of the refusal of planning permission for cases where permitted development rights have been reduced or taken away following the making of an Article 4 direction.

The claimant need only be a "person interested in the land". It is a nice point whether – for compensation purposes under section 107 – there is any difference between a person "having an interest in land", and one who is "interested in the land". The last is the form of words used in section 107: see *Pennine Raceway Ltd* v *Kirklees Metropolitan Borough Council* (1982) 263 EG 721, CA. In this case the owner of a licence with an enforceable right against the freeholder was able to claim compensation. A prospective purchaser who owned an enforceable option to purchase would be entitled to claim. A contractor employed by a person entitled to claim may have to proceed against the latter for damages for breach of contract rather than against the planning authority concerned, but in this event the damages would be added to the compensation claim itself as being "loss or damage which is directly attributable to the order".

The most reliable method of calculating depreciation is to value the interest before and after the revocation or modification of consent. If consent has been revoked, the after value assumes consent for Schedule 3 development only: see section 107(4) and *Burlin* v *Manchester City Council* (1976) 238 EG 891, LT. If the revoked consent had itself been for Schedule 3 development, the after value is nevertheless to be prepared on the assumption that consent is available for Schedule 3 – see *Colley* v *Canterbury City Council* [1993] 1 EGLR 182, CA – even though this strict interpretation of section 107(4) eliminates the claim for depreciation. Other losses and fees can, however, still be recovered. The remedy is to follow the revocation claim with a purchase notice. See study 7 and note (2) to study 5.

Compensation for the revocation, modification and discontinuance of mineral working is complex. A detailed explanation is given in the author's *Handbook of Land Compensation*[15]. See also chapter 11.

Study 5

Revocation of planning permission
A straightforward case where planning permission is revoked before the land is sold.

The owner of a house and orchard employed a surveyor to obtain planning consent for a house on part of the orchard plot and to sell that land with the benefit of the consent.

The surveyor obtained planning permission and received a genuine and fair offer of £40,000 for the plot, but before this could be accepted the consent was revoked and the offer withdrawn. The surveyor charged £300 for obtaining the consent and £1,100 in connection with the attempted sale.

The freeholder's claim would be as follows:

	£
1. *Depreciation in value because of the revocation of consent*: (1)	
Site value with permission for one house	40,000
Less: value subject to section 97 order (2), Say	7,500
Depreciation	32,500

---

[15] Please see Further Reading at the end of this chapter.

2. *Plus*: Abortive expenditure	
Cost of obtaining consent	300
Agent's fee for the attempted sale	1,100
Total claim	£33,900

Plus: professional fees for preparing the claim.

Notes

(1) See section 107 for heads of claim.

(2) The value subject to the order is to be calculated on the assumption that consent would be given for Schedule 3 development, subject to the provisions of Schedule 10. This is the only consent to be assumed in this case. If some other consent remained alive, it is arguable that its value should be brought in as reflecting the true after value. One would do so in a modification case if the surviving consent was the remnant of the original. A truly independent consent is another matter. If it and the revoked consent were not for mutually exclusive developments, then its value would also be present in the before valuation and might be left out of both the before and after calculations, provided there was no financial or physical linkage between the two schemes. If, however, the alternative consent is a true alternative rather than a companion, to include its value in the after valuation would seem contrary to section 107(4) and would reduce the depreciation below that actually attributable to the revocation.

(3) Note that any compensation received for the depreciation in the value of an interest is registered as a compensation notice and may be repayable on the grant of a subsequent planning permission: see sections 107 to 111.

(4) The costs of all professional fees for plans and obtaining planning consent are recoverable under section 107(2) even though these may precede the grant of the planning consent: see *Burlin* v *Manchester* (above) and earlier cases, in particular *Holmes* v *Bradfield Rural District Council* [1949] 1 All ER 381.

(5) Had the owner of the plot, instead of attempting to sell the land, set about having the house built for his own use, then upon revocation he would be entitled to claim such direct losses as compensation to a builder for breach of contract. This would be in addition to compensation for the depreciation in the value of his interest.

(6) Losses claimed, including depreciation and/or loss of profits,

need not themselves have arisen on the site that is the subject of the revocation or modification order, provided they were directly caused by the order. The test is that the loss must not be too remote: there must be causation: see *Cawoods Aggregates (South Eastern) Ltd v Southwark London Borough Council* [1982] 2 EGLR 222 in which the Member said, ". . . There seems . . . to be nothing in section 164 of the 1971 Act which limits loss to the land which is the subject of a revocation order."[16]

Study 6

Modification of planning consent

This shows the calculation of a claim for a developer who had started work but whose planning permission was modified before the development had been completed. Loss of profit is taken into account.

The developer bought 1 ha of land with outline consent for 28 houses some years ago for £370,000. He later obtained full approval and started work. An estate road was built, a sewer laid and construction started on 10 houses. The local planning authority, responding to intense local pressure, revoked the consent on 0.25 ha where no work had been done. The district council has offered to buy this part of the site for £6,000 to add to the village green, otherwise the land must remain undeveloped as amenity land.

The freeholder's claim would be as follows:

Value before revocation order:

	£	£
GDV 0.25 ha with consent for 7 houses (1), 7 houses at £200,000		1,400,000
Less: costs of development:		
Construction, 7 houses at £108,000	756,000	
Site works, sewer, road etc. Say	55,000	
	811,000	
Architect's and quantity surveyor's fees	107,400	
	918,400	
Interest on £501,200, 2 years at 11%	106,580	
Legal and agent's fees at 3% of GDV	42,000	
Developer's risk and profit, 5% of GDV (2)	70,000	

[16] Section 164 is now section 107.

	1,136,980
	263,020
PV £1 in 2 years at 11%	0.8116
Value before order, Say	213,467
Less: Value subject to the order	6,000
Direct loss	£207,467

Plus: abortive expenditure (see Note (3)), professional fees in connection with plans (Note (4)) and professional fees in connection with the preparation of the claim.

Notes

(1) The 0.25 ha could be valued either as here or by valuing the 1 ha in its totality both before and after the order. A problem could arise if there is some abortive expenditure to identify, as care must be taken to avoid double counting: see note (2), Study 7.

(2) The compensation under section 107 is to be full compensation for loss or damage. Subject to the rules against remoteness and double-counting, this can include profits, provided they are reasonably close at hand and certain. Generally the development value of land reflects the possibility of profit from a purchaser's point of view.

   The profits in this example are close at hand, and this is allowed for by deducting only 5% for developer's risk instead of the 15% used elsewhere.

(3) The sewer laid on the 0.75 ha part of the site may now be larger than is strictly necessary. Whether or not this extra cost could be recovered is arguable. It could probably be if none of the developer's profit came into the compensation, but since the profit on the 0.25 ha could only be made by providing the services, and as that profit has been included in the compensation, it is not thought reasonable to claim for the excess capacity in the services as well: see section 107(3). There would be a valid claim for abortive expenditure if the sewer had to be relaid.

(4) If the 0.25 ha was no longer capable of agricultural or other beneficial use, and had the council not wanted to buy it, a purchase notice could be served.

Compensation for the refusal of permitted development

Section 108(2) imposes a 12-month time-limit for claiming compensation following the revocation or modification of a planning consent originally granted by a development order. Subsection (4) allows regulations to exclude permitted development for demolition from the requirements of section 108.

A planning authority may, with the consent of the Secretary of State, make a direction under Article 4 of the GPDO to strike out some or all classes of permitted development. The direction may be made applicable to an entire district or to a single property. Similar powers may be attached to Special Development Orders. The effect is to remove the relevant consent(s), so requiring those whose properties have been affected to seek specific planning consent. If that consent is refused or is granted subject to conditions different from those imposed by the development order, the applicant has been disadvantaged by comparison with those whose properties do not come within the scope of the direction. Compensation may, therefore, be claimed under section 107 (modification or revocation of planning consent). Compensation is calculated in exactly the same way as for any revocation or modification case, as shown in studies 5 to 7. This basis is the same regardless of whether the lost consent had been for development within or beyond Schedule 3: the development order originated a consent which has been revoked or reduced.

Schedule 18 to the 1991 Act provides for the payment of interest from the date of the adverse planning decision which denied or reduced the permitted development.

Study 7

Compensation where a purchase notice follows a revocation order
The revocation of planning permission, the land being rendered incapable of reasonably beneficial use.

The owner of a piece of scrubland sought planning permission for one house. This was granted, but, before offering the land for sale, the owner, in the hope of receiving a better price, spent £6,750 on demolishing a derelict shed and partly laying a drain. The consent was, however, revoked before this work was completed.

Comparable unimproved building plots are worth £45,000. The owner paid £500 to the land agent who obtained the consent. Deprived of its consent the land is worthless, save perhaps as a car park for the parish hall.

A purchase notice has been served on the local council and accepted. The purchase price is agreed at £4,500, the claim is to be prepared.

Compensation under section 107(1) (see Note 1):

	£
Value with consent for one house (2)	
Unimproved site value, Say	45,000
Less: Value subject to the revocation order (3)	4,500
Depreciation	40,500
Plus: Compensation under section 107(2), fees	500
Section 107(1) and (3), expenditure on work rendered abortive	6,750
To be received from the local planning authority	47,750

Compensation under section 137 (see Note 4)	£	
Purchase price (4)	4,500	
Less: if compensation is paid for expenditure on abortive work the value of that work (5),		
Say	1,000	
To be received from the district council	3,500	3,500
Total claim		£51,250

Plus the costs of preparing the claim

Notes
(1) See sections 107, 137, 144.
(2) This is unimproved site value. The use of the improved site here would lead to double counting if the cost of the works (£6,750) is also claimed. The work would not be abortive if it raised the site value from unimproved to improved value.

    In practice, evidence of value based on comparables would have to be carefully scrutinised to distinguish between improved and unimproved sites.

    It would be open to the claimant to opt for whichever calculation gave the most compensation. The choice here is between unimproved site value plus abortive expenditure on one hand and improved site value only on the other. While it might be expected that when completed the site works would add more than their cost to the value of the site, it is more likely that in a partially completed state the cost incurred would exceed the value added.

(3) See section 107(4). In addition to any real planning permission in hand, planning consent for Schedule 3 development (and only Schedule 3) is to be assumed. In this study the assumption of consent for Schedule 3 does not help. The existing use value is established at £4,500.

(4) A purchase notice valuation follows the ordinary compulsory purchase rules, modified as directed by the Act. Section 144 is the starting point for the valuation rules. This brings in section 117 if an alternative consent is given: see subsections (2) and (5). See Chapter 9 for details of compulsory purchase valuations.

(5) Where compensation is paid for expenditure on work under section 107, compensation on acquisition is reduced by the value of the works in respect of which compensation was paid under section 107[17]. Section 107 pays compensation based on the expenditure, whereas section 144(1) deducts the value of the works. There is no reason to suppose that cost and value are equal.

Whether fees paid for under section 107(2) are to be deducted from the purchase price is not clear from section 144(1).

The key question may be whether the "value with consent" used in the calculation of depreciation under section 107 takes account of the value of the work, for if it does it may be possible to argue that the work was not abortive. In many cases, however, it is likely that the cost of preliminary work will not be met by an increase in value. Carelessness here in the preparation of the claim could leave the claimant out of pocket when the total received is compared with the apparent intention of section 107, which is to pay full compensation. Perhaps the intention of section 144(1) is merely to avoid double payment.

Compensation for discontinuance orders, including those requiring some reduction in the intensity of a use

Section 102 authorises the making of a discontinuance order, sections 115 to 117 deal with the compensation, and section 137 with the possibility of serving a purchase notice. Schedule 18 to the

---

[17] See section 144(1).

1991 Act provides for the payment of interest from the date the damage is suffered or the expense is incurred.

If a purchase notice is resorted to, section 144(7) bars the recovery of section 115 compensation for the discontinuance order if the notice results in the purchase of the claimant's interest. It is also barred if the availability of an alternative consent is indicated and compensation is paid under section144(5) and 117. A purchase notice cannot be used if a valuable consent is attached to the discontinuance order: see under purchase notices below.

Where a planning permission is included with the discontinuance order, a distinction may have to be made between a consent which authorises the existing use but in a restricted way and one which is for some development that is entirely different. If the value of a permission attached to the order is less than the value prior to the discontinuance, there is no difficulty: one has suffered depreciation and the claim would allow both that and disturbance. If, however, the alternative consent produces a higher land value than the established (pre-discontinuance) use, then there is an appreciation in the development value. Whether this could bar disturbance in a case involving a total cessation of the business and its forced removal elsewhere is open to doubt, bearing in mind that discontinuance is one occasion when a landowner may have a planning permission thrust at him without having made an application. One relevant consideration may be whether the enhancement in land value actually covers the costs of disturbance: one goes back to the basic principle that compensation is to put the claimant in the position, so far as money can do it, of not having been interfered with. One thing is clear: the totality of the compensation is a single entity, no matter how that total is arrived at, and this suggests that disturbance is not immune from downward adjustments.

There are special rules for the assessment of compensation in the case of mineral workings which are the subject of a discontinuance order. Details are given in *The Handbook of Land Compensation*.

Study 8

Discontinuance Order requiring a reduction in the use
In which a discontinuance order requires a reduction in the intensity of use which in turn depreciates the value of both freehold and leasehold interests.

The tenant and the freeholder of a factory near a residential area have been served with a section 102 order requiring a reduction in the factory's working from 24 to 15 hours per day. The tenant is willing to comply, but the loss of night production will necessitate alterations to some of the plant at a cost of £270,000 and profits will fall by 25% from their present average of £1,100,000 pa before tax.

The tenant holds on a lease having five years to run at a net rent of £216,000 pa. The current net full rental value is £396,000 but with reduced operating time this will fall to £324,000 pa.

The compensation claims for freeholder and tenant are set out on page 348.

Notes

(1) A lower rate is used because the income is secure.
(2) If, as a result of the order, the investment was sufficiently changed to justify a change in the yields, this should be done. The term here is still secure, although secured on a reduced profit rent.
(3) Any other losses that the claimant could show would also be recoverable. An example might be interest on the excess capacity of plant and machinery now used less intensively.

Study 9

Discontinuance order with alternative consent
In which a discontinuance order extinguishes a use, requires work to be done and grants planning permission for some alternative development on the site.

A coal yard adjoining a residential area and in use for many years is operated by the freeholder. Part of the work done in the yard includes the mechanical grading and bagging of coal.

A discontinuance order has been confirmed, requiring the cessation of the business on the site and the removal of the old machinery and stacks of coal. It also gives planning permission for four houses on the land.

In its present use the site is worth £243,000. If cleared for housing it is worth £180,000. A suitable alternative site for the business could be bought for £324,000. This site is two miles away.

The freeholder's compensation claim is as follows (page 349):

Study 8

Freeholder's Claim
Valuation of freehold interest prior to the order

	£ pa	£	£
Net Income	216,000		
YP 5 years at 10½% (1)	3.7	799,200	
Reversion to full net rental value	396,000		
YP perp deferred 5 years at 11%	5.4	2,138,400	2,937,600
Less: Value of freehold interest after Order (2)			
Net Income	216,000		
YP 5 years at 10½% (1)	3.7	799,200	
Reversion to full net rental value	324,000		
YP perp deferred 5 years at 11%	5.4	1,749,600	2,548,800
Freeholder's total claim			388,800

Plus Professional fees for preparing the claim

Tenant's Claim
Value of leasehold interest prior to the order

	£ pa	£
Rent received, full rental value	396,000	
Less: Rent Paid	216,000	
Net Profit Rent	180,000	
YP 5 years at 13% and 4% (Tax at 40%)	2.285	411,300

Thereafter Landlord and Tenant Act lease at FRV

	£ pa	£
Less: Value after the order		
Rent received, full rental value	324,000	
Less: Rent Paid	216,000	
Net Profit Rent	108,000	
YP 5 years at 13% and 4% (Tax at 40%)	2.285	246,780

Thereafter Landlord and Tenant Act lease at FRV

	£ pa	£
Depreciation		164,520
Plus:		
Loss of profit (3)	275,000	
YP Say	3	825,000
Cost of complying with the order, alteration to plant		270,000
Tenant's total claim		1,259,520

Plus Professional fees for preparing the claim

Study 9

	£	£
Depreciation (1)		
Value prior to the order	243,000	
Less: value subject to the order (2)	180,000	
Depreciation		63,000
Plus: Cost of complying with the order (3)		
Demolition of buildings etc.	36,000	
Removal of fuel stocks	6,500	
		42,500
		105,500
Less: salvage value received: (4)		
Hardcore etc. from demolition of buildings		4,700
		100,800
Plus: Disturbance: (1) (5)		
(a) Loss on forced sale of plant and machinery: (5)		
Value to incoming tenant	36,000	
Less: scrap value	5,400	
		30,600
(b) Other Losses, including temporary and/or permanent loss of profit, any reasonable increase in distribution costs (capitalised), duplicated expenditure, *Crawley* costs, etc. (6) (7) (8) Say		72,000
Total compensation claim		£203,400

Plus professional charges incurred because of the order.

Notes

(1) See section 115(2).
(2) Cleared site value with planning permission for seven houses.
(3) See section 115(3).
(4) See section 115(4). Note that the salvage is to be deducted form the total claim and not just from the costs of compliance.
(5) See Chapter 9 for details of disturbance claims.
      So far as the plant and machinery is concerned, the loss is the difference between the scrap value achieved and what would have been paid for it by an entrepreneur coming in to operate the business on the site. It has been assumed that it would not be economic to remove the machinery to the new site. A claimant must always minimise his loss.

(6) For these items of claim, only an indication of some likely ones has been given and a spot figure assumed. It is important to avoid double counting. The cost of removing the coal stocks, required by the order, has already been claimed. Any other removal expenses, however, could be claimed here.

It is also important to exclude items which are not allowable. The £45,000 difference in site values between the two coal yards would not, for instance, be recoverable. The reason for this is that the claimant will be deemed to have received value for money in the purchase of the new yard – it is up to him if he buys a better one.

(7) See *K&B Metals Ltd* v *Birmingham City Council* (1976) 240 EG 876 regarding compensation for activities between the making and the confirmation of the order. A section 102 order does not take effect until confirmed.

(8) For a further example of a valuation following a section 102 order, in which one use was extinguished and an alternative consent given: see *Raddy* v *Cornwall County Council* (1967) 230 EG 976 (or *Valuers Casebook* (R W Westbrook), Vol. 2 p 65).

Study 10

Compensation where a purchase notice follows a discontinuance order
In which a use is extinguished and the land is rendered incapable of reasonably beneficial use.

0.25 ha of land on the edge of a heathland nature reserve has been used since before 1964 as a dump for scrap metal (see note 1), a dealer paying the freeholder the full rental value of £1,000 pa for the use of the land.

A discontinuance order requiring the cessation of the use and the removal of the scrap has been confirmed. No alternative consent has been given. Neither the freeholder nor the dealer has other land in the vicinity. As bare heathland the site is worth £1,000.

The freeholder's claim is as follows, but see note (2):

	£ pa	£
Existing use value		
Open storage: (4)		
Net income	1,000	
YP in perp at 14%	7.2	7,200
Purchase price if the purchase notice is upheld		£7,200

Plus professional fees.

Notes
(1) The use is established regardless of whether planning permission was obtained or not: see section 191. If a limited consent had been granted the effect of the limitation would have to be taken into account.

(2) The freeholder has the choice of whether to claim under section 115 or to serve a purchase notice under section 137. The latter would seem the better choice. This would preclude a claim by him under section 115, but the scrap dealer's rights to claim under section 115 would not be affected. Note the word's "of that order" in section 144(7).

(3) The scrap dealer may claim under section 115 regardless of whether his interest is sufficient for him to satisfy section 137(2)(b) or not. Section 115 allows a claim to be made by any person who has suffered damage as a consequence of the order; section 137(1)(c) and (2)(b) allow a purchase notice to be served by any person entitled to an interest in the land.

(4) The valuation rules under section 137 are those of the Land Compensation Act 1961, sections 5, 14 to 16 etc.

One of the values thereby allowed is Schedule 3 value, plus disturbance if applicable.

In this study a "nil" section 17 certificate of appropriate alternative development is assumed. See chapter 9.

(5) If the freeholder had been running the business on the site, he would have the same choice between section 115 and section 137. If he chose the latter he could not recover the cost of complying with the order, but would sell as if his interest were being compulsorily acquired. The disturbance compensation would depend upon the availability or otherwise of alternative accommodation. The value of the site as a scrap dump would form part of the total compensation.

Compensation for listed buildings

The owner of a listed building has two hurdles to jump before work may go ahead: to obtain listed building consent, and to get planning permission if the work amounts to development. There is no compensation for the refusal of listed building consent nor for the refusal of planning permission. Compensation is recoverable for the revocation or modification of listed building consent, but no claim may be made for abortive expenditure on work undertaken in advance of the consent. It is also available if a building preservation notice is served but the building is not subsequently listed.

Schedule 18 to the 1991 Act provides for the payment of interest from the date of the section 23[18] order or from the date a building preservation notice is served.

The LB Act provisions apply to conservation areas, but much is then deleted by the LB Act section 75 and by regulations. The overall result is that demolition in conservation areas requires conservation area consent, and compensation is available if that consent is subsequently revoked or modified (unless the revocation is unopposed).

Any building or structure within the curtilage of a listed building is treated as part of the listed building provided it is attached to the land and has been so since before July 1 1948[19]. Listed building consent is, therefore, needed for work to these buildings

If a building is listed, section 7 of the LB Act requires that specific consent be obtained for any work of demolition, alteration or extension which would affect its character as a building of special architecural or historic interest. This is all-embracing and wider than the definition of development contained in section 55. It means that specific consent must be obtained even if the proposed work fits within section 55(2)(a) or if consent might already appear to have been given by Schedule 2 to the GPDO.

So far as dwellinghouses are concerned, the development permitted by Schedule 2, Part 1, Class A, of the GPDO is excluded if it would consist of or include the erection of a building within the curtilage of a listed building: in these cases the right to erect a building within the curtilage would derive from Class E, which limits the volume to no more than 10 m³.

Apart from the above restrictions, there is no general exclusion of permitted development rights for the enlargement, alteration or improvement of a dwellinghouse which is a listed building. Those works still require specific listed building consent under the LB Act, section 8, and permitted development rights cannot be exercised in its absence without giving rise to an offence under section 9.

If the specific refusal of permitted development rights (via the making of an Article 4 direction) ran with a refusal of listed

---

[18] LB Act, revocation order.
[19] LB Act, section 1(5).

building consent, some scope for argument might exist about remoteness and the true cause of the loss.

Compensation for the revocation or modification of listed building consent

See LB Act, section 28. The provisions are similar to those of section 107 except that there is no equivalent to section 107(4).

There is no compensation if LB consent is revoked or modified by agreement. Such modification is commonplace where it becomes necessary to change the terms of an original consent in the light of fresh discoveries about the building made during the course of the works: what was originally envisaged and approved may turn out to be impossible or too damaging, so that an alternative solution has to be agreed. The best advice is to foster close co-operation between the owner, the builder and the local planning authority.

Schedule 18 to the 1991 Act provides for the payment of interest from the day of the order.

Study 11

The modification of listed building consent
Planning permission was obtained for the division of a listed building into three houses and for the construction of three garages. The conversion work involved changes to the facade of the building. The consent was modified to allow the internal division of the building into two units with a common front door.

The freeholder's claim for compensation:

Value with the original listed building consent: (1)

	£	£	£
3 houses at, say, £144,000			432,000
Less:			
Costs of conversion		170,000	
Cost of 3 garages		18,000	
Architect's fees		23,000	
			211,000
Value			221,000
Less:			
Value with consent as modified: (2)			
2 houses at say, £180,000		360,000	

Less:			
Costs of conversion	145,000		
Cost of 3 garages	18,000		
Architect's fee	20,000		
		183,000	
Value		177,000	177,000
Depreciation			44,000

Plus: professional fees rendered abortive, calculated by apportionment of the fees actually incurred (3):		
Abortive element, Say		2,200
Total claim		£46,200

Plus: professional fees for preparing the claim.

Notes

(1) It is irrelevant whether the consent is a listed building consent or otherwise. In either case a planning consent has been revoked or modified, and the consequential loss is recoverable. Compare section 107(1) with LB Act, section 28(1).

(2) Note there is no mention of Schedule 3 in LB Act, section 28. The after value will be the value of the interest on the open market subject to the revocation or modification of the listed building consent – which may or may not have involved the removal of development lying within Schedule 3.

(3) The modification would involve further work by the architect, but not all his preliminary work would be wasted. Compensation is due for the extra fees involved if this is the true measure of the cost of the modification to the owner.

(4) For simplicity, interest charges, developer's profit etc. have been omitted. These issues are discussed in earlier studies to do with the revocation of planning consent.

Compensation for the imposition of a building preservation notice

A building preservation notice under the LB Act, section 3 has the effect of temporarily listing a building for six months, after which it will lapse. It also lapses when and if the building is listed or the Secretary of State notifies the local planning authority that he does not intend to list it.

If the building is not subsequently listed, compensation under section 29 becomes payable for all direct loss or damage. This includes any compensation paid by the claimant to a third party for breach of contract caused by having to discontinue or countermand works: see section 29(2) and (3).

If the building is listed, there is no compensation under section 29. Listed building consent would have to be sought in the normal way. The owner of an unlisted building who, acting properly, started to carry out permitted development under the GPDO, but then received a building preservation notice would have to stop work and seek listed building consent – or wait. If the building is subsequently listed, the owner would suffer a loss – perhaps for breach of contract, perhaps for depreciation, or both. Had the loss resulted from a planning refusal following the making of an Article 4 direction, compensation would have been recoverable as if consent had been revoked. However, sections 29 and 28 do not appear to provide for compensation under these circumstances.

Schedule 18 to the 1991 Act provides for the payment of interest from the date the building preservation notice is served.

Study 12

The imposition of a building preservation notice
In which a building preservation notice is imposed but the building is not subsequently listed.

The owner of an old house obtained approval under the Building Regulations for extensive internal alterations at an estimated cost of £10,000. He was served with a building preservation notice forbidding any alteration, but the building was subsequently not listed. By the time the building preservation notice ceased to have effect the builder's estimate had risen by 15% and house values had fallen by 3%.

The claim may be contentious if the parties cannot agree what is – or is not – the direct and reasonable consequence of the order. The most basic claim may appear thus:

Cost of work at time notice ceased to have effect	£11,500
Less Cost of work at time of revocation	£10,000
Direct loss	£1,500
Add Other direct loss or damage	
Losses attributable to delay, Fees etc.	

Compensation: the additional cost of the work, £1,500 (15% of £10,000) plus other direct loss or damage, including fees.

Notes
(1)   The compensating authority may argue that the fall in house values is in no way attributable to the imposition of the building preservation notice. LB Act, section 29(2) awards compensation for any loss or damage directly attributable to the effect of the notice.

   See *C&J Seymour (Investments) Ltd* v *Lewes District Council* [1992] 1 EGLR 237: the claimant buys and sells in the same market and is disadvantaged only if deemed to sell in a lower-priced market than that in which the replacement or reinvestment is purchased – as happened in *West Midland Baptist (Trust) Association (Inc)* v *Birmingham Corporation* [1970] AC 874.

(2)   A contrary view may be that because the compensation is for all direct loss or damage, the fall in the value may be contended for – but the market change is not a directly attributable consequence.

(3)   If the alterations reduced the running costs of the building, then the additional running costs incurred as a result of the enforced postponement would come within section 29(2) provided they were not too remote. Losses included in the claim would have to pass the test of being a direct consequence of, or not too remote from, the service of the preservation notice.

Conservation area consents

The listed building provisions are transferred to apply to conservation areas but many have been deleted by LB Act, section 75, and by regulations.

   The result is that demolition in conservation areas requires conservation area consent, and compensation is available if this is subsequently revoked or modified, unless that change is unopposed. The basis of compensation is the same as for the revocation or modification of a listed building consent. This is the only conservation area compensation.

Purchase notices and listed buildings

See the LB Act, sections 32 to 37. The general rules are similar to other purchase notice cases – see studies 7 and 10. The test is that the building and land have become incapable of reasonably beneficial use in their existing state and cannot be rendered capable by the execution of works for which listed building consent has been granted or promised and that the use of the land is subsequently inseparable from the building (so that the two should be treated as a single holding). When considering what might be regarded as a reasonably beneficial use, prospective development outside Schedule 3 is to be disregarded, and so is any work requiring listed building consent if that consent has not been promised by either the local planning authority or the Secretary of State.

The LB Act, section 48 restricts the service of a listed building purchase notice if a repairs notice has been served but does not utterly prevent it.

Sections 49 and 50 set out the valuation rules. Under normal circumstances, listed building consent may be assumed for the alteration of the building or for its demolition in connection with Schedule 3 development, except in so far as compensation may have already been paid under section 27 for any refusal. (The 1991 Act has repealed section 27.)

If the building is being acquired under the minimum compensation rules of section 50, which apply where the building has been deliberately allowed to fall into disrepair, the planning assumptions are that planning permission would not be granted for any development or redevelopment and that the only listed building consent available would be for the restoration and repair of the building.

Conservation area purchase notices

These are as for listed building purchase notices where the incapacity is caused by refusal, revocation or modification of consent in prescribed circumstances.

Compensation derived from tree preservation orders

Trees are dealt with in Part VIII of the Act. Sections 197 to 202 are concerned with the powers and duties of the local planning authority to preserve trees and woodlands in the interests of amenity, including the making of tree preservation orders.

Trees in conservation areas are protected: sections 203 to 205.

Schedule 18 to the 1991 Act provides for the payment of interest from the date of the refusal of consent or its grant subject to conditions.

Requirements as to replanting, enforcement and penalties are in sections 206 to 210; the availability of injunctions in section 214A, of rights of entry in sections 214B to 214D.

There are two relevant sets of regulations. Both require, *inter alia*, that a tree preservation order is made substantially in the form of the Model Order set out in their schedules. Both govern what may or may not be done, and deal with procedures.

The Town and Country Planning (Trees) Regulations 1999 (SI 1999 No 1892) apply only to tree preservation orders made on or after August 2 1999, whether confirmed before or after it. They substantially reduce the measure of compensation, but do not affect orders made before this date.

The old regulations, which apply to all tree preservation orders made before August 2 1999 are The Town and Country Planning (Tree Preservation Order) Regulations 1969 (SI 1969 No 17). The Town and Country Planning (Tree Preservation Order) (Amendment) and (Trees in Conservation Areas) (Exempted Cases) Regulations 1975 (SI 1975 No 148) also apply.

A tree which is the subject of a tree preservation order may not be topped, lopped, felled or otherwise destroyed without the specific consent of the local planning authority unless it is dying, dead or has become dangerous. A tree preservation order does not survive the grant of planning permission for development if work to the tree is immediately required for the purpose of carrying out that development. There are other exceptions: see sections 198 and 200 (forestry operations on land subject to a forestry dedication agreement) and the Model Order.

Trees in conservation areas are similarly protected, in effect by a blanket tree preservation order, but specific consent need not be obtained from the authority before a prohibited act is undertaken – one must give the authority six weeks' notice of intent. The authority may then either consent, impose a specific tree preservation order, or say nothing. In the latter event, protection will lapse for two years. Exceptions are made to this protection by the Town and Country Planning (Tree Preservation Order) (Amendment) and (Trees in Conservation Areas) (Exempted Cases) Regulations 1975 (SI 1975 No 148).

Compensation is available under section 203 to anyone who suffers loss or damage as a consequence of the refusal of consent under an order or the grant of conditional consent. It is also available under section 204 where an order is made requiring the replanting of woodland which was felled in the course of permitted forestry operations, and the Forestry Commissioner decides, on forestry grounds, that no grant shall be made. The calculation of any compensation should take account of any money already paid for the same trees and of injurious affection (if any) which would have been caused to the value of the land had the trees been felled.

Exceptions to the availability of compensation are where the authority certified that:

- its decision was made in the interest of good forestry, or
- the tree(s) have a special or outstanding amenity value.

Compensation under section 203: Refusal of consent to fell

Where a tree preservation order is in force, compensation for loss or damage caused by the refusal of consent (or by the imposition of conditions attached to a consent) may become payable, but subject to any exceptions or conditions specified in the order. The revocation or modification of consent would also be compensatable.

Compensation under the 1999 regulations, section 203 cases

The new arrangements are very different from the old, are much less generous and no longer cover all loss or damage. The claimant must prove that the loss is caused by the refusal, or conditional grant of consent.

That done, the rules are different for refusals to fell in connection with forestry operations and for everything else.

In connection with forestry, only the owner of the land may claim[20], and the compensation is restricted to the depreciation in the value of the trees attributable to any deterioration in the quality of the timber because of the refusal[21]. Section 11(3) to (5) of the Forestry Act 1967 applies to the assessment of compensation as it does to the refusal of felling licences under section 10 of that Act.

In all other cases, the claim must exceed £500. No claim may be made for the loss of development value, nor for other depreciation in the value of land. This reverses a number of Lands Tribunal decisions, and cases decided under the old regulations must be treated with caution. Development value is defined as any increase

---

[20] This need not be the freeholder.
[21] Model Order Art 9(3) and (5).

in value due to the prospect of development, and development here includes land clearance, so compensation ceases to be available for the loss of any increase in the value of land by converting woodland to some agricultural use, as well as the more obvious losses which may be caused where the refusal of consent reduces the density of development.

There is more. An applicant must give a statement of reasons for making the application, and where the refusal – or conditional grant – of consent results in a loss which was not reasonably foreseeable at the time of the decision (having regard to the statement of reasons and any evidence put forward to support it), no claim may be made.

No claim may be made for foreseeable loss or damage where the claimant fails to take reasonable steps to mitigate.

Compensation under the old regulations, section 203 cases

The measure of the compensation is the loss or damage caused by the adverse decision. Its calculation by means of a before-and-after valuation would be a reasonable starting point if the land value has been affected, but losses and costs unrelated to land value must be assessed separately.

In *Bell* v *Canterbury City Council* [1988] 1 EGLR 205 the Court of Appeal ruled that loss or damage included damage to the value of the claimant's interest in the land and that the loss arose at the moment the consent was refused rather than at the time the order was made. In *Bell* the loss was the difference between the value of the land as woodland and as grazing. The council also argued, in the context of what is now section 203, that if they had granted consent they would also have imposed a replanting requirement. This would not, however, affect the overall compensation, since section 204 also carries compensation rights.

In *Fletcher* v *Chelmsford Borough Council* [1991] 2 EGLR 213, LT a claim followed the refusal of consent to fell a mature tree which was considered to be endangering the foundations and stability of a nearby house. The opinions of an aboricultural consultant and the planning authority were in conflict, so the owner commissioned a more detailed report which involved measuring the movement of a wall for a year. Although insurers paid for the subsequent underpinning, the authority denied liability for the cost of the report on the ground that it was too remote because the report had been deliberately commissioned, and months after the refusal of consent to fell at that. The Lands Tribunal held that the cost of the report was properly incurred as part of the process of counteracting the effects of a known hazard on the land.

In *Deane* v *Bromley London Borough Council* [1992] 1 EGLR 251, the claimant was given permission to reduce the height of 26 mature boundary trees after a large limb had fallen from one on to adjoining land. The council imposed a condition which required that the work be carried out by an approved contractor, which was done at a cost of £1,437.50. The claimant sought compensation on the basis that he and his wife maintained their garden and, but for the condition, would have done the work for nothing. The council argued that the claimant benefited from the work, being under an obligation to see that the trees were safe, and denied that there was any loss. The Tribunal awarded £977.50, being the total cost less what the claimant would have spent hiring additional equipment and the wear and tear thereon (£499 plus VAT).

In *Bollans* v *Surrey County Council* (1968) 209 EG 69, the Lands Tribunal discussed several important issues and concluded that:

1. Damage or expenditure consequential on a refusal under a tree preservation order is analogous to loss or damage directly attributable to a revocation of planning permission.
2. Loss of profit may be too remote if the use for the profitable purpose had still to be started at the time of the revocation or order.
3. It would not be double counting to allow a claim which included both the value of the tree as timber and the profit from its subsequent disposal if the claimant operated both the growing and retailing businesses from which those profits came.

Study 13

Refusal of consent to fell (old regulations)

In which consent is refused for the conversion of woodland, the subject of a tree preservation order, to pasture.

The freeholder of a farm sought consent for the felling of six acres of woodland, the subject of a tree preservation order made under section 198, in order to convert the land to pasture. A ministry grant would be forthcoming. The local planning authority have refused consent. The land would be worth £2,000 per acre as pasture or £500 per acre as woodland, inclusive of the value of the trees. Compensation is payable under section 203.

Value had consent been given (1)

	£	£
6 acres pasture at £2,000	12,000	
PV £1 for 3 years at 4% (2)	0.8889	
		10,667
Less: cost of reclamation	2,700	
PV £1 for 1 year at 4% (3)	0.962	
		2,597
Less: proceeds from sale of trees	1,400	
		1,197
Value with consent for reclamation		9,470
Less: Value subject to refusal of consent		
6 acres woodland at £500		3,000
Depreciation in value		6,470
Plus: consequential losses and disturbance		
Value of any grant towards cost of reclamation (4), Say	800	
PV £1 for 1 year at 4% (5)	0.962	
		769
Total claim		£7,239

Plus: professional fees for preparing the claim.

Notes
(1) See *Bell* v *Canterbury City Council* (1986) 279 EG 767 for a case in which many of the principles are discussed and on which this study is based.
(2) The deferral reflects the time which must elapse following grubbing out before the pasture is fully established.
(3) The Lands Tribunal considered that the reclamation costs should be deferred for one year in *Bell's* case but did not defer the sale of timber. The report does not discuss the reasoning, but presumably the timber would be sold in advance of reclamation.
(4) The cost of reclamation used (£2,700) is the full cost. If a grant towards the cost of the work is likely to be forthcoming – for example, from the Ministry of Agriculture, Fisheries and Food – then its amount should be brought into the calculation, either as here or by using the net-of-grant cost of reclamation (£1,900) in place of £2,700 in the earlier part of the valuation.
(5) The Tribunal did not show deferral of grant for one year in *Bell's* case, adding it directly as a consequential loss, perhaps because the grant figures were not disputed. It would, however, seem consistent to discount this for the same period

as the reclamation costs, since the grant would not be paid until the work had been carried out. Failure to discount would result in different final totals between the two approaches.

Replanting Directions: Compensation under section 204

Where a local planning authority direct that land cleared in the course of forestry operations be replanted and the Forestry Commission decides not to make a grant or loan under section 1 of the Forestry Act 1979 because the direction frustrates the use of the land for commercial forestry, then compensation is payable for any loss or damage due to the direction to replant. If under the 1999 Regulations, the claim would have to be for at least £500, and no claim may be made for any diminution in the value of the land.

Schedule 18 to the 1991 Act provides for the payment of interest from the date the direction requiring replanting is given.

There is, of course, no right to compensation attached to the automatic obligation, contained in section 206, to replant trees removed in contravention of a tree preservation order.

Trees and Purchase Notices under section 198

Where a tree preservation order is in force and refusal of consent concerning the trees renders the land incapable of reasonably beneficial use, a purchase notice may be served: see section 198(3)(c) and (4)(b).

Tree felling does not come within the definition development in section 55. When considering whether land has beneficial use, one must disregard the prospect of any unauthorised prospective use of the land[22]. Once it has been established that following the refusal of consent the land has become incapable of reasonably beneficial use, the normal compulsory purchase code applies.

Purchase notices are considered later in this chapter.

Study 14

Purchase Notice following refusal of consent to fell

Land rendered incapable of reasonable beneficial use in its existing state by the imposition of a tree preservation order.

---

[22] Defined in section 138(2).

A freehold site within the confines of a village's residential area contains sufficient gnarled pine trees to prevent its use. The trees are attractive and are the subject of a tree preservation order. Consent for felling has been refused.

Land surrounding the site is occupied by good-class housing, some set among trees and some with trees confined to the plot boundaries.

If limited felling had been allowed it would have been reasonable to expect planning permission for four houses. If only the boundary trees were kept, eight houses could be fitted on to the site. The plot values would be £26,000 and £16,000 respectively, both based on comparable but cleared sites.

It is accepted that the land has been rendered incapable of reasonably beneficial use by the decision. It could, if cleared, have been used without carrying out any development lying beyond the scope of Schedule 3, either for horticulture or planted up as an orchard. A purchase notice has been accepted.

Assessment of the compensation will follow the normal compulsory purchase rules, sections 14 to 17 of the LCA governing the planning assumptions. These are examined in Chapter 9. If planning permission for residential development could have been expected the claimant will receive residential site value.

The claim would be calculated on the site value disclosed by the planning assumptions, disregarding the tree preservation order. If consent could be expected for houses set among trees the price would be £104,000 (four plots worth £26,000), but £128,000 if it would be reasonable to expect consent for housing with a boundary screen of trees (eight plots worth £16,000).

See *Eardisland Investments Ltd* v *Birmingham Corporation* (1970) 214 EG 1579 in which a purchase notice was served and accepted following the imposition of a tree preservation order and the refusal of planning permission for residential development.

The inference of the report seems to be that the tree preservation order (at least in this case) made no material difference to the density of development that might be expected having regard to the development plan and with a section 17 certificate substituted for the private open space shown. The report is not as clear as one would wish. What does seem clear is that the valuation might be undertaken in two stages: first, the calculation of the site value subject to the normal planning expectations but as if the tree preservation order did not exist and then the calculation of the value subject to the restrictions of the order. The difference between these two figures would be the loss attributable to the order. The unencumbered site value would, of course, have regard to the development plan. This would leave the valuer with the figure for depreciation because of the order – which might be recoverable under section 203 (depending upon the wording of the order) – and a value on which to base a claim for compensation if a purchase notice should be the appropriate remedy.

Compensation for restrictions on advertising

Regulations are made under sections 220 and 221. Section 222 gives certain planning permissions, section 224 deals with enforcement, section 223 with compensation for advertisements in place before August 1 1948, and section 220(2)(c) and (3)(b) with purchase notices. The regulations are the Town and Country Planning (Control of Advertisements) Regulations 1992 (SI No 666, amended).

Control over advertisements is exercised by the regulations. Advertisements which satisfy their requirements do not need planning permission, but the regulations may specify classes of advertisements which do require consent. There is no compensation for the outright refusal of consent.

Compensation is available in only two cases, as set out below.

1. The revocation or modification of an express consent for an advertisement

Express consent may be revoked or modified under regulation 16 and is compensatable under regulation 17. An express consent is one which has been granted by the local planning authority or the Secretary of State, as distinct from one which is granted by the regulations.

Compensation is available to any person suffering loss or damage. The measure is the amount of expenditure rendered abortive (including expenditure on the preparation of plans, etc), plus any other direct loss or damage, but excluding depreciation in the value of any interest in land.

The claim must be made within six months of the date on which the order is approved.

2. The removal of a prohibited advertisement or the discontinuance of an advertising station.

Regulation 8 allows the local planning authority to serve a discontinuance notice affecting advertisements having a deemed consent.

Under section 223 and regulation 20, the forced removal of an advertisement which was on display on August 1 1948, or the discontinuance of the use of land as an advertising station if so used since that date, is compensatable.

Compensation is limited to the expense of removing the advertisement or of discontinuing the use of the land. It is available to any person carrying out works to comply with the regulations. It must be claimed within six months of completing the work. Schedule 18 to the 1991 Act provides for the payment of interests from the date the expenses are incurred.

Compensation for stop notices

Sections 183 and 184 deal with the power of a local planning authority to serve a stop notice and section 186 with a claimant's right to compensation. Sections 185 and 187 deal with service and penalties. Schedule 18 to the 1991 Act provides for the payment of interest from the date the stop notice is served.

There is no prescribed form for a claim under section 186, but the claimant's letter must, if it is to count as a claim, make it clear that a claim is being made, not that one will be made at some future date. Neither the amount of compensation nor other details need be given at this stage: see *Texas Homecare Ltd* v *Lewes District Council* [1986] 1 EGLR 205.

Compensation becomes available only in the following circumstances:

1. The withdrawal of the stop notice, for whatever reason.
2. If the supporting enforcement notice is quashed on the ground that planning permission ought to have been granted.
3. If the enforcement notice is varied or withdrawn in similar circumstances.

There is, therefore, no compensation if planning consent is subsequently given or if planning conditions are lifted. Compensation is also excluded for the prohibition of an activity which was, or contributed to, a breach of planning control at the time the notice was in force. Neither is it available for loss or damage which could have been avoided had the claimant responded properly to the local planning authority's notices. Where this happens compensation may be reduced to the extent that the damage was avoidable.

Where entitlement to compensation is established, any person having an interest in, or occupying, the land at the time the notice was first served may claim for:

1. Any loss or damage directly attributable to the prohibition contained in the notice, or
2. If the enforcement notice is varied, for loss or damage due to the prohibition of any activity which ceases to be prohibited on the variation of the enforcement notice – but subject to the exclusions described above.

Claims may include sums paid for any breach of contract caused by the need to comply with the stop notice's prohibitions.

See *Robert Barnes & Co Ltd* v *Malvern Hills District Council* [1985] 1 EGLR 189 for a case which raised a number of heads of claim, including interest charges. Interest on the purchase of the affected site was allowed. A claim for interest on the postponed receipt of development profits was disallowed because the claimant failed to show that the profits were any less than they would have been without the stop notice. It was the claimant's failure to establish the extent of the loss, not the concept of claiming interest on postponed profits, which was defective.

See *J Sample (Warkworth) Ltd* v *Alnwick District Council* [1984] 2 EGLR 191 for a review of the question of remoteness of loss. The meaning of "directly attributable" is not qualified by the concept of "reasonable foreseeability".

See *Graysmark* v *South Hams District Council* [1989] 1 EGLR 191. In this case the Lands Tribunal (upheld by the Court of Appeal) allowed claims for interest on the cost of the development which had been stopped and for the deferment of the profit which would have been made had there been no stoppage. Claims for worry, loss of repute with bankers and the like were held to be too remote.

Study 15

Stop Notice compensation
Compensation for a stop notice attached to an enforcement notice which is quashed on appeal.

A householder enlarged his house to the full extent permitted by Schedule 2 Part 1 Class A of the GPDO. Some time later, and without planning permission, he contracted with a local builder for the erection of a brick-built hobbies room. This was partly constructed when an enforcement notice, coupled with a stop notice, was served on him. He appealed against the former under section 174(2)(c) on the ground that the room came within Part 1 Class E. Two years later he won his appeal but the cost of the work had risen from the original £10,000 to £13,500. In addition he spent £425 protecting the part-built structure.

Assessment of the compensation:

Direct loss or damage (1)	£
Increase in the cost of construction	3,500
Expenditure on protective work	425
Compensation	£3,925

Plus: Interest from the service of the Notice (2) and professional fees for the preparation of the claim, but not of the appeal (3).

Notes

(1) Section 186(2) specifies that compensation is for "any loss or damage directly attributable to the prohibition . . .". Subsection (4) specifically allows the recovery of payments made in respect of breaches of contract caused by the obligation to comply with the stop notice.

(2) Schedule 18 to the 1991 Act provides for the payment of interest from the date on which the stop notice is served.

(3) The costs of the appeal against the enforcement notice are not recoverable, since that notice does not stem from the stop notice but precedes it. See *Barnes* v *Malvern* and *Sample* v *Alnwick*, above.

(4) It is arguable that interest should be claimed for the loss of any money spent on the work prior to the stop notice, since this has effectively been tied up without either interest or use for the length of time taken by the appeal.

Compensation for the closure of a highway to vehicles

A local planning authority may apply to the Secretary of State for an order banning vehicles from a highway (other than a principal or trunk road) so as to improve the amenity of the area[23]. The status of the highway is then reduced to that of bridleway or footpath, as happens when a street is turned into a pedestrian precinct.

---

[23] Under section 249.

Subsections (3) and (4) allow for exceptions to the ban, in terms of vehicle type, particular persons or time of day.

Section 250 deals with compensation for anyone having an interest in land which has lawful access to the highway.

The measure of compensation is the depreciation in the value of the interest directly attributable to the order, plus any other direct loss or damage. This would include the costs of making the claim.

Schedule 18 to the 1991 Act provides for the payment of interest from the date of the section 249 order.

A simple before-and-after valuation would determine the depreciation. Direct loss or damage could include increased operating costs, loss of profits (if any), and so on, provided these were directly attributable. If the scheme improved the amenity of the area, any claim for losses would need careful preparation. There appears to be no provision for set-off, nor for the repayment of compensation if the scheme is subsequently done away with.

In *Ward* v *Wychavon District Council* [1986] 2 EGLR 205, LT no compensation was awarded, the new access arrangements being found, as a matter of fact, better than those which were stopped up. In *Saleem* v *Bradford City Metropolitan Council* [1984] 2 EGLR 187, the claimant argued that falling profits were due to the closure of the highway to vehicles, but the valuation evidence was poorly assembled and inconsistent and the Lands Tribunal found that the failure of the business was due to inexperience and excessive borrowing. See also *Leonidis* v *Thames Water Authority* [1979] 2 EGLR 8 for a successful claim for damage to a business following the closure of one end of a road for a year (albeit under the Public Health Act).

Compensation for the preservation of ancient monuments

The Ancient Monuments and Archaeological Areas Act 1979 has been considerably amended by the National Heritage Act 1983. The 1979 Act deals with the protection of ancient monuments, guardianship agreements, the acquisition of monuments (including by compulsory purchase), adjoining land and easements, and with compensation.

A compensation claim may arise in one of three ways:

1. *Following the refusal of scheduled monument consent or its grant subject to conditions*[24]:

---

[24] 1979 Act, section 7.

To give rise to a claim the loss must stem from the refusal of scheduled monument consent for work within one of three categories:

(a) Work in connection with development for which express planning permission was given (ie not by way of a development order) prior to the scheduling of the monument. This permission must itself be effective when the scheduled monument consent is applied for. A permission that has time-expired will not do.

(b) Work which is not development at all[25] or which is permitted development under a development order. Even so, no compensation is payable if the work would result in even the partial destruction or demolition of the monument, unless that work is incidental to agriculture or forestry[26].

(c) Work that is necessary for the continuation of the existing use(s) to which the monument was being put immediately before the application for scheduled monument consent. Such use(s) must not be in breach of any legal restrictions.

Where an express planning consent is frustrated by the refusal of scheduled monument consent (or its grant subject to conditions), compensation is not recoverable if the planning consent was granted after the monument was scheduled. A planning authority may, of course, refuse planning consent in order to protect a monument.

So far as compensation is concerned, the link between monument protection and planning control is akin to that between listed building consent and planning permission in general. In *Hoveringham Gravels Ltd* v *Secretary of State for the Environment* (1975) 235 EG 217 it was held that the value of the claimant's interest had been damaged chiefly by the failure to obtain planning permission to work minerals rather than by the imposition of a preservation order which merely frustrated agricultural permitted development under the GPDO. The compensation was, therefore, severely limited. This position has now been reinforced, in that compensation is not payable if the implementation of a planning permission is frustrated by the refusal of scheduled monument consent if that planning permission was granted after the monument was scheduled.

---

[25] See section 55 of the 1990 Act.
[26] Section 7(2) and (4).

The measure of any compensation under section 7 is the amount of expenditure or other loss or damage suffered by a person having an interest in all or part of the monument. It will include the depreciation in the value of the interest – a before-and-after valuation in accordance with section 5 of the LCA, for which see chapter 9, and professional fees.

Section 8 of the 1979 Act deals with the recovery of compensation if consent is subsequently granted.

2. Compensation following the cessation of authorisation of works affecting a scheduled monument following the modification or revocation of scheduled monument consent.

Section 9 of the 1979 Act deals with this, the compensation being for expenditure on works rendered abortive because further work has ceased to be authorised, and for any other loss or damage directly due to this cessation, including the depreciation in the value of an interest in all or part of the monument. The entitlement to compensation is based on sections 107 and 108 of the 1990 Act (revocation orders). Professional fees would be recoverable.

3. Compensation for damage caused by the exercise of certain of the powers contained within the 1979 Act.

A claim may be made if the exercise of the powers of entry and investigation contained within sections 6, 26, 38, 39, 40 and 43 results in damage to land or to any chattels thereon.

Compensation in connection with hazardous substances

The HS Act 1990 has been amended by the Environmental Protection Act 1990 and by the 1991 Act.

Compensation may arise out of either of two events: the revocation or modification of any hazardous substances consent under section 14(1) and, in defined circumstances, the revocation or modification of consent following an application for its continuation under sections 17 to 19.

The HS Act admits other powers of revocation, but these do not carry any entitlement to compensation: see HS Act, sections 13 and 24.

Revocation or modification of hazardous substances consent by order under section 14(1)

Section 14 allows a hazardous substances authority to revoke or modify consent. The reason for so doing must lie within either 14(1)

– that, having regard to any material consideration, it is expedient
– or 14(2), which specifies four cases that may be summarised as
change of use or abandonment. There is no compensation if section
14(2) is used.

The use of the expediency provision of section 14(1) opens the
way for a claim under section 16. There are three possible
components: depreciation, disturbance and the expense of works
necessary to comply with the order. The calculation of any claim
will be under section 117 and 118 of the TP Act as if the compensation was being calculated for the purposes of a discontinuance order
under section 115.

Any person who has suffered damage as a consequence of the
order may claim for either or both:

1.   The depreciation in the value of an interest;
2.   Disturbance.

Any person who carries out works of compliance may recover the
reasonable expenses so incurred from the authority.

Schedule 18 to the 1991 Act provides for the payment of interest
from the date of the section 14(1) order.

The revocation or modification of hazardous substances consent following an application for its continuance under section 17

Sections 17 to 19 refer. The revocation power is contained within
section 18, and compensation is payable under section 19. Schedule
18 to the 1991 Act provides for the payment of interest from the
date of the revocation or modification.

There are no specific rules for claiming compensation – as to
either time-limits or method.

Section 6(2) states that a hazardous substances consent enures for
the benefit of the land to which it relates unless the terms of the
consent provide otherwise. Such consent is, however, automatically
revoked if there is a change in the person in control of part of the
land, unless prior application is made by him to the hazardous
substances authority for the continuation of that consent after the
change. If such an application is made, the authority – in addition
to agreeing to the change – have the right to modify or revoke the
consent, but if they do other than agree they must then pay to the
person who was in control of the whole of the land before the
change, compensation for any loss or damage sustained by him
which is directly attributable to the modification or revocation of

the consent. If the person who is in control of the land omits to apply, in advance, for the hazardous substances consent to continue after the change of control, then no compensation will be payable for the automatic revocation of consent which will follow that change of control.

Compensation for restrictions on mineral workings

The 1991 Act, the T&CP (Minerals) Regulations 1995 (SI 1995 No 2863) and the T&CP (Compensation for Restrictions on Mineral Working and Mineral Waste Depositing) Regulations 1997 (SI 1997 No 1111) have made significant changes. Matters are complicated. Details can be found in *The Handbook of Land Compensation*. See also chapter 11.

Compensation for statutory undertakers

Sections 262 to 283 provide for compensation to be paid to statutory undertakers on a special basis under a variety of circumstances which include the refusal, conditional grant, revocation or modification of planning permission on operational land, the imposition of a requirement to remove or re-site apparatus, the extinguishment of a right of way, and compulsory purchase.

In the case of compulsory purchase, a statutory undertaker may elect to waive the special basis of section 280 and instead to invoke section 281 and have matters dealt with in the ordinary way.

Not all planning refusals are compensatable.

A distinction is made between operational and non-operational land. The list of those deemed to be statutory undertakers varies: it depends upon the circumstances as to whether one is "in" or "out". For example, telecommunications operators are not statutory undertakers but are intermittently included as such.

The compensation and valuation rules are set out in sections 279 to 282.

For claims coming within section 279(1), Schedule 18 to the 1991 Act provides for the payment of interest from the date of the decision under section 266 (planning refusals) or of the order under section 97 (revocation or modification of consent). For claims under section 197(2) (extinction of a right or imposition of a requirement under section 271) interest is payable from the date the right is extinguished or the requirement imposed.

For the refusal, or conditional grant, of planning consent to be compensatable, it must have been for what would have been permitted development but for the making of a direction and is not development which has received specific parliamentary approval within the meaning of section 264(6).

The right to compensation is established by section 279 and the special basis for calculating it is in section 280. The basis is that compensation is recoverable for:

1.   The reasonable expenses of any acquisition made necessary.
2.   The same, of works made necessary.
3.   The loss of profits, calculated in accordance with subsection (3).
4.   The reasonable expenses of removing apparatus as required by an order under section 279(2) or (3) – extinguishment of a right vested in a statutory undertaker or the imposition of a requirement. Subsection (2) deals with statutory undertakers and flows from section 271. Subsection (3) applies to tele-communications operators and flows from section 272. Claims for interest appear to be excluded so far as the tele-communications operators are concerned, there being no mention of section 279(3) in Schedule 18 to the 1991 Act.

Compensation for losses under the Wildlife and Countryside Act 1981

Areas of Special Scientific Interest (SSSI)

Section 28 of this Act provides for the Protection of areas of special scientific interest, notice being served on the owner or occupier. The effect is that any operations specified in the notice must not be carried out without the Nature Conservancy Council having been given prior notice – the latter may then enter into an agreement, which may involve making payments to the owner or occupier. There is no right to compensation under section 28.

Nationally important SSSIs: Nature Conservation Orders

Section 29 provides additional protection for sites of special scientific interest which are of particular national or international importance, designation being by way of a Nature Conservation Order prohibiting specified operations. Compensation is available under section 30(2) to any person having an interest in land comprised in an agricultural unit for depreciation in the value of that interest. The compensation is calculated by reference to the value of the interest subject to, and free from, the order: see *Cameron v Nature Conservancy Council* [1992] 1 EGLR 227 – the first reference under the Wildlife and Countryside Act 1981.

Section 29 permits the council to prevent any specified operations, so that the owner or occupier seeking to execute them must notify the council. Subsections (5)–(7) extend the period for making an agreement from three to 12 months and permit compulsory acquisition. Compensation is available under section 30(3) to anyone with an interest in the land at the time the notice was given who has, as a result of the barring of specified operations under subsection (5), as modified by subsections (6) or (7), either:

1. Reasonably incurred expenditure which has been rendered abortive, or spent money on work which has been rendered abortive, or
2. Has suffered direct loss or damage (excluding damage to the value of an interest in land).

Interest is recoverable from the date of the claim until payment. The relevant regulations are the Wildlife and Countryside (Claims for Compensation under section 30) Regulations 1982 (SI 1982 No 1346).

Compensation for Planning Blight

The law is to be found in sections 149 to 171 and Schedule 13.

Where property is adversely affected by the prospect of an impending acquisition for public works, the owner may serve a blight notice on the appropriate authority requiring it to acquire early. Generally, it must be shown that no sale could be achieved on the open market except at a substantially reduced price. The events which permit the use of a blight notice are specified in Schedule 13 to the 1990 Act.

The recipient authority may serve a counternotice challenging the validity of the blight notice for one or more specified reasons. The terms of the authority's counternotice are definitive and the date of that notice is the material date. Thus, in *Burn v North Yorkshire County Council* (1991) 63 P&CR 81, the authority could not avoid a blight notice where the scheme came to be abandoned after the date of the counternotice.

If a counternotice is served the matter will – unless resolved – go to the Lands Tribunal. If the Tribunal upholds the blight notice, the valuation rules are as if the relevant compulsory purchase procedure had arrived at the notice to treat stage. The reader is referred to chapter 9.

The claimant must have a qualifying interest. These are defined as:

1. Resident owner-occupier (freeholder or tenant with three years' unexpired term) who has been in occupation for not less than six months.
2. Owner-occupier of any property with a net annual value of not more than £24,600. (The limit was raised from £18,000 on April 1 1990)[27].
3. Owner-occupier of an agricultural unit with six months' occupation of the whole.

The six months' rule can also be satisfied if property has been unoccupied for not more than 12 months, and six months' occupation occurred prior to the property becoming vacant.

With two exceptions, the property must first be tested on the open market by attempted sale. No sale need be attempted where a compulsory purchase order is in force, or if compulsory acquisition is authorised by a special enactment.

Blight notices do not bar home loss or farm loss payments. A valid blight notice is a deemed notice to treat and attracts the full compensation provisions of the appropriate acts: see chapter 9. A disturbance claim under section 5(6) of the Land Compensation Act 1961 is not excluded. The argument sometimes put forward, that the claimant brought the disturbance on himself and should not be compensated, it not accepted. The blight was brought on the claimant by the inclusion of the property in the specified descriptions and became the source of loss to hand when the owner decided to sell – whether selling by choice or of necessity. The statute does not provide for the exclusion of disturbance, and accordingly the vendor under a blight notice is in no different a position than one who sells at the instigation of the authority itself.

Section 157 provides for compensation to be reduced for listed buildings in disrepair and for slum clearance cases. This stops blight notices being used as a way round these provisions.

Full details and a review of the leading cases can be found in *The Handbook of Land Compensation*.

---

[27] T&CP (Blight Provisions) (England) Order 2000 (SI No 539).

Purchase notices and "reasonably beneficial use"

Section 137 enables an owner to serve a purchase notice on the district council if, following either the refusal of planning permission or its grant subject to conditions, his land is "rendered incapable of reasonably beneficial use in its existing state".

Purchase notices may also be served following revocation or discontinuance orders and in several other situations, including in connection with listed buildings[28]. In discontinuance cases, the claimant need only be entitled to an interest as distinct from being the owner.

Once a purchase notice has been served and not contested, or confirmed by the Secretary of State, the procedure is the same as if a notice to treat has been served and compensation follows the compulsory purchase rules – see chapter 9. Home loss, farm loss and disturbance claims are not excluded.

The refusal of consent does not have to reduce the value of the land, indeed an outright refusal can never do this, since planning permission need never be sought merely to continue what is already lawfully established. (An exception is if planning permission was originally given for a limited time and that time has expired. Apart from this one case the only way in which a lawful or established use can be upset is by the service of a discontinuance order.) It follows, therefore that since the refusal of consent cannot reduce value, neither can the refusal itself render the land incapable of reasonably beneficial use. The worst that a refusal can do is to prevent an owner putting his land to some more advantageous use.

So far as entitling an owner to serve a purchase notice is concerned, it must be because events other than the planning refusal have so upset the status quo that the current use has ceased to be of reasonable benefit, thus forcing the owner to seek permission to carry out some form of development. If this permission is refused, or is granted subject to conditions, so that the owner is pinned down and unable to shift to a beneficial use, then he may serve a purchase notice. This in effect says to the council: "If you like my property so much the way it is, then buy it from me or find someone else to do so, for it is useless to me". However, a purchase notice is not intended to provide a remedy merely because an owner is refused permission to release development value. It is to deal with the case where an owner is trapped with land that has become incapable of reasonably beneficial use.

---

[28] For details of this, see earlier in this chapter.

It has, by some, been considered a nice point as to whether a purchase notice could be used to help an owner whose land has always been useless. That it may be so used was decided in the case of *Purbeck District Council* v *Secretary of State for the Environment* [1982] 2 EGLR 156, where it was held that it is only necessary to look at the situation as it has become following the planning decision, there being no need to look at the history of the land, nor to be concerned with what brought the present situation about: section 137 refers only to the land's existing state, not its former condition. This applies even if it is the owner's or occupier's activities which have rendered the land useless: there is nothing in the legislation which states that the cause has to be involuntary. It should, however, also be noted that in this case the Secretary of State's decision to uphold the purchase notice was quashed because the occupier, the claimant's tenant, had rendered the land useless by persistent defiance of conditions attached to a planning consent, despite the council's best efforts to force compliance. It was held that a claimant should not be allowed to take advantage of his own wrongdoing to foist the land on an unwilling local authority, and that section 137 excludes situations where the land has become incapable of reasonably beneficial use because of (in planning terms) unlawful activities on the land: 'It would be monstrous if a local planning authority were not entitled to say "We should not have to buy because the reason the land is incapable of reasonably beneficial use is because conditions we imposed in giving planning permission were not complied with".' A breach of planning control is unlawful: see *LTSS Print and Supply Services Ltd* v *Hackney London Borough Council* [1975] 2 EGLR 148, CA.

This decision follows that in *Adams & Wade Ltd* v *Minister of Housing and Local Government* (1965) 18 P&CR 60 at p67: "The purpose of the section is to enable an owner whose use of his land has been frustrated by a planning refusal, to require the local authority to take the land off his hands. The reference to beneficial use must therefore be a reference to a use which can benefit the owner (or prospective owner) and the fact that the land in its existing state confers no benefit or value upon the public at large would be no bar the service of a purchase notice".

If part of a parcel of land is rendered incapable of reasonably beneficial use and part is not, the owner can oblige the authority concerned to purchase only the part which is not capable of such use: see the Court of Appeal decision in *Wain* v *Secretary of State for the Environment* [1982] 1 EGLR 170.

When deciding whether land is capable of reasonably beneficial use, section 138 states that no account is to be taken of any unauthorised prospective use. Note the word "unauthorised", which here means carrying out development other than that specified in Part 1 of Schedule 3. If the notice is served following the refusal (or conditional grant) of planning permission, paragraph 1 of the Schedule (rebuilding) is subject to the limitations of Schedule 10 (restrictions on increases in floorspace).

Authorised prospective uses, which should be taken into account, include activities and changes which do not amount to development at all – section 55 refers.

The claimant must also satisfy section 137(3) or (4) by showing that the land cannot be rendered capable of reasonably beneficial use by carrying out development for which planning permission is already in existence, whether by a specific grant or by way of a development order consent. In cases deriving from the refusal of permission or the conditional grant of permission, or from revocation and modification orders, one must also show that the land cannot be rendered capable by carrying out development for which consent has been promised by either the Secretary of State or the local planning authority. Such promises need not be considered in discontinuance cases: compare section 137(3) with section 137(4) – the latter has no equivalent to subsection (3)(c).

Section 138(1) is clearer than its predecessor, section 180(2) of the 1971 Act. The key word in section 138 is "unauthorised". Section 180 had it that prospective new development must be disregarded, which was in conflict with section 180(1) so far as available planning consents were concerned, a conflict which was resolved in *Gavaghan* v *Secretary of State for the Environment* [1989] 1 PLR 88 where it was held that such consents should be considered.

There was also some inherent difficulty when the refusal of consent was for what is now Schedule 3 development itself – did one have to assume its continuing availability when considering whether the land had reasonably beneficial use? Clearly, to do so would make a nonsense of the intention behind the purchase notice provisions. Again, the word "unauthorised" resolves the problem: if Schedule 3 development has been refused it can no longer be an authorised prospective use and, being unauthorised, would have to be disregarded.

The Secretary of State has gone so far as to say that Schedule 3 is relevant as a guide to compensation once a purchase notice has been accepted, but has no relevance to the question of whether or not the land has reasonably beneficial use.

In summary, when deciding whether a purchase notice can be served, the existing use must be looked at. If the land has become of no reasonable benefit and if it cannot be made beneficial without carrying out development which requires planning permission (and this has not been promised), then a purchase notice can be served. It is essential to realise that "incapable of reasonably beneficial use" is not just another way of saying "less valuable" or even "less useful". So far as agricultural land is concerned, it will be incapable of reasonable beneficial use if its size, shape or location is such that farming is not practicable and any other use requires development and, hence, planning consent.

It is wrong to say that a purchase notice can only be served once permission for Schedule 3 development has been refused. One can be served whenever land is left incapable of reasonably beneficial use, including after the refusal of Schedule 3 development. Once a purchase notice has been served and accepted, the valuation rules are those of the Land Compensation Act 1961. The planning assumptions to be made are those of sections 14 to 17 of that Act: see chapter 9.

There are two distinct stages to be thought about. The first is whether the land has been rendered incapable of reasonable or beneficial use, and here one ignores the possibility of any unauthorised prospective use of the land. Once that stage has been passed, and a valid purchase notice resorted to, one moves on to the second stage, which is to prepare the valuation for compensation. At this stage one takes account of the planning assumptions, which may bring in the value of development lying both within and beyond Schedule 3.

The following is a summary of paragraphs 12 to 19 of Circular 13/83, which relate to reasonably beneficial use.

1. The question in each case is whether the land in its existing state, and taking account of operations and uses for which planning permission or listed building consent is not needed, is incapable of reasonably beneficial use.
2. No account shall be taken of:
    (a) The prospect of any unauthorised use.
    (b) Works for which listed building consent would be needed, unless such consent has been promised by the local planning authority or the Secretary of State.

    (c) A use which would be beneficial to someone other than the owner or prospective purchaser of the land. For the latter there must be a reasonably firm indication that such a person really exists.

3.   Relevant factors when considering the land's capacity for use are:

    (a) The physical state of the land.

    (b) Its size, shape and surroundings.

    (c) The general pattern of uses in the area.

    (d) A use of relatively low value may be beneficial if such a use is common for similar land in the vicinity.

    (e) It may be possible to render a small piece of land capable of reasonably beneficial use by using it in conjunction with some larger parcel, provided, in most cases, that the latter is owned by the owner or prospective purchaser of the small piece.

4.   Valuation evidence:

    (a) Profit from the land may be a useful test, but to point to an absence of profit is not conclusive evidence that the land has no beneficial use. The notion of reasonably beneficial use is not specifically identifiable with profit.

    (b) When considering whether the land has become incapable of reasonably beneficial use, a relevant test – in appropriate cases – is to consider the difference between the annual value of the land in its existing state and the annual value if Schedule 3 Part 1 is disregarded.

5.   "Land cannot be rendered capable of reasonably beneficial use by carrying out development for which planning permission has been granted or promised": section 137(3) and (4).

    Such permission or promise must have been given before the service of the purchase notice. If neither permission nor promise has been given, but the local planning authority thinks that some type of development not sought in the application ought to be allowed, and that this would render the land capable of reasonably beneficial use, the authority should ask the Secretary of State to direct[29] that such a permission would be granted if asked for. Where such a direction is made, the Secretary of State will not confirm the purchase notice.

---

[29] Under section 141(3)–(5).

6. The Secretary of State would normally expect to see evidence that the claimant had tried to dispose of the interest, by sale or letting, before being satisfied the land had become incapable of reasonably beneficial use.

7. The whole of the land which is the subject of the purchase notice must be incapable of reasonably beneficial use – if part is capable of such use, the notice will fail.

    The following cases are of particular relevance:

    R v MHLG, ex parte Chichester Rural District Council [1960] 2 All ER 407.

    General Estates Co Ltd v MHLG (1965) 194 EG 201.

    Trocette Property Co Ltd v Greater London Council (1974) 231 EG 1031.

Compensation where an alternative consent is given

Under section 141(3) the Secretary of State may, instead of confirming the notice, direct that some specified planning consent be given if sought. If the value of this permitted development is less than the Schedule 3 value, compensation is available for the difference: see section 144(1) and (6), and Studies 2 and 4.

Studies which involve Purchase Notices

Studies 1, 2, 3, 4, 7, 10 and 14 involve purchase notices.

Further Reading

Brand, Clive *Encyclopedia of Compulsory Purchase & Compensation* Sweet and Maxwell *

Davies, Keith *Law of Compulsory Purchase and Compensation* Tolley

Denyer-Green, Barry *Compulsory Purchase and Compensation* Estates Gazette

Grant, Malcolm *Encyclopedia of Planning Law and Practice* Sweet and Maxwell *

Hayward, Richard *The Handbook of Land Compensation* Sweet and Maxwell*

---

* Looseleaf publication with updating service.

Chapter 11

Minerals

Introduction

There are a number of factors to be considered in the valuation of mineral properties which distinguish them from valuations of other types of property.

One of these is that the mineral to be valued is hidden from view and techniques must be employed to elucidate the extent and quality, which will have a bearing on the value of the mineral.

In early mining and quarrying times the extent and quality of a mineral was very much established by trial and error until local knowledge was built up giving indications of its extent, quality and nature of disposition. With technological advancement the techniques for evaluating a mineral reserve have now reached a very advanced stage. For example, the evaluation of oil reserves will employ a detailed geological and geophysical investigation prior to drilling. The use of computers has greatly assisted the interpretation of geophysical information and therefore increased the reliability of these non-intrusive investigation techniques. Even so the majority of oil wells drilled are still unproductive. Even after extensive exploration work the detailed nature of minerals remains largely unknown until they are worked. Certain mineral deposits are relatively easily understood while others are extremely complex. With underground mines its is more difficult, and hence more expensive, to undertake exploratory work and the points of geological reference, for economic reasons, become more widely spaced than with surface deposits. The geological interpretation therefore becomes less reliable and underground mining is particularly prone to encounter unforeseen geological and mining problems. The valuer must have a good understanding of the geology and its effect on the viability of the mining/quarrying operations. A knowledge of the methods of working the mineral is also essential.

The mineral must be capable of being economically extracted and processed to a state to meet market requirements. The quality and specification requirements are now very important and British,

European and International standards have to be considered, as well as customer requirements. The valuer should have a good knowledge of mineral processing techniques and the particular market conditions.

The market determines the price of minerals and any individual mineral working must of essence be capable of economic operation ie, unit working plus investments costs must be less than selling price. The selling price is determined by what the customer is prepared pay at the point of use. Therefore a mineral working's location with respect to the market becomes of paramount importance with low value bulk minerals such as aggregates.

Mineral deposits vary considerably in their size, quality, geological complexity and ease of working and there can, therefore, be wide variations in operating costs and profitability from one site to another. The valuer should have a good understanding of the economics of mineral working in order to be capable of independently assessing, in the broadest terms, the viability of individual operations.

Another important consideration in the valuation of minerals within the United Kingdom is the Town and Country Planning legislation. The development of new mineral reserves is now controlled by development plans and supporting Town and Country Planning legislation. If a mineral deposit does not have planning permission then no value other than hope value can be attached to it. Again it is essential that the valuer has a sound knowledge of planning law and its effect on mineral working.

Additionally, mineral extraction is now controlled by Environmental Legislation and there is other specific legislation such as the Mines (Working Facilities and Support) Act 1923 as amended by the Mines (Working Facilities and Support) Act 1966 which may need considering. The list of specific legislation affecting minerals is too great to detail here and valuers will have to refer to text books on law and mineral land management. The valuer should have good knowledge of environmental and other legislation so far as it affects mineral workings.

The ownership of minerals is far from simple and while many minerals are still vested in the surface owner it is not uncommon for the surface and mineral ownerships to be severed. Several minerals have been taken into public ownership over the past 60 years or so and the ownership of gold and silver has, since time immemorial, been vested in the Crown. Former copyholders have retained their interest in minerals and there are still traditional rights to work lead in the Peak District of Derbyshire and coal and iron in the Forest of Dean. The valuer should have a good knowledge of mineral ownership.

While underground mining can take place without destroying the surface of the land, open pit mining or quarrying by necessity remove the surface of the land, thus precluding the use of the land for any other purpose while mining takes place. The mineral working is a temporary use of the land but the character of the land is changed forever. The same minerals cannot be worked twice and as a result the working of minerals is often referred to as a "wasting asset", rather similar to the reducing term of a leasehold interest.

Surface mining in the past was often very destructive, the surface of the land suffering high levels of dereliction. Modern restoration techniques now offer beneficial afteruse and in some cases a greater development potential can be achieved. Valuation of a mineral working may now require input from a number of specialists to comment on the problems of each development phase, ie the mineral extraction, the landfilling and the afteruse development potential. The valuation of land containing a mineral reserve may therefore involve a multi-disciplinary approach.

As in any valuation, it is essential to establish the purpose of the valuation and the basis on which it is to be prepared.

Valuations fall into two broad categories:

Valuations Statutory for Purposes
Compulsory purchase
Compensation for planning decisions
Estate Duty
Inheritance Tax
Rating
Capital Gains Tax
Mining Code

Market Valuations
Valuations for sale
Valuations for purchase
Asset valuations

The valuer will need to look at the basis of valuation. The basis for statutory purposes is usually defined.

Non statutory valuations need a wholly different approach and the basis of valuation and the methods to be adopted in arriving at an opinion of value are usually in accordance with the *Appraisal and Valuation Manual* (The Red Book) issued by the Royal Institution of Chartered Surveyors. The Red Book defines a variety of valuation bases: Open Market Value, Existing Use Value, Estimated Realisation Price et al. Practice Statement 22 of the Red Book is concerned with minerals and sets out the considerations which have to be taken into account.

Open Market Value is an opinion of the best price at which the sale of an interest in property would have been completed unconditionally for cash consideration on the date of valuation, assuming:

(a) a willing seller;

(b) that prior to the date of valuation, there had been a reasonable period {having regard to the nature of the property and the state of the market} for the proper marketing of the interest, for the agreement of the price and terms and for the completion of the sale;

(c) that the state of the market, level of values and other circumstances were, on any earlier assumed date of exchange of contracts, the same as on the date of valuation;

(d) that no account is taken of any additional bid by a prospective purchaser with a special interest; and

(e) that both parties to the transaction had acted knowledgeably, prudently and without compulsion.

Whatever valuation basis is required the valuer will have to take into account the fact that the value of mineral properties is reduced with each tonne of mineral extracted, processed and removed from the land.

The essence of any valuation is that where possible it should be achieved by way of comparison with similar recent transactions in similar circumstances. This simple principle becomes more complicated when, as in the case of minerals, the assets being valued are of a type which come on to the market relatively infrequently and comparable transactions may be scarce or non-existent. Mineral valuers attempt to overcome the problem by looking for comparable evidence such as rents, royalties and profits, etc. When valuing mineral properties the traditional approach is to capitalise the rent or royalty. An addition may then be applied to take into account any plant which might exist and the benefit of any trading agreements or market share which the quarry might enjoy. Such an approach can give a good indication of the value which might be considered prudent for the purposes of security for a loan or for an asset valuation but such a valuation is likely to be well short of the selling price which an attractive quarry might achieve if sold on the open market. It is therefore considered necessary in many cases to take into account the profit made from working the minerals.

Practice Statement 22 of "The Red Book" is concerned with the valuation of mineral properties and recognises the difficulty in obtaining reliable comparables. Thus it states that:

> In valuing mineral properties, it may be appropriate to reflect the trading potential of the property in terms of volume (expected outputs) and profitability (usually reflected in the royalty value). Thus in arriving at values the Valuer should have regard to market transactions providing evidence of capital and royalty values and to the profitability of the mineral and other related assets. Where the level of profitability influences the final valuation it should be discussed with the Directors.

There are more lease transactions than sales of mineral properties and these are sources of evidence for comparison when preparing mineral valuations. Landlords and tenants will within a mineral lease have agreed the appropriate royalty for the right to extract, process and sell the mineral and this will have taken market conditions and other factors into account. There may well be other payments within the lease and these will need consideration in order to arrive at the total amount agreed for the right to extract the mineral.

Mining leases usually provide for three kinds of rent:

1 A minimum rent (sometimes referred to as a dead or certain rent) is the rent payable regardless of the quantity of minerals extracted. This often merges with the royalty rent whereby minerals to the value of the minimum rent can be worked before royalty payments become due.

2 A royalty rent, which is payable for each tonne (or some other measure) of mineral extracted, processed and sold.

3 A surface rent in respect of surface occupied. This can vary in the case of surface mineral extraction according to the extent of land occupied.

Study 1 shows an example of the rents payable to a landlord under a mining lease incorporating the usual rent elements of a mining lease. Although the valuation of minerals require special consideration they are not divorced from the general economic considerations and, as with most commodities in the 21st century, require consideration not only of national but also European and world markets.

Study 1

The freeholder of an area of land containing limestone has been approached by the lessee, who wishes to purchase the freehold interest. The lessee is the operator of a quarry which is capable of producing 400,000 tonnes pa and has proven reserves of 5 million tonnes with possible further reserves of 3 million tonnes. The expected output over the next four years will be 150,000; 200,000; 250,000 and 350,000 tonnes respectively, building up to full production of 400,000 tonnes in the fifth year. Approximately 2 ha of surface land are occupied for the purpose of processing limestone.

The terms of the lease are:

Total area included in lease	50ha
Rent for surface land occupied	£250 per ha/pa
Minimum rent	£40,000 pa
Shortworkings	
clause	
Royalties:	first 200,000 tonnes pa 25p/tonne
remainder	20p/tonne

Term 25 years unexpired with break clause if minerals become exhausted.

Restoration: area to be restored by the lessee to agricultural use.

If there are shortworkings* to date of £20,000 advise the lessor of the value of his freehold interest in the 50 ha.

---

*Shortworkings are where the royalties less than the minimum rent due for any year and some leases allow this shortfall (shortworkings) to be recovered from "overworkings" in subsequent years. Overworkings are the royalties in excess of the minimum rent.

Table 1

Life of Quarry	tonnes	tonnes	Years
Proven reserves		5,000,000	
Deduct next 4 years workings	150,000		
	200,000		
	250,000		
	350,000		
Reserves when full production commences		4,050,000	
Divided by annual output of	400,000		
Life on full production			10.125
Possible additional reserves		3,000,000	
Divided by annual output	400,000		
Possible further quarry life			7.5

Table 2 Royalty values

	Output Tonnes	Royalty/ Tonne	£s	Amount Due £s
Year 1	150,000	@ 25 pence		37,500
Year 2	200,000	@ 25 pence		50,000
Year 3	200,000	@ 25 pence	50,000	
	50,000	@ 20 pence	10,000	60,000
Year 4	200,000	@ 25 pence	50,000	
	150,000	@ 20 pence	30,000	80,000
Year 5–21	200,000	@ 25 pence	50,000	
	200,000	@ 20 pence	40,000	90,000
Year 22 (1)	200,000	@ 25 pence	50,000	
	50,000	@ 20 pence	10,000	60,000

Table 3 Recoupment

Year	Royalty Value £	Minimum Rent £	Short-workings £	Over-workings £	Shorts to date £	Amount due £
0					20,000	
1	37,500	40,000	2,500		22,500	40,000
2	50,000	40,000		10,000	12,500	40,000
3 (2)	60,000	40,000		20,000		47,500
4	80,000	40,000		40,000		80,000
5	90,000	40,000		50,000		90,000
6 to 20	90,000	40,000		50,000		90,000
21	90,000	40,000		50,000		90,000
22	60,000	40,000		20,000		60,000

Table 4 Lessor's interest

		Present Value
	£s/pa	£s
Minimum Rent		
Minimum rent	40,000	
YP for 22 years @ 13% (3)	7.1695	286,780
Add Royalties		
Year 3 (royalty in excess of minimum rent)	7,500	
YP for 1 year @ 16% (4) 0.8621		
PV £1 in 2 years @ 16% 0.7432	0.6407	4,805
Years 4–22 (royalty in excess of minimum rent on first 200,000 tonnes pa)	10,000	
YP for 19 years @ 17% 5.5845		
PV £1 in 3 years @ 17% (5) 0.6244	3.487	34,870
Year 4 (royalty in excess of 200,000 tonnes pa) (6)	30,000	
YP for 1 year @ 18% 0.8475		
PV £1 in 3 years @ 18% 0.6086	0.5158	15,474
Years 5–21 (royalty in excess of 200,000 tonnes pa)	40,000	
YP for 17 years @ 19% 4.9897		
PV £1 in 4 years at 19% (7) 0.4987	2.4884	99,536
Year 22	10,000	
YP for 1 year @ 20% 0.8333		
PV £1 in 21 years @ 20% (8) 0.0217	0.0181	181
Add rent for surface land		
50 ha @ £250 pa	12,500	
YP for 23 years @ 10% (9)	8.8832	111,040
Add reversion of land to agricultural use		
50 ha @ £200 pa	10,000	
YP in perp @ 6% 16.6667		
PV £1 in 23 years @ 6% (10) 0.2618	4.3633	43,633
Total		596,319
Present Capital Value, Say		£600,000

Notes

(1)   Reserves when full production commences   4,050,000 tonnes
       Possible reserves   3,000,000 tonnes
       Total   7,050,000 tonnes

Less

17 years working @ 400,000 pa	6,800,000 tonnes
Final year	250,000 tonnes

(2) Amount due, being minimum rent plus the difference between excess workings and short workings to date, ie £40,000 plus £20,000 minus £12,500.

(3) Minimum rent considered at risk rate of 13% as less than 50% of expected total income and is considered to be more assured than royalty rent.

(4) Royalties over and above the minimum rent but not yet paying full dues.

(5) The minimum rent will be paid whether the quarry is producing or not, but royalties are paid on output, which from the lessor's viewpoint is not so secure as the minimum rent; therefore the royalties are given a higher risk rate.

(6) The difference between the amount due and the minimum rent plus royalties paid at 25p when merged, ie £80,000 less £50,000.

(7) The royalties at 20p are not as certain as those at 25p and therefore given a higher risk rate.

(8) The final year's working is less certain.

(9) Surface rent for land occupied.

(10) Reclaimed agricultural land.

Rating

The purpose of this study is to provide advice to a client on the expected rating assessment for a new quarrying operation. Advice will be in the form of an estimate based on the planned quarrying operation and mineral output from the quarry. It is not overlooked that the ratepayer could seek the advice of the Valuation Office Agency. However, there will be occasions in the early stages of project planning when the company may want to restrict the details of the scheme to company personnel.

In order to arrive at the estimated rating assessment it is necessary to find the correct rateable value for the "hereditament" according to law.

The rateable value is defined by paragraph 2(1–1B) of Schedule 6 of the Local Government Finance Act 1988 (as amended):

> The rateable value of a non-domestic hereditament none of which consists of domestic property and none of which is exempt from local non-domestic rating shall be taken to be an amount equal to the rent at

which it is estimated the hereditament might reasonably be expected to let from year to year if the tenant undertook to pay all the usual tenant's rates and taxes and to bear the cost of the repairs and insurance and other expenses (if any) necessary to maintain the hereditament in a state to command that rent.

The generally accepted method of valuing a quarry for rating purposes is to adopt a royalty per tonne which a tenant would be prepared to pay for the right to work the mineral. The right to work the mineral often, but not always, includes ancillary rights such as the right to process and make merchantable without the payment of additional rent. The royalty is by convention multiplied by the tonnage of mineral extracted from the quarry during the preceding year. To this sum is added the rent of the surface or other land in the hereditament and also an appropriate percentage of the capital value of the buildings and rateable plant and machinery. Too often this is a carried out as a mathematical exercise. It must be remembered that the process is merely one method of arriving at the rent which the yearly tenant might be expected to pay. The important stage often overlooked is a "stand back and look" where all the peculiarities of the hereditament which might affect the bid of the "hypothetical tenant" are considered.

Study 2

Rating
Sherwood Farm lies adjacent to a main road 12 miles from Bigoak, a major commercial centre. The farm which extends to 100 ha is owned by Sherwood Quarrying Company Ltd. The quarry was purchased by means of a capital sum many years ago. A sand and gravel deposit about 4 m thick underlies an average of 1.5 m of overburden. The base of the sand and gravel is above the water table. The underlying solid strata is a sandstone which is a local acquifer providing water for a nearby brewery. The sand and gravel contains approximately 60% sand and 40% gravel. Planning permission has recently been granted, and a contract has been signed, for the construction of a sand and gravel processing plant capable of producing 75 tonnes per hour. Current quarry output is expected to be in the order of 90% of plant capacity. Restoration of the quarry is to low level agriculture.

Much Aggregates Ltd have a quarry which they hold under a lease. A haul road 50 m long has been constructed to give access to a main road. The quarry is 5 miles from Sherwood Farm and 11 miles from Bigoak. The geology of the Much Aggregates Quarry is similar to Sherwood Farm, the sand and gravel being some 4.5 metres thick which underlies overburden some 1.2 metres thick. The silt content of the sand and gravel deposit is slightly higher. The quarry is subject to a lease which was signed in October, three years prior to the new Rating List coming into force. The term is for 21 years, the Certain Rent is based on an output of 100,000 tonnes pa. The opening royalty was 80 pence per tonne. The royalty is reviewed annually to the Retail Price Index (RPI) and there is an open market review every three years. There is a termination clause on mineral exhaustion. The quarry output is around 200,000 tonnes pa. Restoration of the quarry is to low level agriculture.

Big John Quarrying Co have a sand and gravel quarry which is situated 22 miles from Bigoak and is reached by narrow country lanes. The sand and gravel is some 5 m thick underlying overburden which is approximately 1 m thick. The geology differs from Sherwood Farm in that the bottom 4 m of sand and gravel are below the water table. The material is processed at a nearby, but not adjacent, plant site. There is a planning restriction which prohibits the erection of a mineral processing plant within the mineral deposit area. The plant site is leased at an annual rent of £20,000 pa. The sand and gravel is held on lease which was signed two years before the Rating List was due to come into force. The term is for 18 years with annual reviews to RPI and open market reviews every three years. The opening royalty was 60 p per tonne. The output of the quarry is of the order of 200,000 tonnes pa. The landlord is keen to develop water based activities such as fishing after the quarrying is complete.

Analysis of Information – Assumptions
The valuation date for the Rating List is two years before the List comes into force (1).

Inflation is currently running at 3% pa.

Big John Quarrying Co – Analysis
Lease details are compatible with the statutory definition of rateable value. (2)

	£ per tonne
Royalty payable	0.60 (3)
Adjust for remote plant site – transport	+ 0.12 (3)
Adjust for distance from market say 5	+ 0.03 (4)
Adjust for part of deposit being below water table say 5%	+ 0.03 (5)
Equivalent royalty for Sherwood Farm say	£0.78

Much Aggregates Ltd – Analysis
The lease is compatible with the statutory definition of rateable value (2)

	£ per tonne
Royalty payable	0.80
Adjust to the valuation date	+ 0.015 (6)
Equivalent royalty for Sherwood Farm	£0.815 (7)

Valuation of Sherwood Farm
In terms of "tone of the list" (8) royalty, the evidence from Much Aggregates is preferred because:

1. It is closer to Sherwood Farm than Big John Quarry, and a similar distance to the market.
2. There are fewer adjustments to be made to the royalty to make the evidence comparable to the situation at Sherwood Farm.
3. Although the evidence is six months prior to the valuation date, it is not so remote that adjustment would make it unsafe as comparable evidence.

An appropriate "tone of list" royalty for Sherwood Farm would therefore be, say 82 pence (rounding of the actual figure of 81.5 p) and the evidence from Big John Quarry would give a general level of support for this "tone of list" royalty.

Valuation of Buildings, Plant and Machinery
Buildings rateable plant and machinery situated on mines and quarries are valued using the contractors test method of valuation (9). This method is adopted because it is rare to find evidence of rental value for buildings plant and machinery on mineral producing hereditaments. The fundamental theory behind the contractors test method is that the hypothetical tenant would be prepared to pay as a rent a figure, which represents a reasonable

percentage of the effective capital value of the buildings, plant and machinery. Effective capital value is determined having regard to the cost of construction (10), deductions being made as necessary for age, functional and technical obsolescence.

Table 5

Ref	Item	N/R etc	Size	Unit Cost (10)	Effective capital value/ allowance	Adjusted capital value
	Buildings					
B1	Weigh Office/Office	m²	16.1	220	1	3,542
B2	Mess Room	m²	15.5	240	1	3,720
B3	Stores	m²	20	210	1	4,200
B4	Workshop	m²	300	180	1	54,000
B5	Control Cabin	m²	13.3	209	1	2,779.7
	Supports	m³	15.96	35	1	558.6
B6	Steel Security Store	m²	14.64	75	1	1,098
B7	Steel Security Store	m²	14.64	75	1	1,098
	Plant & machinery					0
P1	Primary Feed Hopper supports/grizzly	m³	40	40	1	1,600
	Foundations		90	250	1	22,500
P2	Conveyor (800mm) & ww	m	20	250	1	5,000
P3	Conveyor (700mm) & ww	m	14.4	220	1	3,168
P4	Barrel wash supports	m³	280	40	1	11,200
	Foundations	m³	60	250	1	15,000
P5	Conveyor (800mm) & ww	m	25	250	1	6,250
P6	Surge Bin	N/R			1	0
	Foundations/feed ramp		3.6	250	1	900
P7	Conveyor (700mm) & ww	m	25	220	1	5,500
	Trestle	m	6	150	1	900
P8	Bins	N/R			1	0
P8	Supports	tonnes	14	1,200	1	16,800
	Foundations	m³	65	250	1	16,250
P9	O/size conveyor	m	5	150	1	750
P10	Crusher Foundations	m³	7.5	250	1	1,875
	Crusher supports	tonnes	11	200	1	2,200

P11 Return conveyor	m	25	180	1	4,500
Supports	tonnes	3.5	1,200	1	4,200
P12 Radial conveyor	m	12	17	1	204
P13 Radial conveyor	m	12	75	1	900
P14 Radial conveyor	m	12			0
P15 Radial conveyor	m	12	75	1	900
P16 Bund Wall to oil tanks	m²	14	35	1	490
P17 Bund Wall to diesel tanks	m²	14.5	35	1	507.5
P18 Aggregate Bays		85	120	1	10,200
P17 Concrete		1,550	15	1	23,250
P18 Electrics		1	12,000	1	12,000
Services		1	10,000	1	10,000
Total					£248,040

Legislation prescribes the decapitalisation rates to be used in contractors test valuations. For the 1990 and 1995 Rating Lists these are 6% and 5.5% respectively. For the 2000 Revaluation the rates are 3.67% for defence, educational and healthcare hereditaments, and 5.5% in all other cases (11).

Total Buildings rateable plant and machinery	£248,040
Decapitalistion @ 5.5%	
Rateable value	£13, 642

Table 6 Valuation of Sherwood Farm Quarry

	Unadjusted Rateable Value (12)	Rateable Value
180,000 tonnes /annum (say) @ 82 pence	147,600	73,800
Buildings, rateable plant and machinery	13,462	13,462
Total	£161,062	£87,262

It would then be appropriate to advise Sherwood Quarrying Co Ltd that:

Rateable Value × Unified Business Rate = Rates payable

The rates would be payable on a pro rata basis for the likely period of occupation during the rate year.

Notes
(1) Local Government Finance Act 1988 Schedule 6 (3)(b).
(2) The royalty was agreed on the valuation date and therefore no adjustment is necessary to the date of valuation.
(3) Many mining leases as well as granting rights to extract the mineral give the right to erect a processing plant. It would have to be assumed that the royalty agreed reflected the fact that the tenant had to pay an additional rent for the remote plant site and the cost of transporting the unprocessed sand and gravel to the remote processing plant.
(4) The analysis aims to arrive at the rent which would be payable at Sherwood Farm which is closer to the main market. It is therefore necessary to make an adjustment to the rent at Big John Quarry to compensate for the difference in distance to the main market.
(5) There is a further disability because part of the deposit is below the water table with increased working difficulty and the same principle will apply as in (4) above.
(6) In the absence of any further information it has to be assumed that sand and gravel royalties are moving in line with inflation. The adjustment of royalties by an appropriate index, in the absence of market evidence, has been accepted by the Lands Tribunal to be an appropriate method of adjusting mineral royalties to the valuation date: see *Hodgkinson (Valuation Officer) v ARC Ltd* (RA 342–343/93).
(7) In terms of a mineral royalty, the construction of a 50 metre road for access to the public highway and the slight variation in silt content are not considered sufficient disabilities to require further adjustments to the market royalty for Sherwood Farm Quarry.
(8) "Tone of List" is not a term recognised in rating law as the law requires the valuation of each hereditament (see *Ladies Hosiery and Underwear Ltd v West Middlesex Assessment Committee* 2 KB 679) 1932. However, "tone of list" is accepted by rating practitioners as a means of comparing similar properties by adopting similar unit rates to arrive at rateable value.
(9) Buildings, rateable plant and machinery in mines and quarries are valued using the contractors test because rental evidence for them is rare. The method is detailed in *Ryde on Rating*, is included in RICS Guidance Notes, and has often been considered by the Lands Tribunal. See *Dawkins (VO) v Leamington Spa Corporation and Warwickshire County Council*

[1961] RVR 291. The theory is that the tenant would be willing to pay a reasonable percentage of the effective capital value of the buildings, plant and machinery as rent. This capital value is found from the cost of construction (10), with deductions for age, functional and technical obsolescence and suchlike. See *Monsanto plc* (now *Solutia UK Ltd*) v *Farria (VO)* (1968) RA 107, and *Shell Exploration and Production Ltd* v *Assessor for Grampian Valuation Joint Board LTS/VA/1998/47.*

(10) If appropriate the valuer should consider professional fees, location factor and scale of contract either within the unit rate or as additional items to arrive at the total adjusted capital value for the buildings, rateable plant and machinery.

(11) The following provide regulations as to the decapitalisation rates to be adopted: the Non-Domestic Rating (Miscalleneous Provisions) (No 2) Regulations 1989 SI No 2302; the ditto (Amendment) Regulations 1994 SI No 3122; the ditto (No 2) Regulation 1989 (Amendment) (England) Regulations 2000 SI No 908 (W39). See also DOE Consultation Paper 30/3/94.

(12) There is a special provision regarding mines and quarries which allows for the capital element which is part of the royalty payment. Regulation 5 of the Non Domestic Rating (Miscellaneous Provisions) Regulations 1989 as amended applies to any hereditament which consists of or includes a mine or a quarry. The Regulation provides that for the purposes of assessing rateable value of a mine or quarry no account shall be taken of sums payable in respect of extraction of mineral from such land in so far as sums are attributable to the capital value of minerals extracted. The allowance also applies to land occupied for the purposes of winning, working, grading, grinding and crushing of minerals. There is no statutory term before the application of the 50% reduction but it is generally known as the Unadjusted Rateable Value.

Study 3

Compensation for Compulsory Purchase of Mineral Bearing Land

A mineral operator owns an area of land with the underlying gypsum some 80 ha in extent. Approximately 16.5 ha have been worked out by surface extraction methods and 4 ha are used for processing the gypsum, leaving 59.5 ha to be worked.

The gypsum is 4.5 m thick and it is envisaged that 5 ha of land will be excavated each year.

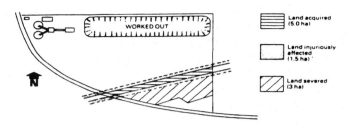

A motorway is to be constructed over the unworked portion of land and a compulsory purchase order has been made in respect of 5ha of land. The sketch above shows the details of the scheme. What is the owner's claim for compensation?

Table 7

	£'s	Present Value (£'s)
Land acquired		
Area 5.0 ha		
5.0 ha at £20, 000 (1)		100,000
Injurious affection		
Area 1.5 ha		
1.5 ha at £20,000 (2)	£30,000	
Less agricultural value 1.5 ha at £3,000 (3)	£4,500	25,500
Severance		
Area 3.0 ha		
3.0 ha at £20,000 (4)	£60,000	
Less agricultural value 3.0 ha		
at £3,000 (3)	£9,000	51,000
Disturbance		
Annual value of plant (5)	£120,000	
YP for 2 years @ 16% and 2.5% Tax 40%	1.02	
Value		£122,400
Deferred 10 years PV £1 of 10		
years @ 16%	0.227	
Add		£27,785
Break up value at closure (5)	£80,000	
PV of £1 of 10 years @ 16%	0.227	£18,160
Break up value at closure (5)	£80,000	
PV £1 in 12 years @ 16%	0.168	£13,440
Difference in Present Value (deduct)		£4,720
Disturbance claim (6)		23,065
Total claim		£199,565

Notes
(1) Comparable transactions have been analysed and adjustments made for differences between them and the land being considered. A figure of £20,000 per ha is indicated, including the minerals.
(2) The motorway must be supported on either side, which means that a strip of minerals either side cannot be worked. These minerals are not acquired but their value is injuriously affected by the acquisition to the extent that it is totally eliminated.
(3) On the south side of the motorway the land injuriously affected and the land severed (see note 4) will continue to be used for agricultural purposes. The loss suffered, therefore, for those portions is the mineral value less the agricultural value.
(4) The piece of land to the south of the motorway. Neither the planning authority nor the economics of extraction will permit the working of this small area in isolation and there are no prospects of it being worked with other adjacent land. The cost of providing a bridge over/under the motorway is in excess of £450,000.
(5) The loss of 9.5 ha of land means that the processing plant will reach the end of its useful life two years earlier than it would have done but for the motorway. That is, at the end of the 10th year instead of at the end of the 12th year. The acquisition is, therefore, the direct cause of the loss of the annual value of the plant for a two year period. However, the break-up value of the plant will be realised two years earlier, so the actual loss suffered for disturbance will be reduced by the gain made on the earlier realisation of the break up value of the plant.
(6) Present annual value of the plant for years 11 and 12 less the difference between the present break up value in 10 years and 12 years time respectively.

Compulsory Purchase and Compensation Under the "Mining Code"

The Railway Clauses Consolidation Act 1845 sections 77 to 85, known as the 'Mining Code', introduced a code of practice regulating the right for support of railways from the underlying minerals.

Briefly, the code requires a mine owner, who can be the owner, lessee or occupier of the minerals, to serve a notice of approach to a railway company when any workings reach 40 yds distance from any railway, buildings or works. If the railway company requires the minerals to remain unworked to support the railway then they must serve a counternotice requiring support, and pay compensation to the mine owner for the minerals unworked. The amount of compensation is the loss suffered by the owner by virtue of leaving the minerals unworked, plus any additional expense incurred.

Similar "Mining Codes" are incorporated in other enactments such as the Waterworks Clauses Act 1948, Public Health Act 1875 (Support of Sewers) Amendment Act 1883, Public Health Act 1936, Pipelines Act 1962, Water Act 1989 and Water Resources Act 1990. There is statutory provision for the "Mining Code" to apply to land and buildings that are included in compulsory purchase orders. The 40 yds has been substituted with 37 m in the more recent legislation.

The "Mining Code" was amended by Part II of the Mines (Working Facilities and Support) Act 1923, so far as it applies to railways, providing for the mine owner to serve the "notice of approach" and determines an area of protection, being the width of the protected works plus a distance of 40 yds or half the depth of the seam, whichever is the greater. The area of protection is divided into an inner area and an outer area, the inner area being the width of the protected works plus 40 yds, the remainder being the outer area. The compensation is to be assessed separately for the mine owner and the royalty owner and is specified at a rate per tonne for the total area and payable at 100% for the inner area and 33⅓ % for the outer area. Subsequent legislation prescribes a statutory distance of 40 yds (37 m) but also allows the designating authority to substitute other distances. It is, however, common for the provisions of the 1923 Act to be incorporated.

It should be noted, however, that if the mine owner has a right to work and withdraw support then all the minerals are deemed to be within the inner area.

Study 4

Mining Code
A gas company wishes to obtain a pillar of support for a length of pipeline and has served a counternotice to a notice of approach served under the Pipelines Act 1962 by the operator of a fluorspar mine. The lessee is a substantial operator making a net profit of £10 per tonne. The order authorising the pipeline construction

designated the distances prescribed in the Mines (Working Facilities and Support) Act 1923. The relevant terms of the lease are – royalty £0.50 per tonne of crude ore; minimum rent £4,000 pa; there is an average clause; prior to the acquisition of the land by the gas company there was a right to work the minerals but not to let down the surface.

The mineral vein is restricted to the carboniferous limestone strata which in this area is abut 60 m thick and this overlain by about 100 m of non-productive shales. The vein varies considerably in width but averages 2.5 m. The vein is for all intents and purposes vertical. Fluorspar is the major constituent of the vein, with a grade of about 50%, the remaining constituents are mainly barytes, calcite, silica and galena. The fixed royalty of £0.50 a tonne is paid irrespective of variations in grade. It has been agreed that the loss of minerals over the next four years will be as follows:

Year 1   13,560 tonnes within the inner area and
         3,130 tonnes within the outer area of the proposed pillar of support.
Year 2   16,000 tonnes within the inner area and
         8,000 tonnes within the outer area
Year 3   8,000 tonnes within the inner area and
         7,040 tonnes within the outer area
Year 4   8,000 tonnes within the inner area and
         7,040 tonnes within the outer area of the proposed pillar of support.

In addition, it has been agreed that there will be 5,480 tonnes of unworkable fluorspar of which 2,740 tonnes would have been worked during the second year and 1,370 tonnes in the third and fourth years.

Leaving the pillar of support unworked will interrupt production from the mine while access roads are driven in the country rock to reach the vein beyond the pillar.

Advise the freeholder and the operator of the fluorspar of their claims for compensation from the railway company for the proposed sterilisation of the fluorspar.

Lessee's Claim	£s	£s
Year 1		
3,560 tonnes @ £10 a tonne (1)	35,600	
3,130 tonnes @ £3.33 a tonne (2)	10,423	
	46,023	
PV of £1 in 1 year at 12%	0.893	41,098

Add Year 2

16,000 tonnes @ £10 a tonne (1)	160,000	
8,000 tonnes @ £3.33 a tonne (2)	26,640	
2,740 tonnes @ £3,33 a tonne (3)	9,124	
	195,764	
PV of £1 in 2 years at 12%	0.797	156,024

Add Year 3

8,000 tonnes @ £10 a tonne (1)	80,000	
7,040 tonnes @ £3.33 a tonne (2)	23,443	
1,370 tonnes @ £3,33 a tonne (3)	4,562	
	108,005	
PV of £1 in 3 years at 12%	0.712	76,900

Add Year 4

8,000 tonnes @ £10 a tonne (1)	80,000	
7,040 tonnes @ £3.33 a tonne (2)	23,443	
1,370 tonnes @ £3,33 a tonne (3)	4,562	
	108,005	
PV of £1 in 4 years at 12%	0.636	68,691
Compensation for Minerals		£342,713

Compensation for Premature Development Work

Leaving the pillar of support unworked has meant that development work has had to be brought forward so that the vein beyond the pillar can be accessed. The cost of this work is estimated to be £210,000 and this needs to be expended immediately instead of over a period of three years. We assume that the development work would have been spread equally over the three-year period and that the cost would have been £70,000 pa.

	£s	£s	£s
Immediate Development Cost		210,000	
Less Previous Annual Development Costs	70,000		
Years' Purchase 3 years @ 12%	2.402		
Present Value		168,140	
Compensation for Premature Development Work (4)			41,860
Lessee's Total Claim			£384,573

Freeholder's Claim	£s	£s
Year 1		
3,560 tonnes @ £0.50 a tonne (1)	1,780	
3,130 tonnes @ £0.167 a tonne (2)	523	
	2,303	
PV of £1 in 1 year at 10% (5)	0.909	2,093
Add Year 2		
16,000 tonnes @ £0.50 a tonne (1)	8,000	
8,000 tonnes @ £0.167a tonne (2)	1,336	
2,740 tonnes @ £0.167 a tonne (3)	458	
	9,794	
PV of £1 in 2 years at 10% (5)	0.826	8,090
Add Year 3		
8,000 tonnes @ £0.50 a tonne (1)	4,000	
7,040 tonnes @ £0.167 a tonne (2)	1,176	
1,370 tonnes @ £0.167 a tonne (3)	229	
	5,405	
PV of £1 in 3 years at 10% (5)	0.751	4,059
Add Year 4		
8,000 tonnes @ £0.50 a tonne (1)	4,000	
7,040 tonnes @ £0.167 a tonne (2)	1,176	
1,370 tonnes @ £0.167 a tonne (3)	229	
	5,405	
PV of £1 in 4 years at 10% (5)	0.683	3,692
Freeholder's Total Claim		£17,934

Notes
(1) The inner area at full value.
(2) The outer area at one third value.
(3) Additional mineral made unworkable.
(4) Additional compensation under the Mining Code.
(5) The risk is not as high for the royalty owner; he would receive a minimum rent irrespective of whether the minerals were worked.

Study 5

Computer Spreadsheets and Discounted Cash Flow
Computers are ideally suited for valuation calculations and any modern spreadsheet programme can be used for this purpose.

Most spreadsheets can undertake simple calculations so that the valuations can be calculated from first principles while the more sophisticated spreadsheets have direct functions for calculating Present Value and Years' Purchase as well other financial formula. There are also several specialist valuation programmes commercially available.

Computers allow easy computation and it becomes relatively simple to carry out numerous complex calculations and undertake sensitivity analyses.

The basic feature of any spreadsheet is that it is divided into a series of rows and columns and the junctions between rows and columns are referred to as a cells. The columns are usually referenced: A, B, C, etc whereas the rows are referenced 1, 2, 3, etc. Each cell on the spreadsheet can then be referred as: A1, A2, B1, B2, etc. Numbers can be placed in the boxes and then boxes can be summed, subtracted, multiplied, divided, raised to the power, etc. The possibilities are almost endless.

When constructing a valuation spreadsheet it is necessary to have a good working knowledge of valuation formulae. Once the valuation spreadsheet has been constructed it will repeat the calculations as often as required but it should be remembered that if the mathematical formulae are entered incorrectly it will always calculate the wrong answer. It is essential therefore after constructing a spreadsheet that the mathematics are independently checked either by reference to tables or by calculating from first principles.

Perhaps the simplest way of undertaking a valuation using a spreadsheet is by using the Discounted Cash Flow (DCF) method. This method consists simply of the summation of the Present Value of £1 of the income stream for the years to be valued. It should be remembered that DCF gives exactly the same results as using Years' Purchase.

The following tables shows how a valuation calculation can be undertaken. Each row shows the calculation for each year of the term and the columns show the following:

Column	Content
A	Year
B	Annual output
C	Unit profit or Royalty
D	Annual Income
E	Discount/risk rate
F	Present Value of £1

Table 8

	A	B	C	D	E	F	G	H
1	Year	Output	Royalty	Income	Discount /Risk Rate	Present Value of £1	Present Value	Present Value Running Total
2	1	1,000	1	B2*C2	0.1	1/((1+E2)^A2)	D2*F2	G2
3	1+A2	B2	C2	B3*C3	E2	1/((1+E3)^A3)	D3*F3	H2+G3
4	1+A3	B3	C3	B4*C4	E3	1/((1+E4)^A4)	D4*F4	H3+G4
5	1+A4	B4	C4	B5*C5	E4	1/((1+E5)^A5)	D5*F5	H4+G5
6	1+A5	B5	C5	B6*C6	E5	1/((1+E6)^A6)	D6*F6	H5+G6
7	1+A6	B6	C6	B7*C7	E6	1/((1+E7)^A7)	D7*F7	H6+G7
8	1+A7	B7	C7	B8*C8	E7	1/((1+E8)^A8)	D8*F8	H7+G8
9	1+A8	B8	C8	B9*C9	E8	1/((1+E9)^A9)	D9*F9	H8+G9
10	1+A9	B9	C9	B10*C10	E9	1/((1+E10)^A10)	D10*F10	H9+G10
11	1+A10	B10	C10	B11*C11	E10	1/((1+E11)^A11)	D11*F11	H10+G11

Note: * means multiply and ^ means raise to the power of.

G	Present Value of the income
H	Running total of yearly present values

The income shown in column D is the product of multiplying columns B and C ie B1*C1. The Present Value of £1 shown in column F is the reciprocal of one plus discount rate raised to the power of the appropriate year ie $1/((1+E1)^{\wedge}A1)$. Column G is the product of multiplying columns D and F ie D1*F1. Column H shows the running total and is simply the previous total shown on column H added to the figure in column G for the appropriate year ie H1+G2. By typing a cell number into a different cell will repeat the value.

Table 8 shows how the spreadsheet was constructed with the formula in each of the boxes whereas Table 9 (page 408) shows the calculated values.

It will be seen from Column G how the Present Value, for any individual year, decreases the further it is deferred. Column H shows how the value accrues and the annual rate of increase reduces as the deferment period increases.

Although it is not obvious, a valuation calculated using this method takes into account a Sinking Fund at the same rate as the discount rate (ie single rate). In the example above we have used an income of £1,000 pa for a period of 10 years at a discount rate of 10% and this gives a Present Value of £6,144.57. We assume that the investor has purchased the property for this sum and therefore receives an income of £1,000 pa for the next 10 years. The investor requires a return on investment of 10% and this equates to £614.46 pa. As the income is £1,000 that leaves £385.54 pa to be invested. If this is done at 10% then the initial investment sum will be recouped. See following example:

Capital Invested	£6144.57	
Return on investment @ 10%	£614.46	pa
Annual income	£1000.00	pa
Excess income to be invested	£385.54	pa
Amount of £1 pa for 10 years @ 10%	15.9374	
Accrued Sum	£6144.57	
Initial Investment	£6144.57	

If the income from the property is to be deferred or no income is to be received for certain years then the calculations are achieved by

Table 9

1	A Year	B Output	C Royalty	D Income	E Discount .Risk Rate	F Present Value of £1	G Present Value £s	H Present Value Running Total £s
2	1	1,000	1	1,000	0.1	0.9091	909.09	909.09
3	2	1,000	1	1,000	0.1	0.8265	826.45	1,735.54
4	3	1,000	1	1,000	0.1	0.7513	751.31	2,486.85
5	4	1,000	1	1,000	0.1	0.683	683.01	3,169.87
6	5	1,000	1	1,000	0.1	0.6209	620.92	3,790.79
7	6	1,000	1	1,000	0.1	0.5645	564.47	4,355.26
8	7	1,000	1	1,000	0.1	0.5132	513.16	4,868.42
9	8	1,000	1	1,000	0.1	0.4665	466.51	5,334.93
10	9	1,000	1	1,000	0.1	0.4241	424.1	5,759.02
11	10	1,000	1	1,000	0.1	0.3855	385.54	6,144.57

Table 10

	A Year	B Output	C Royalty	D Income	E Discount .Risk Rate	F Present Value of £1	G Present Value £s	H Present Value Running Total £s
1								
2	1							
3	2							
4	3	1,000	1	1,000	0.1	0.7513	751.3	751.3
5	4	1,500	1	1,500	0.1	0.683	1,024.5	1,775.8
6	5	1,500	1	1,500	0.1	0.6209	931.35	2,707.15
7	6	1,500	1	1,500	0.1	0.5645	846.75	3,553.9
8	7	1,500	1	1,500	0.1	0.5132	769.8	4,323.7
9	8	2,500	1	2,500	0.12	0.4039	1,009.75	5,333.45
10	9	2,500	1	2,500	0.12	0.3606	901.5	6,234.95
11	10	2,500	1	2,500	0.12	0.322	805	7,039.95
12	11	2,500	1	2,500	0.12	0.2875	718.75	7,758.7
13	12	2,500	1	2,500	0.12	0.2567	641.75	8,400.45

simply not entering an income for the appropriate years. Again if there is variable income and perceived greater risk in later years then this can be easily calculated. Table 10 shows a spreadsheet calculation using varying income and discount rates and deferred for 2 years.

Study 6

Valuation Matrices

Where there is a large number of variables it may be appropriate to construct a valuation matrix. This will assist the valuer in arriving at the valuation figure and will demonstrate to the client the volatile nature of the parameters and their effect on value.

An example of where a matrix may been appropriate is with underground coal mines. Modern capital intensive longwall mining can be highly productive but requires relatively large areas which are free from geological faulting, even minor faulting can severely disrupt production. Relatively small shortfalls in budgeted output can turn a profit into a loss. Many mines operate with only broad outline knowledge of the geology in front of the workings and the detailed geology is proved by the driving of development roadways. There may only be a year or so of "measured" coal reserves* but "indicated" coal reserves may be sufficient to last several decades. The mine could therefore work profitably for the foreseeable future but it could hit geological problems which could close it in, say, less than two years.

The following matrix is prepared on the assumption that there is sufficient proved coal to last about two years and possibly sufficient coal to last at least a further 10 years where detailed geology has yet to be proved. The financial risks associated with underground coal mining are high and we have accordingly used an interest rate 18% for the first two years and interest rates varying between 25% and 40% for the remaining ten years.

It will be seen from the matrix that similar present values are arrived at using lower outputs and lower interest rates to that using higher output and higher interest rates. The matrix shows a range

---

*Reserves which are potentially suitable for longwall mining. The geological structure should be well known from boreholes, in seam headings and over or underlying workings and seam should be measured and analysed on at least three sides of a block less than 0.5 km across. The quantity and quality of information is such that any shortfall should be small.

Table 11

Annual Production (tonnes)	Estimated Profit £s/tonne	Estimated Annual Profit £s	Interest Rates 18% & 25% Present Value £s	Interest Rates 18% & 30% Present Value £s	Interest Rates 18% & 35% Present Value £s	Interest Rates 18% & 40% Present Value £s
700,000	0	0	0	0	0	0
750,000	1	750,000	2,888,000	2,546,000	2,292,000	2,098,000
800,000	2	1,600,000	6,161,000	5,412,000	4,889,000	4,475,000
850,000	2.5	2,125,000	8,183,000	7,214,000	6,493,000	5,944,000
900,000	3	2,700,000	10,397,000	9,166,000	8,250,000	7,552,000
950,000	3.5	3,325,000	12,804,000	11,288,000	10,159,000	9,300,000
1,000,000	4	4,000,000	15,403,000	13,580,000	12,222,000	11,188,000

of values between £2.1 million and £15.4 million but for all practical purposes the range is from £2.9 million to £11.2 million since one would not combine a low risk rate with high output levels. The valuer would have to carry out a detailed analysis of the circumstances at the mine before deciding on the appropriate value.

Compensation for Mineral Planning Decisions

The current provision for compensation for mineral planning decisions can be found in the following legislation

1   The Town and Country Planning Act 1990.
2   Planning and Compensation Act 1991. Guidance can be found within Mineral Planning Guidance (MPG) 9.
3   The Environment Act 1995: Review of Mineral Planning Permissions. Guidance can be found within MPG14.
4   The Town and Country Planning (Compensation for Restrictions on Mineral Workings and Mineral Waste Depositing) Regulations 1997, SI 1997 No 1111, which came into force on March 25 1997. Guidance can be found within MPG4.

There are three broad areas which can give rise to compensation following mineral planning decisions.

(i)   The review of mineral planning permissions under the Environment Act 1995 (the 1995 Act). The 1995 Act provides for initial and subsequent periodic reviews of mineral planning permissions to maintain mineral planning permissions in line with changing environmental practices and standards. Compensation may be payable if new conditions, other than restoration and aftercare conditions, restrict working rights. Compensation can either stem from the Mineral Planning Authority (MPA) serving notice and including within the notice a statement that the new conditions would adversely affect, to an unreasonable degree, either the economic viability or the asset value of the site. Alternatively, the land or mineral owner may be of the view that the new conditions will affect the viability of the site. In these circumstances Parts IV and XI of the Town and Country Planning Act 1990 (the 1990 Act) have effect as if a modification order had been made and confirmed under sections 97 and 98 of the 1990 Act. Persons having an interest in the land or the minerals are able to claim compensation, eg the mineral and/or land owner and the mineral operator.

(ii) The Planning and Compensation Act 1991 (the 1991 Act) introduced new procedures for dealing with permissions for the winning and working of minerals or the depositing of mineral waste, originally granted under Interim Development Orders (IDO's). The 1991 Act refers to the IDO's as old mining permissions. The 1991 Act required certain actions to be taken in respect of the old mining permissions if they were to continue to have effect. In the case where a person with an interest in the land or minerals has submitted a scheme of working including restoration conditions and the MPA impose conditions different from those submitted in the application, then there is a right of appeal to the appropriate Secretary of State. The Government, when introducing the 1991 Act, made it clear that they only envisaged conditions applying to environmental and amenity aspects of working sites and it did not anticipate that any new conditions applied would affect the asset value or economic structure of the operation. The guidance notes indicate that conditions which would significantly affect the asset value are more appropriate for MPA reviews introduced under the Town and Country Planning (Minerals) Act 1981. Compensation does not seem to be a concept envisaged under the IDO registration and agreement of conditions. It is assumed that compensation would arise from a review of planning permissions under the Environment Act 1995 or by the MPA serving an order under the 1990 Act as detailed in the following paragraph.

(iii) MPA's still retain their powers to make orders revoking, modifying, discontinuing, prohibiting or suspending mineral workings. In summary the following may attract compensation from the MPA if they are confirmed by the Secretary of State:

   (a) Revocation and modification orders (section 97 and Part II of Schedule 5, 1990 Act);
   (b) Discontinuance orders (section 102 and paragraphs 1 to 3 of Schedule 9, 1990 Act);
   (c) Prohibition orders (paragraphs 3 and 4 of Schedule 9, 1990 Act); and,
   (d) Suspension orders and supplementary suspension orders (paragraphs 5 to 9 of Schedule 9, 1990 Act).

A claim must be made under the appropriate provisions of the 1990 Act. The provisions are section 107 (or 279 as appropriate) in the case of revocation and modification orders, or section 115 (or 280 as appropriate) in the case of other orders. It should be noted that the Town and Country Planning (Compensation for Restrictions on Mineral Workings and Waste Depositing) Regulations (the 1997 Regulations) define the circumstances in which compensation is not to be payable following the making of a modification or discontinuance order. The 1997 Regulations also modify section 115 of the 1990 Act in its application for claims for compensation following the making of prohibition, suspension or supplementary orders. The 1997 Regulations revoked the Town and Country Planning (Compensation for Restrictions on Mineral Working) Regulations 1985 and the Town and Country Planning (Compensation for Restrictions on Mineral Working) (Amendment) Regulations 1990.

Compensation

MPG14 states the Government's intention to bring the compensation entitlement following a revocation, modification or discontinuance order into line with the compensation provisions for periodic reviews, ie there would be no compensation for orders which imposed restoration or aftercare conditions, nor for orders which imposed conditions which did not affect working rights. Working rights are affected if any of the following are reduced or restricted:

(a) The size of area which may be used for the winning and working of minerals or the depositing of mineral waste;
(b) The depth to which the operations for the winning and working of minerals may extend;
(c) The height of any deposit of mineral waste;
(d) The rate at which any particular mineral may be extracted;
(e) The rate at which any particular mineral waste may be deposited;
(f) The period at the expiry of which any winning and working of minerals or depositing of mineral waste is to cease; or
(g) The total quantity of minerals which may be extracted from, or of mineral waste which may be deposited on site.

The circumstances where *no* compensation is payable following the making of orders are set out in the 1997 Regulations and summarised in MPG4.

It is not proposed to repeat all the circumstances here as the conditions where compensation will arise and the amount of compensation due varies under the orders. Any valuer dealing with a claim for compensation will need to refer to the 1990 Act and the 1997 Regulations (guidance from MPG4). As an example it may be useful to consider the question of compensation where a review is made under the Environment Act 1995.

The scenario being that revised conditions have been submitted for an active Phase 1 site under the provisions of the 1995 Act (Phase 1 sites are detailed under the 1995 Act but in brief refer to active sites where the predominant planning permission/s was/were granted after June 30 1948 and before April 1 1969). The MPA has not accepted the conditions submitted by the applicant but has substituted amended ones. Some of the conditions will affect the economic viability of the site. The MPA has served a notice (see Schedule 13 paragraph 15 to the Environment Act 1995) stating that it is their opinion the revised conditions could affect the economic viability of the site. The same scenario would arise where the applicant has made a successful appeal (see Schedule 13 paragraph 11 to the Environment Act 1995). Parts IV and XI of the 1990 Act will have effect as if a modification order had been made and confirmed under sections 97 and 98 of the 1990 Act. The applicant will be able to make a claim for compensation under section 107 of the 1990 Act, as modified by the Town and Country Planning (Compensation for Restrictions on Mineral Working and Mineral Waste Depositing) Regulations 1997.

First, dealing with the modification under the Town and Country Planning (Compensation for Restrictions on Mineral Working and Mineral Waste Depositing) Regulations 1997.

Regulation 3 provides that no compensation is payable where the following conditions are satisfied:

(a) the order does not impose any restriction on working rights other than restoration or aftercare conditions (see paragraphs 2 (a)(i)&(ii));

(b) that either the permission was granted not less than five years before the grant of the order, or the planning permission was granted before February 22 1982 (see paragraphs 2(b)(i)&(ii)); and

(c) the order was made more than five years after any previous order or orders in respect of the same land, and more than 5 years after an application for determination of conditions under Schedule 2 to the Planning and Compensation Act 1991 or under Schedule 13 or 14 to the 1995 Act was finally determined (see paragraphs 2(c)(i)&(ii)).

If these conditions are not satisfied then Section 107 of the 1990 Act applies and unmodified and unabated compensation is payable.

Section 107 of the 1990 Act sets out that a person with an interest in the land or minerals can make a claim, within the prescribed time and prescribed manner, to the local planning authority if it can be shown that they have:

(a) incurred expenditure in carrying out work which is rendered abortive by the order;

(b) otherwise sustained loss or damage which is directly attributable to the order.

Study 7

Compensation for Modification of Mineral Planning Permission
A quarry operator has an active Phase 1 site and has submitted an Application for Determination of New Conditions (1). All the correct procedures have been adopted but both parties agreed to a time extension to determine conditions. After lengthy discussions the MPA has determined conditions different from those submitted by the applicant.

The revised conditions relate to the construction of a new access (which avoids residential dwellings and a school). The new access includes a screened entrance but will sterilise 180,000 tonnes of limestone. The limestone would have been worked in five years time as shown on working plans previously submitted to the MPA. The current profits on sales are £1.80 per tonne. Some £15,000 was spent on a new weighbridge and wheelwash at the existing access some two years ago. It will be possible to relocate these items to the new access at a cost of £8,500. The cost of the new access will be £95,000. Consultants' fees for revising the restoration plans and construction details for the new access including negotiations with the Highways Authority and supervision of the works will be £3,750. The quarry production has averaged 90,000 tonnes over the last eight years and this is supported by statutory returns.

Stripping costs of adjacent land containing reserves will now take place in four years time instead of six years time. Some 10,500 cubic metres (m³) of soils and 9,000m³ of overburden is involved. The costs of moving the soils is £0.70/m³ and the cost of moving the overburden is £0.60/m³.

There is no opportunity for landfill. The proposed restoration of the whole site is to low level restoration with afteruse for agriculture, with some amenity woodland planting. The restoration plan along with a five-year aftercare scheme has been agreed with the MPA. The conditions within the original planning permission in respect of restoration required all plant and machinery to be removed from the site at the cessation of mineral extraction. The soils stripped from the site will be respread over the quarry floor and the land returned to agricultural use. The quarry operator has estimated that on current costs the agreed restoration plan and aftercare scheme will cost a further £15,000 above the cost envisaged for the proposed restoration under the original planning permission (2).

Advise the quarry owner on the amount of compensation to be submitted in a claim to the MPA.

Table 12

		£s	Present Value £s
Cost of constructing new access			95,000
Relocation of wheelwash and weighbridge			8,500
Loss of limestone year 5			
– 90,000 tonnes × £1.80	£162,000		
PV of £1 in 5 years @ 8%	0.6806	110,257	110,257
Loss of limestone year 6			
– 90,000 tonnes × £1.80	£162,000		
PV of £1 in 6 years @ 8%	0.6302	102,092	102,092
Fees incurred in plan preparation etc			3,750
Cost of moving soils in 4 years time			
10,500m³ × £0.70	£7,350		
9,000m³ × £0.60	£5.400		
	£12,750		
PV of £1 in 4 years @ 8%	0.735	9,371	
Deduct			
Cost of moving soils in 6 years time			
10,500m³ × £0.70	£7,350		
9,000m³ × £0.60	£5,400		
	£12,750		
PV of £1 in 6 years @ 8%	0.6302	8,035	
Additional cost of moving soils		(9,371 – 8,035)	1,336
Total claim			£320,935

Notes
(1) See Planning and Compensation Act 1991 and MPG8 for procedures and time scale.
(2) The claim for compensation for additional restoration and aftercare costs is excluded under the Town and Country Planning (Compensation for Restrictions on Mineral Workings and Mineral Waste Depositing) Regulations 1997: see Regulation 3.

© PR Deakin & MC Jervis 2000

Chapter 12

Rating

As was the case for the last edition of *Principles into Practice*, the rating system has changed yet again. This time the changes are not so far reaching and are confined mainly to the community charge being replaced by the council tax in 1993 and changes to the Plant and Machinery Regulations.

Chargeable Dwellings

The definition of a dwelling is contained in section 3 of the Local Government Finance Act 1992. Basically a chargeable dwelling is a "hereditament" that certifies certain conditions set out in the Act. There is one bill per dwelling, based on bands related to capital value, as at April 1 1991. The tax has both property and personal elements. The personal element is based on an assumption that the dwelling is occupied by two adult residents and the bill is discounted by 25% where there is only one. The basis of valuation is the amount that the property would have sold for on the "open market" by a willing vendor on April 1 1991 on assumptions contained in Regulation 6 of The Council Tax (Situation and Valuation of Dwellings) Regulations 1992.

The main valuation assumptions are:

- That the sale was with vacant possession.
- That the interest sold was freehold except for flats when a lease for 99 years at a nominal rent is to be assumed (the actual lease term is ignored).
- That the dwelling had no potential for any building work or other development requiring planning permission.
- That the size, layout and character of the dwelling and the physical state of its locality which existed on April 1 1991 were the same as those which actually existed on April 1 1993 or a later date depending on why the alteration is being made.
- That the dwelling (and any common parts such as an entrance or hallway shared with other dwellings) was in a "state of reasonable repair".

Size Layout, Character and Locality

While the dwelling is valued at April 1 1991, the size, layout character and locality of the dwelling are required to be considered as they actually existed at April 1 1993 or, at a later date, depending on the facts:

If there needs to be a correction of an error in the valuation list when the Council Tax came into effect, the date would be when the list became inaccurate. The dwelling is valued as if all the physical factors existed when the valuation first came into force.

If there is a division or merger of dwelling (s), the date would be the date of the split or merger. Dwellings are valued having regard to all the physical factors as they existed at the date of the completion of the works.

If the dwelling was improved, the date would be the date when the building is subsequently sold as a dwelling cannot be rebanded due to increases arising from improvements until it is next sold. The dwelling is revalued having regard to all the physical factors as they existed at the date of sale.

If there is a change in the balance of the business/domestic use then the date would be that when the change occurred, with the whole dwelling valued having regard to the physical factors as they existed at the date when the change of use occurred.

If part of a dwelling is demolished with no intention of rebuilding, then the date is the date of demolition. If however the demolition is part of planned future works then no alteration may be made. Regard is given to the physical factors as they existed at the date of demolition.

If the value is reduced due to alterations for making them suitable for a physically disabled person, then the date would be the date of adaptation with the physical factors considered as at that date.

Where there is a change in the physical state of the dwelling's locality, the date would be the later of:

(a) date of previous alteration,
(b) date of any previous sale giving rise to an alteration
(c) April 1 1993.

Note that if a reduction in value has been caused by external factors, these cannot be offset by improvements executed by the current taxpayer. The state of the locality would be taken as that when the changes to the locality took place.

Non-domestic hereditaments

For non-domestic rating the basic essentials of rateability remain unchanged and the term hereditament has the same meaning that it always had. To be included in a local rating list a hereditament has to be either used wholly for non-domestic and non-exempt purposes or there has to be at least some part of it which is neither domestic property nor which is exempt. Details of exemptions are to be found in Schedule 5 of the Local Government Finance Act 1988 and include among other exemptions, agricultural premises, places of religious worship, parks and property used for the disabled. The legislation provides that property is domestic if it is used wholly for the purpose of living accommodation. This may be taken to be accommodation or property used for sleeping, eating and associated purposes. Included within the specific definition of domestic property are private garages used wholly or mainly for the accommodation of a private motor vehicle and private storage premises used wholly or mainly for the storage of articles of domestic use. Excluded from the definition of domestic property is certain "short-stay" accommodation; "property" is not domestic if it is wholly or mainly used for the purposes of a business for the provision of "short-stay" accommodation, ie: *inter alia*, accommodation which is provided for short periods to individuals whose sole or main residence is elsewhere. It is however generally considered that a minor non-domestic use of an otherwise wholly domestic hereditament will not give rise to liability to non-domestic rates, if that use is *de minimus* and does not materially detract from the domestic use. The boundary between domestic and non-domestic use is a grey area. All domestic property comprises or is ancillary to, living accommodation, but some living accommodation comes within the definition of rateable non-domestic property because it is used for the purposes of a business for the provision of short stay accommodation. Hotels for example comprise living accommodation but are rateable because they provide short stay accommodation for persons whose sole or main residence is elsewhere. Rooms occupied by the long stay residents and accommodation occupied by the proprietors or staff, as their sole or main residence, does come within the definition of domestic property. The value attributable to the use of these parts has to be excluded from the rating valuation. (see composite hereditaments below). Single units of self-catering holiday accommodation are rateable as are second homes made available for letting commercially for more than 140 days in a year. A particular difficulty is the seasonal hotel which reverts to wholly domestic use out of season. A detailed discussion of these problems is beyond the scope of this chapter.

Composite Hereditaments

The valuation basis for composite hereditaments is provided in Schedule 6 paras 1A and 1B as inserted by Schedule 5 para 38(4) of the Local Government and Housing Act 1989. Hereditaments which comprise both rateable non-domestic property and domestic property are known as composite hereditaments. For such hereditaments it is only the non-domestic use which is to be valued but for rating the valuation must reflect the benefit of the presence of the domestic accommodation within the larger hereditament. The extent of the non-domestic use which is to be valued, is either the actual physical extent of that use, or possibly a notional part of the hereditament consistent with the prevailing pattern of domestic/non-domestic use in the locality. This concept of "notionality" is intended to overcome the bizarre and atypical distributions of uses which are sometimes encountered. Notionality applies only to distributions of uses within composite hereditaments. It cannot be used to bring into rating premises built as business premises, but which are used wholly as domestic accommodation, against the prevailing local pattern of uses.

Valuation

The underlying valuation philosophy remains virtually the same, ie to find the annual value of the hereditament. Previously hereditaments were valued to Gross Value or Net Annual Value as defined by section 19 of the General Rate Act 1967. Now, all non-domestic hereditaments are to be valued to Rateable Value by virtue of The Local Government Finance Act 1988 Schedule 6, 2–(1). This definition is identical to the old section 19(1),(3) Net Annual Value of the 1967 General Rate Act and is effectively the annual rent with the tenant carrying out all repairs, insuring and paying all outgoings. Rateable Value is essentially a full repairing and insuring rental and follows current market practice. The valuation date is defined by Schedule 6,2–(3), and for the purposes of a new rating list is the day on which the list must be compiled, or such day preceeding that day as may be specified by the Secretary of

State. For the 2000 list the date will be April 1 1998. However, Schedule 6,2–(7) states that the state of the hereditament and the environment are to be taken as at the date the list was compiled. This in effect means that the assessment of a hereditament must be based on April 1 1993 values, modified to take account of physical circumstances as they were expected to be on April 1 1995. The old principle of making two valuations, one at value levels at the date of the alteration of the list on the proposal, and a valuation at "tone" levels has gone. All valuations are now to the "tone of the list" and "absolute" falls in value after April 1 1993 do not give rise to reductions in rateable value between revaluations. The matters to be considered at the date the list was compiled are set out in Schedule 6, 2–(7) and are:

(a) Matters affecting the physical state or physical enjoyment of the hereditament,

(b) The mode or category of occupation of the hereditament,

(c) The quantity of minerals or other substances in or extracted from the hereditament,

(cc) The quantity of refuse or waste material which is brought onto and permanently deposited on the hereditament,

(d) Matters affecting the physical state of the locality in which the hereditament is situated or which, though not affecting the physical state of the locality; are nonetheless physically manifest there, and

(e) The use or occupation of other premises situated in the locality of the hereditament.

Where the rateable value is determined to make an alteration to a list which has been compiled, these matters are to be taken as they existed at the date of the proposal. (Schedule 6,2–(6).

Hypothetical Tenancy

All valuations are of course in essence the valuer's opinion of the outcome of imagined negotiation between a purchaser/lessee and a vendor/lessor. Valuations for different purposes require a variety of scenarios for the hypothetical bargain. Rating valuation requires the valuer to determine a rent, which is the result of a negotiation for a hypothetical tenancy. The terms of the tenancy are based in the legislation as interpreted by decisions of the Lands Tribunal and higher courts. In considering the likely rental bids all possible occupiers, including the actual occupier and owner, must be taken into account: *R* v *London School Board* [1886]17 QBD 730. However, where there is only one possible tenant, this would place that tenant in a strong negotiating position which must be taken into account in determining the assessment: *Tomlinson* (VO) v *Plymouth Argyle Football Club Ltd* [1960] EOD 330. In these circumstances the ability of the actual tenant to pay the rent is to be considered. Usually there is more than one potential tenant and the hypothetical tenant is to be taken as being neither an exceptionally good nor an exceptionally bad business person. Rather he/she is of more middle-of-the-road commercial capabilities.

The landlord must not necessarily be assumed to be a local landlord with special interests: *Coppin (VO)* v *East Midlands Joint Airport Committee* [1971] RA 49, but will generally let to the highest bidder. Both parties are to be considered reasonably minded but commercially prudent. Although the letting is assumed to be from year to year it would normally have a reasonable prospect of continuance: *R* v *South Staffordshire Waterworks Co* [1885] 16 QBD 359. The state of repairs to be envisaged for the valuation is not necessarily the existing state of repair at the valuation date. Following the decision in the case of *Wexter* v *Playle (VO)* (1960) 2 WLR 187; (1960) 1 All ER 338 before the Court of Appeal, the party having the liability to maintain premises in repair or to keep premises in repair, has the obligation to put the premises in repair. This assumption has been put beyond doubt by the Rating (Valuation) Act 1999, reversing the Lands Tribunal decision of *Benjamin (VO)* v *Anston Properties Ltd* (1998) RA 53. The rent required by the rating valuations is that which the tenant would bid to reflect the obligation to first put the hereditament into, and then to maintain it in an appropriate standard of repair. The standard of repair to be assumed in determining a rating assessment was considered in the Lands Tribunal case of *Brighton Marine Palace & Pier Co* v *Rees (VO)* (1961) 9 RRC 77, RVR 614 and will differ from hereditament to hereditament depending upon such factors as the age of building, the quality of the locality and type of tenant. As is the case elsewhere in the rating hypothesis the landlord is to be taken to be a reasonably minded person. The facets above are considered to provide an underlying philosophy behind the hypothetical tenancy concept. There are many more aspects which have been explored by the Lands Tribunal.

Valuation Methods

The valuation methods of the rating valuer remain unchanged with rental comparison being the preferred approach wherever possible. Whichever method is used, the end result should theoretically be the same when correctly applied, but the least appropriate methods require unacceptable degrees of adjustment. All buildings should be measured in accordance with RICS/ISVA guidelines.

Rental Comparison

Market rental evidence when it is available, is always to be preferred, as being the most direct evidence and hence rental comparison is for the majority of hereditaments the usual valuation method. Ideally the letting should be grants of fresh tenancies entered into at the April 1 1993 and on rateable value terms. In reality such evidence will rarely be available and the rents will need considerable adjustment to be "reconciled" with the statutory definition. This makes demands upon the valuer's skill and is greatly aided by the backup of an in-house research department.

The rental adjustment process involves considering the actual bargain at the date it was entered into, to establish the rent which would have been paid at that time, had the letting been in the terms of the hypothetical tenancy. Assisted by the evidence, a view is then taken of the rent which would have been paid on those terms at the valuation date. The general process of adjustment of rents to terms of rateable value is well established. However, it must always be remembered that the greater the extent of the adjustments which have to be made the less reliable is the rent as evidence.

Tenants frequently execute works of improvement and extension to their property during the tenancy. The total cost to the tenant of the occupation of the property, the virtual rent, includes both the rent reserved and the cost of those works. The valuer's task is to consider the actual economic decisions of the tenant to determine the rent that would have been paid for the improved property, in terms of rateable value, had it been so improved or extended at the date the current rent was fixed. These improvements will potentially have a value to the tenant only for so long as they are not reflected in the rent he pays to his landlord. Thus the valuer will have to consider the Landlord and Tenant Acts and the terms of the tenancy to establish when the value of the improvements may become included in the rent. This represents the maximum period for which the improvements will have any value to the tenant. The period may be shorter if the "life" of the improvements, either physical or economic, comes to an end earlier, see *Edma (Jewellers) Ltd* v *Moore (VO)* [1975] RA 343 LT.

If the rent is not on full repairing and insuring terms the adjustment for these items must be carried out at the final stage. It is customary for a percentage adjustment to be used for this purpose but a more refined approach, based upon the actual costs, would be preferable. The rents so adjusted to terms of rateable value, will require analysis to unit prices for the comparison purposes. This may entail the use of end allowances which are discussed later. Review periods greater than one year may require adjustment to take account of overage (the amount by which the rent fixed for several years may differ from that fixed on an annual basis), although, in practice, such annual lettings may not always be reviewed on a yearly basis. Various formulae exist which attempt to quantify this figure. However, it may well be that evidence does not exist in the market in a suitable form to substantiate such an approach for rating assessments.

It is to be noted that not all evidence of lettings is necessarily of equal value. Rent reviews during the course of a lease or renewal of a lease determined under the Landlord and Tenant Act 1954, may require a market rental. This does not mean that a market rent is determined. Many factors may influence the eventual figure, not least the relationship of the actual landlord and tenant. The only true market rent is that achieved when a property, with the benefit of vacant possession, is freely and efficiently exposed to the market with adequate time allowed for negotiation.

Profits or Accounts (Receipts and Expenditure) Method

This method is based on the premise that the rental value will depend upon the profit earning capability of a hereditament. The greater the estimated profitability of a business conducted at the premises the more the tenant can afford to pay. Certain hereditaments are valued by this method rather than rental comparison. This is usually because rental evidence does not exist in a form that allows for direct comparison. This would typically be where a hereditament enjoys some form of uniqueness, normally resulting from a factual or legal monopoly. A hotel opposite the only railway station in a provincial town with few other hotels, would be a typical example of a factual monopoly. The license required by a public house to sell intoxicating liquor, would similarly be representative of a legal monopoly.

The method entails an analysis of the trading figures to determine the level of profit the hypothetical tenant might expect from the business conducted at the premises and from that, the amount of rent that he could reasonably afford to bid. This would naturally be on rateable value terms with the occupier assumed to be liable for all repairs and insurance. The gross income is estimated having regard to the actual accounts. From this are deducted the purchases and working expenses (excluding rent but including rates) to give the divisible balance. This figure is then divided between the tenant's share and the amount available for payment of rent. It is to be noted that the tenant's share is a most important constituent and must be sufficient to induce him/her to run the business. An adequate return on the capital tied up in the business must be allowed before considering the sum available for rent and rates.

This method calls greatly upon the skill of the valuer to interpret the accounts which will not have been prepared for the purposes of a rating valuation. The trading accounts to be considered are those for the (three) years preceeding the valuation date as these are deemed to be available to the hypothetical tenant.

Contractor's method

This is often described as the "method of last resort". It is certainly true to say that it is only used where the previous methods are not applicable. It is based upon the premise that cost approximately equates to value. If a hereditament were not available to rent then a prospective tenant must construct one. The land would need to be acquired and the building constructed. The occupier would have to borrow the capital in order to construct the building and pay interest on this loan. The amount of the annual interest payable would be equivalent to the maximum rent he might pay. Briefly the process is as follows. The building costs as at the valuation date are determined. They are then adjusted to take account of any age and/or obsolescence disabilities the actual building may suffer by comparison with a new building, or its equivalent, from which the costs will have been derived. The resulting adjusted replacement cost of the building(s) plus the value of the site works and fees is then "rentalised" or decapitalised. The rate percentage at which the total cost is decapitalised represents the actual costs of making the money available, adjusted to reflect the essential differences between the benefits to be derived from owning the property rather than renting it, on the terms of the hypothetical tenancy. While the philosophy of the decapitalisation rate is interesting, for the 1995 and subsequent rating lists it is somewhat academic as the rate percentage are now prescribed by law as being, 3.67% for hospitals and educational establishments and 5.5% for other hereditaments. (The Non-Domestic Rating (Miscellaneous Provisions, (No 2) Regulations 1994 SI 1994 No 3122.)

The fifth stage of the approach, and possibly the most vital, requires the valuer to stand back and check the results of the calculation against known values of properties as similar in type as possible. Adjustments at this fifth stage may be either downwards or upwards.

The contractor's method was recently scrutinised in *Imperial College of Science & Technology* v *Ebdon (VO)* [1985] 1 EGLR 209, LT and [1987] 1 EGLR 164, CA, and more recently in *Monsanto plc* v *Farris (VO)* (1998) RA 107.

It is interesting to note that since the revaluation in 1973, the statistics produced by the British Costruction Information Service show a divergence between the costs of construction and the tender price for a contract. This has been brought about largely by the economic climate. In fact construction costs appear to have increased approximately twice as fast as the tender price. If a potential occupier were to have a building constructed, would he/she construct the building or put it out to contract? The decision would have an impact on the capital involved.

"Phasing" or "Transitional arrangements"

For the 1990 revaluation, the government enacted legislation to ease the burden on businesses facing high increases in payments of local taxation. They placed limits on proposed increases and these were "funded" by similar limits placed on reductions. These were known as transitional arrangements (also known as phasing) and they have been retained and developed in the last revaluation and amended since. Transitional arrangements are not part of a valuation and consequently are not considered in detail here. However, their implications for the actual liability of a ratepayer must be taken into account. Failure to consider the consequences may result in a client paying more than they otherwise would which, at the very least, would lower the valuers' esteem in the eyes of their client. For a more detailed examination of phasing arrangements see the Non-Domestic Rating (Chargeable Amounts) (Amendment) Regulations 1996 SI 1996 No 911. Explanations may be found in *Phasing pain or relief* [1995] 24 EG 137 and *Resolving the problem of the RxJ/S syndrome* CSM July/August 1996 pp 40–41.

1995 revaluation issues

The 1995 revaluation created particular problems due to the nature of the property market, particularly in the south of the country where rental values were falling. (This is in contrast to the 1990 revaluation where values had increased.) This drop in values was not always reflected in the headline rent for a number of reasons. For instance, there was a general reluctance to reduce the actual rent paid as this could have a knock on effect when these transactions were used as comparables. Reverse premiums were sometimes paid by landlords to tenants to maintain the rent apparently passing. Similarly, rent-free periods, common in normal market conditions were sometimes extended, effectively reducing the rental value while keeping the actual rent paid "artificially" high. Sometimes landlords contributed to the costs of fitting out premises to the same effect. All these factors needed to be taken into account when analysing rents for rating purposes. The following examples are given to illustrate the various concepts and suggest one of the various approaches to solving the problem.

Study 1

The letting of a shop unit is agreed at a rent of £35,000 pa on FR&I terms for the next five years. The landlord paid a reverse premium to the tenant of £ 30,000.

As a reverse premium was paid to the tenant, the true rental value would be higher than the agreed rent passing of £35,000. To adjust the rent, an annual equivalent of the reverse premium would be added.

		£
Rent passing		35,000
Less		
Reverse premium	30,000	
Decapitalise at YP 5 years 8%(1)	3.99	7,519
		27,481
Actual rental value say		27,500

Notes

(1) These have all been undertaken using single rate figures, ie from a freehold landlord's position. It could be argued that the figures should also be analysed from the tenants perspective, using dual rate tax adjusted figures and that the actual rent paid would reflect these two valuations.

The concept is similar to that commonly used for valuations in connection with surrender and renewals.

Study 2

A letting has been agreed of a shop unit at a rent of £45,000. The landlord has contributed £25,000 towards the costs of fitting out the unit. There will be a review to full rental value in five years time.

This is very similar to the above example. The actual rent paid reflects the fact that the landlord is contributing towards what would normally be a tenants expenditure. Consequently the true rental value would be less than the £45,000 pa agreed.

		£
Rent passing		45,000
Less		
Reverse premium	25,000	
Decapitalise at YP 5 years 8%(1 above)	3.99	6,265
		38,735
Actual rental value say		38,750

Study 3

The rent of shop is agreed at £75,000 pa until it is reviewed to the open market rental in five years time. A rent-free period of two years has been granted.

Again the actual rent agreed does not reflect the actual rental value as a rent free period has been agreed and the real rental value will be less. In order to calculate the real rental value, the value of the rent-free period must be taken into account. On the basis that the landlord is not losing anything of value, the capital value of the term before review, must be the same for the actual agreed terms as they would be if the actual market rent was being paid, ie $V tar$ = Vtomr where $V tar$ = the value of the term on the actual rental and $V tmor$ = the value of the term at an open market rental.

Value of Term on the actual rental (*Vtar*)		£
Rent passing		75,000
YP for 3 years at 8% (1 above)	2.577	
Deferred 2 years at 8%	0.857	2,208
Capital Value of term		165,600

As this must = the value of the term at the open market rental (*Vtar* = *Vtom*), we can decapitalise it to find what that rental should have been:

Capital Value of term		165 600
Decapitalise YP 5 years at 8% (1 above)	3.99	41 503
True open market rental Say		£41 500

The following case studies relate to England and Wales only.

Shops

See also chapter 7

These are normally valued by direct rental comparison as comparable rental information is reasonably common. The method most often used to compare one shop with another is that of Zoning. This has not always been the case and the method has a chequered history. It is probable that the method arose from the valuation of "parlour" shops in the earlier part of this century, when shops commonly comprised the front room of a converted house, for sales space, with the rear room for storage[1]. The rear room (storage) was often valued at half the rate of the front room (sales), the depth of the rooms determining the "zone" (commonly 15 or 20 ft). Over the years the dividing walls were removed but the process of using 15/20 ft zones and halving back continued.

The method is not without its problems. In some circumstances great sophistication is required in analysis and valuation. Slavish adherence to the principle of halving back is not to be recommended. "Zoning should be subservient to the valuation and not valuation subservient to zoning." Whatever the valuation entails, it should be remembered that "as you devalue so must you value."

---

[1] Shops – Valuation for Rating, Rating and Valuation Association, 1983, p6.

Essentially a shop is zoned back from the main frontage in strips. These are usually 20 ft, 6 or 7 m. However it is possible to use natural zones where the shops used in analysis and valuation have some obvious common depth. This could be the case in a modern shopping development. There are many advantages in using a natural zone for a specific valuation. The main disadvantage is that by definition, the natural zone will not be common to many other shops. Consequently any analysis will be of limited use. However, if the more conventional zone pattern is adopted, the analysis can be used for any valuation for which the comparable is itself valid.

The first zone is considered the most valuable, with value decreasing with distance from the main frontage. It is established valuation convention to halve back. Hence the first zone (Zone A) will have a value of £X. The next zone (Zone B) will have a value of .5X and so on. However, the valuer must be watchful for local market rents which may show this to be inappropriate. The reduction used should follow the local market evidence. The other areas of the shop which have value are similarly assessed "in terms of zone A, (ITZA)" in order to obtain a total picture as to the likely value.

The resulting fractions, or "X" factors as they have sometimes been known, may seem very odd if not interpreted in monetary terms: *Lewis's* v *Grice (VO)* [1982] RA 224. Hence market analysis might show the zone A to be £250, and some other part of the shop to be worth £16. The X factor would therefore be 16/250 or 0.064. Where a significant differences in area occurs between rented shops and those to be valued, the figures may be reconciled using an adjustment known as a "quantity allowance". Where the quality of shops is significantly different, a quality allowance may be used. Where the layout of a shop has significant disadvantages over another a disability allowance may be used.

Any rateable plant and machinery may be valued by an annual equivalent on the depreciated replacement cost (Contractor's Method) if not more conveniently incorporated into the value of the building directly.

While most shops can reasonably be analysed using zoning methods, it is sometimes appropriate to use an overall basis. This depends not only on the subject property but also the nature of the comparables. In *Lewis's* v *Grice* 1982, a large department store in Liverpool comprising over 37,000 m^2 and seven floors was valued in terms of Zone A due to the nature of the comparables. However in *Argyll Stores Propertiess Ltd* v *Edmonson (VO)* (1989) RVR 113 the Lands Tribunal accepted a valuation of a supermarket on an overall basis, as there were acceptable comparable supermarket style shops capable of comparison on this basis. It should also be remembered that the needs of the hypothetical tenant of a large store may be quite different from the hypothetical tenant of small lockup shop.

When analysing rents of shops, it is useful wherever possible, to start with a simple unit free from disabilities, return frontages and similar peculiarities. This should enable the differences to be highlighted in order to construct an overall picture.

Study 4

Analyse 1 The High Street, a city centre lockup shop 5 m wide by 12 m deep, situated in the south east of the country. Use the information to assess a similar shop 5, The High Street, which is 6 m wide by 12 m deep. No 1 was let in March 1993 on a 15-year lease with 5 yearly rent reviews at £45,000 pa on FRI terms.

Analysis
Taking 6 metre zones and halving back,

	Size	Area	"x" factor	Area (ITZA)
Zone A	5m × 6m	30	A	30
Zone B	5m × 6m	30	A/2	15
Total area ITZA				45

£45,000/45 = Zone A rent of £1,000 per m ITZA in RV terms (1)

Valuation (i)

	Size	Area	"x" factor	Area (ITZA)	at £	£RV
Zone A	6m × 6m	36	A	36		
Zone B	6m × 6m	36	A/2	18		
Total area ITZA				54	1,000	54,000
					RV =	£54,000

OR

Valuation (ii)

	Size	Area	at £/m	£ RV
Zone A	6m × 6m	36	1000 (A)	36,000
Zone B	6m × 6m	36	500 (A/2)	18,000
				54,000
			RV =	£54,000

Notes

(1) Although there are five yearly rent reviews, in this case there is no market evidence to support a deduction for overage.

Study 5

Using the comparable evidence above, value 7, High Street, which is 6 m wide by 17 m deep. To the rear there is a store comprising 30 m².

Valuation

	Size	Area	at £/m	£
Zone A	6m × 6m	36	1,000 (A)	36,000
Zone B	6m × 6m	36	500 (A/2)	18,000
Remainder	6m × 5m	30	250 (A/4)	7,500
Store		30	100 (A/10)	3,000
				64,500
			RV =	£64,500

Notes

(1) "x" factor from analysis of rents used to derive price per m².

Study 6

Using the above comparable value 11 High Street which has a return frontage. The unit is 6 m wide by 20 m deep. There is a basement stock room of 15 m², and you consider the basic layout to have an inherent trading disability compared with comparable properties.

Valuation

	Size	Area	at £/m	£
Zone A	6m × 6m	36	1,000 (A)	36,000
Zone B	6m × 6m	36	500 (A/2)	18,000
Remainder	6m × 8m	48	250 (A/4)	12,200
				66,000
Add:				
Return frontage 5% (1)				3,300
Basement stock		15	50 (A/20)	750
				70,050
Deduct: disability 3% (2)				2,100
				67,950
RV Say				£68,000

Notes
(1) From an analysis of similar shop units with return frontages. There are a number of ways that return frontages can be analysed : per metre run, a spot figure or a percentage addition to the value of the ground floor. The most appropriate method will be determined by the nature of subject property and the comparables.
(2) To allow for the disability and enable the assessment to be reconciled with similar properties excluding the disability. The percentage should again comes from market analysis. Other end allowances such as quantity allowances would be deducted at this stage of the valuation as appropriate.

Study 7

Value a supermarket which is situated on an island site to the south of the High Street, with similar sized units of similar quality. The units share a large shoppers car park, but have their own loading access.

Valuation

Description	Area in m^2	Price/m^2	£
Sales (1)	2,750	70 (2)	192,500
Meat preparation	100	52	5,200
Canteen	88	43	3,784
Warehouse	582	35 (3)	20,370
Cloaks	12	17	204
Store	23	17	391
Switch/motor room	51	15	765
Gas main room	3	10	30
			223,244
		Say RV =	£223,200

Notes
(1) An overall treatment for the sales space, rather than zoning is used, as this is a large unit with good comparable evidence of similar sizes and there is no direct relationship with small "zoned" shops.
(2) Evidence from analysis of similar units.
(3) One would expect the parts directly ancillary to the sales space to have a value relative to that use and the space ancillary to

the warehouse to have values relative to the adopted warehouse value.

Study 8

Sometimes buildings can span different uses. In the following example a large versatile space is used as a trade warehouse. This space could easily be used solely for warehousing or out of town retail. The valuation contains elements of both valuation approaches and is increasingly common with the growth in town retailing.

Value a trade warehouse which is situated on the edge of town. The accommodation comprises both high and low bay warehousing, offices, enclosed loading bays, entrance and exits and car parking. Unlike the comparables a modern sophisticated sprinkler system has been installed.

Valuation

Description	Area in m²	Price/m²	£
High bay warehouse			
Full height (1)	5,850	20	117,000
Under offices (2)	500	18	9,000
Low bay warehouse	7,500	18	135,000
First floor offices	800	25	20,000
Enclosed loading area	250	14	3,500
Enclosed entrance and exit	50	15	750
			285,250
Add sprinklers at 2.5%			7,131
			292,381
Add car parking			
850 spaces at £35			29,750
			322,131
RV Say			322,100

Notes
(1) Height can be an important factor see study 11 also.
(2) In this case the height is restricted because of the first floor offices. In some cases restrictions may be imposed by a mezzanine floor. In either case a distinction has to be made in value which depends upon the individual circumstances of the case.

Offices

See also chapter 6.

Many types of offices exist but they should all be capable of assessment by rental comparison. The valuation may require assessment of a single office, suite of offices or a whole office block, depending on the extent of the rateable occupation. In any event the basic principles are the similar. Modern offices are measured to net internal area (NIA), which excludes common areas such as staircases, toilets, fire corridors, etc. In a multiple occupation building , reception areas, high quality atria, etc, which benefit all tenants will generally be reflected in the value of the separate occupations. Occasionally some residual value may remain in the hands of the landlord as a separate assessment.

The nature of the demountable partitioning needs to be considered to determine whether it is part of the setting within which the business is carried out: *British Bakeries Ltd* v *Gudgion (VO) and Croydon London Borough Council* [1969] RA 465. If this is the case then the partitioning is rateable and must be reflected in the assessment. This is usually done by increasing the unit price or taking a percentage of the depreciated replacement cost of the partitioning. If on the other hand the partitioning is necessary for the purposes of the particular business it is considered to be plant, and so long as demountable partitioning remains unlisted in the relevant statutes, will not be rateable. Older office buildings, frequently the product of conversion of residential accommodation, will often be measured and valued on an effective floor area basis (EFA).

Study 9

This is an analysis and valuation of offices on first and seconnd floors above a bank in the High Street of a country town let to a solicitor in 1989 at £2 400 pa, the tenant being responsible for internal repairs. EFA of the first floor 50 m²; seconnd floor 48 m² (1) and there is no lift. Rents are considered to have fallen by 10% between the letting date and the April 1 1993. There are no landlord's services.

Adjustments	£
Rent Passing	2,400
Deduct for external repairs and insurance(2)	150
Rent in RV terms	2,250

Analysis

Either

£2,250 for 98 m² = £22.96 per m²

Or in terms of first floor space

1st floor 50 m² at	X =	50
2nd floor 48 m² at	0.75X =	36
		86

2,550/86 = £29.65 per m² 1st floor

and hence £22.24 per m² 2nd floor (0.75 of £29.65)

This second analysis involves "valuer judgment" of the relative values of floors and would preferably be made with the benefit of some evidence. The analysis provides evidence of rental values in terms of RV at April 1989. With knowledge of the amount of the general value change to April 1993 in the locality the valuer may decide to apply equivalent prices less 10%.

Assessment

1st floor	50 m² at	£26.70	1,335
2nd floor	48 m² at	£20.02	961
			2,296
RV Say			£2,300

Notes

(1) EFA would be probably be used due to the nature of the accommodation.

(2) It would be necessary to estimate the likely expenditure as at April 1 1993. The calculation and apportionment of costs would not be without problems.

(3) This calculation has been produced for illustration purposes. While the letting of the hereditament itself is usually very good evidence, it is not the only, or necessarily the best evidence. If sufficient evidence exists to assist the valuer's judgement of the ratio of values on the first and second floor, there will probably be additional rental comparable evidence to help with the assessment.

Study 10

"Zenith Buildings" is a large, good quality office block of recent construction situated in a Midlands city centre. The accommodation comprises 11 floors each of 500 m²; lifts; central heating; 20 car parking spaces; occupied by one tenant.

Rental Evidence

"Nadir House" is a similar office block in a similar location. Eight floors each of 465 m²; atrium; lifts; central heating; 20 car parking spaces separately let at £110 per space, all floors let during March 1993.

Ground and first floors: 15 year lease; reviews at 5 yearly intervals. Rent £30,000 pa, tenant doing internal repairs; separate service charge reviewed annually, currently £ 330 pa.

Second to 6th floors inclusive each let separately on leases as above; rent for each floor £15,500, current service charge per floor £820 pa.

Seventh and 8th floors: let as one unit on lease as above; rent £32,000 pa, current service charge £1,450 pa.

Analysis of "Nadir House"

	Area m²	£/m
Ground and 1st floor		
lease rent	30,000 (1)	
Area 930		
Thus rent per m² =	30,000/930	= £32.25
2nd–6th floors (5 separate floors)		
lease rents	15,500	
Area 465		
Thus rent per m² =	15,500/465	= £33.33 (each floor)
7th and 8th floors		
lease rent	32,000	
Area 930		
Thus rent per m² =	32,000/930	= £34.40

Notes

(1) The tenant already undertakes internal repairs and pays for external repairs and insurance by way of the service charge, making the rent full repairing and insuring.

(2) It has been assumed that there is minimal partitioning on each floor.

(3) It is not uncommon to see a higher rent on the top floors of an office building. Initially it might seem to be appropriate to apply the prices per m² directly as follows:

Valuation of "Zenith Buildings"

		at £/m	£ RV
Ground floor	500 m² at	32.25	16,125
First floor	500 m² at	32.25	16,125
2nd–9th floors	4,000 m² at	33.33	133,320
10th floor	500 m² at	35.00	17,500
			£183,070
Add partitioning (1)			
£150,000 at 5.5%			8,250
			191,320
20 car parking spaces at			
£100			2,000
			193,320
RV Say			£193,300

The tenant for the whole building is however more likely to see the value in terms of an overall price per m². The hypothesis requires that the valuer considers the rent of the whole block (the rateable hereditament) which may or may not equate to the sum of the values of the industrial floors. From the evidence we have the valuer may decide that the value of the whole building, 5,500 m² is £33.50 per m² to include partitioning.

Valuation

Area m	at £/m	£ RV
5,500	33.50	184,250
20 car parking spaces	100.00	2,000
		186 250
RV Say =		£186 250

Notes

(1) It is assumed that there is clear evidence of demand for office blocks of such size and hence no quantity allowance is appropriate.

Factories and Warehouses

See also chapters 5 and 16.

　　Factories have always been valued to net annual value so there is virtually no change in the valuation process. Many factories are similar to each other in the basic facilities that they provide. These can normally be assessed using comparable rental assessments. However, certain factories may be specialised with much plant and machinery, as in the case of oil refineries. These will need special treatment which requires specialist knowledge which is outside the scope of this chapter.

Warehouses were previously valued to gross value and must now be assessed to rateable value. Modern warehouses are more than just boxes, being designed for efficient storage and retrieval. Regard must be paid to the working height and units with a height less than 7.6 m are not likely to be so well suited to this purpose. The unit assessment must be reduced accordingly. Where good comparable evidence of a similar style exists locally, it may be that valuation can take place on an overall or gross internal area basis. In other cases a more sensitive approach may be needed. Some parts of the country have their own "custom and practice", but whichever method is adopted the RICS/ISVA measuring codes are to be preferred.

Study 11

Value the Enterprise warehouse which is situated in the north of England on its own site and with its own loading facilities. It comprises 700 m^2 of which 70 m^2 are offices. The structure is steel framed with metal PVC covered cladding. Eaves height is 6.5 m and the roof is lined. In addition there is also a sprinkler system. You have a comparable warehouse of similar construction with an eaves height of 4 m and sprinkler system. It comprises some 600 m^2 including 60 m^2 of office accommodation. It was let in March 1993 on FRI terms at £15,750 pa.

Analysis

	Area in m^2	"x" factor	Total
Main Space			
(Work/storage)	540	1.00 (1)	540
Offices	60	1.25	75
			615

£15,750/615 = £25.60 m^2 Main Space in RV terms

Valuation of "Enterprise Warehouse"

	Area in m²	"x" factor	Total	£/m	£
Work/storage	700	1.00	700		
Offices	70	1.25	87.5		
			787.5	29.5(2)	23,231
Say RV =					£23,225

Notes

(1) The work area determines the value of the unit and other values are related to it. Hence the weighting of 1.00.

(2) From other evidence and valuer judgment the increased height is considered to increase the value in main space terms by about 15%.

Hotels

See also chapter 18.

Hotels often have some form of monopoly, eg by virtue of their position and are not very frequently let. They are generally assessed using the profit's method. In some cases there may be a sufficient number of comparables to enable a rating assessment using comparable rental evidence. This would be the case with a number of hotels in such a seaside resort as Blackpool.

Study 12

A good class hotel with excellent all the year round trade. Three years accounts have been examined and show that the trading pattern is reasonably consistent. The following figures are based on the accounts for the year ending December 31 1992.

Gross Receipts	£
Bedroom lettings	
Restaurant	
Bar	490,000
Less :	
Purchases	53,000
Gross Profit	437,000

Working Expenses
salaries and wages (including National Insurance)
electricity, gas, water (metered) stationery, printing
postage, telephone and advertising, laundry and
cleaning materials, repairs and renewals of furnishing
and equipment (1)
repairs to structure (2)
Insurance; structure and 3rd party, rates payable and

sundry expenses	360,000
Net Trading profit	77,000
Less :	
Interest on Capital (3)	
Chattels	
Cash	
Stock 250,000 at 8%	20,000
Divisible Balance	57,000
Less :	
Tenants share at 40% (4)	22,800
Amount left for rent	34,200
RV	34,200

Notes

(1) This figure should represent the annual amount necessary to maintain the trade at a consistent level.

(2) Repairs and insurance of property are required to bring the rent to NAV terms.

(3) In the rating hypothesis the tenant has to provide these items from their own capital. They are therefore entitled to the interest on the "opportunity cost" of the capital.

(4) This share should be sufficient to induce the tenant to trade. It is the amount that they would require for running the business and would normally be in addition to a managers salary. It is common for a percentage to be used.

(5) A benefit of the antecedent valuation date is that the actual rate liability at April 1 1993 may be adopted.

The RV as determined by the profit's method can be analysed on a bedroom basis and applied to other hotels where accounts are not available or are unreliable. A standard unit basis is adopted taking the "best" double bedrooms as 100% or a factor of 1.00. This is often referred to as the "Reduced Bedroom Basis".

Study 13

An average quality hotel, typical of its type, in a large seaside town with good comparable evidence.

Accommodation
Ground floor: lounge, bar (full on-licence), residents' dining room.
First floor: 10 double, 3 single bedrooms all with sea views
Second floor: Similar to 1st floor
Third floor: 2 double bedrooms with no sea views, 2 staff bedrooms (1) Comparable rental evidence shows a rent per reduced bedroom between £1,200 for the best to £600 for the poorest.

Valuation

Floor	Bedroom	Reduced Bedrooms
1st floor	10 double at 100%	10
	3 single at 75%	2.25
2nd floor	10 double at 75%	7.5
	3 single at 50%	1.5
3rd floor	2 double at 50%	1.0
		22.25

Reduced bedroom rent from similar comparable say £1,000 (2)
$$22.25 \times 1,000$$
$$= \underline{22,250}$$
RV   £22,250

Notes
(1) At the time of writing, the staff bedrooms are subject to the Council Tax, ie this is a composite hereditament
(2) The valuation could be undertaken by the profit's method but comparable assessment is always to be preferred where good rented comparables are available.
(3) Were there to be extensive non-resident use of the bar and the dining room these might require separate treatment in the valuation.

Schools and Nurseries

Most schools, both private and those occupied by Local Education Authorities, are now assessed on a contractors basis. However, some private schools may have sufficient comparable rental evidence to be valued by rental comparison. This may be by comparison with other schools or with similar premises which are let for other uses. Hence, a large Victorian dwelling-house might be let for use as an office or a private school at a similar rental if planning permission were available.

In most cases public schools and purpose-built independent schools will have to be valued by the contractor's method. Particular problems will be experienced if the buildings are old, of great historical and architectural merit but functionally obsolete. In such circumstances a valuer would be well advised to have regard to *Eton College* v *Lane (VO)* [1971] RA 186 considered below.

Study 14

A leading case regarding a public school is *Eton College* v *Lane (VO)* [1971] RA 186 and this study is based upon the valuation principles approved by the Lands Tribunal. Eton College comprises buildings that are old, of great intrinsic merit and interest but which are less than ideal for use as a large boarding school. The valuation approach can interestingly be compared with the Scottish example of *Aberdeen University* v *Grampian Regional Assessor* (1990) RA 27, a rather detailed consideration of the problems encountered in the valuation of a university.

Valuation

Buildings	£	£
Estimated replacement costs	16,000,000	
Less:		
Allowances for age and obsolescence		
factors say 63%	10,080,000	
Effective capital value		5,920,000
Site		
22.67 ha at £ 40,000 per ha	906,800	
Less: 63%	571,284	335,516
Effective capital value of land and buildings		6,255,516
Playing fields		
80 ha at £5,000 per ha		400,000
Total effective capital value		6,655,516
Decapitalise at 3.67% (NB see note 1)		244,257

Deduct: end allowance 10%	24,426
	219,831
RV Say	£219,830

Notes
(1) For the 1995 rating list there is the statutory decapitalisation rate as provided for, by 1994 S1 3122 The Non-Domestic Rating (Miscellaneous Provisions) (No2) (Amendment) Regulations 1994.

Public Buildings

These are best assessed by comparable rental evidence where such evidence is available. This would normally be the case with local government offices. In other cases such as public museums or libraries it would be unlikely that suitable rental evidence exists. In these cases it will usually be appropriate to use the contractor's method. This might also be the case with municipal sports facilities which at first sight may seem suitable candidates for the profit's method. However, these facilities may not be run on a commercial basis but rather to provide the facilities to the local community and as such the trading accounts would prove unsuitable. (At the time of writing, the valuation approach for these types of property is the subject of dispute.)

Study 15

A modern dry ski slope owned and operated by a local district council in an area of relatively low land values. The unit comprises a dry ski slope, mechanical lift, car parking and ancillary buildings.

Valuation

	£
Site works including fencing, services, floodlighting, car parking and mechanical lift (1)	200,000
Building costs including all fees	80,000
Total building costs	280,000
Add:	
Land costs 9 acres at £16,000	144,000
Total effective capital value of land and buildings (2)	415,000
Decapitalise at 5.5% (3)	22 850
RV Say	£22 850

Notes
(1) The extent to which certain equipment will be considered part of the rateable hereditament will be governed by the Plant and Machinery Regulations as disused at the end of this chapter.
(2) As the ski slope is of modern construction obsolescence is minimal.
(3) Statutory decapitalisation rate as provided for, by The Non-Domestic Rating (Miscellaneous Provisions) (No 2) (Amendment) Regulations 1994 SI 1994 No 3122.

Study 16

A modern crematorium built by a district council; at the time of valuation there are some 2,750 cremations annually but the premises has a full working capacity of 7,500 cremations annually; thus some of the facilities are surplus to requirements.

Valuation	£
Estimated buildings costs	
Crematorium	650,000
Cremators (furnaces) (1)	120,000
Site works	50,000
	820,000
Add:	
Land	30,000
	850,000
Deduct: for surplus capacity at say 40% (2)	340,000
Effective capital value	510,000
Decapitalise at 5.5%	28,050
RV Say	£28,000

Notes
(1) Such items are rateable as plant even though they do not occur in an industrial building.
(2) This item is estimated, *inter alia*, by reference to the ratio between the actual use and potential capacity.

Cinemas

When the 1973 valuation was undertaken cinemas were declining in number. Today the birth of the multiplex cinema has given this entertainment form a new lease of life. The method of assessment usually used is a form of direct profits valuation by taking a percentage of gross receipts. The number of seats available and "sold" are used to calculate the number of full houses per week. This does not mean actual full houses but the number of "customers" divided by the number of seats in order to give a notional number of full houses per week.

Study 17

A modern cinema with 1,250 seats; average admission charge is £4 per seat and there are 7.5 full houses per week throughout the year.

	£
Box office receipts per full house £5,000 × 7.5	37,500
	52
Estimated annual box office receipts	1,950,000
Receipts from sundries; ice cream etc.	104,000
	2,054,000
RV at 6%	123,240
RV Say	£123,250

Petrol Filling Stations and Garages

See also chapter 14.

Forecourt values could perhaps be assessed by a direct form of profits valuation based on the throughput. It is however more usual for a rental per thousand litres to be established to include the tanks, forecourt, kiosk, canopy, rateable pits for tanks, etc. Ancillary accommodation may be assessed using rental comparison with similar units. Lubrication bays and sales space may be valued by reference to rental evidence within the locality.

These days oil companies are rarely interested in car sales and lubrication bays. Their main concern is the sale of petrol and they are increasingly disposed towards associated car wash and shop sales. High volumes of petrol sales are vitally important and where an annual throughput of four to five million litres or more exists, oil companies are likely to compete with each other. Where there is no tie competition may produce an overbid. If the throughput is much below four million litres the station may be part of a secondary market with significantly lower rentals. During the 1990s super-stores and hypermarkets developed a considerable presence in petrol sales and were estimated to account for 23% of retail sales in 1997. Discounting and price matching caused the number of petrol sales operators to decline from 18,500 in 1992 to 14,800 in 1997 and with reducing profit levels on petrol sales operators are seeking to maximise income from other profit centres. Consequently, shop sales are increasingly important.

Location is vitally important with the left hand side of a commuter route, going out of town, likely to be the most profitable. However, throughput is fundamentally important. The gross profit to the occupier produced by the petrol sales may be calculated by multiplying the throughput by the dealers margins. In such circumstances it is worth noting that the profit per litre made from say four star petrol, unleaded petrol and derv may be different. Care should accordingly be exercised when comparing stations with broadly similar throughputs, as the make up of those throughputs may affect the profitability.

Car wash plant is not rateable and is notionally cleared from the site. It is the potential of the site for car wash use which is valued by reference to potential turnover.

Study 18

A filling station primarily serving a rural village. The filling station stays open from 0700 to 1900 hours each day of the year, but is closed on Bank Holidays (1).

	£
Petrol sales	
1,820,000 litres at £5 per 1000 litres (2)	9,100
Workshop	
100 m² at £15	1,500
RV	£10,600

Notes

(1) It may be claimed that these hours are unreasonably restricted and that as a consequence the operator may be able to increase throughput and profitability by extending the opening hours. Valuation *rebus sic stantibus* would suggest that this should not be considered. Furthermore the tenant-operator is reasonably efficient and should not be considered to be "super-efficient".

(2) The petrol sales would normally cover the rental element of the forecourt, pits for tanks, canopy and associated kiosk sales.

Any excess area over that necessary to sell petrol may be valued separately.

A leading case dealing with the circumstance governing the rateability of petrol storage tanks is *Shell Mex & BP* v *Holyoak (VO)* [1951] 1 WLR 188.

Study 19

A busy filling station on a main commuter road out of a city centre. The pump layout is a modern starting grid style, and the facilities include a car wash and sales area additionally to the kiosk.

	£
Petrol Sales	
5,687,500 litres at £9 per 1,000 litres	51,187
Sales area 25 m² at £50	1,250
Car Wash (1)	52,437
RV Say	£52,425

Notes
(1) Car wash plant is not rateable, the value is reflected in the potential of the site.

Plant and Machinery

While the general principle has not changed since the previous edition of this text, certain amendments have been made. Plant and machinery may form part of the rateable hereditament in many types of property, not only industrial hereditaments. However, it must be capable of classification within one of the generic types specified in The Valuation for Rating (Plant and Machinery) Regulations 1994 (SI 1994 No 2680). Detailed tables of the plant and machinery to be considered are included within these regulations and consequently are not stated here. However, the main elements are:

- Class 1 Power plant
  (eg steam boilers, dynamos, storage batteries, transformers, cables, water wheels, compressors, windmills, wind turbines, solar cells)
- Class 2 Services
  (eg heating cooling and ventilation, lighting, draining, supplying water, protection from hazards)

- Class 3 Conveyers & Pipelines
  (eg railways, lifts, cables, poles and pylons, towers, pipe lines)
- Class 4 Plant which is on the nature of a structure.
  Class 4 contains two tables of plant and machinery.

Table 3 is automatically included. Items in table 4 are not included if their "total cubic capacity (measured externally and excluding foundations, settings, supports and anything which is not an integral part of the item) does not exceed four hundred cubic metres and which is readily capable of being moved from one site and re-erected in its original state on another without substantial demolition of any surrounding structure." Four hundred cubic metres is about the size of a modest two-storey house. .

Plant and machinery in class 4 must be considered to be in the nature of a structure and reference should be made to the relevant case law. In all other cases the plant and machinery is automatically rateable by virtue of inclusion in this order.

The value of plant and machinery may be included by an increase in the unit price as with central heating. In other cases it may be included by taking an annual equivalent of the depreciated replacement cost. This usually requires specialist knowledge. The current decapitalisation rate is 5.5% but is subject to amendment.

This chapter was drafted prior to the 2000 revaluation. Since then a number of changes have occurred. Petrol stations have stopped selling four star petrol and the new rateable values came into effect on April 1 2000, with an antecedent valuation date of April 1 1998.

There have also been some changes in the basis of valuation. The most notable of these arose from *Benjamin (VO)* v *Anston Properties Ltd*, (1998) RA 53. Here, the Lands Tribunal held that if a hereditament was in a state of disrepair that would depress the rental value, then the rating assessment should also reflect this fact. This ran counter to the Schedule 6 valuation basis that existed at the time. The result would have been entries in the valuation lists which varied according to the state of repair. Since then, The Rating (Valuation) Act 1999 has amended the Local Government Finance Act 1988 so that according to Scedule 6 the estimated rent takes into account the following assumptions:

(a) the first assumption is that the tenancy begins on the day by reference to which the determination is to be made;

(b) the second assumption is that immediately before the tenancy begins the hereditament is in a reasonable state of repair, but excluding from this assumption any repairs which a reasonable landlord would consider uneconomic;

(c) the third assumption is that the tenant undertakes to pay all usual tenant's rates and taxes and to bear the cost of repairs and insurance and other expenses (if any) necessary to maintain the property in a state to command the rent mentioned above.

Hence at the actual date of the rateable value assessment, the property is considered as being in a reasonable conditon consistent with the market rent achievable, with the property being maintained by the tenant thereafter to the same standard.

Transitional relief has been maintained. The transitional arrangements remain with the property and consequently pass on to new purchasers although there are still calls for it to be abolished.

© Michael Jayne 2000

Chapter 13

Development Properties

In a competitive land market development can ensure the efficient use of land resources. If a property can be used more profitably or possesses latent development value, capital investment can realise that development potential. The evaluation and appraisal of individual development projects is essential to ensure that the purchase and development costs are reasonable in terms of the final purpose and value on completion of such developments. Such evaluation of development potential may encompass market conditions and prices, site capabilities, needs and problems along with legal and planning issues. What is an apparently simple and straightforward valuation task is arguably the most susceptible of error, and one, moreover, which is heavily reliant upon individual judgment. Because of the sensitivity to fluctuation by the internal elements within a development scheme, the results of an appraisal or viability study can never be wholly accurate but merely reflect informed professional opinion. Nevertheless, despite the speed of legislative change, the vagaries of political policy and economic volatility, there remains a need to underpin entrepreneurial flair with some degree of analytical rigour.

Probably the most important part of the valuation process when considering development properties is the appraisal of all the determining factors which underlie and condition the various components of the valuation itself. Rental income or capital value, costs of construction and other development charges, initial yield and thus capitalisation rate, building contract period and the total time taken to complete and dispose of the project, and the cost of finance and method of funding the scheme are all dependent upon a number of critical surveys and investigations. Some of these can be listed briefly as follows:

(i) *Planning Policy*
     Reference should be made to any approved statutory development plans, relevant planning and development briefs, design guides, zoning designations, density standards,

user constraints, conservation policies, highway proposals and general attitude towards planning gain agreements.

(ii)   *Planning history*

Inquiries in respect of past and present planning decisions relating to the site in question and to surrounding properties should be made. It also pays to have some regard to the personalities and politics involved in local planning issues and be aware of the influence of various interest groups and amenity societies.

(iii) *Statutory undertakers*

In addition to any discussions held with the local planning authority it will usually be necessary to consult with certain statutory undertakers to ensure the availability of such services and facilities as gas, water, sewerage, drainage, electricity, telephones and transport.

(iv) *Market analysis*

It should almost go without saying that research into prevailing market conditions is a prerequisite to valuation and will normally take account of such matters as capital values, rental levels, past rental growth, achieved yields, vacancy rates, outstanding planning permissions, developments in the course of construction, cleared sites and other opportunities, comparable transactions, rates and some conception of future trends.

(v)   *Locality*

Special consideration would be given to the immediate vicinity of the site to be valued; in particular, the general environment, communications, transport, labour, local access, services, facilities and adjoining premises.

(vi) *Site*

A thorough and comprehensive survey of the site is, of course, essential and will cover aspects such as ownership, acquisition, boundaries, area, topography, landscape, stability, access, layout, buildings, services, archaeological remains and any other physical factors likely to affect development potential.

A development valuation or viability study can be undertaken for a number of different purposes, which include:

(i)   Calculating the likely value of land for development or redevelopment where acceptable profit margins and development costs can be estimated.

(ii) Assessing the probable level of profit which may result from development where the costs of land and construction are known.

(iii) Establishing a cost ceiling for construction where minimum acceptable profit and land values are known.

(iv) Establishing the optimum intensity or quality of a proposed development.

(v) Estimating a required minimum level of rental income to justify the development decision.

A combination of these calculations can be conducted to explore alternative levels of acceptable costs and returns, but all valuations for development purposes require an agreed or anticipated level of income or capital value with which to work.

Several methods of assessing viability can be employed, with the appropriate choice of technique largely resting upon the particular circumstances and objectives of the developer concerned. The principal method used is that of *capital profit*, by which total development costs are deducted from gross development value and a residual profit is established. An alternative approach is the estimation of the *yield* produced by a development scheme. This can be a simple comparison of the expected initial income expressed as a percentage of the likely development costs; or it can be a more refined relationship between estimated income allowing for rental growth and the attainment of a specified yield by a selected target date. Further related viability assessments between alternative development projects include sensitivity analysis, ground rent appraisals and long-term funding appraisals.

In conjunction with all these methods of appraisal it is possible to use a *discounted cash flow* approach towards analysis so as to explore more closely the effects through time. It is also becoming normal practice to use two or more techniques of appraisal both as supplementary checks and in order to avoid any inherent anomalies. Whatever the result of the appraisal it is none the less essential that the *opportunity cost* of the prospective development should be identified in respect of alternative areas of enterprise. Although the completed building becomes an investment property, development appraisal is essentially concerned with gauging viability; it does not produce valuations for either investment or sale.

The foundation of all valuations for development purposes is the *residual method*. This produces a capital sum representing either property or land value, depending upon the situation, which may be applied in reaching a particular development decision. It should be recognised that the Lands Tribunal are extremely sceptical about adopting the residual approach as a primary method of valuation in the context of their jurisdiction because of the variability of the constituent factors. A small adjustment in a certain item can disproportionally affect the final residual value and consequently the technique is open to manipulation in the interests of a particular case. A justification for the use of the method is that it may represent the only approach possible in the appraisal of a potential development scheme. Many schemes have unique characteristics in terms of location, funding, design and disposition. Comparable market transactions may not be available and the deductive process of moving towards a site value based upon the value of the potential development and the costs of achieving that development reflects the bids of purchasers in the development land market. The existence of higher and lower values is a result of the efficiency of alternative land use proposals, varying individual finance costs and desire to acquire the site, and not of inherent flaws in the methodology used.

It should be noted that all rents and costs quoted in the following studies are for illustrative purposes.

Conventionally, interest rates applicable to property finance are some few percentage points above base rates. Recent years have seen base interest rates move substantially ranging from 7¼% to rates of over 10%. The rates used in the following studies are for illustrative purposes and reflect the higher end of that spectrum.

The incidence of VAT is ignored in the calculations on the assumption that a developer would be able to reclaim all tax payable, although it is accepted that there may be a delay between payment and recovery which can affect cash flow adversely.

Note: Practitioners involved in the valuation or appraisal of development properties should at all times be aware that their professional duty to adhere to good valuation practice requires them to be conversant with relevant guidance notes and practice statements such as those contained in the RICS Appraisal and Valuation Manual. Essential reading in this regard includes Guidance Note 17 which covers the choice of method, stages in the valuation and guidance on the inputs involved.

Study 1

The purpose of this study is to set out a conventional format for preparing a simple residual valuation to find a land value and, in doing so, to identify the basic components of a development scheme and illustrate a method of treatment.

A prospective developer finds it necessary to ascertain how much he can afford to offer for a small, prime suburban site in London which has planning permission for 2,000 m^2 of offices producing 1,600 m^2 of lettable floor space. The projected development period is expected to be 24 months and the building contract period 12 months. Six months have been allowed before building works start to take account of detailed design, estimation and tendering. Six months following practical completion have been allowed for any possible letting voids which may occur. Finance can be arranged at 1.2% per month, comparable schemes have recently yielded 6%, rents of around £185 per m^2 net of all outgoings have currently been achieved on similar properties, and a developer's profit of 15% on capital value is required. Construction costs have been estimated at £800 per m^2. A development valuation to assess the residual value of land can be conducted as follows:

Valuation	£	£
A. Capital value after development		
Anticipated net rental income (1)	296,000	
YP in perp at 6% (2)	16.67	
Estimated gross development value		4,934,320
B. Development costs		
Building costs (3)		
2,000 m^2 gross floor area at £800 per m^2	1,600,000	
Building finance (4)		
Interest on building costs 1.2% pm for		
18 months × ½	191,606	
Professional fees (5)		
12.5% on building costs	200,000	
Interest on fees (6)		
1.2% pm for 18 months × ⅔	31,934	
Promotion and marketing (7)		
Estimated budget (including interest)	50,000	

Contingency (8)		
5% on costs (including interest)	103,677	
Agents' fees (9)		
Letting at 10% on initial rent	29,600	
Sale at 3% on capital value	148,030	
Developer's profit (10)		
15% on capital value	740,148	
Total development costs		3,094,995

C. Residual Land Value (11)

Sum available for land, acquisition and interest	1,839,325

Let x = land value = $1.00 x$

Finance on land (12)
1.2% pm for 24 months = $0.33/x$

Acquisition costs (13)
0.04 at 1.2% for 24 months = $0.053 x$

$1.384 x$ = £1,839,325, $\therefore x$ =	1,328,992
Residual land value now (14)	
Say	£1,330,000

Notes

(1) Rental income is assessed by reference to net internal areas, sometimes known as net lettable or net usable areas and even effective floor area. In this particular scheme the gross to net floor area relationship, called the "efficiency ratio" is 80%, which is an average figure for small new office buildings. All measurements should be undertaken in accordance with the current Code of Measuring Practice (RICS/ISVA).

(2) The initial yield used to find the rate at which the rental income is capitalised to calculate the gross development value is market derived. It is selected either by reference to other similar transactions or by negotiated agreement with a financial institution willing to enter into a forward purchase commitment. The rate will vary according to the nature of any funding arrangement and the exact needs of the long-term investor involved.

(3) Building costs are normally calculated by reference to the gross internal area of a building and based upon an analysis of the principal elements such as foundations, basement, superstructure, services, external envelope, internal division and finishes. The reasoning behind this is that it allows for a

general comparison derived from other schemes without being too specific about varying external finishes. At this stage in a development appraisal it is quite usual to use slightly adjusted figures applied to gross external measurements because detailed plans have not yet been prepared and only schematic drawings are available. An additional uncertainty is that when tenant demand is high it is probable that the building will be completed to a "developer's finish" standard, leaving the ingoing tenant to provide wall, floor, ceiling finishes etc. In a poor market, the developer may well have to bear this expenditure so as to provide accommodation available for immediate occupation.

(4) To facilitate preliminary appraisal in the absence of an actual building contract it is normally assumed that the regular payments made to the contractor are equal throughout the contract period. As a developer will only borrow funds as liabilities occur it is reasonable at this stage to calculate the cost of building finance in an approximate way to reflect cash flow. This can be done in a number of ways. The full rate of interest can be applied for the entire construction period and half the result taken as the finance charge, as here; or half the rate of interest can be applied for the funding period; or the full rate can be applied to half the period; or a percentage figure other than 50% can be applied to the full rate for the whole period if an obviously irregular cash flow or compounding is likely. In practice, the phasing and breakdown of costs typically follows an "S" shaped curve showing a gradual build-up of expenditure normally reaching a peak after about 60% of the contract has elapsed, with something of a tailing-off towards the end. The profile of the curve varies between sectors. In this study the building costs are funded for the 12-month construction period and for the possible subsequent six-month letting void. Strictly speaking, the total costs should be compounded at the full rate of 1.2% for the final six months rather than being halved to reflect cash flow, or a higher percentage than 50% adopted.

(5) Professional fees can vary considerably depending upon the nature of the work, the size of the scheme and the problems encountered. While they are normally made in accordance with the relevant professional scales conventionally observed within the building industry, the more complex the project the higher the level of fees or, alternatively, the larger the scheme

the more likely it becomes that a negotiated fee arrangement below 13.5% can be agreed between the parties. In very broad terms these fees may be broken down among the various contributing professions so that the architect usually receives about 6% on construction cost, the quantity surveyor 2.5%, the consulting engineer 2.5% and other specialist engineering services a further 2.5%.

(6) Payments to professionals are usually made at prescribed points in the process such as upon the receipt of contract documents and then at various stages during construction until completion. It has been shown, however, that an acceptable measure of the cost of financing these payments can be taken at between two-thirds and three-quarters of the full rate of interest over the construction period. Frequently, however, professional fees are added to building cost, and the finance charges calculated together.

(7) The amount of the budget allowed for promotion and marketing can vary considerably depending upon the nature and location of the project concerned. Because marketing needs differ so widely between developments it is unwise to adopt a simple percentage of cost or value. Rather, a figure related to the probable costs of promoting the individual development must be estimated.

(8) Although the developer's profit margin makes some allowance for risk, and many would argue that there should be no provision for miscalculation, a contingency sum of about 5% on building cost is advisable.

(9) Again the level of letting fees is frequently negotiable and may be reduced where either a sole agent is appointed or a regularly retained agent employed. In the residential sector a figure of between 2% and 3% of sale price is used. Where two agents are jointly instructed, as frequently happens in the commercial sector, fees are often taken at around 15% of initial rent. The sale fee assumes that the developer will sell on the development to an investor and, though negotiable, is usually pitched at around 3% to 4% of the agreed price or, at this stage, the gross development value.

Depending upon the financing arrangements, it may also be necessary to include fees for negotiating the funding facility. Typically this might be ½–1% of the sum borrowed. It is becoming increasingly necessary for developers to effect insurance against defects in the building. Insurance policies, covering everything from the main structure to services, can be obtained for a one-off premium of about 1% of the construction costs.

(10) Developer's profit reflects the required return for enterprise, organisation, overheads and risk. Where the value of land is known it is commonly expressed as a percentage of total capital expenditure, and may be anything between 15% and 25%. When deriving a residual land value, however, and total development cost is uncertain, it is frequently pitched at a slightly lower percentage of capital value – between 10% and 20%. Experience shows that there is a very rough relationship between, say, 20% on cost and 15% on value among different schemes.

The residual method assumes a sale of the completed scheme with the developer hoping to gain profit by way of a capital sum. If the developer is likely to hold the development as an investment he may be more concerned as to whether rental income will cover long-term finance costs.

(11) The method described in the study for calculating the residual value of the land is theoretically correct. In practice, however, some valuers still apply the relevant present value of £1 to the sum available for land, acquisition and interest. While unsound, the result is little different.

(12) The cost of financing the land is "rolled-up" and calculated over the whole development period because all acquisition costs usually occur at the moment of purchase and recoupment does not commence until some form of disposition.

(13) Acquisition costs are taken at an average of 4% comprising typically estate agents' fees of 1%–2%, conveyancy costs of about 1% and stamp duty of 1%–2%.

(14) In this way £1,330,000 is available to purchase the land. Any price in excess of this sum erodes the developer's profit and vice versa.

(15) Throughout this and all subsequent studies, it should be recognised that in practice a valuer will frequently rely upon direct comparable evidence to assess development value. If similar land has recently been sold for the same purpose in the locality and a general level of £x per m², per ha or per plot can reasonably be established, then the residual method should be checked against market comparables.

Study 2

This study shows how development profit can be assessed when the value of the land is known or can be assumed.

A vacant and partially derelict deconsecrated church building in the centre of a large provincial town is being offered for sale at £2.5 m². A local property development company is interested in converting the building into a small speciality shopping centre on two floors. The reconstructed building will be approximately 3,000 m² gross in size providing about 2,000 m² of net lettable floorspace divided into 18 units of between 50 and 250 m². Rental income is predicted to average out at around £300 per m². An investment return of 7.5% is sought. Building costs are estimated at £550 per m². Bridging finance is available at 1.4% per month and the development will probably take 21 months to complete and let. The development company are anxious to know what will be the likely level of profit.

Valuation	£	£
A. *Gross development value*		
Estimated rental income at £300 per m² on 2000m	600,000	
YP in perp at 7.5%	13.33	
Gross development value		7,998,000
B. *Development costs*		
Building costs at £550 per m² on 3,000m (1)	1,650,000	
Professional fees at 15% (2)	247,500	
Contingencies at 5% on (i) and (ii) (3)	94,875	
Promotion, Say	50,000	
Finance on (i) to (iv) at 1.4% pm for		
18 months × 0.65 (4)	377,486	
Letting fees at 10% of rent	60,000	
Sale fee at 3% of GDV	239,940	
Land cost	2,500,000	
Acquisition at 4%	100,000	
Finance on land and acquisition at 1.4%		
for 21 months	881,532	
Total development costs		6,201,333
C. *Development profit*		
Residual value in 21 months		1,796,667
PV of £1 in 21 months at 1.4% pm		0.747
Value now		£1,342,110

Profit on cost in 21 months

$$\frac{1,796,677 \times 100}{6,201,333} \qquad = \qquad 28.97\%$$

Profit on cost now (5)

$$\frac{1,342,110 \times 100}{6,201,333} \qquad = \qquad 21.64\%$$

Profit on value in 21 months

$$\frac{1,796,667 \times 100}{7,998,000} \qquad = \qquad 22.46\%$$

Profit on value now

$$\frac{1,342,110 \times 100}{7,998,000} \qquad = \qquad 16.78\%$$

Investment return on cost

$$\frac{600,000 \times 100}{6,201,333} \qquad = \qquad 9.68\%$$

Notes

(1) Building costs are more difficult to estimate accurately for reconstruction or renovation work.

(2) Professional fees are slightly higher than normal owing to the special and complex nature of the project and the need for additional structural engineering advice.

(3) A reasonable contingency budget is essential in work of this kind.

(4) A proportion of 65% rather than the usual 50% approximation has been allowed to take account of the front loading of fees and the rolling up of the entire debt charge during the period of letting voids following completion.

(5) In theory it is correct to discount the residual profit upon completion and sale to a present value now. In practice this is not always done. Both the profits on cost and value in 21 months and those expressed as a return now are presented. The difference between them will obviously depend upon the length of the development period and the finance rate used for discounting.

Study 3

Ground leases and ground rent calculation.

The owner of a site may not wish to dispose of his interest outright, preferring to grant a developer a long lease at a rent reflecting the development value of the land.

Historically, and currently, much development has been undertaken on a ground lease basis. Although the owner of the land might be foregoing receipt of a capital sum (not necessarily the case, as the ground lessee may be required to pay a substantial premium as well as rent), an important consideration is that the owner can retain some control over the development (particularly significant before the introduction of planning control and building legislation). In addition, modern ground leases will enable the ground lessor to share in the rising value of the development and, of course, at the end of the ground lease the land, together with any buildings, reverts to the freeholder.

Initial ground rents may be calculated in a number of ways and, as with funding arrangements, ground leases have become very complex.

A simplified approach can be illustrated using the facts as in study 1. Suppose that instead of selling the freehold for £1,330,000 the owner wishes to know the ground rent that might be expected if, as an alternative, a developer is granted a 120 years' ground lease, commencing upon completion of development, when the building agreement (which covers the construction period) ends.

		£	£pa
Expected income from scheme (ERV)			296,000
Less:			
Long-term finance			
£2,354,807 at 7% (2)		164,836	
Developer's risk and profit			
£2,354,807 at 1.5% (3)		35,323	200,158
Amount available for ground rent			95,842
Return on site value to freeholder (4)			

$$\frac{95,842 \times 100\%}{1,330,000} \quad = \quad 7.2\%$$

Gearing of ground rent to ERV (5)

$$\frac{95,842 \times 100\%}{296,000} \quad = \quad 32.38\%$$

Notes
(1) Total development costs less developer's profit.
(2) Long-term finance can be provided in many ways – mortgage, debenture, sale and leaseback etc, and the finance rate will depend upon the method adopted, the general level of interest rates and the status of the developer.
(3) As always, the return a developer is prepared to accept (here in the form of annual income) will reflect the nature of the scheme, letting potential and possible movement in rental values. The ground lease will almost certainly provide for periodic refurbishment/redevelopment and a developer would argue in negotiation that an additional allowance should be made, perhaps by way on an annual sinking fund, for this expenditure.
(4) This is a useful check for the freeholder so as to compare the options of sale and leasing.
(5) To avoid problems at review, ground leases will often provide for ground rents to be reviewed to an agreed proportion of future rental values of the scheme.
(6) These calculations are based upon estimates and assumptions and the actual rental value of the scheme, development costs, etc will change. Provision is usually made for a recalculation of the ground rent when the position is clear but the freeholder will normally be guaranteed a minimum ground rent. Pre-development agreements will also provide for sharing surplus income (if the scheme is very successful for example) although the developer will normally bear deficits.

Study 4

To show that the residual method can be modified to take account of phased development projects.

Consider the freehold interest in a cleared site which has planning approval for the construction of 90 detached houses. It is thought probable that any prospective purchaser would develop the site in three phases, each of 30 houses. The total development period is estimated at three years with separate contract periods of 12 months for each of the three phases. The sale price of the houses is set at £150,000 each. Construction costs are estimated at £85,000 a house and the cost of site works and the provision of services is assessed at an average of £8,000 per house plot. Finance is available at 1.5% per month and a developer's profit of 10% on gross development is considered likely to be sought.

A phased residual valuation can be conducted as follows.

Valuation	£	£
Phase 1		
A. *Sale price of house*		150,000
B. *Development cost*		
Building costs	85,000	
Site works and services	8,000	
Professional fees at 10%	9,300	
Advertising, Say	2,000	
Finance on £104,3000 at 1.5% for 12 months $\times$ ½	10,201	
Disposal fees at 3% of sale price	4,500	
Developer's profit at 10% of sale price	15,000	
Total development costs		134,001
Balance per plot		15,999
C. *Site value phase 1*		
Number of plots		30
Amount available in 12 months		479,970
Let site value =	1.000 x	
Acquisition costs =	0.040 x	
Finance at 1.5% on 1.04 x for 1 year =	0.203 x	
	1.243 x	
1.243 x = 479,970		
∴ x = 386,138		
Site value of phase 1, Say		386,000
D. *Site value of phase 2*		
Site value phase 1	386,000	
PV of £1 in 12 mths at 1.5%	0.836	322,696
E. *Site value of phase 3*		
Site value phase 1	386,000	
PV of £1 in 24 mths at 1.5%	0.700	270,200

978,896
F. *Value of entire site*
Say £980,000

Notes

(1) It can be seen that the value of each phase is in the order of £386,000 if development on all three phases started immediately. Allowance has been made, however, for the cost to the developer of holding phase 2 for a further 12 months and phase 3 for a further 24 months. Even if the three phases were sold separately to different developers it is assumed that the local demand for new housing would not hold up sufficiently to permit consecutive development of all 90 houses within the first year. An alternative calculation with adjusted selling prices to reflect increased supply could with advantage be performed.

(2) Although the method suggested above is an improvement to a global residual it still gives only a broad indication of value. A detailed discounted cashflow analysis would provide a much better approximation of value.

Study 5

This study demonstrates that the phased residual valuation can be taken a stage further so as to produce a residual cashflow valuation. Consider a 10 ha site on a motorway location some 12 miles west of London. It has planning permission for "high technology" industrial development and a site coverage of around 30% is envisaged. Rents of £120 per m² are forecast and a major financial institution has shown interest in buying the scheme once completed if it can show an initial yield of 7%. The total development period will be about three years, but is considered possible to develop and let the equivalent of one ha every three months starting in month nine. Construction costs are estimated at £500 per m² and short-term finance can be arranged at 16%. The site is on offer for £12 m. A residual cashflow valuation to take account of the effects of expenditure and revenue during the development period and show the likely level of profit can be performed as follows (see table on p468).

Preliminaries

(i)   10 ha = 100,000 m².
(ii)   Site coverage at 30% = 30,000 m² gross.
(iii) *Less* 5% = 28,500 m² net lettable floorspace.
(iv)   Building costs of £15,000,000 averaged at £1,250,000 a quarter.

Residual cashflow (000s)

| Timescale | | Cash outflows | | | | | Cash inflows | | |
Year	Month	Land	Building	Promotion	Fees	Rent	Sale	Cumulative cashflow	Interest at 3.78% pq
0	0	(12,480)		(30)				(12,510)	
0	3		(1,250)	(30)	(125)			(13,915)	(479)
0	6		(1,250)	(30)	(125)			(15,793)	(544)
0	9		(1,250)	(30)	(159)	85		(17,691)	(618)
1	0		(1,250)	(30)	(159)	171		(19,577)	(692)
1	3		(1,250)	(30)	(159)	256		(21,452)	(766)
1	6		(1,250)	(30)	(159)	342		(23,315)	(840)
1	9		(1,250)	(30)	(159)	427		(25,167)	(913)
2	0		(1,250)	(30)	(159)	513		(27,006)	(986)
2	3		(1,250)	(30)	(159)	598		(28,833)	(1,058)
2	6		(1,250)		(159)	684		(30,616)	(1,130)
2	9		(1,250)		(159)	769		(32,386)	(1,200)
3	0		(1,250)		(159)	855		(34,140)	(1,270)
					(1,466)		48,872	11,996	
Total		(12,480)	(15,000)	(300)	(3,306)	4,700	48,872	11,996	(10,490)

(v)   Promotion costs of £300,000 spread over the first 2½ years at an average of £30,000 a quarter.

(vi)   Professional fees taken at 10% of building costs and paid an average rate of £125,000 a quarter.

(vii) Letting fees taken at 10% of initial annual rents and paid as quarterly rents are received.

(viii) Sale fee of £1,466,154, being 3% of gross development value payable upon completion and sale.

(ix)   Rental income at 10% of the total quarterly rent roll to commence in month 9 and grow by a further 10% each quarter until month 36 when the full quarterly rent roll of £320,625 is reached.

(x)   Gross development value of £48,871,800 calculated by capitalising rental income of £3,420,000 a year by 14.29 YP and received in three years' time.

(xi)   Annual interest of 16% to be taken at 3.78% a quarter.

(xii) All cashflow figures rounded to the nearest thousand.

	£	£	£
A. *Gross development value*			48,872,000
B. *Land and construction costs*			
Land	12,480,000		
Building	15,000,000		
Promotion	300,000		
Fees	3,306,000		
Interest	10,490,000		
Total		41,576,000	
Less:			
Revenue during development		4,700,000	
C. *Total development costs*			36,876,000
D. *Development profit*			
Development profit in 3 years			11,996,000
PV of £1 in 3 years at 16%			0.641
Development profit now			7,689,436

Profit on cost in 3 years

$$\frac{11,996,000 \times 100}{36,876,000} = 32.53\%$$

Profit on cost now

$$\frac{7,689,436 \times 100}{36,876,000} \quad = \quad 20.85\%$$

Profit on value in 3 years

$$\frac{11,996,000 \times 100}{48,872,000} \quad = \quad 24.55\%$$

Profit on value now

$$\frac{7,689,426 \times 100}{48,872,000} \quad = \quad 15.73\%$$

Return on cost

$$\frac{3,420,000 \times 100}{36,876,000} \quad = \quad 9.27\%$$

Study 6

This compares the conventional residual method of valuation with a discounted cashflow analysis in order to demonstrate the need to be more conscious of the effects of time and the incidence of costs and revenue in the valuation of development properties.

A local property development company has been offered a prime corner site on the high street of a prosperous provincial town for £1.5m and is anxious to establish the probable viability for development with a view to disposing of it to an institution once fully completed and occupied. The site is currently used as a builder's yard with some vacant and near-derelict shops and has a high street frontage of 120 m and a depth of 45 m. It lies within an area allocated for shops and offices with an overall plot ratio of 1.5:1 and a general height restriction of three storeys. Preliminary discussions indicate that the usual parking standards of one space to every 200 m² of office floorspace and five to every 100 m² retail floorspace could be relaxed if 30 spaces are provided on-site and a section 106 agreement under the Town and Country Planning Act 1990 is entered into whereby a further 50 spaces are funded by the developer in a nearby local authority car park to be constructed in one or two years' time. A condition limiting a substantial portion of any office floor space to local firms will almost certainly be imposed on a planning permission. all ground floor development must be retail and rear access to shops is considered essential. A small supermarket of approximately 1,000 m² is thought likely to attract support and because the local planning authority are concerned that some form of suitable development takes place as soon as possible, negotiations should be relatively straightforward with permission probably granted in three months.

The quantity surveyor retained by the company has supplied the following information:

(a) A 6 m grid to be used throughout with 5 m ceiling heights for retail space and 3 m ceiling heights for offices.
(b) Building costs for shops to be taken at £400 per m² excluding fitting out and shop fronts and equally phased over nine months. Standard shop units to be 6 m × 24 m.
(c) Building costs for offices to be taken at £800 per m² for letting, including lifts and central heating and equally phased over 15 months.
(d) Demolition and site preparation to be allowed for at £20 per m² across the entire site.
(e) External works, including landscaping will cost £200,000.
(f) All payments to be made three months in arrears.

The property development company's knowledge of the area indicates rental levels of £150 per m² pa for supermarket space, an average of £29,000 pa or approximately £200 per m² pa for a standard shop unit and £100 per m² pa for offices. Given three months to obtain planning permission and prepare the site and nine months to construct the shops, it is envisaged that a further three months should be time enough to allow for a successful letting campaign and sufficiently complete the superstructure of the building so that the shops could be let at the end of 15 months. In view of the probable local user condition the letting climate of the offices is slightly more uncertain, and it is considered appropriate to allow a full six months following completion before they are fully let and disposition to an institution can be effected. Professional fees to the architect and quantity surveyor have been negotiated so that £60,000 is paid as a lump-sum following planning permission and the remainder calculated subsequently at 10% of building cost on a three-monthly basis. Short-term finance has been arranged with a merchant bank at 15% (3.56% per quarter).

It is well reported that institutions are interested in schemes of this nature, scale and location, but seek yields of between 7% and 8%. It is therefore necessary to establish whether a yield can be accomplished which provides for this and also additionally allows for the developer's risk.

Site Layout

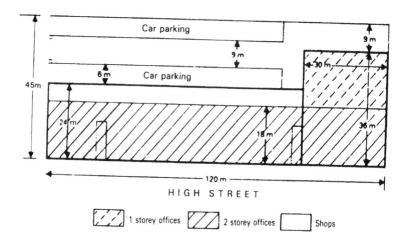

Preliminaries

	m^2
(a) *Site*	
Site area = 5,400 m^2	
Plot ratio = 1.5:1	
Therefore gross permitted commercial floor space	8,100
(b) *Shops*	
Supermarket 30 m × 36 m	1,080
Standard units 90 m × 24 m less 2 ground floor	
office entrances 3 m × 12 m	2,088
Total Gross retail floor space	3,168
(c) *Offices*	
(Gross permitted floor space commercial floor space	
– gross retail floor space) 8,100 m^2 – 3,168 m^2	4,932
Taking account of 6m grid constraint: 2 storey above	
shops 120 m × 18 m	4,320
Single-storey extension above supermarket 18 m × 24 m	432
Add – 2 ground-floor entrances 3 m × 12 m	72
Total Gross office floor space	4,824

(d)   *Gross areas to net*

Gross	Deduction	Net m²
Supermarket 1,080 m²	10%	972
Standard units 2,088 m²	15%	1,775
Offices 4,824 m²	20%	3,859

(e)   *Income*   £pa

Supermarket
972 m² at £150 per m² pa   145,800
Standard units
1,775 m² at £200 per m² pa   355,000
Offices
3,859 m² at £100 per m² pa   385,900
                                                                     £886,700

Conventional residual	£	£
A. *Capital value after development*		
Estimated net rental income	886,700	
YP in perp at 7.5%	13.33	
Gross development value		11,819,711
B. *Development costs*		
Building costs		
Shops – 3,168 m² at £400 per m²	1,267,200	
Offices – 4,824 m² at £800 per m²	3,859,200	
Site preparation	108,000	
External works	200,000	
Total	5,434,400	
Professional fees		
Arch and QS by negotiation	603,440	
Contingencies		
5% on (i)	271,720	
Finance on building		
3.56% pq for 7 quarters × ½ on		
(i) + (ii) + (iii)	875,297	
Agents' fees		
Letting fees 15% on initial rent	133,005	
Sale fees at 3% on GDV	354,591	
Land costs		
Land	1,500,000	
Acquisition	60,000	
Finance at 3.56% for 8 quarters	503,768	
Total development costs		9,736,221

C. *Residual capital value*

Capital value in 24 months' time	2,083,490
PV of £1 in 2 yrs at 15%	0.756
Net present value (developer's profit)	£1,575,118

D. *Profit*

Profit on GDV	$\dfrac{1,575,118 \times 100}{11,819,711}$	=	13.33%
Profit on cost	$\dfrac{1,575,118 \times 100}{9,736,221}$	=	16.18%
Development yield	$\dfrac{886,700 \times 100}{9,736,221}$	=	9.11%

Notes on residual and tables A, B and C

The present value of the profit indicated by the residual is £1,575,118. This may be compared with an appraisal by discounted cashflow (see table A p475). It is commonly thought that a discounted cashflow analysis produces a more accurate result (in this case £1,572,170, virtually the same as the profit produced by the residual method).

The principal advantage of using cashflows is that there is considerably more flexibility in the timing of payments and receipts. Generally, the same information (costs and values) are used initially in residuals and cashflows such that inaccuracies in these inputs will lead to errors whichever method is adopted. For investment and development appraisal, two discounted cashflow approaches are used commonly. The net present value (NPV) approach involves the discounting of all inflows and outflows. The sum of the discounted inflows and outflows produces the net present value of the developer's profit. Individual developers have their own "target" discount rates but for initial appraisals it is logical to use the short-term finance rate. A positive NPV indicates that, potentially, the scheme is profitable and a negative NPV that a loss is likely.

Table A on p475 shows how the present value of developer's profit may be calculated. All the costs used are as in the residual, but an attempt has been made to indicate how these costs might be spread throughout the scheme.

If the finance rate is used for discounting, the difference between the discounted residual profit and the NPV derived from the

Table A

Item/Quarter	0	1	2	3	4	5	6	7	8
Land cost	(1,500,000)								
Acquisition costs	(60,000)								
Site preparation		(108,000)							
Building costs:									
shops		(422,400)	(422,400)	(422,400)					
offices		(771,840)	(771,840)	(771,840)	(771,840)	(771,840)			
other				(25,000)	(25,000)	(150,000)			
Arch and QS fees		(190,224)	(119,424)	(121,924)	(79,684)	(92,184)			
Contingency		(65,112)	(59,712)	(60,962)	(39,842)	(46,092)			
Agency & Legal fees						(75,120)			(412,476)
Shop income						125,200	125,200	125,200	
Sale proceeds								11,819,711	
Cash flow	(1,560,000)	(1,557,576)	(1,373,376)	(1,402,126)	(916,366)	(1,010,036)	125,200	11,407,235	
PV of £1 at 3.56%	1	0.966	0.932	0.900	0.869	0.840	0.811	0.783	0.756
NPV	(1,560,000)	(1,504,618)	(1,279,986)	(1,261,913)	(796,322)	(848,430)	101,537	98,032	8,623,870
Cumulative NPV	(1,560,000)	(3,064,618)	(4,344,604)	(5,606,517)	(6,402,839)	(7,251,269)	(7,149,732)	(7,051,700)	1,572,170

Discounted Cashflow Analysis (NPV)

Table B

Quarter					Internal Rate of Return				
	0	1	2	3	4	5	6	7	8
Cashflow	(1,560,000)	(1,557,576)	(1,373,376)	(1,402,126)	(916,366)	(1,010,036)	125,200	125,200	11,407,235
PV at 6% p qtr	1	0.943	0.890	0.840	0.792	0.747	0.705	0.665	0.627
NPV	(1,560,000)	(1,468,794)	(1,222,305)	(1,177,786)	(725,762)	(754,497)	88,266	83,258	7,152,336
Total NPV									

Quarter	0	1	2	3	4	5	6	7	8
Cashflow	(1,560,000)	(1,557,576)	(1,373,376)	(1,402,126)	(916,366)	(1,010,036)	125,200	125,200	11,407,235
PV at 8% p qtr	1	0.926	0.857	0.794	0.735	0.681	0.630	0.583	0.540
NPV	(1,560,000)	(1,422,315)	(1,176,983)	(1,113,288)	(673,529)	(687,835)	78,876	72,992	6,159,907
Total NPV									(342,175)

Table C

Item/Quarter					Cashflow Analysis				
	0	1	2	3	4	5	6	7	8
Cashflow	(1,560,000)	(1,557,576)	(1,373,376)	(1,402,126)	(916,366)	(1,010,036)	125,200	125,200	11,407,235
Cumulative costs	(1,560,000)	(3,117,576)	(4,546,488)	(6,061,577)	(7,143,819)	(8,375,552)	(8,512,564)	(8,694,868)	2,398,373
Quarterly interest		(55,536)	(112,963)	(165,876)	(221,697)	(262,212)	(307,504)	(313,994)	(320,715)
Cumulative cost at end of each quarter	(1,560,000)	(3,173,112)	(4,659,451)	(6,227,453)	(7,365,516)	(8,637,764)	(8,820,068)	(9,008,862)	2,077,658

The cashflow is taken from Table A p475

discounted cashflow results entirely from the treatment of construction finance in the residual and the inclusion of income in the discounted cashflow approach enables the calculation of the internal rate of return (IRR) of the scheme. The IRR may be defined as the discount rate which when applied to inflows and outflows produces a net present value of £0.

The necessary calculations are shown in Table B on p476, which involve discounting the net cashflows at two trial rates, hopefully producing one negative and one positive NPV. The IRR is then found by interpolation (see p478), in this case 31.6% pa, confirming that the return exceeds the finance rate of 15% pa (3.56% per quarter). A free discussion of discounted cashflow theory is beyond the scope of this chapter but it is felt necessary to repeat that measures of returns produced are not automatically more accurate than those given by the residual.

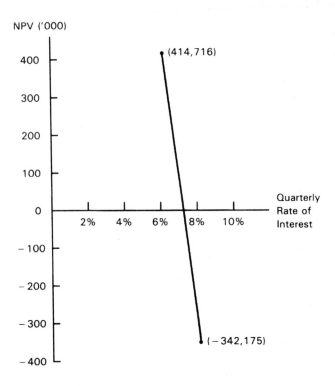

Graphical estimation of internal rate of return (see Table B p476)

Formula calculation of IRR

$$\text{IRR} = 6\% + 2\% \ \frac{414,716}{414,716 + 342,175} \quad \text{p qtr}$$

$$= 6\% + 2\% \ \frac{414,716}{756,891} \quad \text{p qtr}$$

$$= 6\% + 1.096\% \quad \text{p qtr}$$

$$= 7.096\%, \text{Say } 7.1\% \quad \text{p qtr}$$

$$\text{Effective annual IRR} = (1.071)^4 - 1$$
$$= 0.316$$

ie 31.6% pa.

Table C, p476, shows a cashflow analysis. With this method an accounting process is undertaken, ie interest is added to expenditure on a period-by-period basis. The interest in each period is calculated at 3.56% on the cumulative cost in the previous period. This analysis produces a developer's profit of £2,077,658 comparable approximately with the residual profit of £2,083,490. (Once again, the difference between these figures is attributable to finance charges and inflows of rent before the development is completed.)

Cashflows are useful in that they enable a developer (or the provider of finance) to assess the likely financial commitment throughout the scheme (ie in quarter 4, total spending amounts to £7,365,516). This facility is not available with residual or discounted cashflow methods. As a check, the discounted cashflow method should produce the present value of the cashflow profit (subject to rounding).

Study 7

This study is intended to illustrate how probable changes in costs and values during the development period can be accommodated within residual cashflow valuation.

Consider the conversion of a terrace of eight Victorian houses, which are vacant and somewhat dilapidated, into a block of 24 luxury flats. The houses are situated in a desirable part of north London and similar flats to those planned are selling for around £100,000 each, but demand is such that prices are confidently expected to rise by about 1% per month over the next year (12.68% pa). Building costs to effect the conversion are estimated at £25,000 per flat but are considered likely to increase by 1.25% per month for the next year (16.08% pa). Building costs are to be evenly spread over a 12-month development period. It is intended that six flats will be sold in months 9 to 12 inclusive. Professional fees are assessed at 10% of building costs, and will be paid by equal instalments of 2.5% a quarter. Agents' and solicitors' fees will be charged at 2% of the sale price of each flat. A development profit of 25% of the sale price is sought and bridging finance is available at 1% a month (12.68% pa).

A prospective developer wishes to know how much he might have to pay for the eight houses now.

Preliminaries
Receipts

Month 9 sees six flats sold at £100,000 each × 1.0937	=	£656,220
Month 10 sees six flats sold at £100,000 each × 1.1046	=	£662,760
Month 11 sees six flats sold at £100,000 each × 1.1157	=	£669,420
Month 12 sees six flats sold at £100,000 each × 1.1268	=	£676,080

Building costs + fees + profits
Total costs are 24 flats at £25,000 = £600,000
Average of £50,000 a month
Increase by 1.25% per month

	£	
1	50,625	
2	51,258	
3	51,899	+ 10% fees
4	52,548	
5	53,205	
6	53,870	+ 10% fees
7	54,543	
8	55,225	
9	55,915	+ 10% fees + 2% fees + 25% profit
10	56,614	+ 2% fees + 25% profit
11	57,322	+ 2% fees + 25% profit
12	58,038	+ 10% fees + 2% fees + 25% profit

It is assumed that surpluses in the latter few months will be reinvested at the same rate of interest. The resulting residual represents an absoluted maximum bid.

Valuation

Month	Receipts	Costs	Net cashflow	Capital outstanding	Interest
1		(50,625)	(50,625)	(50,625)	(506)
2		51,258)	(51,258)	(102,389)	(1,024)
3		(67,277)	(67,277)	(170,690)	(1,707)
4		(52,548)	(52,548)	(224,945)	(2,249)
5		(53,205)	(53,205)	(280,399)	(2,804)
6		69,832)	(69,832)	(353,035)	(3,530)
7		(54,543)	(54,543)	(411,108)	(4,111)
8		(55,225)	(55,225)	(470,444)	(4,704)
9	656,220	(249,662)	406,558	(68,509)	(686)
10	662,760	(235,559)	427,201	357,925	3,579
11	669,420	(238,065)	431,355	792,859	7,929
12	676,080	(257,777)	418,303	1,219,091	

Sum available in 12 months =	£1,219,091
Let the value of the houses =	$1.0000 x$
Acquisition costs at 4% =	$0.0400 x$
Finance at 12.68% on $1.04 x$ =	$0.1319 x$
	$\therefore 1.1719 x = 1,219,091$
$\therefore x$	$= 1,040,269$

The developer might offer £1m for the eight houses.

Study 8

This study shows how it is possible to explore the effect of changing levels of performance during the development period among such factors as yield, cost, rent, finance and time. To begin with, a simple sensitivity analysis know as a Mini-Max Evaluation is demonstrated, then the sensitivity of a scheme is tested by the Extinguishment of Profit Method and finally the individual components are measured by the Percentage Change Method.

Consider the position of a developer who has paid £500,000 for a plot of land which has planning permission to build a discount cash and carry warehouse of 4,000m² gross. Rents of around £85 per m² overall have been quoted in the vicinity. Initial yields in the region of 8.5% are reported for similar properties. Construction costs are estimated at £500 per m². Finance can be arranged at about 1.2% per month and the development is considered to take approximately 18 months to complete and let.

I. Median or realistic valuation
A. *Gross development value*

	£	£
Rental income at £85 per m² on 4,000 m²	340,000	
YP in perp at 8.5%	11.76	
Gross development value		3998,400

B. *Development costs*

Building costs at £500 per m² on 4,000 m²	2,000,000	
Professional fees at 10% of (i)	200,000	
Finance at 1.2% pm for 18 months × ½	263,458	
Letting at 10% of rent	34,000	
Land	500,000	
Acquisition at 4%	20,000	
Finance at 1.2% pm for 18 months	124,544	
Total development costs		3,142,002

C. *Development profit*

Sum available in 18 months	856,398
PV of £1 in 18 months at 1.2% pm	0.807
Profit now	£691,113

Profit on cost

$$\frac{691,113 \times 100}{3,142,002} \quad = \quad 22\%$$

Return on cost

$$\frac{340,000 \times 100}{3,142,002} \quad = \quad 10.8\%$$

II Maximum or optimistic valuation
Rents predicted at £100 per m²
Initial yield taken at 8%
Building costs estimated at £450 per m²
Finance at 1.0% pm
Development period and letting 15 months

A. *Gross development value*

	£	£
Rental income at £100 per m²	400,000	
YP in perp at 8%	12.5	
Gross development value		5,000,000

B. *Development costs*

Building costs at £450 per m²	1,800,000
Professional fees at 10%	180,000
Finance at 1.0% pm for 15 months × ½	159,359

Letting at 10% of rent	40,000	
Land	500,000	
Acquisition	20,000	
Finance at 1.0% pm for 15 months	83,704	
Total development costs		2,783,063

C. Development profit

Sum available in 18 months		2,216,937
PV of £1 in 15 months at 1.0% pm		0.861
Profit		1,908,783

Profit on cost

$$\frac{1,908,783 \times 100}{2,783,063} \quad = \quad 68.6\%$$

Return on cost

$$\frac{400,000 \times 100}{2,783,063} \quad = \quad 14.4\%$$

III. Minimum or pessimistic valuation
Rents down to £70 per m[1]
Initial yield up to 9%
Building costs up to £550 per m^2
Finance at 1.4% per month
Development period 21 months

A. Gross development value

	£	£
Rental income at £70 per m^2	280,000	
YP in perp at 9%	11.11	
Gross development value		3,110,800

B. Development costs

Building costs at £550 per m^2	2,200,000	
Professional fees at 10%	220,000	
Finance at 1.4% pm for 21 months $\times$ ½	410,251	
Letting at 10% of rent	28,000	
Land	500,000	
Acquisition	20,000	
Finance at 1.4% pm for 21 months	176,306	
Total development costs		3,554,557

C. Development profit

A loss of		(£443,757)

Note

It can, therefore, be seen that changes to the various components of rent, yield, cost, finance and time can have dramatic effects upon the profitability of a scheme when acting in concert.

Sensitivity analysis

Initial development valuations and appraisals must be based upon a considerable number of "guestimates". The more complex the scheme, the more likely it is that inaccuracies will occur.

However, with experience and up-to-date market knowledge, it is possible to arrive at an acceptable approximation. The reliability of the "answers" depends not so much on the method as the reliability of data used.

Changes in rents, yields (capital values), building costs and, with longer schemes, finance rates can all affect values or profit.

Reworking calculations to demonstrate the influence on land values of rental value for example is relatively tedious and the increasing use of computer programs is, in part, because most commercial and general purpose (spreadsheet) packages enable sensitivity testing to be undertaken. The effects on profit of rising or falling rental values can be determined almost instantaneously.

Recent Trends

In particular, the major involvement of financial institutions and advisers in property investment have led to the demand for more explicitly analytical evidence for property development appraisals. While the basic approach remains to establish whether a scheme is likely to prove viable, based upon assumptions made, this is usually accompanied by feasibility studies, discounted cash flow analysis and scenario testing based upon ranges of probable costs and returns.

Detailed analysis is greatly assisted by the increased utilisation of PC based spreadsheets involving the use of discounted cash flow methods as the property industry is increasingly willing to draw upon material and knowledge from outside traditional professional experience. No standard residual model spreadsheet is used throughout the profession because precise criteria for inputs and results sought differ from project to project. The common use of such analysis is in project development appraisal often based upon in-house spreadsheets developed and refined through review of previous projects. Dedicated software packages often based on the Microsoft Excel package are also available for busy professional practices requiring ease of use as well as speedy and accurate assessments. Such packages contain features such as traps to prevent input of unreasonable data, deletion of relevant data, to prevent errors and provide automatic file back-ups.

Typical uses of such systems include the appraisal of complex development options with cash flows on completed schemes over investment periods. Individual appraisals may be assessed, aggregated and used as wished. Project tailored appraisals may be developed in a user friendly manner and options in terms of sensitivity analysis, phasing and financing can be looked at in greater detail.

Chapter 14

Garages and Petrol Filling Stations

In the past the two types of property, garages and filling stations, were often on the same site and run by the same proprietor. This is now becoming less common. The two types of business have been moving steadily away from each other in the last 20 years and it might well be that in future editions of this book they are dealt with under entirely separate chapters.

Both types of property have one thing in common – they have seen a substantial reduction in their numbers in the last 30 years. The traditional garage has been steadily losing ground to a growing number of specialist fast-fit businesses which now dominate the market in tyres, exhausts, shock absorbers, batteries and windscreen replacement. In 1975 there were 10,500 franchised garages in the UK. A number of motor manufacturers are presently sharply reducing these numbers so that by the turn of the century there could be fewer than 6,000 at which stage they could possibly be outnumbered by fast-fit centres.

Similarly the number of petrol filling stations in the UK has fallen sharply and the survivor sites tend to be larger, well laid out and capable of generating substantial non-fuel revenues and the oil companies are eagerly setting about transforming themselves into general roadside retailers. This might be a long-term, defensive strategy as the arrival of new forms of power generation for the car becomes more possible with pressure for zero-emission increasing.

A surveyor may be required to carry out valuations and conduct negotiations in connection with the purchase or sale of both categories of property and deal with new lettings and rent reviews. In the past it was rare that petrol filling stations were bought for investment purposes. Because their value is related to annual throughput they have been regarded as vulnerable to bypasses and traffic management schemes which can rob them of important traffic flow. However, more conventional types of investment properties suffered greatly in the recession of the early 1990s and the financial standing of tenants became much more important to investors so that there is more interest now in petrol filling stations as investments, particularly when the tenant is a major oil company. The same cannot really be said of garages. Even the best of the listed motor groups have market capitalisations which are tiny by comparison with the oil companies. Some motor groups have nevertheless arranged sale and lease-backs on some of their garages. This has been achieved usually by accepting artificial rent review provisions so that the rent at review is raised in a formula related to the Retail Price Index, or to another type of property such as off-centre retail stores.

Petrol Filling Stations

The industry

The valuer specialising in petrol station properties should have a working knowledge of the oil industry. Petrol is a standard product and it is marketed internationally, most of the larger wholesalers in the United Kingdom being part of multi-national oil companies. Shell, BP and Exxon, Esso's parent company, are among the largest companies in the world. Petrol is one by-product of crude oil and in the industry oil is described as "equity crude" to mean the oil produced by an oil company itself, and "crude" which is oil purchased from a third party in the open market. Equity crude as well as crude itself has to be valued at international market prices and not in relation to production costs, if for no other reason than that it could be sold at such prices, but also because it is hardly possible to identify the production costs of petrol alone as part of the oil-refining process. Bulk trade in north-west Europe is referred to as the "Rotterdam market" but the Rotterdam market is not an organised market with prices established in open-floor trading but rather a market where direct bargains are struck, although they are then reported as "Rotterdam prices". Refiners themselves sometimes buy or sell from Rotterdam when they have supply difficulties. It should be noted at this stage that the price of crude oil is now quoted in terms of spot market prices rather than the official listed prices and this is because of the supply situation.

For the purposes of definition the functions of the international oil industry can be divided between "upstream activities" and "downstream activities". The former description covers exploration and production. However, in this chapter we are concerned only with downstream activities, or refining and marketing and chiefly with the latter, though the activities of an international, vertically-integrated oil company will have considerable influence on the market place for petrol filling stations. Petrol is one of a range of products produced by refining crude oil, the others including fuel oil, lubricants, kerosene, naphtha, etc, and in the course of the process a decision has to be made regarding the proportions of the various products demanded by prevailing market conditions, and as already remarked it is for this reason that it is difficult to specify the cost of refining petrol alone.

History of the petrol business

Petrol was first imported from the USA and marketed in Britain at the turn of the century and at that time it was sold in sealed two-gallon cans. After the first world war, private motoring and commercial road transport developed quickly and there was a consequent increase in the number of petrol retailers. In 1920 hand-operated petrol pumps came into use in conjunction with underground storage tanks of about 500 gallons capacity and premises where petrol had been sold in cans then became pump sites. By 1938 there were over 35,000 outlets with an average throughput of 24,000 gallons a year and there was only a fraction of today's traffic on the roads. After the first world war Shell and Esso controlled the market, but between the wars various other suppliers established themselves. There was a rapid growth in demand and the establishment of refineries in the UK quickly followed.

The Petroleum Board, which controlled petroleum distribution and prices during the second world war, was dissolved in 1948 and Anglo-American (Esso) introduced the tied garage or solus trading system from the USA. Considerable capital investment by oil companies was needed for the construction of refineries and it was therefore particularly necessary for the companies to secure assured outlets for their products. When petrol rationing ended, Esso, followed by Shell-Mex and BP and then by others, started to acquire outlets to secure ties and within a very short period all the leading suppliers had introduced similar arrangements, so that by 1953 eight out of 10 stations were subject to exclusive supply arrangements of one sort or another. There were by then five supplying groups, namely Shell-Mex, BP and National Benzole; Anglo-American (Esso) and Cleveland; Regent; Vacuum (Mobil) and Fina (Petrofina) and they controlled virtually the whole market. These companies invested heavily in the improvement of existing petrol stations and the development of new stations, either directly or through their dealers. The then existing system would not have been adequate to cope with the increased demand and the oil companies provided the basis and the finance for change; indeed at the beginning of this period many petrol stations had no forecourts as we know them today and were simply prepared to sell petrol, over the pavement as it were, as an adjunct of their garage and repair businesses.

At the end of the last war there were about 30,000 stations still in existence, mostly of poor quality, and the number of stations hardly altered over the next 20 years, although average sales per station doubled more or less in line with the increase in motor traffic. For some years pump prices were fairly stable, but Jet entered the market in 1958 and began to cut prices, to be followed by others.

By 1968 there were 39,958 stations in the UK with an average annual throughput of 85,789 gallons. The number of stations was virtually halved by 1989 and the average annual throughput rose to 343,158 gallons.

By 1997 the number of stations fell to 14,748 and the average annual throughput rose to 435,547 gallons. In the period from 1968 to 1997 the demand for motoring fuel had virtually doubled.

Present market

While the total number of stations has been reducing steadily in the last 30 years, a major factor in the recent closure rate has been the arrival in the market of the leading grocers like Tesco, Sainsbury and Asda. In a decade this sector has gone from nowhere to operating approximately 800 stations and they now account for approximately 25% of the UK market. Earlier superstore stations were not usually planned to pick up business from the passing motorist. They relied instead on their grocery customers and throughputs were often price-induced. However, a number of recent superstore developments have stations planned so that they can also be used easily by the passing motorist.

BP have merged their downstream interests in the UK with Mobil and they with Shell and Esso can be described as the majors. Between them they own or supply 5,650 stations in the UK and enjoy approximately 40% of the market.

The Burma chain has been sold to Save and Shell have now acquired the Gulf chain which earlier in 1997 was involved in an abandoned merger with Elf and Murco. Total and Fina have recently joined forces in the UK.

There is a healthy demand for high throughput stations, but the supply is limited. It is estimated that in a typical year there are no more than 15 opportunities to acquire sites with a proven or projected annual throughput of a million gallons.

Petrol tie and the law

The simple early method of obtaining a tie was to enter into a short-term solus agreement under which the dealer agreed to sell the oil company's petrol to the exclusion of other petrols, usually in return for special wholesale price rebates on petrol sold and sometimes in return for loans. These were sometimes secured by a mortgage but the type of arrangement that developed very considerably was a purchase and lease-back transaction, because in that way the dealer-owner could obtain the highest price for his freehold and then lease back for a tied rent which was low in relation to value, the lower because a rebate on petrol purchases was usually arranged as well and on the other hand because the oil company could obtain a secure tie. On top of these transactions the petrol companies themselves acquired successful stations by outright purchase or they purchased greenfield sites or run-down stations and then carried out development. From a legal point of view all these arrangements seemed at the time to be suitable ways of securing a tied outlet but the position has been conditioned by case law and if the valuer has to deal with a tied station he is well advised to consider the precise nature of the tie and the question of whether it is binding or not.

The security of tenure provisions of the Landlord and Tenant Acts apply and a tenant is able to claim a new lease at a rent suitably adjusted to allow for the continued imposition of a tie, although of course petrol companies are able to oppose claims for new leases on the grounds specified in the Acts. As between the 1927 Act and the 1954 Act there is no great alteration in the provisions relating to the terms of a new tenancy and, while the court has a wide discretion, it is difficult to imagine a case where the petrol company would be required to grant a new lease without the tie, since the tie was the company's only object in acquiring the property in the first place. On the other hand, and having regard to one aspect of the reasoning in the restraint of trade cases discussed below, it is just conceivable that the courts might choose not to reimpose a tie on renewal, but in such circumstances it would seem to follow that the new rent would have to be at the free rental value level and would thus be beyond the reach of a company other than an oil company. Rent review clauses in tied tenancy agreements should make the tied basis of valuation clear but often do not.

Covenants on the part of a tenant to deal with his landlord alone and to take his goods have been accepted for many years, subject only to the implied term that the landlord will supply the goods at a fair and reasonable price. Such covenants have been held to be legal and binding in equity on an assignee.

The practice of arranging for the operation of oil company stations by licensees or by wholly-owned operating companies is now fairly common, though there has been a tendency in the late 1990s for oil companies to take stations back into direct management. In the case of licensee operations an extra-statutory code is applied, which does give the dealer some protection and security. The normal practice is to grant a licence for a three-year period, renewable on terms and with provision for compensation in the alternative. For obvious reasons the valuer will not be greatly concerned with this type of arrangement

Contracts contrary to public policy are unlawful and agreements in restraint of trade had been held to be contrary to public policy and against the public interest. In the past the term has been taken to include arrangements restricting business activity and limiting competition. Over a period of seven years or so starting in the mid-1960s four important petrol tie cases were decided: *Petrofina (GB) Ltd* v *Martin* [1966] CH 146; *Esso Petroleum Co* v *Harper's Garage (Stourport) Ltd* [1968] AC 269; *Cleveland Petroleum Co* v *Dartstone* [1969] 1 WLR 116; and *Total Oil (GB) Ltd* v *Thompson Garages (Biggin Hill) Ltd* [1972] 1 QB 318. The valuer should study these cases. *Petrofina* v *Martin* more or less destroyed the simple solus agreement method of obtaining a tie. *Esso* v *Harper's Garage* largely destroyed the mortgage coupled with a sales agreement method. The *Dartstone* case tended to validate lease-and-leaseback or purchase-and-purchase back types of tie arrangements and the position was strengthened by the *Total* v *Thompson* case.

In a normal rent review situation the valuer will find that he is required to arrive at the open market rent on the assumption that the premises are offered to let with vacant possession and he therefore values on a "free of tie" basis. From the point of view of the tenant this may produce unfortunate results because, as will be appreciated, the free rent could be much higher than the tied rent, although in such circumstances the tenant's subsequent remedy would probably be to sublease the premises to an oil company at the free rent that he would otherwise have to pay himself, arranging to lease the premises back at a tied rent.

Direct management, licensees, leases and dealers – the modern retailing operation

Sites owned by wholesalers are either operated by them under a form of direct management, run by licensees or operated by tied tenants. On the other hand, independent sites are normally the business of solus dealers who operate under exclusive supply agreements with particular oil companies. Nowadays these agreements usually run for three or five years. Hypermarkets and supermarkets are also supplied under solus agreements and the larger the operation the greater the bargaining power of the retailer in terms of prices paid for supplies.

Oil companies prefer to lease out premises where there are workshops and car sales businesses, or even to sell off such parts, but simple petrol stations are more often found to be under direct management. It should be noted of course that a lessee operation gives the tenant the benefit of the lease renewal provisions of the Landlord and Tenant Act 1954, although section 30 of that Act and in particular subsections (*f*) and (*g*) enable an oil company landlord to obtain possession at the end of a lease for reconstruction or direct occupation, so the protection of the Act is to some extent illusory. The licensee type of occupation deserves examination case by case because it is well established in law that the mere use of the word "licensee" does not preclude the possibility that the occupier is in fact a tenant with all the advantages that flow from that.

Major petrol stations, while usually the property of oil companies, are sometimes found to be owned by the major petrol retailing chains. Smaller wholesalers might take on lesser sites in rural areas, but low volume and badly located sites are not attractive and are often sold for alternative development, sometimes with restrictive covenants against petrol sales where the vendor fears redevelopment and resulting competition. Of course low-rated sites can be improved, particularly in terms of layout and the provision of forecourt shops and indeed are improved if the potential for growth exists.

When petrol stations are acquired by oil companies it is, as has already been remarked, for the purpose of securing a tied outlet for fuel, but oil companies do not enter the market where the petrol sales potential is too low to make acquisition worthwhile. In these latter circumstances the market is restricted to dealers who, as tenants or as owners, in the normal course of business make solus agreements with oil companies for the supply of particular brands of petrol; they are the solus dealers.

Under a solus agreement a station proprietor agrees to buy his fuel from a wholesaler – a refiner or a supplier – at scheduled prices less a rebate and in addition he will sometimes receive what is known as selective price support on a temporary basis, this being a form of support enabling a dealer to reduce his prices to meet or to try to beat local competition. A recent development has been the introduction of supply agreements which give the dealer what has been called "a guaranteed margin". With the exception of hypermarkets there are usually found to be only small overall variations in retail prices, although prices do tend to be higher in areas where there is little competition and lower where there is a real competition – from hypermarkets for example.

Modern filling station

Thirty years ago most filling stations had small sales kiosks where the motorist could pay for his petrol and at the same time buy from a very limited range of goods, mainly cigarettes and confectionery. This has been changing steadily since then so that today oil companies and petrol retailers are striving to improve their retailing skills and enlarge their facilities in order to offer a much wider range of goods to the motorist.

Shell UK Ltd has recently published a study called "Night-time Convenience Shopping and the 24 hour Society". The key facts are revealing. It is estimated that the UK forecourt convenience market is worth £2.8 billion. Shell's own retail business, "Select", does 20% of its business between 10.00pm and 6.00am and 80% of the 850 shops are open 24 hours a day and seven days a week. There are 2,500 to 3,000 product lines and Select is the country's fourth largest "own-label" sandwich supplier. Forty four of the shops have off-licences. Sixteen per cent of customers arrive on foot and 65% of all customers buy something other than fuel.

A recent E K Williams survey for "Forecourt News" suggests that typical profit margins vary from 35.1% on ice-creams down to 8.1% on cigarettes. The margin of 20.7% for groceries will be the envy of grocers generally. Since cigarettes will account typically for over a third of takings, average overall margins should be in the region of 15–20%. It is now becoming more common for a successful filling station to have non-fuel shop takings of £8,000 a week and many will have commission from lottery ticket sales which will be worth typically £5,000 a year.

It is likely that the oil companies will continue to look for new sources of profit and new ways of attracting customers. In late 1997 Texaco signed a five-year contract with the Bank of Scotland to launch a nationwide cash machine service at 200 of its stations by the end of 1998. It will be seen therefore that non-fuel activities are rapidly becoming a fundamental source of revenue at a station. Since the motorist is encouraged to buy non-fuel items before paying for his petrol the lingering time at the pump will be longer. This will tend to make greater demands on ideal site areas so that a station standing on much less than half an acre will not have enough space to extend the facilities to include a convenience store. Planning applications by oil companies to extend their shops to 1,000 sq ft or even 2,000 sq ft have caused some concern among Planning Officers, who fear that established high streets and shopping centres close by could be adversely affected by the growth of convenience stores attached to filling stations. However, resistance to the increasing public demand for this type of retailing and the trend for larger shops will be difficult to justify particularly at a time when many councils are looking hard at ways of reducing the amount of traffic in the centre of their towns and cities.

Adding to the demands on space is the importance to a successful station of a car wash where gross receipts might typically be £40,000 a year and for a simple hand-held lance wash £10,000 a year. The profit margins are remarkable. Returns of approximately 30% are not uncommon. Strong traffic volumes are essential to a successful station. The valuer can usually obtain traffic counts from the highways departments of local authorities or the Highways Agency. In regard to an established station with a proven throughput the valuer must have regard to factors which might affect traffic volume and therefore trade. He will make careful enquiries to establish that there are no road improvements proposed which could affect the traffic volume. He should also discover whether there are any proposed developments of new stations in the area and not necessarily on the same route. This latter point is particularly relevant in the case of superstores with stations. He should look too at other stations which are already trading in the area, and compare their prospects for survival. With new petrol stations the valuer will be called on to estimate the potential mature throughput. In some cases the mature throughput can be reached within a year of the start of trading but in other cases it is only reached after about three years. He should look carefully at competing stations in the area which are trading poorly and are unlikely to survive. This calls for some fine judgement but is more apparent in the case of stations which have severely restricted site areas or difficult access and egress. Such a station might have a modest annual throughput of say 300,000 gallons but its closure could lead to most of its throughput going to the next station along the route and will rarely benefit more than two other stations in the area.

As a rule of thumb on trunk roads the turn-in proportion is usually about 3 to 4% of passing motorists. On busy roads most of the throughput would come from traffic which is passing on the same side of the road as the station. The average purchase in 1998 was approximately 5.5 gallons.

By way of illustration, in the case of a good site with a frontage of 150 ft to 200 ft, a depth of no less than 100 feet, and a two-way 24 hour traffic flow of 40,000 private vehicles and vans, the estimated throughput calculation would look like this:

Assessment of Throughput

Total two-way 24 hour traffic flow for cars and vans		40,000
Less half for those vehicles on the side of the road opposite to the station		20,000
Effective daily number of potential motoring customers		20,000
Average turn-in rate		3%
Estimated number of daily customers	(vehicles)	600
Average fuel purchase	(gallons)	5.5
Average daily throughput	(gallons)	3,300
Effective number of days in the year		313
Estimated annual throughput	(gallons)	1,032,900

It is important to note that this example assumes that the traffic flow only applies to six days of the week. This will vary enormously from one road to another. It is never the case that a road as busy as the one in the example will be totally without Sunday traffic and a large residential population in the immediately vicinity of a station would have a considerable bearing on the calculation. The 24-hour day seven-day week contemplated in the Shell study is of special relevance to the future. So too is the increasing number of major sporting events taking place on Sundays.

In 1995 the Department of Transport estimated traffic growth of 58–92% between then and 2025. Despite this, in late 1996 the Department of Transport abandoned over 100 longer-term road schemes though they did say that they would consider smaller-scale improvements to tackle local problems. It is possible that in the longer term the rate of petrol sales growth may decrease as saturation point is reached in terms of car ownership and because motor manufacturers have made great progress in terms of economic fuel consumption. These matters will have a considerable bearing on the future growth of throughputs and non-fuel turnover.

Basis of valuation

As is the case with the valuation of hotels, public houses, restaurants, cinemas and theatres, a close knowledge of the trade carried out on the premises is a necessary concomitant to this type of work.

As has been said, oil companies do not purchase for investment but to obtain an outlet for their products. Their petrol profits arise from the refining and wholesaling process – and to some extent from rents from tied outlets – whereas dealers' profits arise from retailing. The concept of a tied rent and a free rent must be understood. The free rent is the rent which an oil company is prepared to pay and is the total of the tied rent that it can expect to receive from its dealer plus an overbid, or special value element, which in fact is met out of refining and wholesaling margins. The dealer pays his rent out of his retailing profits, whereas the oil company can and does add to the rent received from the dealer a proportion of its own earnings. In practice oil companies will be found in most cases to purchase rather than lease their outlets.

It must be borne in mind that this special aspect of property valuation involves projections in an imperfect market; this in the sense that where substantial petrol throughputs are involved the oil companies and petrol distributors are virtually the only buyers and, furthermore, their interest varies from month to month because of marketing or budgeting factors rather than for property investment reasons. There is a limited number of possible purchasers and they are not all in the market at the same time; nor are they necessarily in the market in the same region of the country. In the circumstances consistency will be found to be lacking and it is probably more true to say in this case than most that the value of the property is what it will fetch.

The valuer must be cautious in terms of analysis and synthesis when dealing with individual transactions because it is occasionally the case that excessively high prices are paid for large volume outlets. The oil companies usually operate within budget parameters and the availability of funds as well as marketing needs will condition offers. It is of vital importance that the valuer should consider all factors likely to affect petrol sales and the trade generally such as possible road diversions and a variation in traffic volume, the possibility of new stations being established, of old stations being enlarged and improved and of stations being closed. Valuable information can usually be obtained from the planning and highway authorities.

Directly comparable evidence is of great assistance, but it is not always available and valuations may well have to be constructed from first principles. To be useful, a comparable must be directly analogous and, by way of example, if in search of such evidence the valuer analyses the sales at, say, six stations on a busy 30-mile stretch of road with little variation in traffic volume, he will nevertheless find great variations in sales owing to such things as comparative accessibility, visibility, lay-out, dealer operation standards and, of course, pricing policy. It is after all from throughput and other turnover, and the consequent profit, that forecourt rent derives. Competitive pricing is vital but standards of operation are equally so and, surprising though it is, a good dealer might possibly sell twice as much petrol as a bad dealer at a given station. National advertising and special promotions by major oil companies can also affect throughputs.

It must be made clear that stations on other routes are no more directly comparable than would be rental evidence taken from shops in positions other than the one to which a valuation relates. It is true though that there is a general pattern of values and it is this pattern that the specialist valuer must build up in his mind in the course of practice. With particular reference to forecourts, it is clear that oil companies apply specific rates per gallon on a sliding scale adjusted for throughput when buying or leasing.

Forecourt values

Tied rents and licence fees, on the one hand, and open market rents, on the other, are to be distinguished from each other, tied rents being artificial or closed market rents paid to suppliers and conditioned by retail profit margins. Private landlords – non-oil company landlords – will let premises at open market rents but these rents will be affected by oil company demand only if the throughput is sufficient to attract it. The oil company wholesaler will of course be found to pay an open market rent and then, if he subleases, to do so at a lower tied rent.

Open market forecourt rents, that is to say free rents or the rents an oil company might be expected to pay, are related to the cost of producing petrol, Derv and lubricants and the profits of marketing them, whereas tied rents are related to the profits derived from retailing these products. Dealer rents, in the sense of rents paid to private landlords – non-oil company landlords – or low throughput stations of insufficient interest to attract oil companies are directly derived from likely trading profits.

Oil company interest is clear at about 800,000 galls pa or above, but there is an area between that and about 600,000 galls pa where the interest of oil companies is variable and depends on a variety of marketing and extraneous factors. It can be taken for granted that at the present time even a minor oil company would probably not be interested in the acquisition of a station selling 600,000 galls pa or less. This does not mean that there is no demand for small throughput stations but it is a demand from dealers, retail chains or speculators. Whether for capital value or rental value the valuer will use great caution when dealing with a station with a throughput of less than 400,000 gallons and particularly so if the throughput is falling

By and large the property investor acquires a hedge against inflation and his yield, in the case of freeholds anyhow, is a net yield in the sense that he should not have to build up a sinking fund to compensate for the deleterious effects of inflation on the capital investment. However, petrol filling stations are bought by oil companies and oil companies are not investors in the sense in which the word has been used here; they buy property to secure an outlet for their fuel products. Of course, good, well-sited stations will still be a hedge against inflation but the central reason for the purchase is to obtain a property from which petrol can be sold or a property that can be let subject to a tie to secure the sale of petrol.

It will thus be apparent that normal investment considerations are irrelevant.

The great majority of petrol station transactions are purchases, leasing being a least preferred alternative, and it follows from this that most real evidence of value arises from capital transactions. The price paid for petrol filling stations can be closely correlated with throughput and rises steeply in terms of the amount paid per gallon as throughput increases.

The following table is an attempt by way of synthesis to give an approximate indication of free of tie capital values in relation to throughput:

Annual throughput in galls	Capital value per gall
500,000	70p
600,000	80p
700,000	100p
800,000	120p
900,000	140p
1,000,000	150p
1,250,000	160p
1,500,000	170p
1,750,000	190p
2,000,000	200p

It is important to stress that these values apply to the whole station and assume non-fuel and car-wash receipts usually found at stations with the sort of fuel throughputs indicated.

Helpful to the specialist valuer is the *UK Pump Price Report* which is published twice a month by the *Petroleum Times*. This lists current pump prices for all forms of motor fuel at 39 towns and cities throughout the United Kingdom. There is a difference of approximately 25p a gallon between the dearest and the cheapest unleaded gallon. There is evidence too of a well established trend for cheaper fuel along a line following the M62 between Liverpool and Hull. There is a similar band of cheaper fuel following the M8 between Edinburgh and Glasgow. Logically, the valuer should experience corresponding variations in value when dealing in the open market. In fact, he will find it difficult to discern such a pattern. He could find that he gets a better price for a site in Edinburgh than he does for a site in Inverness even though

Inverness pump prices are stronger. This is not entirely explained by the fewer number of oil companies active in the Highlands.

From capital value to rental value

Capital values being the yardstick, rental value often has to be derived from the level of figures given in the last section, or rather from their market equivalents, and this is the reverse of the normal investment valuation process. The distinction between an investment purchase and a purchase made to secure an outlet for petrol sales will have become clear.

It follows that since the determination of the rental value of a petrol station often involves the ascertainment of capital value as a first step – the opposite of the normal investment valuation process – the years' purchase used for decapitalisation purposes must be constant. Decapitalisation at a higher years' purchase for a better station would produce an incongruous result in trading terms, since the more valuable station would then have a rental value proportionately lower than the less valuable station, and as a matter of strict logic it could be suggested that the better a petrol filling station is in trading terms the lower should be the decapitalisation years' purchase figure, although this would result in the initially surprising conclusion that, for example, a 500,000 galls pa station's capital value should be decapitalised on an 8% basis and a 1,000,000 galls pa station on a 10% basis. If it is argued that varying rates should be employed as in the conventional investment valuation process then it would follow that the rate of values per gallon must rise at the lower end of the scale and this would not represent the real nature of the market. The oil company's target is throughput and the higher the throughput the more valuable the station. It cannot be the case that an oil company would be prepared to pay a higher rent in proportion to capital value for a smaller station.

Of course the conventional investor might be prepared to pay a proportionately higher price in relation to rent for the better station but that would follow the initial free-rent equation and the increment could be regarded as the premium a conventional investor was prepared to pay for security of income in terms of covenant and perhaps for growth prospects, although he would have to possess a specialised knowledge of the oil industry to judge the latter. An investor might also take into account the possibility of redevelopment for alternative uses, a factor usually ignored by oil companies, though that could change in the long term.

It is now necessary to turn to the question of rental yield in relation to capital value and as far as possible to base the conclusions on empirical evidence. In the normal investment market the valuer's analysis is derived from evidence of lettings followed by sales, or from sales followed by lettings, so the relationship of capital and rental value in terms of yield is immediately clear, whereas in the case of petrol stations the property will have been bought by an oil company and, if let, let at a tied rent related to the tied tenant's expected retail profits not to the wholesale profits that fix the capital value of a free station, and the same conditions apply if an oil company leases a station. In short, in the equation one factor, rent or capital value, is likely to be hypothetical or at least not derived from the same transaction, simply because oil companies do not buy a station and then lease it at a free rent.

From the evidence available it would appear that in terms of today's market the inherent rent yield lies between 7% and 9% and can fairly be said to average 8%, and it will have become evident from the foregoing that the decapitalisation rate must be constant throughout the range of capital values.

In practice it is suggested that in determining rental value the practitioner should start from capital value unless, that is, he has equally good or better evidence of rental value, in which case he should use both approaches. A table of indicated free rental values derived from the table given above and based on a decapitalisation rate of 8% follows:

Annual throughput in galls	Rental value per gall
500,000	5.6p
600,000	6.4p
700,000	8.0p
800,000	9.6p
900,000	11.2p
1,000,000	12.0p
1,250,000	12.8p
1,500,000	13.6p
1,750,000	15.2p
2,000,000	16.0p

As with capital values these rents are for the whole station and the same assumptions about non-fuel and car-wash receipts are made.

These figures are perhaps unduly precise but, if nothing more, they will establish the fact that a free rent could hardly be paid out of the retailing profits of a tied tenant.

The values of petrol stations have risen more or less in line with commercial and industrial property values, possibly doubling between 1984 and 1989, and it is believed that it was in the period from 1986 to 1987 that capital prices of £1 per gall were first realised for prime stations. Major oil companies claim that their return on the capital investment in service stations is as low as 6%, but this must reflect the fact that tied rents as an element in the return are artificially low, the difference between the tied rent and the open market rent, or the annual equivalent of the purchase price in the open market, being the premium or special value element that petrol companies are prepared to pay for the tie.

The Harewood case

If there is a lack of evidence about petrol throughput in lease renewal arbitrations and in new tenancy cases under the Landlord and Tenant Act 1954 – for example if the tenant or supplier refuses to supply throughput figures – the valuer has in reserve the possibility of applying for an order for disclosure1 as part of the arbitration or court process. Such an order can usually be obtained if a valuation depends on evidence of trade done or of profits as is invariably the case with petrol stations, throughput being a vital element in determining the rental value of a forecourt. Disclosure might include valuation reports and general accounts as well as petrol sales figures of course. The leading case is *Harewood Hotels Ltd v Harris* (1957) 170 EG 177, which was referred to in *WJ Barton Ltd v Long Acre Securities Ltd* [1982] 1 EGLR 89. The *Harewood* case made it clear that evidence of the financial results of a business was admissible if needed to show what rent premises were likely to fetch and the second case in limiting this judgment nevertheless made it clear that in the case of petrol stations trading evidence was relevant and could be called for.

It would seem to follow from the foregoing that if in negotiations a valuer acting for a wholesaler or for a retailer declines to produce petrol throughput figures he is encouraging the lessor to go to arbitration or to the court and that this might as well be recognised from the beginning. Petrol stations can only be valued on the basis of actual or potential throughput, although, of course, surveys and checks can be carried out to determine throughput and indeed to estimate potential throughput. There is at least one firm specialising in this type of work.

Fast Food Caterers

There is a growing trend for oil companies to acquire sites where they can trade alongside fast food caterers, particularly those with strong brand images. The two activities are complementary and a popular caterer can typically improve throughputs by as much as 10%.

Shops and car washes

Shops and car washes are usually valued with the forecourt and as far as shops are concerned it would appear on analysis that rents charged by oil companies to their tenants often relate directly to throughput in a ratio of something in the region of £2 per sq ft per 100,000 galls pa, in other words £20 per sq ft at 1,000,000 galls pa, but this would not apply in an urban situation where there was non-motorist trade as well, this probably justifying a higher rent. The size of the shop is critical of course and would normally be in the region of 500 sq ft to 800 sq ft though in appropriate locations shops will be much larger.

In the last few years the sector has seen remarkable changes in the oil companies' approach to retailing at filling stations. Shops have become larger and larger and the amount of trade which walks onto the station can be remarkably high in places where the filling station is surrounded by houses or offices. Shops with areas of close to 2,000 sq ft are in appropriate cases the objective of the oil companies. They are called convenience stores. The range of goods carried in these shops has expanded dramatically, so that one can buy fast food, groceries, freshly baked bread and fresh flowers. The businesses have expanded to include lottery ticket sales, electricity tokens and few new filling stations presently being planned are without a cash dispensing machine. Often premium pricing policies are applied to certain goods such as cigarettes.

The oil companies have clearly aimed at establishing themselves as roadside retailers. Esso have joined forces with Tesco and BP with Safeway in an attempt to develop this type of business.

It might well be that the valuer would do well to deal with the rental value of shops or convenience stores on a profits basis and expect a rent of approximately 15% to 25% of the net profitability of the shop. He will need to be satisfied that he has all the necessary information about the tenant to carry out this task. It should be noted that as shop sizes become bigger then the profitability per square foot will fall as the range of goods is extended to include items with lower profit margins.

A fully equipped car wash in a permanent building would cost in the region of £100,000 to build and might produce a gross return of £60,000 pa with running costs of £20,000 pa, leaving a net return of £40,000 pa. An oil company landlord would probably look for something in the region of a third of the net profit as rent, so in the case quoted the rent would be around £13,333 pa.

Study 1

By way of example the following is a rental valuation of a modern petrol filling station free of tie, with an annual motor fuel throughput of 1,000,000 gallons, annual shop sales of £400,000 excluding VAT and car wash annual sales of £60,000 and in an area of average pump prices.

Annual throughput		1,000,000 gallons
Rent per gallon		9 p
		£90,000
Non fuel turnover	£400,000	
Profit Margin	20%	
	£80,000	
Attributable to rent	25%	£20,000
Car wash turnover	£60,000	
Cost of running	£20,000	
Return	£40,000	
Attributable to rent	33%	£13,333
Annual Rental Value		£123,333

Cost of construction and site values

At present the cost of constructing a typical fully-equipped petrol service station or petrol filling station stands between £550,000 and £900,000. A good modern station will be found to have a forecourt with a frontage of at least 150 ft and a depth of about 150 ft as well and will be equipped with three or perhaps four multi-hose pump islands under a canopy. The storage tank capacity of a station selling upwards of 1,000,000 galls pa usually consists of two or often three 12,000-gall tanks or six 5,000 gall tanks. Deliveries are nowadays made by 6,000-gall tankers and if three deliveries a week are assumed it will be seen that it is possible to calculate the required tank capacity when peak sales are taken into account.

A typical station construction cost would relate to a station comprising a forecourt with storage tanks and interceptors, a canopy, a shop and an area of hard-standing, together with electrical and mechanical services. The provision of a car wash would increase the cost by between £50,000 and £100,000.

The residual method of valuation is commonly used by oil companies to determine site value and, speaking in very general terms, the value of a prime site with a potential throughput of 1,000,000 galls pa would be in the region of £600,000 to £1,000,000, this for a site with an area of about half an acre.

It appears to be the case that certain oil companies which follow USA accounting principles prefer short-term building leases, for example leases for 20 years, and are prepared to pay rents equivalent to those normally paid under long-term leases. This follows from the usually fallacious reasoning that rent is a liability, this conclusion ignoring the fact that a lease is usually regarded as an asset. It will sometimes be found that building leases for normal long terms contain break clauses in favour of the tenant at, for example, 20-year intervals and the valuer will perceive the reason for this in the circumstances just outlined.

Motorway Service Areas

Prior to 1992 the Department of Transport was responsible for identifying, acquiring by compulsory purchase where necessary and promoting the development of motorway service areas.

These sites were leased, mainly to oil companies and large catering organisations. Originally turnover rents were paid but later the practice has been for the Department of Transport to grant 50-year leases at a premium with nominal ground rents.

In 1992 the Department of Transport opened up this sector to private initiative so that oil companies, catering companies and developers have since then identified what they saw as potentially good sites, often filling in large gaps between existing sites.

Obtaining planning permission for these service areas is a daunting and expensive task. Often a multi-site public enquiry is held and can last for many months. It is usual for planning consent to be granted subject to a Signs Agreement from the Department of Transport. This permits the erection of signs on the approach section of the motorway which signs are helpful to the motorist and the business of the service area. In exchange for a Signs Agreement the Department of Transport will create a rent charge on the motorway service area under the terms of which they can continue to enforce the regulations which apply to these sites in connection with car parking capacity, HGV parking capacity, provision of facilities, and a ban on alcoholic drink and operations which could be reasonably described as destination activities.

A valuer will approach this subject with the greatest care and a knowledge of the scarce comparable evidence is fundamental.

The Department of Transport has a general rule that motorway service areas will be available at roughly 30-mile intervals and that additional "infill" services will be allowed, but only exceptionally and where a clear and compelling safety case can be established. For this reason, establishing need is fundamental to the obtaining of planning permission. However, motorway traffic use has increased at a much greater rate than trunk road traffic use and pump prices and catering prices are stronger on motorways so that the profitability is greater. Assuming that a site is large enough to provide a large petrol filling station, an HGV service area, a large amenity and catering building and a Travel Lodge then the developed value of such a site will be in the region of £6 to £8 per annual gallon of throughput, ignoring bunkered DERV throughput. Site values are very difficult indeed to estimate. Infrastructure costs vary considerably from site to site. These can include the provision of services, drainage, road works and even the cost of building crowned roundabouts. The early motorway service areas like Knutsford on the M6 occupied twin sites of not much more than 13 acres whereas the later service areas like Warwick on the M40 occupy twin sites totalling over 50 acres.

The principles applied to the valuation of an ordinary petrol station apply to a motorway service area but the scale is substantially different. This is less marked at the extreme sections of some motorways such as the M6 north of Preston and the M4 west of Newport. In areas like these, and particularly in Scotland, values are more easily related to values on trunk roads. The difference is explained entirely by the great difference in traffic flows.

Garages

In 1975 there were approximately 31,000 petrol filling stations in Great Britain. By 1999 there were less than half that number. The change has been dramatic.

The change in the motor trade has been just as great. The traditional business of a garage has been greatly eroded by the development of Fast Fit Centres which have enjoyed a meteoric rise. Thirty years ago the traditional motor trade captured 80% to 90% of all car tyre sales. In 1998 this share had fallen to 8% to 10%. In the same period the trade's share of exhaust sales has fallen from 60% to 30%.

The Fast Fit businesses were able to succeed because they concentrated on jobs that could be done quickly and easily by semi-skilled staff. Motorists could go in without making an appointment and sit down and wait while the job was done. They did not need to book an appointment for three days' time. These businesses were also perceived by the public as being cheaper, although this was not always the case. Their advantage came from menu pricing so that the motorist knew in advance what he was going to pay.

The valuer will usually be faced with arriving at a capital value of a franchised garage – that is to say a garage which holds a franchise from a motor manufacturer with a prescribed dealer territory in which to operate. A franchise agreement ties the proprietor of the garage to sell only the new products and parts of a particular manufacturer. In approaching the valuation of a franchised garage or dealership, the most important thing is to distinguish between the value of the property and the value of the franchise. Although the Block Exemption Rules are at present a little kinder to garage proprietors, and may become even more so in 2002, a sound property valuation should ignore the advantage or disadvantage of a particular franchise.

Retaining a franchise is not guaranteed. Neither is the popularity of any manufacturer's range of cars. Clearly there are advantages to certain franchises but these advantages should be found in the profit and loss account of the particular business and not in a valuation of the property itself.

It is important too to ignore the profitability of a business carried on from the property to be valued. It is not uncommon to see businesses in poor property with less than desirable franchises but nevertheless making good profits. In such a business the proprietor will have special strengths and skills like used car sales. So the

profitability of a business should not be confused with the value of
the property.

Valuation

Study 2

By way of example the following valuation is of a provincial
dealership in a good position with attractive buildings erected 20
years ago.

	Sq ft	Rent £s per sq ft	Rent £s pa
Showroom	3,000	12.00	36,000
Sales Office	300	6.00	1,800
General Office	1,000	6.00	6,000
Reception	200	4.00	800
Workshops	6,000	3.00	18,000
Parts Storage	200	3.00	600
Mezzanine	500	1.50	750
Rental Value of buildings			63,950
Open used car stances	No 25	400	10,000
Car parking spaces	No 40	100	4,000
Total Rental Value			77,950
Year's purchase @ 9%			11.11
			866,024
Value Say			860,000

The valuation of a franchised dealership is based on the capitalised
estimated rental value as shown in the example. Considerable skill
and experience is required in the valuation of the whole and the
constituent parts.

Showroom

There is a relationship between the rental value in the example of
£12 per sq ft and good quality, off-centre retail rents in the area. Had
the showroom been 4,000 sq ft and not 3,000 sq ft it would still have
been reasonable to adopt the same rent. Had the showroom been
10,000 sq ft the rent to be adopted would have been considerably
reduced. Showrooms are a manufacturer's requirement. Unless

proportionately increased sales can be made, having showrooms which are too large merely reduces the amount of retained profit per vehicle sold. The showroom has to bear rent, or occupancy costs, rates, cleaning, lighting and decorating. The number of new cars a garage proprietor can sell is influenced to a large extent by the market share of the particular franchise he holds.

Offices

It is sometimes the case that a garage is also the head office of a larger business with other branches. In the example shown the offices of 1,000 sq ft are appropriate to a garage of the size shown. Had the general offices been 3,000 sq ft it would have been sensible to adopt a lower rent on the assumption that a typical willing buyer would be unlikely to have a need for so much office space. There is little relationship between the value of ancillary offices in a garage and the value of purpose built offices in the same town. Typically an ancillary office rent should not exceed two-thirds of the value of showroom space and is more usually half.

Workshops

The rent of £3 per sq ft allocated to the workshops in the example represents the value of stand alone workshop space of a comparable quality in the same town. The principle of marginal utility applies to workshops just as it does to showrooms and offices. The servicing and repair of vehicles has changed enormously in the last 20 years so that much less workshop accommodation is needed.

Used car stances

Valuing used car stances calls for some fine judgment. The annual rental value of one stance should be related approximately to the average retained profit on one vehicle. A good used car operation as part of a franchised garage should turn over in a year approximately 10 times the number of cars displayed. The difficulty for the valuer is that some businesses do far better and many more do far worse. Here again the principle of marginal utility is important. Had there been 100 used car stances it would be wrong to assume that the willing purchaser might sell 1,000 cars a year.

Such a volume could only be achieved by a specialist used car operation because a typical franchisee in the type of property in the example would be hard pressed to generate more than 300 used car sales a year from part exchanges and the only way to get to a much higher number would be to have a very skilled used car buyer. It would be wrong for the valuer to assume that the typical willing buyer would have those skills.

If the garage in the example had no used car spaces or a very limited number of stances then the valuer's approach to the other constituent parts of the valuation would be different. The property would lack a vital source of profit. The increased reliability of used cars has changed the business considerably and is likely to lead to an increase in the establishment of massive, specialist used car operations.

Most garages are owner-occupied and open market lettings are rare so good comparable evidence is difficult for the valuer to find.

Capitalising the rent

This is the most difficult stage for the valuer. In other types of property there is an investment market in which institutions, property companies and individuals are prepared to be landlords. There is only a tiny investment market in garage properties.

Some sale and lease backs have been achieved by garage groups, though usually they have guaranteed rental growth and/or artificially related the rental value for review purposes to other types of property, like off-centre retail space.

The yield adopted for motor trade property should have a closer relationship to the actual cost of money at the time. This can vary considerably. For example, a substantial motor group listed on the London Stock Exchange might buy the garage in the example for £860,000. As we start the 21st century the trend is for consistently lower inflation and much lower interest rates, the motor group might borrow the money to buy the property and pay 7% interest which would be £60,200 a year, considerably less than the estimated rental value.

Location

The location of a garage is very important to the valuer. As the number of franchised dealerships has been falling steadily for the last 25 years, the trend is likely to be the establishment of larger, flagship dealerships in big towns and cities enjoying much larger dealer territories. It will be increasingly difficult to value garages in smaller towns. Some will survive as unfranchised garages, but many will go out of the trade completely.

Fast Fit Premises

While franchised garages generally tend to be owner-occupied, this is less so in the case of Fast Fit properties and the valuer will find it comparatively easier to obtain evidence of comparable transactions in the area. Rental values for such premises are not directly related to the value of workshop space in the area. Rents can often be as much as twice the level of ordinary workshop rents, particularly when the property is in a prominent location.

Recommended Reading

"Forecourt" the monthly magazine of the Petrol Retailers Association published monthly by the AFL Deeson Partnerships Ltd, Ewell House, Graveney Road, Faversham, Kent, ME13 8UP.

"Petroleum Times Energy Report" published twice monthly by Nexus Media Ltd, Nexus House, Azalea Drive, Swanley, Kent, BR8 8HY.

"Automotive Management" published fortnightly by Leading Edge Publishing, 2 Oxted Chambers, 185/187 Station Road East, Oxted, Surrey, RH8 0QE.

"The Motor Industry of Great Britain" published annually by the Society of Motor Manufacturers and Traders Ltd of Forbes House, Halkin Street, London, SW1X 7DS.

"Appraisal & Valuation Manual" The Royal Institution of Chartered Surveyors.

Chapter 15

Lease Renewals and Rent Reviews of Commercial Property

There can be little doubt that it should be the aim of professional advisers to keep their clients out of litigation and to obtain a proper settlement by negotiation, if so instructed. In this connection it may be advantageous to use the services provided by PACT which applies to disputes relating to unopposed lease renewals under the Landlord and Tenant Act 1954 – see the appendix to this chapter.

In spite of the best efforts by their respective professional advisers, negotiations between landlords and tenants of commercial premises break down. If a lease has expired, the final arbiter is the county court or the High Court. The costs are clearly high and there is therefore an added inducement for the matter to be settled.

However, where the parties have fallen out over the terms of a rent review, then the matter will fall to be settled either by an independent surveyor or by an arbitrator, according to the lease. He tends to be a Chartered Surveyor (RICS) and is normally appointed by the President of the Royal Institution of Chartered Surveyors, rather than by agreement between the parties. The settlement of the dispute by one of these methods should normally be quicker and cheaper than any court proceedings.

A significant number of references arise because the tenant is either reluctant or cannot afford to pay the increased rent and therefore reference itself may give some temporary relief to cash flow problems.

Increasing need to keep up to date

The necessity to speed up proceedings and keep down costs is reflected in the recent Acts, the Civil Evidence Act 1995 and the Arbitration Act 1996.

The increasing volume of work dictates that the private practitioner must keep fully abreast of any reported Landlord and

Tenant Act cases. He needs to be able to give evidence, to prepare proofs of evidence and, if the matter is going to be dealt with by a written representation, to be able to draft and write them on behalf of his client. If he is an arbitrator, then he must have or acquire the ability to write an award which, by the time he has set out the various arguments and recorded his conclusions in writing, may well be a document of several pages.

In January 1997 the RICS issued a practice statement and guidance notes for surveyors acting as expert witnesses.

Practice Statement and Guidance Notes – RICS January 1997

Expert Witnesses have been getting a bad press, especially with the Judiciary. Lord Woolf in his report *Access to Justice* devotes a whole chapter to Expert Witnesses. In his report of July 1996, he said

> Those responsibilities (of the expert witness) should be made clear in the guidance issued by the relevant professional bodies.

The RICS reacted by publishing a Practice Statement (which became mandatory in March 1997) and Guidance Notes (RICS Byelaw 19(7). Conduct Regulation 23).

The general thrust is that a witness has a primary duty to the court, is to be truthful, honest as to opinion, independent, objective and unbiased.

A Chartered Surveyor may only accept instructions to appear as an expert witness:

1. If he advises the client *in writing* that the Practice Statement applies and offers to supply a copy on request (RICS bookshops).
2. If he keeps a *written record* of matters on which expert evidence is required.

Any changes in instructions or supplemental instructions must be recorded *in writing*.

Where an inspection of any property is required it must always be carried out to the extent necessary to produce an honest opinion which is professionally competent, having regard to its purpose and the circumstances of the case.

In producing his report, the surveyor must:

*   personally sign and date it and;
*   make a declaration that:

(a) he believes it to be accurate
(b) includes all facts relevant to his opinion (ie no weeding out of comparables)
(c) state that the report complies with the 1997 Practice Notes and Guidelines.

The publication of some 26 pages contains much guidance and common-sense suggestions. It includes a suggested layout of a witness report.

It remains to be seen how the RICS is to police compliance with the document. Some changes are detailed above and there is no substitute for reading the document, having it to hand, and having copies for clients. Bear in mind that the changes are mandatory and that failure to do certain things may result in a disciplinary hearing at Great George Street.

A section is written on the dual roles as expert and advocate. These frequently arise in Valuation Tribunals, Arbitrations, Rent Assessment Committees and Planning Appeals.

> Members are recommended to consider the desirability of the roles. Members are under a duty, when undertaking these two roles, to make it clear to the Judicial Body which aspect of the two roles they are fulfilling at each and every stage.

Also quoted is the case of *Multimedia Productions Ltd* v *Secretary of State for the Environment* [1988] EGCS 83 when it was said:

> The Expert, who has also played the role of Advocate, should not be surprised if his evidence was later treated by the Court with some caution.

Many surveyors appear in the dual role. They need to be especially aware, not only of their duties and to whom they are owed, but also of the Practice Statement and Guidance Notes.

Arbitration Act 1996

The great significance of rent in terms of a tenant's outgoing is not generally realised. Commerce and industry in general terms pay as much in corporation tax as they do in rates; rent clearly can be more significant than both rolled together. The open market place is where rents are settled by supply and demand. The arbitrator's task is to determine what the open market is but there is often conflicting evidence. An expanding chain of shops such as fast food shops may well pay more than the market in order to expand business. Other trades may be artificially depressed due to recession. Finding a balance is never easy and the landlord and tenant are entitled to expect a high level of professional competence. An arbitration award is usually more satisfactory to all parties because the professionals see their arguments advanced and any legal points disposed of, hopefully for all time; the client may well have an award of one page which of itself gives no indication whatsoever of the considerable work that has gone into the presentation of his case. The losing client would not be human if he did not take a jaundiced view.

Under the Arbitration Act 1996 an arbitrator must give a reasoned award unless the parties agree otherwise. In contrast an independent expert does not have to give reasons, and rarely does so; he is open to be sued for negligence. An arbitrator is immune from suit unless an act or omission is shown to be in bad faith.

Lease Renewals

Previously, in chapter 4, the statutory rights of a business tenant to a new lease have been set out in summary. There is a considerable body of case law as to who or what amounts to a business tenant. The definition is far wider than plain English would indicate, and a tennis club has been held to be a business tenant as defined under the Act. There has also been a considerable amount of case law arising from a landlord successfully, or unsuccessfully opposing the application by his tenant for the grant of a new lease. There are seven grounds specified under section 30 of the Landlord and Tenant Act 1954 Part II (see chapter 4). Clearly a valuer has to have some knowledge of these, and even if the grant of a new tenancy is opposed the question of an interim rent, and therefore values, may well arise if the matter proceeds to court and the hearing is after the lease has expired. It is possible today for a business lease to be outside the terms of the Landlord and Tenant Act 1954 Part II. In the 1960s it became apparent that landlords were unwilling to grant short leases of property which they intended to redevelop, or perhaps needed for their own occupation, either because tenants could successfully apply for new leases or could delay the landlord's plans by lodging an application for a new lease. Under the Law of Property Act 1969 a business lease can now be registered in court as being outside the terms of the Landlord and Tenant Act 1954 Part II.

The valuer normally first gets involved with a lease renewal with the necessary service of statutory notices. The valuer should point out to his client that a lease is coming to an end and that certain necessary statutory notices and counternotices should be served. As the final arbiter of the new lease and its rent will be either the county court or, in the case of more valuable property, the High Court, services of a statutory notices are best left to solicitors. A valuer concerned with a case going to the county court or to the High Court will not only be concerned with the appropriate value that the court will be invited to determine under section 34, but must also enquire whether either the landlord or the tenant is seeking to vary any of the terms of the original lease. Very often the original lease may either be old or expressed in an out-of-date language that may nor reflect current market conditions or current legal thinking, particularly relating to clarification or extension of the law in reported cases. Section 35 of the Act gives the court the right to impose such terms and conditions as it thinks fit, having had regard to the current tenancy and all relevant circumstances. The valuer giving evidence must therefore be prepared to give evidence in chief or to be cross-examined upon any changes that might be made by the court and to express an expert view as to what extent such changes should be reflected in the rent determined under section 34.

In February 1982 the House of Lords, in *O'May v City of London Real Property Company Ltd* [1982] 1 EGLR 76, gave a judgment that is some indication as to how far the courts should go in varying the terms of the lease. The facts of the case were that the landlords had sought to impose a full repairing lease upon the tenants and the High Court had thought this to be a reasonable variation. The High Court was overturned by the Court of Appeal and the House of Lords upheld the judgment of the Court of Appeal. It was an agreed fact between the valuers that the transfer of the burden could be adequately compensated by a reduction in rent from £10.50 per sq ft pa to £10.00 per sq ft pa. The High Court set out some tests and these do not appear to have been materially dissented from in the Court of Appeal or the House of Lords. There appear to be three main tests:

1. Has the party seeking a variation in the terms of the lease shown any reason for doing so?
2. If such a change in term is granted can it be adequately adjusted by a reduced rent determined under section 34?
3. Will the proposed change materially impair the tenant's security in carrying out his business or profession?

With all three questions in mind, the court should consider whether each one is fair and reasonable between parties.

In his judgment Lord Hailsham said:

> Obviously it is to the advantage of the landlord to transfer the financial risk of fluctuation to the tenant and there can be no possible reason why, if the tenant agrees (and the evidence was that many do) he should not do so. But the crucial question is, if the current lease does not so provide and the tenant does not agree, by what possible reasoning the court should impose the burden on the tenant against his will as a condition of his receiving a new tenancy under Part II of the Landlord and Tenant Act 1954. It may be granted that the transfer of the risk from the landlord to the tenant is a perfectly legitimate negotiating aim for the landlord to entertain. But the argument is two-edged. It is an equally legitimate negotiating aim of the tenant to resist the change. Granted that a reduction in rent of 50p from £10.50 per ft to £10 per ft is, in the limited sense described, an adequate estimate of the rent payable to the landlord if the risk is to be kept where it is under the current lease. But neither of the two statements assists to answer the question, where in the new lease, is the risk of fluctuation to lie. If I am correct that the inference from the authorities is that the language of Section 35 requires that the party (whether landlord or tenant) requiring a change must justify as reasonable a departure from the current lease in case of dispute about its terms, the answer must be that *prima facie* it must lie where the current lease provides and that a mere agreement about figures based on either or both of two rival hypotheses does not shift the burden in any way.

It has been suggested that the court's reluctance to vary the lease was because the lease being renewed was itself a short lease. It is not considered that this is necessarily the case as the maximum length of lease the court can award is14 years. The acid point of the case was the transfer of a risk. While it was agreed that the market would reflect the risk in financial terms, this did not mean that the transfer of the risk itself was fair or reasonable to the tenant.

Rent Reviews

The valuer will either be acting for the landlord or the tenant. Whether he be negotiating or whether he be preparing a proof of evidence he first of all has to determine the basis of the valuation. The basis of his approach should be contained within the civil contract, that is the lease between the two parties. It is a basic rule of construction and interpretation of leases that they must be read strictly. The court will usually only add words missing if they are necessary to make commercial sense of the document.

In 1985 and 1986 there was considerable litigation over rent review clauses which in some instances were held, when read strictly, to require future rent reviews to be ignored on a rent review. The result in some of those cases was that what had to be valued was the unexpired period of the lease ignoring future reviews (and therefore increasing the hypothetical tenant's bid).

These results were almost certainly never intended by those drafting the leases. As each lease is often different as to the exact wording applied, it is not helpful to comment in general terms on such a series of cases. Recent judgments indicate that commercial sense should be applied in construing such clauses. No doubt there will be further cases before it is possible to give general, as opposed to particular guidance.

The rent review clause will normally require the value to be determined by reference to such expressions as "the full rental value of the property" or "the open market value of the property". Rarely are two clauses similar, though the Royal Institution of Chartered Surveyors and the Law Society have suggested standard appropriate wording. The whole basis of the valuation must reflect not only the rent review clause but the lease as a whole. There may be points to be disregarded, there may be points to take into account; questions of improvements, inflation and the user clause may all be relevant. There are three basic methods of approach:

Rental Values

With such standard business premises as a shop, office or factory, there is normally evidence of rent. The amount that a hypothetical tenant might pay, or a hypothetical landlord might accept, must, in the long run, reflect whether the tenant expects to make a profit out of his occupation. It must also reflect the fact that such a landlord would fall short of asking a rent so high that the tenant would be unwilling to pay because he could not continue trading with sufficient profit.

Profit's Test

There are a number of properties where reference to the profitability of the tenant may be the correct approach. Examples are public houses, cinemas, theatres, clubs, nightclubs and football clubs. The valuer, preferably by reference to certified accounts, should be able to analyse those accounts and calculate the true profits of the operation, then apportion those profits between the landlord and the tenant.

The certification of accounts as a true and correct record by a chartered accountant is only the point at which the valuer begins. Rarely will a valuer be told, as I was on one occasion by another chartered surveyor to his eternal credit, to ignore the certified accounts that had been put in as his client was "well known to the Inland Revenue".

Capital Values

Finally, there is a small category of rent reviews where the valuer can obtain no assistance by way of market rental and the property is not an appropriate one to be valued by reference to profits. In this last resort the valuer has to consider capital values. The type of property where this arises is normally a building erected but not necessarily used for a particular purpose, such as a church or a school, and where the tenant is tied to a very restrictive user clause that otherwise makes the property, if not unmarketable, only marketable to a very limited extent. The valuer would only use this approach as a last resort and among many other matters would have regard to the age of the building and its maintenance and running costs.

Evidence of Value

A major part of preparing a client's case, whether it be to negotiate with another valuer, to present the case in the High Court or the county court, to an arbitrator or to an independent surveyor, is the quality of the evidence put forward as opposed to the quantity. There is at the moment no authoritative text book on the question of evidence as applied in arbitrations. For evidence generally and the rules of evidence, there are several excellent books currently in print but none of these cover the points relevant of a rent review. All other things being equal, which they rarely are, the valuer producing and proving the best evidence serves his client well.

The Best Evidence

The following is a suggested order of preference:

(a)   Market evidence within the same building
(b)   Market evidence in the immediate vicinity
(c)   Rent review in the same property
(d)   Rent reviews in the immediate vicinity
(e)   Independent Surveyor's determination
(f)   Hearsay evidence

Under the law of evidence an arbitrator's award is not admissible in court as evidence without the consent of both parties[1]. However, an arbitrator's award can only be as good as the evidence given to that arbitrator and upon which he made his award. An independent surveyor's determination should also for very practical reasons be treated with considerable reservation. An independent surveyor can be sued for negligence and therefore the determination is rarely more than a page long and will contain no reasoning. Thus the determination may not even indicate whether the parties to the dispute were able or were invited to put forward evidence, what their evidence was, or whether the independent surveyor relied entirely upon his own practical experience or knowledge.

Evidence generally

The Civil Evidence Act 1995 came into force in 1997.

Essentially the Act allows hearsay evidence to be given if the party wanting to produce hearsay evidence has given notice. If requested the party must give particulars of or relative to the evidence.

Usefully the Act lays down what considerations an arbitrator should bear in mind in weighing hearsay evidence and deciding its weight and importance. Section 4 provides:

(1)   In estimating the weight (if any) to be given to hearsay evidence in civil proceedings the Court shall have regard to any circumstances from which any inference can reasonably be drawn as to the reliability or otherwise of the evidence.

(2)   Regard may be had, in particular, to the following:

(a)   Whether it would have been reasonable and practicable for the party by whom the evidence was adduced to have produced the maker of the original statement as a witness;

---

[1] *Land Securities* v *Westminster City Council* [1993]. But in arbitrations under the Agricultural Holdings Act 1986 there is a statutory direction to have regard for rents fixed by arbitration under that Act.

(b) Whether the original statement was made contemporaneously with the occurrence or existence of the matters stated;

(c) Whether the evidence involves multiple hearsay;

(d) Whether any person involved had any motive to conceal or misrepresent matters;

(e) Whether the original statement was an edited account, or was made in collaboration with another or for a particular purpose;

(f) Whether the circumstances in which the evidence is adduced is hearsay are such as to suggest an attempt to prevent proper evaluation of its weight.

Proof of Evidence

The parties can agree that the strict rules of evidence do not apply to a dispute. Otherwise a surveyor putting forward the rent payable on another property as evidence must be in a position to prove the transaction. In the High Court and before an arbitrator the proof required is production of the actual documentation or certified copies. As a practical point in an arbitration the arbitrator is normally perfectly happy to accept comparable evidence as being proved and fully admissible if both the parties agree this as a fact. At one time it was thought that post review evidence could only be admissible as evidence of trend in values and not as evidence of values as such.

In *Melwood Units PTY Ltd* v *Commissioner of Main Roads* [1979] AC 426 PC, an Australian case, the Privy Council held the Land Appeal Court wrong to reject evidence of a June 1966 sale in relation to a compulsory purchase order of September 1965.

Some later cases, such as *Duvan Estates Ltd* v *Rossette Sunshine Savouries Ltd* [1982] 1 EGLR 20, appear in conflict with the Privy Council case.

In *Segama NV* v *Penny Le Roy Ltd* [1984] 1 EGLR 109, it was held, following *Melwood*, that an arbitrator was entitled to admit evidence as to rents agreed after the relevant date. However, Slaughton J said that other judges might take a different view.

The weight that an arbitrator or independent surveyor attaches to evidence, be it pre-review or post-review date, is a matter of both judgment and experience. Post-review evidence is therefore probably fully admissible not just of trend in values but of actual values. But the longer the period between the valuation date and the date of the transaction, the less weight the evidence is likely to carry with the arbitrator. If there has been some political or economic or local event which has significantly affected market values, this too will reduce – perhaps to zero – the weight to be attached to the evidence.

The Arbitrator's Award

If a rent review is being decided the arbitrator has to come to a decision by a logical process of thought. An arbitrator is bound in general terms by the evidence put before him, whether it be evidence of comparable properties or evidence of opinion. In either event he is much assisted by the logical presentation to him of each side's case. As stated above an arbitrator now must issue his award in the form of a reasoned award unless the parties agree otherwise. In the appendix to this chapter is an example of an award that might be made by the arbitrator. This is not put forward as definitive as to how an award should be written; it does, however show how the arbitrator might marshal the facts and evidence and deal with certain of the subjects covered in this chapter, such as improvements, inflation and evidence.

Appeals

The view has been expressed that appeals to the court against an award will be more difficult to make since the Arbitration Act 1996. It is true that a challenge cannot be made without the aggrieved party having first exhausted any available arbitration process of appeal or review, and any available recourse by way of correction of Award or additional Award. But subject to that, challenges can be made upon:

(a) the substantive jurisdiction of the arbitrator (section 30); or
(b) grounds of serious irregularity, formerly known as misconduct (section 68); or
(c) a question of law arising out of the Award, unless otherwise agreed by the parties (section 69).

A party overlooking the requirement that an appeal must be lodged within 28 days of the date of the award does so at its peril.

Appendix 1

The arbitrator's award

There is no standard or approved format other than in the case of agricultural holdings*. The award may be a lengthy document depending on the arguments advanced and the complexity of the case. The true test of the quality of the award is that not only should the parties to the arbitration understand it but also that is should contain sufficient information for a judge to follow the arguments advanced and the reasons for arbitrator's conclusions.

In *Gleniffer Finance Corporation* v *Guardian Royal Exchange Assurance Ltd*, the Hon Mr Justice Goff (High Court, April 14 1981) concluded his judgment with the following words:

> The Arbitrator's Award was a full and accurate one in which he reviewed all the evidence and on the basis of that exercised his judgement based on his skill and experience. The exercise of an Arbitrator's judgement on this basis is not an appropriate case for review under the Arbitration Act 1979.
>
>   I therefore dismiss this application with costs.

While there is no approved method it is suggested that the award should start with four main preliminaries.

1. A recital of the appointment of the arbitrator, his meetings, directions, inspections and other factual events leading to the hearing or written representations.
2. A list of what matters have been agreed between the parties prior to or during the arbitration.
3. A brief summary of the main differences between the parties.
4. Reproduction of the rent review clause and other clauses referred to by the parties.

As an award is a privileged document between the parties it is not possible to reproduce an actual award.

The award that follows is an attempt to show how an arbitrator in a fictitious reference might summarise arguments put to him and come to reasoned conclusions when presented with difficult problems.

---

* In the Agricultural Holdings (form of award in arbitration proceedings) Order 1990 (SI 1990 No 1472) there is a prescribed format.

AWARD
in the matter of Rental Valuation
of
736 High Street, Balham
as at
December 25th 1999

Daniel Parr, FRICS   27th May 2000
Arbitrator

Graham Kimber, ARICS for the Claimant
Philip Forwood, FRICS for the Respondent

1.   WHEREAS The president of the Royal Institution of Chartered
      Surveyors appointed me Daniel Parr, FRICS, as Arbitrator on the 16
      January 2000 following a dispute arising under the lease between
      Balham Refurbishments Ltd and The Agile Trading Company Ltd.
      My appointment was made under clause 6 of the lease dated 7th
      January 1977.

      I directed that a preliminary meeting be held. At that meeting on
      February 2nd the landlords, Balham Refurbishments Ltd, were
      represented by Mr Kimber and the tenants, The Agile Trading
      Company, were represented by Mr Forwood.

      I designated the landlords as claimants and the tenants as
      respondents. Both parties requested an oral hearing and I issued
      directions for such a hearing to commence on March 25 2000 and
      confirmed these directions in writing the same day.

      The hearing took place on March 25, 26 and 27 2000. I took evidence
      on oath. On April 2nd I inspected 736 High Street, Balham and
      various other comparable properties put forward by the parties,
      accompanied by Mr Kimber and Mr Forwood.

2.   Prior to the hearing the parties had at my request agreed certain
      factual information as to the areas and other facts concerning 736
      High St, Balham and certain of the comparables. Those matters of
      agreement were, as directed, handed to me as an agreed bundle of
      documents at the commencement of the hearing.

I DO NOW HEREBY MAKE AND PUBLISH THIS MY FINAL
REASONED AWARD

3.   *The main differences between the parties.*
      The major differences between the parties can be summarised as
      follows:

A.   Should there be a reasoned award or not.
B.   Was certain evidence admissible or not.
C.   How were certain improvements to be valued.
D.   What was the effect of the restrictive user clause in the lease.
E.   What was the rental value of 736 Balham High Street.

4.   I set out the relevant clauses in the lease relating to the rent review, the restrictive user clause and improvements.
(These would be set out at this point.)

5.   I now deal in detail with the differences as listed above.

A.   *Should there be a reasoned Award?*

This is a simple point. One party wants a reasoned award, the other party does not. Under the Arbitration Act 1996 I am bound to issue a reasoned award unless the parties agree otherwise.

Regardless of the merits of the claimant's case, the Arbitration Act 1996 makes it mandatory in this instance to issue a reasoned award.

I THEREFORE, AS A PRELIMINARY POINT, DETERMINE THAT THIS BE A REASONED AWARD.

B.   *Was certain evidence admissible or not*

During his evidence Mr Forwood put forward a letter from Mr Kimber dated 17 November 1997. This two page letter was marked "Without Prejudice". I advised Mr Forwood that I regarded this letter as being inadmissible and that I would record this determination in my award but I would have no regard whatsoever to the contents of that letter.

Mr Kimber submitted as hearsay evidence a schedule of shops available to let in South London. Whilst admissible, these "comparables" are only evidence that certain shops are on the market. I attach little weight to this evidence.

C.   *How were certain improvements to be valued*

I heard evidence at length on the cost to the tenant of putting in a new shop front and improving the toilet facilities.

The fact that these improvements had been made was not disputed.

In my opinion the lease dated 7th January 1977 is a full repairing lease. I therefore dismiss Mr Forwood's lengthy contention that

I must value and take into account the condition of the property as at January 1998.

With regard to the shop front I accept the argument of Mr Kimber that this does not increase the rental value of 736 High Street, Balham as most, if not all, tenants of comparable property install their own shop fronts, normally in a "house style".

The improvements to the toilets are in my view "tenant's improvements" that I am required to exclude under the terms of the lease, I therefore accept Mr Forwood's argument and reject that of Mr Kimber.

The cost in 1977 of providing internal male and female WCs, to replace a single external WC in the yard was £2,000. Mr Forwood argued that this sum should be adjusted upwards for inflation to 1999 costs and written off over the 5 year review and justified as a reduction in rental value of £850 p.a. In my view Mr Forwood's approach is wholly wrong.

The correct method is either to write the expenditure off over the total period of the 30 year lease or to assess the increased rental value of the shop resulting from these improvements.

I have no evidence of fact or opinion and I therefore have to adopt a robust view which is that the effect of the improvements is to increase the rental value by 5%. This must therefore be deducted from the rental value.

D.   *The Restrictive User Clause*

Under the lease the shop may only be used for the retail sale of sports equipment.

It is, I believe, a fair summary of the opinions of the two experts that Mr Kimber's expert opinion is that no discount should apply, whereas Mr Forwood for the tenant has put forward a figure of not less than 25 per cent and has suggested that this is a sensible figure relative to the reductions of 31.62 per cent in *Plinth Property Investments Ltd v Mott, Hay & Anderson* [1979] 1 EGLR 17 and the 10 per cent and 11.66 per cent which can be found or deducted from the decisions in *UDS Tailoring v BL Holdings* [1982] 1 EGLR 61 and *Clements (Charles) (London) Ltd v Rank City Wall* [1978] 1 EGLR 47.

In my view one fundamental factor has been overlooked in the cases presented to me. In the *Plinth* case the Arbitrator had stated a case in alternative figures. He had already determined that a reduction of 31.62 per cent might apply, but he was uncertain as to whether a restrictive user clause applied to a rent review or not.

The restricted user clause was held to apply, and it therefore followed that the rent was discounted by 31.62 per cent. As a matter of law an Arbitrator's award, as such, is inadmissible in relation to a third party, save as hearsay evidence. In my view there must be in practical terms a more common-sense approach as to how I regard the decision in *Plinth* in relation to 736 High Street, Balham. The expert who determined the 31.62 per cent was not an independent surveyor but was an Arbitrator. the figure that he arrived at in the absence of any reasoned award I must presume to be only as good as the evidence, be it evidence of fact or evidence of opinion, that was submitted to him. Mr Forwood suggested to me his figure of 25 per cent is in line with the *Plinth* case (and with the two cases in London – the *UDS Tailoring* case and the *Charles Clement* case). I think I would be wrong to place any reliance at all on the 31.62 per cent for the reasons I have already given, and it must therefore be wrong to have any regard to whether it was or was not "in line" with the *Plinth* case.

I must also say that the *Plinth* case, whilst most certainly one concerning a restrictive user, was of an office block in Croydon, which in valuation terms must be considered remote when compared to a shop in Balham.

I now refer to the other two cases cited. I do not consider the *Charles Clements* case to be particularly relevant because, although a reduction of 11.66 per cent can be inferred, Mr Justice Goulding said in his judgement that this had been referred to in argument but there was no agreement on the point. The *UDS Tailoring* case concerned a restriction to a shop in the Edgware Road where under that restriction the tenant was only to use the demised premises as a men's and women's bespoke and ready to wear tailors and outfitters. This restriction was to some extent less strict in that the tenant could apply to the landlord for a change of use but the landlord, at his sole option, determined whether or not that change would conflict with any other trade or business within a block of property formed by 204 to 256 Edgware Road. In *UDS Tailoring* Mr Vivian Price QC said:

" I think that the restrictive conditions do impose a burden upon the tenant over and above that imposed by the general law, and I therefore think it appropriate to make a substantial reduction. I think, however, that 15 per cent is far too high a figure for that and, again in my assessment (which again I must emphasise cannot be a matter of calculation) the correct figure for deduction should be 10 per cent."

I recognise that both these cases relate to shops which are totally different from a shop in Balham. I accept that the *UDS* case is more in point than the *Charles Clements* case, and that the restriction was of a very different nature to the restriction that I have to presume in the hypothetical lease.

I feel however that both cases together with *Plinth* and my own experience are sufficient for me to reject Mr Kimber's contention that there should be no reduction because of the restricted use clause in the hypothetical lease. Mr Forwood's opinion that a reduction of 25% should apply is not substantiated and I determine that the reduction shall be 10%.

E.   *The Rental Value of 736 High Street, Balham*

The parties were in basic agreement that the rental value of the shop was £5,000 but differed as to the effect of improvements and the restrictive user clause.

Mr Kimber additionally argued the wording of the rent review clause requiring determination of the rental value "as might be reasonably demanded" required me to determine a higher figure than that obtainable on the open market. Both Mr Kimber and Mr Forwood advanced legal arguments (referred to earlier). I do not consider those legal arguments are in any way relevant nor do the cases cited assist me. In particular, I do not accept Mr Kimber's argument that "reasonably demanded" justifies an overbid to open market value. The expert's valuations were as follows:

	Claimant £ pa	Respondent £pa
Open market value	5,000	5,000
Add 10% for "demanded"	500	–
	5,500	5,000
Deduction for restricted user clause	–	1,250
	5,500	3,750
Deduction for improvements	–	850
	5,500 pa	2,950 pa

My determination arising from this Award is:

£ p.a.		
Open market value	5,000	
Deduction for Restrictive User 10%	500	
	4,500	
Deduction for improvements 5%	225	
	£4,275	p.a.

I THEREFORD MAKE AND PUBLISH THIS MY AWARD AND DETERMINE THAT THE RENT AS DEFINED UNDER THE LEASE DATED THE 7TH JANUARY 1977 SHALL BE £4,275 P.A. (FOUR THOUSAND TWO HUNDRED AND SEVENTY FIVE POUNDS) AS AT THE REVIEW DATE DECEMBER 25TH 1999.

It is open the me to determine the question of costs. Neither party addressed me on the question although I invited submissions.

My award in the event lies between the figures contended for. My findings on deductions for restrictive user clause and improvements lean towards the claimant whilst the addition claimed for "demanded" falls to the respondent.

In my view therefore neither side should obtain costs and I so order. Both parties shall equally bear the costs of the reference and the costs of the award.

Daniel Parr FRICS
London
December 30th 1999

Appendix 2

P.A.C.T.

Professional Arbitration on Court Terms is a service provided by the Royal Institution of Chartered Surveyors and the Law Society, and has been in operation since July 1997.

It is available as a means of determining disputes relating to unopposed lease renewals under the Landlord and Tenant Act 1954. It offers opportunity for disputes to be resolved without the necessity of going to court. Terms and rent are decided by a surveyor or solicitor acting as either arbitrator or independent expert.

The key issues are summarised in the explanatory booklet* as follows:

Decide with the other party whether it would be advantageous to refer aspects of the lease renewal to a third party solicitor or surveyor rather than a judge.

Identify which aspects of the renewal (if any) are agreed

Decide which aspects to refer to the third party, i.e.

The interim rent, the new rent, other terms of the lease,

the detailed drafting of terms or any combination of these

Choose which aspects are to be resolved by a third party solicitor and which by a third party surveyor

Choose third party's capacity – whether arbitrator or expert

Draft Court application making use of or adapting PACT model orders

Apply to Court for consent

Apply for appointment of person adjudicating (if not agreed)

Proceed with adjudication

Receive decision subject only to cooling off rights and, in arbitration, any right of appeal.

Further Reading

Agricultural Holdings – Muir, Walt and Moss (14th ed) – Sweet and Maxwell
Scammell and Densham's Law of Agricultural Holdings – Butterworths

© P.A.C.T
CW Goodwyn 2000

---

*Obtainable from The Law Society, Arbitration Service, 50–52 Chancery Lane, London, WC2A 1SX (Helpline 020 7320 5698) or
The Royal Institution of Chartered Surveyors, the RICS Dispute Resolution Service, Surveyor Court, Westwood Way, Coventry, CV4 8JE (Helpline 020 7334 3806).

Chapter 16

Asset Valuation

The Royal Institution of Chartered Surveyors (RICS) has been concerned with self regulation of valuation work carried out by its members since the early 1970s, initially with the preparation of definitions of the bases on which valuations could be prepared, and leading on to the production of other guidance notes on what a valuer should do when undertaking a valuation. This regulation, from the start, and right up until today has been concerned with the procedures to be adopted for processing a valuation job, such as accepting the instructions, defining the bases of valuation and the enquiries that a valuer should make, rather than the valuation techniques. Initially, the guidance notes were directed primarily at published valuations that were going to appear in the public domain. The original "Red Book" was voluntary, and the Institution had no powers to enforce compliance.

In 1990 there was a significant change. The then guidance notes were made mandatory for all Chartered Surveyors, ISVA and IRRV members when carrying out valuations that were going to appear in the public domain. There was an "opt in" clause whereby if a client instructed a valuer to follow the Red Book, then compliance also became mandatory. The Institution for the first time had powers to discipline members who failed to comply with the Red Book requirements.

Most recently, in 1995 following publication of the Mallison Committee report, there was a major revision and extension of the Red Book to form the new RICS *Appraisal and Valuation Manual*. This manual now applies to all valuations, subject to a number of exceptions which are listed in the book.

Most assets valuations of the type discussed in this chapter are potentially going to be in the public domain and the objectives of regulations are to:

(a) Produce a consistency of approach in the preparation of valuations which assists shareholders, brokers, analysts and others when comparing one company with another.

(b) Provide a "level playing field", at least in so far as asset valuations in takeover bids are concerned, by ensuring that both sides work to a similar set of principles.

(c) Ensure that information provided to anyone who might rely upon published financial statements incorporating valuations is not misleading.

The *Appraisal and Valuation Manual* is required reading for students and practitioners alike. It not only deals with general principles but also provides the answers to many specific and detailed questions which valuers will meet from time to time. Similar rules are published by The European Group of Valuers Associations (TEGOVA) – "The Blue Book" and by the International Valuation Standards Committee.

In this chapter "asset valuations" are defined, as those valuations which concern fixed assets that are shown in financial statements. The term "financial statement" covers a wide spectrum – from company accounts to circulars issued under the rules of the Stock Exchange and Takeover Panel, to government requirements under the Insurance Company Regulations.

Financial Statements

The principal forms of financial statement with which the asset valuer is likely to be concerned will include the following:

(a) Company accounts

Company financial accounts are published in the form of a balance sheet and a profit and loss account together with a director's report and a chairman's statement. In valuers' terms the balance sheet will show the capital value of the fixed assets and the profit and loss account will show items of depreciation charges, while the directors' report or chairman's statement may make reference to the valuation or to future potential of properties which are not fully reflected in the current values. All these areas of the company's accounts are important to the asset valuer. The importance of a valuer's contribution towards the formation of company accounts has grown with the developing sophistication in preparing them. This has been marked through changing conventions owing to the introduction of revaluation accounting, as a permitted alternative to historic cost. All this has led to requirements for regular valuations of fixed assets.

The RICS and the accountancy profession have the closest liaison and co-operation. Accounting standards now produced by the Accounting Standards Board, which incorporate matters affecting fixed assets, have, in the main, been compiled after consultation with the AVSB (or it's predecessor bodies) and advice has been given to accountants over the appropriate valuation methods and procedures that should be adopted in company accounts.

Any valuer undertaking a valuation for company account purposes should at the outset arrange to meet the directors, their accountants and possibly auditors to ensure the procedures, basis of valuation and presentation of the valuation are fully understood.

(b) Stock Exchange and the Takeover Panel

In the late 1960s and early 1970s much concern was expressed over the activities of the asset stripper. This concern continues and indeed in respect of assets being transferred from the public to the private sector there is new concern that this will be at an undervaluation. The RICS through the AVSB has co-operated with the Stock Exchange and The Financial Services Authority in preparation of detailed requirements for companies seeking a quotation on the London Stock Exchange through their requirements entitled *The Listing Rules*, (the "Yellow Book"). These requirements, laid down in chapter 18 of that book, follow very closely the *Appraisal and Valuation Manual* and while they relate principally to property companies, the rules also apply in general to the property assests of non-property companies where a listing on the Stock Exchange is sought or where classes of transactions are undertaken that need to be reported to shareholders.

The City Panel on Takeovers and Mergers in its book *The City Code on Takeovers and Mergers* (the "Blue Book") lays down rules of conduct and procedures in contested and agreed takeovers. These rules and procedures incorporate the relevant practice statements of the AVSB and need to be followed by valuers.

(c) A number of other forms of financial statement are dealt with later in this chapter. These include for:

*   Insurance Companies
*   Pension and Superannuation Funds
*   Life Assurance Companies

- Property Unit Trusts
- Enterprise Zone Property Trusts

Definition of an asset valuer

The asset valuer needs to be professionally qualified and have the appropriate market knowledge, skills and understanding. Professionally qualified means a corporate member of the RICS or the Institute of Revenues, Rating and Valuation (formerly the Rating and Valuation Association).

An internal valuer is an asset valuer who is a director or an employee and who has no significant financial interest therein. An external valuer is an asset valuer who is not an internal valuer and who has no significant direct or indirect financial interest in the organisation owning the asset. An independent valuer for most purposes is an external valuer who has no other recent or foreseeable potential fee-earning relationship concerning the subject property apart from the valuation fee and who has disclosed any recent or present relationship with any of the interested parties or with the subject property. Practitioners should be particularly careful when accepting an appointment as an independent asset valuer to ensure that they comply with the precise terms of the Red Book and of any statute or regulation under which that appointment is being made.

Bases of valuation

As valuers are well aware, there is not just a single basis of valuation. Those bases of relevance to asset valuation for financial statements are referred to below:

(a) Open market value

The primary basis of valuation for financial statements is to open market value. This will apply to all properties for which there is a general demand with or without adaptation and which are commonly bought, sold or leased on the open market.

The definition of open market value is set out in PS4 as follows:

> Open market value is an opinion of the best price at which the sale of an interest in property would have been completed unconditionally for cash consideration on the date of valuation assuming:
> (a) a willing seller;

     (b)   that, prior to the date of valuation, there had been a reasonable period (having regard to the nature of the property and the state of the market) for the proper marketing of the interest, for the agreement of the price and terms and for the completion of the sale;

     (c)   that the state of the market, level of values and other circumstances were, on any earlier assumed date of exchange of contracts, the same as on the date of valuation;

     (d)   that no account is taken of any additional bid by a purchaser with a special interest;

     (e)   that both parties to the transaction had acted knowledgeably, prudently and without compulsion.

The valuer must state the date of valuation in the valuation certificate. It may be the same as the date of the certificate or an earlier date but it must not be a future date.

In exceptional circumstances, which are more likely to arise in the case of a leasehold property, an open market value may be negative.

The definition envisages properties being bought and sold individually over a reasonable period of time. In recent years it has been argued that the value of a portfolio of properties may be greater than the sum total of the individual valuations. This may, for example, arise in connection with certain trading properties, eg a chain of hotels or petrol filling stations may benefit from economies of scale not available to single units. In the event that this is considered to be the case the valuer may refer to such additional value which may attach to the portfolio if sold as a whole but is required to report in addition the sum total of the values from individual disposals.

Open market value includes such hope value as the property may have for uses other than the existing use, but only to the extent that such value would in practice be reflected in bids made by prospective purchasers in the market. Generally valuations should be made only on such assumptions as the valuer reasonably and realistically could expect to be made by prospective purchasers in the open market.

This definition does not override any statutory definition of market value which may have to be adopted for the purposes of valuation for capital gains tax, compensation on compulsory purchase or other purposes.

(b) Open market valuations having regard to trading potential

Some categories of land and buildings are designed or adapted for particular purposes and are sold fully equipped and trading as an operational entity in the open market. The valuer should clarify in the valuation certificate the basis of the valuation. In the case of an hotel for example, it would be appropriate to use words such as "open market value fully equipped as an operational hotel and having regard to trading potential". Examples of other such properties include public houses, cinemas, theatres, bingo clubs, gaming clubs, petrol filling stations, licensed betting offices and specialised leisure and sporting facilities. The operational entity will include, besides the land and buildings, items such as fixtures, fittings, goodwill and stock at valuation.

Open market transactions which involve the sale of such properties can provide evidence of value for use when valuing this type of property for accounting purposes. When analysing the prices paid for comparable properties and preparing a valuation of the subject property the valuer should have regard to the trading accounts for the previous years, where these are available, and with the help of the existing or proposed management, form an opinion as to the future trading potential and level of turnover and profit likely to be achieved.

Certain operations can only be carried on under statutory consents, permits and licences, and their continuance is an assumption which should be specifically stated in the valuation certificate.

It would be normal to refer to the fact that the fully operational business would include fixtures, fittings, furniture and goodwill with stock at valuation. The new owner will normally engage the existing staff and sometimes the management and would expect to take over the benefit of future bookings which are an important feature of the continuing operations.

Problems can be encountered in differentiating between the value of the trading potential which runs with the property and the value of the goodwill which is personal to the present owner and which may be transferred to other properties in the event of the subject property being sold. In such cases the valuer when assessing future trading potential should exclude any profit which would be available only to the present owner or management but reflect any trading potential that might be realised in the hands of a more efficient operator.

Examples of valuations on an open market basis are carried in the relevant functional chapters in this book. They do not differ from the normal approach to valuations as they are the types of buildings which are commonly bought and sold and leased in the market place and where there is no difficulty in obtaining evidence of the market price.

(c) Depreciated replacement cost

Problems do, however, arise when a valuer is faced with valuing for balance sheet purposes those types of property which rarely, if ever change hands in the open market for their existing use except as part of a sale of the business in occupation. These have been classified as specialised properties.

Their specialised nature may arise from the construction, arrangement, size or location of the property, or a combination of these factors, or may be due to the nature of the plant and machinery and items of equipment which the buildings are designed to house. Examples of specialised properties are oil refineries, chemical works, buildings which are no more than cladding for a special plant and standard building of great size or in isolated or unusual locations. If, because of the lack of demand or market, it is not practicable to ascertain an existing use value of specialised properties, depreciated replacement cost (DRC) is the only satisfactory approach to arrive at the net current replacement cost required for accounting purposes.

For these specialised properties a different valuation approach is required. In the majority of cases they have been erected to give an adequate return on capital to the owners and thus are worthwhile to the business as long as profits can be generated. The basis of valuation of such properties is referred to as depreciated replacement cost (DRC). Such a valuation commences with an estimate of the gross replacement cost of the buildings, which is the estimated cost at date of the valuation of erecting a modern substitute building having the same usable floor area but taking account of current building techniques. This figure is then reduced to take into account physical, functional and economic obsolescence and environmental factors. To this is added the open market value of the land for its existing use. Any valuation prepared on a DRC basis must be made subject to the adequate potential profitability of the business having regard to the total value of the assets employed. It is, however, the directors' responsibility to decide whether the business is sufficiently profitable to carry the property in the balance sheet at the full DRC or whether a lower figure should be adopted. In the Public Sector the equivalent test is prospect and viability of continuance of the existing use.

An undertaking may occupy a property comprising a number of buildings, some of which may be of a specialised nature and others non-specialised. Some undertakings may also carry on from a single property a number of different businesses, each of which is a separate entity. Normally a property should be categorised as a whole. However, if the degree of specialisation is not substantial, either because of the construction of the buildings or because of the number or size of specialised buildings in relation to the whole, it may be appropriate to value some parts of the property on the basis of existing use value and to provide a separate figure, calculated on the basis of depreciated replacement cost, for the specialised parts of the property. If this approach is to be adopted it must be agreed with the directors.

In the case of specialised properties which are in the course of development they should be valued having regard to their existing state and costs current at the date of valuation on the depreciated cost basis and subject to adequate potential profitability on completion.

While these are the only appropriate bases of valuation, there are other definitions of which the asset valuer should be aware.

(i) Existing use value

The standard definition of open market value is adopted but with the additional assumption that the property will continue in its existing use, thus ignoring any possible form of alternative use, any element of hope value, and any possible increase in value due to special investment or financial transactions such as sale and leaseback which would leave the owners with a different interest from the one to be valued. There will, however be a need to take into account possibilities of extensions on developed land or eventual redevelopment of existing buildings.

Existing use should not, however, be interpreted too narrowly. It does not carry the same meaning as in planning law. It does not necessarily mean the particular trade being carried on. Many industrial buildings for instance, would have the same value irrespective of the trade that is carried on. A factory is valued as a factory not as a particular factory and a shop as a shop not as a particular type of shop (unless the market differentiates between the two).

Where there have been special adaptations to suit the particular requirements of the business then any value attributable to those adaptations should be reflected in the value to the business on a DRC basis. If the value of such adaptations is material then that additional value should be expressed as subject to adequate potential profitability.

It is not unusual for a property to have a restrictive covenant preventing assignment in the lease or more rarely the benefit of a planning permission incorporating a personal condition. Both these restrictions should be ignored, as the valuation is for its existing use and thus its continued occupation. Where, however, such a restriction produces a valuation where the open market value of the property is less than the existing use value, then the valuer must report such an occurrence to the owners, who, if the matter is material, will disclose it in the directors' report or as a note to the accounts.

There may well be included in the company's portfolio of property a non-specialised property which is in the course of development. The directors may decide to adopt actual cost to the date of valuation. However, should they elect to revalue then it must be on whichever of the following bases in the valuer's judgment conveys to the reader of the report the most sensible and informative financial appraisal of the property in its state at the date of valuation, having regard to all the circumstances including the purpose of the valuation and the stage of construction reached.

(a) Open market value of the land and buildings in their existing state at the date of valuation; or

(b) Open market value of the land at the date of valuation as a cleared site with the benefit *inter alia* of the planning permission for the actual development plus, stated separately, the costs of development incurred by the date of valuation.

The valuers may be required to justify the basis chosen. In cases where the decision is closely balanced, basis (a) should be adopted.

(ii) Alternative use value

As we shall see later, valuation for incorporation in company accounts for owner-occupied property will be on an existing use basis. The phrase "alternative use valuation" is sometimes used and this is taken to be the open market value having regard to both the existing and any potential alternative uses which might be reflected in market bids. It is therefore usually the same as the open market value.

(iii) Additional valuations made on special assumptions

Open market valuations should be made only on such assumptions as the valuer reasonably and realistically could expect prospective purchasers to make in the open market. The valuer nevertheless should provide the most sensible and realistic appraisal of the actual and potential value of the property or properties. In exceptional circumstances it may be necessary, in order to advise fully upon the potential value of the property, to report an additional valuation on a special assumption. This is an assumption which in the actual circumstances prevailing in the market at the time of the valuation could not reasonably be expected by the valuer to be made by a prospective purchaser. Examples might include the value which would apply in the event of planning consent being obtained for a particular form of development or the price that might be paid by a special purchaser.

In the case of valuations prepared for the purposes of company accounts the normal basis of valuation will be open market value excluding any additional value attributable to special assumptions. If a valuation made on a special assumption is to be provided it should therefore be reported as well as the valuation reported on the normal basis. Such valuations sometimes assume great importance in takeover situations.

Valuations for incorporation in company accounts

Land and buildings are generally held as fixed assets in one of the following categories:

1. For occupation by the business.
2. As investments.
3. As surplus to the requirements of the business.

The basis of valuation to apply will be as follows:

	Existing Use Value	Depreciated replacement cost	Open market value
Owner-occupation	*	*	
Investments			*
Surplus property			*

In respect of properties for occupation by the business it is necessary to bear in mind the concept of company accounts in that they reflect that the business will continue for the foreseeable future. The assets are therefore stated in the accounts at their value to the business as it continues to occupy them and carry on business using those assets. The value of those assets is therefore the value to the owner-occupier and is the net current replacement cost. This is the cost of purchasing, at the least cost, the remaining service potential of the asset at the balance sheet date. So, for those non-specialised properties that are for occupation by the business, the valuation to be incorporated into the company accounts will be on the basis of its existing use. To this will be added the value of adaptation works, if necessary on a DRC basis. Specialised properties will be valued on a DRC basis.

For those properties held as investments, that is being owned for the purpose of letting to produce a rental income, there is not the same concept of the property being required for the continuance of the business and so the basis of valuation to be incorporated into company accounts will be to open market value.

For those properties that are surplus to the requirements of the business or are held for disposal the basis of valuation for incorporation into company accounts is again open market value.

It is important for valuers to show these categories of properties separately in the valuation report and certificate. All these categories of property values will appear in the balance sheet.

The appropriate methods of valuation have been considered in some detail in the sections above.

Categorisation of properties

The problem of deciding which category of property, specialised or non-specialised, is the directors' responsibility, but the valuer should discuss this division with the directors if he considers the category chosen is not appropriate.

Profit and loss account

All companies, with the exception of property investment companies, are required to depreciate all those fixed assets which have a limited economic use or life over the estimated life expectancy of those assets. Property investment companies are in the main exempt, but are required to value their fixed assets on an annual basis as laid down in SSAP 19. In accountancy concepts (both historic and current cost) it is necessary to estimate the amount of depreciation which an asset has suffered over an accounting period and to charge that amount to the profit and loss account. Depreciation is defined as the measure of the wearing out, consumption or other loss of value of the fixed asset, whether arising from use, effluxion of time or obsolescence, through market technology and market changes. Freehold land is not normally liable to depreciation, with the exception of land which has a limited life due, for example to depletion by mineral extraction or which is subject to a limited planning permission. Leasehold land does, however, have a limited length determined by the life of the lease. The figure on which the depreciation charge to be allocated to the profit and loss account will be based is known as the depreciable amount, and it can be expected that valuers will be consulted over such matters, in particular factors such as apportionment of the valuation, degree of obsolescence or life expectancy of the buildings.

As valuations incorporate elements of both land and buildings, there is a need to apportion the valuation so that the depreciable amount can be calculated in respect of the wasting element.

The depreciable amount is calculated by either:

(a) Deducting from the cost or valuation of the asset the value of the land for its existing use; or
(b) By making an assessment of the net replacement cost of the buildings.

Approach (a) follows the logic behind depreciation, namely that depreciation should be charged on the difference between the current value and what will remain to the owner once the buildings have reached the end of their useful life. It is necessary to arrive at the value of the land which is valued as in its existing use. In (b) the net replacement cost is arrived at by deducting from the gross replacement cost factors for physical and functional obsolescence and environmental matters. Gross replacement cost is defined as the estimated cost of erecting the building or a modern substitute building having the same usable area as that existing.

It is important for the valuer to consult with the directors over functional obsolescence as they will be aware of how long the buildings would be needed for their own particular business purposes and of any disadvantages from which they suffer. In those properties which have been categorised as specialised properties, and valued on a DRC basis, the buildings and structures figure will provide the depreciable amount.

In calculating depreciation, the depreciable amount is divided by the years of future economic useful life, thereby providing a figure of consumption or waste during the accounting period in question. It is difficult if not impossible to put a precise life on a building or group of buildings, especially bearing in mind that the life of a building may be extended by expenditure on improvements. It is usual therefore, to adopt a system of banding so that buildings or groups of buildings are identified as having a life of, for instance, 10–30 years or 30–50 years.

It is important for valuers to understand that the depreciation charged in the profit and loss account is not the formation of a sinking fund for the replacement of the asset but is a charge to the profit and loss account of an amount of the measure of wearing out, consumption or loss of value during the period of the account.

Red Book guidance on depreciation issues is being reviewed and is expected to be updated during 2000.

Directors' report or notes to the accounts

The concept of the on-going business in the accounts of a company entails the valuation of the asset occupied by the company for the purposes of the business to be at its existing use value or depreciated replacement cost. However, land and buildings may possess a value different from their existing use where there is a possible use of the property for some other purpose. Normally such values are realised on liquidation or closure and as such are not suitable for inclusion into the accounts. Often, however, the alternative use value may have relevance to the company's overall situation; for instance, in possible defence situations against takeover or in certain cases for security purposes. Where there is a materially different value in alternative use from the existing use it is the duty of the valuer to report the alternative use value to the client. It may also be appropriate to prepare that valuation on the basis of making an additional assumption; for example, as to a planning consent which might be achievable. Such an alternative value might be included in the directors' report or notes to the accountants.

Should a property in a company be declared surplus to trading requirements, then it will be assessed at its open market value which takes into account any possible alternative use at least to the extent that such value would be reflected in market bids. Such a value of surplus land will be shown in the balance sheet. It is important to remember that land and buildings held for investment or development purposes will also be valued on an open market basis, which would take into account any alternative use in so far as that would be reflected in a market bid.

Valuations for Stock Exchange and Takeover Panel

Valuations that are required for these purposes should be carried out in accordance with the rules, regulations and codes of conduct laid down by those bodies which adopt the RICS *Appraisal and Valuation Manual*. In general terms they follow the broad principles outlined in valuations for incorporation into company accounts, although valuers should be especially aware of the following points:

(i) Where there is a likelihood that it would be commercially disadvantageous to provide separate valuations of certain properties, or full information relating thereto, the valuer can probably report in a summarised form, although he should consult both with his client and the Financial Services Authority or City Panel. Similarly the City Panel are likely to permit a summary valuation where there are constraints on time or in practicabilities. Again the City Panel should be consulted.

(ii) In view of the time constraints which often apply, the valuer may in exceptional circumstances issue a qualified certificate based on limited inspections and information. It is, however, important that these restrictions on the normal basis of approach and the information on which the valuer has relied should be clearly stated in the certificate.

Valuations for other purposes

The Red Book contains a number of Practice Statements dealing with valuations for financial statements for a variety of different types of financial institutions and types of investment vehicle, including:

- Practice Statement 16 – Insurance Companies, and in particular the Insurance Companies Regulations
- Practice Statement 17 – Unit Linked Property Assets of Life Assurance Companies
- Practice Statement 18 – Unregulated Property Unit Trusts
- Practice Statement 19 – Property Funds (Authorised Property Unit Trusts)
- Practice Statement 20 – Pension and Superannuation Funds

These various Practice Statements contain important information and additional requirements, including:

- The definition of valuers who may carry out particular types of work.
- Specific additional reporting requirements in relation to valuations for particular purposes.
- Frequency of valuations.
- Additional statutory requirements.
- Background information.

Valuers concerned with valuations for any of these purposes should always consult the Red Book, and appropriate Acts and regulations and rules issued by regulatory bodies for detailed advice when undertaking such instructions.

Working Studies

Set out below are five study valuations:

1. Non-Specialised property with surplus land.
2. Non-Specialised property with adaptations.
3. Non-Specialised development property.
4. Specialised property.
5. Investment property.

The first two studies contain depreciable amount calculations in addition to balance sheet valuations.

Study 1

In this example the property is a 10-year old single-storey warehouse with adjoining five-year old three-storey headquarters office building on a site of three acres with a further 1.1 acres of surplus land which has the potential for a B1 development subject to planning permission and is not required by the business. The property is located on the outskirts of Slough with good access to the M4. The valuation is required for balance sheet purposes.

Lettable floor areas	sq ft	£
Warehouse	40,560	
Offices	15,100	
Estimated rental value		
Warehouse – 40,560 at £8 per sq ft		324,480
Offices – 15,100 at £20 per sq ft		302,000
		626,480
Rental Valuation		
Say		£625.000

Valuation

(a) Property required for purposes of the business – existing use value.

	£
Rental value	625,000
YP in perp at 8%	12.5
	7,812,500
Say	£7,800,000

(b) Property surplus to requirements (open market value)

	£
1.1 acres at £2,000,000 per acre for B1 development	2,200,000
Less: for time to obtain planning consent and risk, say 25%	550,000
Present open market value	£1,650,000

The property is valued by comparison with similar non-specialised properties, adopting the concepts of on-going business, open market value and existing use. Had the adjoining land not been surplus but held for an extension to the warehouse the existing use concept would have required it to be valued for such purposes as follows:

Land for expansion of warehouse
1.1 acres at £1m per acre   £1,100,000

Depreciable amount
So far as the profit and loss account is concerned, the balance sheet
figure of £7,800,000 has to be apportioned into wasting and non-
wasting elements. In this example the warehouse building is
regarded as having a 30–50 year future useful economic life
(average 40 years) and the office block 40–60 years (average 50
years).

Method (a)
Land value – 3 acres at £1m per acre   £3,000,000
Therefore depreciable amount   £4,800,000

Method (b)

Description	GRC £	Future life	Age	Factor	NRC £
Warehouse	2.028,000	40	10	40/50	1,622,400
Offices	1,510,00	50	5	50/55	1,372,300
					2,994,700
Depreciable amount					
Say					£3,000,000

Widely differing results can be achieved from the differing
approaches. Valuers preferring to adopt the logic of approach (a)
but needing to divide the depreciable amount between buildings
with different lives could achieve this by pro-rata adjustment of the
approach (b) with results as follows:

	£
Required depreciable amount	4,800,000
Therefore factor 4,800,000 ÷ 3,000,000 = 1.6	
Warehouse 1,622,400 × 1.6	
Say	2,600,000
Offices 1,372,300 × 1.6	
Say	2,200,000
Residual amount, land (as before)	3,000,000
	£7,800,000

Study 2

This example consists of a purpose-built freehold dairy on a site of 1.5 acres comprising bottling hall, coldstore, covered parking for milk floats, ancillary offices and large yard. It was built over a period covering the last 5–15 years. It is situated about half a mile west of Chelmsford town centre in a predominantly industrial area.

Gross floor area	sq ft
Bottling hall	15,000
Coldstore	2,000
Covered parking	5,000
Offices	1,000
Yard	40,000

The premises have had considerable adaptations carried out comprising special finishes and drainage to the bottling hall and refrigeration in the cold storage area. The rental value of the non-specialised elements is about £85,000 pa, which produces a capital value of £650,000. The net replacement cost of the adaptations is £250,000 with a future economic life of 20 to30 years (average 25 years).

Valuation

The valuation of the property on the basis of the existing use at £900,000 comprises the following elements:

(a)	Value of the standard elements of accommodation – existing use value:	£650,000
(b)	Depreciated replacement cost of adaptation works including special finishes and drainage to bottling area and permanent cold storage	£250,000

As the value attributed to the adaptation works is significant, the total valuation of the property on the above basis would be subject to adequate potential profitability.

The valuation further assumes that the property continues to be occupied by the trading company concerned in connection with the business. Should the property cease to be operational, it should be valued by reference to its current open market value which, in this case, would be at or about £650,000 (in other words, disregarding the adaptations).

There are instances on the other hand where in the case of surplus properties, open market value can exceed value in use; for example, if this site had been in a central office area.

Depreciable amount
The depreciable amount for use in the profit and loss account in this study, assuming continuation in operation and adequate potential profitability, would be £300,000, calculated as follows:

Method (a)
Land value (residual amount)
£600,000 (1.5 acres at £400,000 per acre)

Depreciable amount
£300,000

Method (b)

Description	GRC £	Future	Age	Factor	NRC, Say £
Bottling hall	750,000	35	5	35/40	656,000
Coldstore	150,000	25	10	25/35	107,000
Covered parking	150,000	30	10	30/40	112,000
Offices	60,000	35	5	35/40	52,000
Yard	200,000	35	5	35/40	167,000
					£1,094,000

Method (a) depreciable amount must apply: reduce figure from above pro-rata using ratio 300,000 : 1,094,000 or 27.4%

	£
Bottling hall	180,000
Coldstore	29,300
Covered parking	30,700
Offices	14,000
Yard	46,000
	£300,000

Study 3

This study consist of a 23,000 m² warehouse property development at the half-way stage of construction on an industrial estate near the

M4 and about 10 miles from London Airport. It has been assumed that from commencement to final letting will take a period of two years.

1. Valuation of completed development

	£	£
Estimated rent received	2,000,000	
YP in perp at 9%	11.11	22,220,000
Stamp duty at 3%	666,600	
Agent's fee	250,000	
Legal fees	125,000	1,041,600
		21,178,400
Say		£21,200,000

2. Cost of construction of completed development

	£	£
23,000 m² gross at £430 per m², Say	9,890,000	
Contingencies and extra costs	100,000	
Architects, QS and Eng fees at 12%	1,198,800	
Finance: 1 year's interest at 13%	727,270	
Void period of 1 year's interest at 13%	1,549,100	
Letting, publicity and legal costs	150,000	13,615,170
Developer's profit at 20% of gross cost		
or 12.7% of completed development value	2,723,000	
Total cost including profit		£16,338,170

3. Site value

Gross site value when complete		4,861,830
Deferred 2 years at 13%		0.78
		3,792,227
Stamp duty at 3%	120,750	
Agent's fees	40,500	
Legal fees	21,000	182,500
		3,609,727
Net Value of site at commencement		
Say		£3,600,000

It will be recalled that the site is only half completed. There are, therefore, two possible bases to be adopted. First, open market value of the land and buildings in their existing state at the date of valuation.

Completed value:	£21,200,000	
Deferred one year at 10%	0.9	£19,080,000
Less outstanding expenditure		
(50% of £13,615,170)	£6,807,585	
Less for profit and risk required by a		
developer acquiring half completed scheme:	£1,926,000	£8,733,585
Open market value in the existing state:		£10,346,415
Say:		£10,350,000

The alternative approach is open market value of the cleared site plus costs to date:

Net site value for proposed use:	£3,600,000	
Current cost of development		
50% of £13,615,170(half built):	£6,807,585	
Value		£10,407,585
Say:		£10,400,000

Since it is believed that the property could be sold in its existing state, then there is no reason for departing from the normal presumption that basis (a) is to be preferred, and therefore the value adopted for balance sheet purposes will be £10,350,000.

Study 4

The study used in this example is a purpose-built glass works, held freehold, which comprises interlinking industrial buildings of mainly trussed-roof or portal-frame construction, parts of which are unusually lofty, clad in corrugated asbestos, housing specialised plant and machinery together with extensive single-storey finished product warehousing and ancillary temporary office accommodation.

The property has been developed over a number of years on a site of approximately 10 ha. Market evidence shows that current land values are approximately £250,000 per ha.

A depreciated replacement cost valuation (DRC) has been defined as the "current cost of acquiring the site and erecting the premises, less an appropriate deduction for their present condition". This is again based on existing use, but it is a subjective judgment involving the valuer in a significant knowledge of buildings and industrial processes, cost, physical depreciation and functional obsolescence.

The DRC of a building is the gross replacement cost (GRC) reduced by a depreciation factor (DF), plus the open market value of the land for existing use purposes.

The GRC is the cost of erecting a modern equivalent building.

The DF is used to reflect the physical and functional obsolescence and environmental factors so as to arrive at the value of the building to the business at the date of valuation. The allowance for physical obsolescence can be calculated by dividing the future economic life of a building by the total life expectancy of a modern equivalent, adjusted to reflect functional obsolescence and other factors specific to the building in question. (Compare this approach with the alternative used in the depreciable amount calculations in studies 1 and 2). Other alternatives based on an estimated depreciation curve are also adopted.

Schedule of accommodation

Description	Date built	Life expectancy of modern equivalent	Life expectancy of existing building	GRC £
Main glass works	1967	60	20	6,000,000
Laboratory and toilets	1972	60	20	1,000,000
Canteen and offices	1976	30	5	800,000
Finished-product warehouse	1982	60	40	10,000,000

Valuation
Main glass works
GRC 6,000,000
DF = 40/60 = 66%
but because of changes in process requirements
Say 75% leaving net value at   1,500,000

Laboratory and toilets
GRC 1,000,000
DF = 40/60 = 66% leaving net value at   330,000

Canteen and offices
GRC 800,000
DF = 25/30 = 83% leaving net value at   136,000

Finished-product warehouse
GRC 10,000,000

DF = 20/60 = 33%
but because condition has been allowed to
deteriorate to unreasonable extent,

Say 40% leaving net value at	6,000,000
Depreciated replacement cost	7,966,000
Plus: land value 10 ha at £250,000 per ha	2,500,000
	10,466,000
Less: for poor layout of existing buildings 5%	523,000
	9,943,000
Total value	
Say	£9,950,000

DRC valuations are not so much "valuations" as "costs" and this point should be borne in mind. Cost and value are not the same thing and any valuation of a specialised property by DRC methods must always be subject to the caveat of adequate potential profitability. The directors have the option thereby of reducing the DRC to an acceptable level prior to incorporating the figure into the balance sheet.

The value attributable to land in a DRC calculation is open to much debate and valuers are advised to consider the points discussed in PS4.

Study 5

Cashflow is the basis of an investment valuation. Consequently the valuations tend to differ from those of operational properties where the emphasis lies on use and floor areas.

The following freehold study sets out financial information on a typical investment property comprising a central London building in multi-occupation, refurbished in 1987 with a shop on the ground floor, three office tenants on individual floors above and the top floor vacant.

1. Tenant: Sweets and Things

Next review date 2 years 3 months;
Lease expiring 17 years 3 months.
Rent receivable £15,000 pa

Shop		
Zone A	27 m² at 625	£16.875
Zone B	7 m² at 312	2,184
Basement – storage	45 m² at 100	4,500
Total		23,559
Rental value		
Say		£23,500 pa

2. Tenant: *Shipping Insurance*
Next review date 3 years 3 months;
Lease expiring 18 years 3 months
Rent receivable £35,000 pa

First floor	90 m² at 500	£45,000
Rental value		
Say		£45,000 pa

3. Tenant: *Double O Seven*
Next review date 4 years;
Lease expiring 19 years.
Rent receivable £40,000 pa

Second floor	95 m² at 500	£47,500
Rental value		
Say		£47,500 pa

4. Tenant: *Kay Gee Bee*
Next review date 1 year 6 months;
Lease expiring 16 years 6 months
Rent receivable £27,500 pa

Third floor	90 m² at 500	£45,000
Say		£45,000 pa

5. Tenant: *Vacant*
6-month void period

Fourth floor	95 m² at 500	£47,500
Rental value		
Say		£47,500 pa
Total rental value		
Say		£208,500 pa

Valuation

Rent received		£117,500	
YP in perp at 7%		14.28	£1,677,900
Rent increase		47,500	
YP in perp at 7.5%	13.33		
PV for 6 months at 7.5%	0.96	12.79	607,525
Rent increase		17,500	
YP in perp at 7.5%	13.33		
PV for 1 year 6 months at 7.5%	0.89	11.86	207,550
Rent increase		8,500	
YP in perp at 7.5%	13.33		
PV for 2 years 3 months at 7.5%	0.84	11.20	95,200
Rent increase		10,000	
YP in perp at 7.5%	13.33		
PV for 3 years 3 months at 7.5%	0.79	10.53	105,300
Rent increase		7,500	
YP in perp at 7.5%	13.33		
PV for 4 years at 7.5%	0.75	10.00	75,000
Gross capital value			£2,768,475
Stamp duty at 3%		83,304	
Agent's fee at 1%		27,684	
Legal fees at 0.5%		13,842	124,830
			£2,643,645
Net value			
Say			£2,650,000

© R Baldwin 2000

Chapter 17

Public Houses

Valuations

If a public house is to be offered for sale or to let, one of the first questions that will be asked by prospective takers is: What is the trade? The trade is an influential factor in valuing licensed premises. It does not form the basis of a separate goodwill valuation as in shops and other businesses but is integral to the valuation of the property. The premises have the benefit of a justices' full on-licence and the trade is regarded, in most cases, as being an inherent attribute of the licence. Hence, goodwill, if any, is normally considered to be included in the price.

Trade

In what form can information about trade be obtained? If it is a tied house (that is, tied to one brewer or supplier for liquors) a certificate may be available. The brewers or suppliers keep records of the beers, wines and spirits delivered to each house and issue a statement annually showing the barrelage and gallonage. The beer, including the bottled beer, is shown in barrels (36 gallons) and the wines and spirits in gallons (8 pints or 6 bottles). In a "free" house selling different brews, an alternative source of information is the accounts or the receipts and purchases. If a publican employs a stocktaker periodically, say monthly or quarterly, the stocktaker's statements can be referred to for the receipts and purchases.

Audited and certified accounts are the most reliable form of information and should be sought.

Market

There are several types of purchaser in the market for public houses: brewery companies, retail and catering chains (some of which may be linked to brewery companies), independent retail companies, institutional and other investors and individual traders.

The valuation approach is influenced by who is likely to be the purchaser.

When looking at a public house from the point of view of a brewer, the wholesale profits are a factor. National and regional brewers are mainly interested in houses with large, sound, all-round trades (eg new public houses on residential estates and large catering public houses) and may show little or no interest in buying small- or medium-trade houses in out-of-the-way villages or inferior situations in towns. Indeed, many brewers have been improving the quality of their tied estate by selling off such poor houses as these.

The modern practice of takeovers and mergers among large national brewers has been halted by the effect of legislation consequent upon the Monopolies & Mergers Commission Report on the Supply of Beer. By the Supply of Beer (Tied Estate) Order 1989 breweries with over 2,000 public houses must: (1) dispose of their brewery business, or (2) dispose of the excess of houses over 2,000, or (3) release half of the excess over 2,000 from tie clauses. There are still a number of local and regional brewers who are trying to add to their tied estates by purchasing good rural small-to medium-trade public houses provided they fit into the routes and districts covered by weekly drays.

Examples of valuations with brewery companies as potential bidders are contained in studies 8 to 14.

A new prospective purchaser of modern public houses is the independent brewer who has started up to meet the demand for "real" ale from free trade outlets. Another purchaser or lessee who has recently entered the market is the micro-brewer who can produce enough beer in a converted outbuilding to serve two or three other public houses at the same time. These independent brewers may be a weak force at present but their bids may become stronger if brewing "real" ale for free trade outlets continues to be viable and profitable. It is certainly worthwhile circulating them with particulars of public houses for sale or to lease.

Retail groups and catering chains are often able to outbid the individual retailer by reason of their power to obtain high discounts on food and liquor purchases. A valuation reflecting this factor is contained in study 15.

Individual retailers and small retail groups rely normally on straightforward valuations based on accounts and receipts, and little or no regard is paid to discounts as an additive: see studies 1 and 3.

Investors appear to be interested in retail companies whose pubs are let on leases to sound lessees. Institutional and other investors are searching for stability and growth. Valuations on their behalf will probably depend entirely on a capitalisation of a maintainable rental income, with a prospect of increased rents at future reviews. At present, institutional investors are a minor force in the market.

Landlord and Tenant Reform

Until recently, lessees had no security of tenure under the Landlord and Tenant Act 1954. Public houses were excluded from the protective provisions of Part II. Now all this is changed, owing to recommendations in the Monopolies & Mergers Commission's Report 1989 on the Supply of Beer.

The Landlord and Tenant (Licensed Premises) Act 1990 (which took effect on January 1 1991) repeals section 43(1)(*d*) of the 1954 Act, which excluded pubs, and they now, like other commercial and business premises, have a right to a new lease and, where it is refused, to statutory compensation.

In theory, bringing public houses under the security umbrella of the Landlord and Tenant Act 1954, in line with other commercial and business premises, and the growing practice of granting longer leases should make public houses more attractive to institutional and corporate investors, and after a hesitant hiatus there are now indications of this happening.

Methods of Valuation

There are several methods of capital valuation as follows:

1. Accounts
2. Receipts
3. Capitalisation of Rent
4. Kennedy Method
5. Price per Barrel
6. Overbid Methods
7. Floor Area

They are considered below in this order.

1. Accounts Method

This is a primary method, particularly for free houses. The quality and dependability of the accounts is important. Accounts prepared by an unqualified person, or unsigned accounts, should be treated with reserve.

A valuer should be able to adjust accounts to provide a medium for valuation. A typical set of "free" public house accounts may be adjusted by adding back to, or deducting from, the net profit.

Study 1

Example	£	£
Receipts		406,600
Net profit as shown in accounts		103,600
Add back interest paid on loans		3,700
		107,300
Deduct interest on tenant's capital:		
Trade inventory of furniture, fixtures and fittings valued at ingoing 3 years ago at £9,900. Allow for depreciation and additions since, Say	12,500	
Stock in hand of consumable goods (actual)	8,600	
Cash required to run business, allow one month's purchases of consumable goods, Say	20,000	
	41,100	
Interest at 15%, Say		6,165
Adjusted net profit		101,135
YP		6
		606,810
Capital value, Say		£606,800

= receipts £406,600 × 1.49237 YP

The object of adding back loan interest is to exclude the personal circumstances of the publican. A prospective buyer may be financially independent and require no loans. The deduction of interest on tenant's capital is made so that the "adjusted" net profit represents the profit emanating solely from the licensed premises.

The trade inventory and stock are purchased separately and the end figure will be:

Study 2

	£
Licensed premises – freehold with benefit of licence, Say	606,800
Inventory	12,500
Stock	8,600
	£627,900

There is a wide practice of offering the freehold interest for sale to include the trade inventory.

In all cases, the inventory should be prepared by the vendor's agent and adjustments agreed for deficiencies, if any.

The payment for stock and glassware is always made or adjusted at the "change" on the day of ingoing.

2. Receipts Method

In study 1, it will be seen that the capital value has been devalued as a years' purchase(1.49237) on the receipts. It is a method that valuers are often forced to use in a primary way because the only available evidence of trade is the receipts, the accounts for the previous year not being prepared and certified at the date the property is advertised for sale. Valuations such as this (which take one line) are attractive as time-saving exercises, but they can be used reliably only by valuers with sales experience on this type of property.

Preferably, the receipt's method should be regarded as a check method but as explained above, this is not always its role.

In looking at receipts, valuers should establish whether they include VAT. Receipts extracted from certified accounts will be exclusive of VAT but if they are taken direct from till records or cashbooks, they may include VAT which should be deducted for valuation purposes.

Study 3

Estimated annual receipts based on till records	£477,755	
Deduct VAT (17½%):		
£477,755 ÷ 6.714		71,155 (rounded)
Receipts ex VAT	£406,600	
YP (per study 1)	1,49237	
Capital Value	£606,800	

Years' purchase figures vary generally from 1 to 2 but can be more or less depending on the profitability of the property and its desirability in the market, also upon the valuer's judgment based on his experience and knowledge of comparables.

3. Capitalisation of Rent Method

If an open market rent is being paid and has been recently fixed, a years' purchase can be applied.

Study 4

Example

Rent paid under lease	£43,250 pa
YP, Say	15
	£648,750

Many surveyors are hesitant about valuing licensed premises on a profits basis. Some believe it is traditional to try to reach a rent and then multiply the rent by a years' purchase. Thus, the adjusted net profit might be treated as follows:

Study 5

Adjusted net profit (per study 1)	£101,135
Tenant's share 60% = £60,681, Say	60,680
Rent	40,455
YP	15
Capital value	£606,825

Valuations linked to net profit can sometimes be misleading and it is advisable always to check the answer by other methods, such as a YP on receipts.

In the past it has been considered that the rental values of public houses bore no uniform relationship to the capital values. Possibly, this may still be the position. Certainly, if a lease is terminating within a few years, then it is often the practice to consider a reversion to capital value (if circumstances permit) instead of a revised lease rent.

Study 6

Example

Estimate the capital value of a public house let on a lease at £10,000 pa, the lease expiring in six years' time. Its present capital value with vacant possession is estimated to be £250,000.

	£	£
First 6 years		
Rent under lease (well secured)	10,000	
YP 6 years at 6%	4.92	49,200
After 6 years		
Reversion to capital value at the end of the lease	250,000	
Defer 6 years at 7.5%	0.648	162,000
		£211,200

Compare this on the assumption that a revised rent after 6 years would be £15,000.

Study 7

	£	£
First 6 years as before		49,200
After 6 years		
Revised rent	15,000	
YP of reversion to perp at, Say 7%	9.52	142,800
		£192,000

It may be preferable to consider a reversion to capital value in this situation, not forgetting to give consideration to the possibility of a payment for compensation under section 37 of the Landlord and Tenant Act 1954 if an application for a new lease is refused: see chapter 4 on The Landlord and Tenant Acts.

4. Kennedy Method

Two traditional methods of valuation were in collision in a Development Land Tax Case, *Inland Revenue Commissioners* v *Allied Breweries (UK) Ltd* [1982] 1 EGLR 189, concerning a back-street pub in Birmingham. The Revenue's Licensed Property Valuer based his valuation on an updated version of a brewer's profit basis, called the Kennedy method after the judgment of Kennedy J in an earlier case (*Ashby's Cobham Brewery Co Ltd, re The Crown, Cobham and Ashby's Staines Brewery Co Ltd, re The Hand and Spear, Woking* [1906] 2 KB 754) on redundancy compensation. The valuation in the 1982 case was amended by the Lands Tribunal so that at the end of the day it looked like this:

Study 8

Malt Shovel PH, Birmingham
Current use value at 14.12.1976

Wholesale profits		£
356 barrels draught beer at £3.90		1,388
36 barrels bottled beer at £6.50		234
135 gallons wines and spirits at 90p		122
		1,744
Add: 25% to allow for increase over "tone" of the 1973 valuation list		436
Wholesale profits in 1976 estimated at		2,180
Tied rent (1974)	£1,560	
Add: 15% to update	234	1,794
Gross income		3,974
Deduct: for repairs and insurance		374
Net income		3,600
YP		10
Current use value		£36,000

The brewers' wholesale profits in the valuation were those which formed the basis of rating assessments in 1973 – the Lands Tribunal uplifted them by 25% to reflect the position in 1976, the relevant date in this case.

The reasoning behind the Kennedy Method is that it goes directly to what the brewery is making from the public house: first, the brewery makes wholesale profit from supplying goods and, second, it charges its tenants a tied rent. In the managed houses, which represent about 46% of the brewer's tied estates, the brewer makes a net retail profit instead of a tied rent.

In the Development Land Tax case under discussion, the brewer's valuer depended on a valuation per converted or equivalent barrel (defined in study 11), the price per barrel being derived from a scrutiny and analysis of comparable sales over a wide area. The Lands Tribunal looked selectively at the comparables and reinforced its decision by using a check valuation at per converted barrel. It is interesting to note that it expressed no preference for either method. As it has done in other cases where methods of valuation have come into conflict, it rationalised the evidence and showed how the answers on both methods could confirm each other.

An updated version of a valuation of a public house similar to that in study 8 might be as follows:

Study 9

A current valuation of public house on the amended Kennedy Method

356 barrels of draught beer at £100	£35,600
36 barrels of bottled beer at £120	4,320
270 gallons of wines and spirits (assume tenant not tied for these so brewer receives no profits)	–
Current wholesale profits estimated at	£39,920
Add: current tied rent	30,000
Gross income	£69,920
Deduct: for repairs and insurance	5,000
Net income	£64,920
YP	10
	£649,200
Capital Value on the basis of current use as a public house, Say	£649,000

5. Price per Barrel

The valuation in study 9 devalues as follows in terms of price per beer barrel:

Study 10

356 barrels draught beer + 36 barrels bottled beer = 392 beer barrels
392 beer barrels @ £1,656 = £649,152, Say £649,000

Next in terms of a price per converted or equivalent barrel. Converted or equivalent barrelage is the draught and bottled beer expressed in barrels (36 gallons) plus the gallons of wines and spirits divided by three.

Study 11

356 barrels draught beer + 36 barrels bottled beer + (270 gals wines
and spirits ÷ 3) = 482 converted barrels pa
482 converted or equivalent barrels @ £1,347 = £649,254, Say £649,000

In the past, when 80% to 90% of pubs were tied to breweries, the method valuing at a price per barrel was normal but now it tends to be used more as a check or for analysing comparables. It can no longer be regarded as a reliable method for general use because public houses no longer sell mainly liquor. Many practise extensive catering and others derive significant receipts from amusement machines. Also many purchase guest beers and wines and spirits elsewhere than from the one brewery company and it is difficult or impossible to obtain accurate sums of barrelage.

6. Overbid Methods

Overbid methods are often used for two kinds of valuation (a) in valuing a public house from the point of view of a brewer and (b) from the point of view of a retail group that can command high discounts off purchases.

(a) Overbid by a Brewery Company

A brewery company charges a rent and also receives wholesale profits on the beers supplied to a pub. This was explained earlier in the Kennedy Method. The company is able to use the wholesale profits to substantiate an overbid.

Study 12

Normal lease rental value of public house per annum	£20,000
Brewer's annual wholesale profits:	
300 barrels @ £100 = £30,000	
overbid say 10%	3,000
Rental Value	23,000
YP	15
Capital Value	£345,000

Overbids are likely to be between 5% and 35% depending on the fierceness of the competition between prospective bidders.

This was the range of overbids used by the Valuation Office when it adopted the overbid method in the 1990 Rating Revaluation.

If the barrelage is not known by the valuer but he is aware of the receipts, he might estimate the brewer's overbid thus:

Study 13

Normal rental value as before, pa	£20,000
Receipts £200,000	
Adopt 1½% as overbid	3,000
	23,000
YP	15
Capital Value	£345,000

If the brewery company intended to run a public house under management it would make a net trading profit as well as a wholesale/manufacturing profit, and it might decide to form an overbid from an amalgam of both sources of profit.

Study 14

Normal rental value as before, pa		£20,000
(1) Overbid from wholesale profits (per study 12)	£3,000	
(2) Estimated net trading profit after payment of rent, working expenses, wages and other costs £20,000		
Adopt overbid 10%	£2,000	5,000
		£25,000
YP		15
Capital Value as a Managed House		£375,000

(b) Overbid by a Retail Company or Chain

Assume a retail company obtains a significant discount on its purchases of beer and certain other consumable goods. In order to make a competitive bid for a public house, it uses a part of its discounts to make an overbid.

Study 15

Normal rental value (as before), pa	£20,000
Discounts on beer purchases etc	
amount to £16,000 pa	
Assume the overbid to be 25% of £16,000	4,000
	£24,000
YP	15
Capital Value to Retail Group	£360,000

It is possible for the use of overbid methods to be applied direct to capital values, instead of to rental values and then capitalised.

Tiny retail groups may not generate enough beer trade to qualify for large discounts and this would hamper or limit any overbidding. Similarly, individual retailers would be unable to make an overbid without sacrificing a slice of their net trading profit.

7. Floor Area

Generally, valuations of public houses do not depend on floor area appraisals or zoning except where there is competition for the property for alternative uses such as shops or offices.

Following the extension of permitted hours in the Licensing Act 1988 allowing public houses to open on weekdays from 11:00 AM to 11:00 PM, that is, virtually all day, they can now compete with shops and restaurants for town centre sites.

In such circumstances, regard must be had to demand from other users and a licensed profits valuation should be checked by a retail floor area appraisal. Reference may be made to chapter 7 on "Shops" for methodology.

Rental Valuations

The methods follow those described in the foregoing pages.

For instance, in study 5 (combined with study 1) is an example of a valuation to rental value on the *accounts method*.

A full and detailed accounts valuation will be found elsewhere in the book, in chapter 18 on "Hotels".

The shorter *receipts method* calls for a percentage to be applied to the turnover to produce a rental value.

Study 16

Turnover of urban public house doing a steady trade including a little
catering £185,000
Multiply £185,000 by 10% = £18,500 rent

Rental values vary from 7% to 13% in most cases but can be more
or less depending on the quality of the house and trade and other
factors. Accounts valuations are always preferable, using the
receipts method as a check and a test against comparables.

Overbids will be made by brewery companies and retail chains
in order to lease as well as to buy properties, and the methods have
been outlined (see studies 12 to 15).

Rating

Reference should be made to the Rating chapter 12 for the general
principles of assessment.

For 50 years public houses were assessed on the Direct Approach
Method or Direct Method as it was shortly known. In skeleton it
had this form:

Study 17

Brewer's wholesale profits on tied supplies of beers and wines and spirits	£ . . . . .
Add: tied rent	£ . . . . .
Total brewer's net income	£ . . . . .
Rating assessment reached by taking a brewer's rental bid between 30% and 50%	£ . . . . .

The direct method was introduced following the House of Lords
decision in *Robinson Bros (Brewers) Ltd* v *Durham County Assessment
Committee* [1938] AC 321, where it was held that brewers (who then
owned most of the public houses) should be considered also as
hypothetical tenants of public houses and, furthermore, the effect
on values of competition between brewers to acquire public houses
should be taken into account. Prior to this case, valuations had been
estimated on the basis of what a retailer would pay.

In the 1990 Revaluation, the rateable values were assessed on an
Overbid Method similar to that in study 12 but in the 1995 and 2000
Rating Lists valuations have reverted to a Receipts based method.

Brewery companies now own only about 30% of public houses in the UK. The remainder are owned or leased by large retail chains or free traders. Certain national brewery companies have ceased brewing and converted themselves into giant retail groups. This severe reduction in brewery ownership of pubs is a result of the legislation following the Monopolies and Mergers Commission Report in 1989 on the Supply of Beer (described earlier in this chapter).

Following agreement with the Brewers & Licensed Retailers Association (formerly the Brewers Society) the Valuation Office Agency have issued "Approved Guides" for assessing the rateable values of public houses in the 1995 and 2000 Rating Lists based on a percentage of liquor (or wet) receipts plus a percentage on "dry" receipts from catering, letting rooms, etc.

The Guide for the 2000 Rating List contains graphs shewing the results of rent analysis in terms of percentages which in the provinces commence at 4% on liquor and gaming machine receipts for low earning pubs, with the worse physical characteristics, up to about 12.5% for good quality well located pubs with receipts in excess of £500,000 pa. For dry trade the percentages vary from 3.5% to a maximum of 10.5% with percentages of 7% to 8% for pubs in the middle ranges.

The Guide states that to ascertain the correct rateable value:

> The first and most important consideration is to determine the "fair maintainable receipts" of the property excluding VAT. These will be split between gross receipts for liquor, food, accommodation and other sales and net receipts from gaming machines.
>
> The figure of receipts determined should represent the annual trade considered to be maintainable at 1 April 1998 (the antecedent valuation date) having regard to the physical nature of the property and its location as at 1 April 2000 when the new rating lists come into force (or subsequently following a material change of circumstances) on the assumption that the business will be proficiently carried out by a competent publican responding to the normal trading practices and competition of the locality.
>
> Valuations should not be arithmetical calculations. Each must be considered by a competent valuer within the terms of this guide and having regard to the individual nature of each property, its trading location and style of trade.

Study 18

Assess the rateable value of an average quality provincial public house with a turnover of £430,000 "wet", £20,000 net income from a gaming machine and £10,000 "dry" totalling £460,000 which can be regarded as the fair maintainable receipts.

Wet: £450,000 @ 11%	49,500
Dry: £10,000 @ 4.5%	450
Rateable value	£49,950

Study 19

Assess the rateable value of an average quality catering pub with a wet turnover of £95,000 net income from a gaming machine of £5,000 and a dry turnover of £150,000 which are the fair maintainable receipts.

Wet: £100,000 @ 7.5%	7,500
Dry: £150,000 @ 7.75%	11,625
Rateable value	£19,125

Study 20

Assess the rateable value of a small terrace pub in a poor location with a wet turnover of £80,000 net income from a gaming machine of £1,000 and a dry trade of £2,000, which are the fair maintainable receipts.

Wet: £81,000 @ 6.25%	£5,062
Dry: £2,000 @ 4%	80
Rateable value	£5,142

The antecedent valuation date for the new millennial 2,000 Non-domestic Rating Revaluation is April 1 1998 and the date the List takes effect is April 1 2000.

Notices requesting information of rents and turnover were served by Valuation Officers in 1998. As a result about a quarter of the rateable values have gone down but many have risen, following the market where "the good have got better and the poor have got worse".

It is important to understand that the majority of public houses have living accommodation which is subject to a separate assessment for Council Tax. The government's Valuation Office in analysing rent returns has applied a scale of allowances to the living accommodation. The result is that the rateable values of most rented public houses are less than their actual lease rents because the rateable values exclude the estimated rental values of the domestic accommodation.

There is a new guidance note: *The Receipts and Expenditure Method of Valuation of Non-Domestic Rating* produced by the *Joint Professional Institutions' Rating Valuation Forum and the Valuation Office* which contains useful advice on the Receipts and Expenditure Method (ie the Accounts Method) including a page on the shortened Receipts basis.

It should be remembered that a valuation on receipts based on the Approved Guide is a comparative valuation and is not a valuation on accounts. A valuation on receipts is a short-cut method. If a valuer is acting for a ratepayer it may be advantageous to make an accounts valuation similar in principle to that illustrated for an hotel in chapters 12 and 18. Studies 1 and 5 in this chapter may also provide a guide.

Further reading

Law & Valuation of Leisure Property a book published by Estates Gazette

Rating List 2000 Approved Guide to Valuation of Public Houses published by the Valuation Office. (This may be purchased from Direct Services, Room G33, Chief Executives Office, Valuation Office Agency, New Court, Carey Street, London WC2A 2JE. Telephone 020 7506 1700 for further details.)

Receipts and Expenditure Method of Valuation for Non-Domestic Rating published by RICS Books 1998

Chapter 18

Hotels

Introduction

The term "hotel" encompasses a wide spectrum of property types from larger units having up to or maybe in excess of 1,000 letting bedrooms, to smaller units having as few as 10 or maybe less. They range from modern, purpose-built units, incorporating the latest design techniques, to converted old manor houses or coaching inns. Services offered may be restricted simply to letting bedrooms or include a range of other facilities such as restaurants, bars, conference and banqueting rooms, leisure clubs, golf courses, business centres, etc.

Whichever category a particular property falls in, one factor that most hotels have in common is limited potential for conversion to alternative uses.

Hotels are a class of property which normally changes hands in the open market as fully operational businesses at prices based directly on the trading potential for their existing use. A property will therefore, be normally sold inclusive of all trade fixtures, fittings, furniture, furnishings and equipment, and with the benefit of all licences, permits and trading potential, but excluding stock in trade, leased and badged items, some of which may be subject to a separate agreement. The whole principle of valuation is therefore based upon potential turnover, net profit and the return on capital required by prospective purchasers in the market. However, while larger hotels are generally operated by hotel companies to make a profit, this may not be the primary consideration for some, mainly smaller hotels where the owners are motivated by other reasons such as the provision of a family home, quality of life etc.

The UK hotel market has historically been characterised by it's cyclical pattern, given that the fortunes of the hotel industry are very much a reflection of activity within the economy as a whole. Values have tended to follow these trends.

Corisand

Hotels are a specialist property type and it is essential that the valuer has sufficient knowledge and expertise and understands the market in which he is operating.

Corisand Investments Ltd v *Druce & Co* [1978] 2 EGLR 86, is a leading case on the valuation of hotels and is worthy of study. It concerned the valuation of a hotel in North London for mortgage purposes in September 1973. The valuation was relied on by the plaintiffs to advance a second mortgage, and when the hotel market collapsed shortly afterwards, the borrowers could not repay the loan, and the plaintiffs suffered a loss. Mr Justice Gibson set out in his judgment the matters of principle and fact that an ordinarily competent valuer must have regard to in order to discharge his duty of care in valuing a property such as the subject hotel.

The valuation was made on the basis of the value of the hotel for mortgage purposes with the benefit of vacant possession as at September 28 1973. This was at the end of a property boom, and at a time when the property market was described as "high" or "booming'" The problems of valuing in such a market were considered and reference was made to the judgment of Watkins J in the case of *Singer & Friedlander Ltd* v *John D Wood and Co* [1977] 2 EGLR 84:

> In every case the valuer, having gathered all the vital information, is expected to be sufficiently skilful so as to enable himself to interpret the facts, to make indispensable assumptions and to employ a well-practised method of reaching a conclusion; and it may be to check that conclusion with another reached as the result of the use of a second well-practised method. In every case the valuer must not only be versed in the value of land throughout the country in a general way, but he must inform himself adequately of the market trends and be very sensitive to them with particular regard for the locality in which the land he values lies. Whatever conclusion is reached, it must be without consideration for the purpose for which it is required . . . If a valuation is sought at times when the property market is plainly showing signs of deep depression or of unusual buoyancy or volatility, the valuer's task is made more difficult than usual. But it is not in such unusual circumstances an impossible one. As Mr Ross said, valuation is an art, not a science. Pinpoint accuracy in the result is not therefore, to be expected by he who requests the valuation. There is, as I have said, a permissible margin of error, the 'bracket' as I have called it. What can properly be expected from a competent valuer using reasonable skill and care is that his valuation falls within this bracket.

The unusual circumstances of his task impose upon him a greater test of his skill and bid him to exercise stricter disciplines in the making of assumptions without which he is unable to perform his task; and I think he must beware of lapsing into carelessness or over-confidence when the market is riding high.

Both judgments applied to valuations for mortgage purposes but the duty of care inherent in the valuer's task was made clear. Another aspect of valuation dealt with in Gibson J's judgment was the matter of any speculative content in the market. On this matter he said:

The plaintiffs however, have wholly satisfied me that in September 1973 an ordinarily competent valuer had substantial ground for knowing that any speculative content in his estimated open market price, estimated in that boom market, may well not be maintained in future or be readily realisable on the forced sale of the property. A valuer in fixing in September 1973 a valuation figure for mortgage purposes was not entitled to be optimistic in the sense that he could not treat as a proper basis for a mortgage valuation an open market price containing a substantial speculative content.

A valuer in September 1973 knew that any speculative content in the open market price, attributed by him to a property, might well disappear within a short period of time. There was no rational basis for supposing that such speculation in hotels as had been seen to occur must continue for any particular length of time – it could cease as it had begun. The concept of a speculator's price, as I understand it, is to pay more for a property than other buyers are then willing to pay on ordinary principles of investment return in the belief that the market will rise to and overtake the price so paid. The capacity of speculators in the market to bid for hotels must have been affected by the cost of borrowing money; and the increase in Bank Rate in the summer of 1973, and the letter of the Governor of the Bank of England to which reference has been made, were entirely sufficient to remind any valuer that a boom market may not continue and may not be there when a mortgagee comes to sell his security."

The valuer therefore, must discount any element of speculative value in his valuation or identify this element in his valuation report.

Royal Institution of Chartered Surveyors Appraisal and Valuation Manual 1996

Guidance Note 7 of the Appraisal and Valuation Manual published by the Royal Institution of Chartered Surveyors in 1996, relates to

Trading-Related Valuations and Goodwill and sets out the general approach to be adopted for the capital valuation of properties which are normally sold as fully operational business units of which hotels are a kind.

The full contents of Guidance Note 7, to which the valuer should refer need not be set out in full here, but particular attention is drawn to the following:

GN 7.2.1

The valuation of the operational entity includes:

(a) the land and buildings;
(b) Trade fixtures, fittings, furniture, furnishings and equipment. (Certain items may not be owned and the Report must make clear which are excluded from the valuation.)
(c) The market's perception of the trading potential excluding personal goodwill (see GN 7.2.10), together with an assumed ability to renew existing licenses, consents, certificates and permits (but see GN 7.6. below). This was formerly known as inherent goodwill (see GN 7.2.10).

GN 7.2.2

The valuation method adopted by the valuer should reflect the approach generally used by the market for the particular type of property. Different methods, including the profits method, Discounted Cash Flow (DCF), analysis of comparable transactions, or a combination of these will be appropriate for different types and sizes of property.

GN 7.2.3

In arriving at the valuation, the valuer should analyse and review the trading accounts for the current and previous years, and projections for future years, where these are available; he should then form an opinion, by reference to analysis of the trading accounts of comparable properties, as to the future trading performance and fair, maintainable operating profit likely to be achieved by an operator taking over the existing business at the date of valuation. In doing so, the valuer should consider the economic adaptability of the property to a different and more profitable style of operation. Furthermore, he should have primary regard to the market's perception of maintainable profitability.

GN 7.2.8

The accounts of a particular property will only show how that property is trading under the particular management at the time. The task of the valuer is to assess the fair maintainable level of trade and future profitability that can be achieved by an operator of the business, upon which a potential purchaser would be likely to base his offer.

GN 7.2.17

The determination of the capitalisation multiplier, or the discount rate, to be applied in arriving at the capital value of the property, relies upon the experience and judgment of the valuer. It should reflect the valuer's opinion of the market's perception of the risk or security associated with the subject property and its current and future trading potential, taking account of all available market evidence and economic factors such as interest rates and inflation.

Inspections

The details that need to be taken into account on inspection differ from many other types of property because floor area is not the essential ingredient that determines value. The information required relates to the trading potential of the property.

Location, as with any property, will provide a hotel's most important advantage or disadvantage as the case may be. What is a good location will depend upon the type and class of hotel under consideration, but the valuer must always have regard to the economic prosperity of the surrounding locality and the hotel's proximity and accessibility to sources of business and the target market sector. City centres are always in need of an appropriate supply of hotel bedrooms, and major airports are another prime location where high occupancy can be expected, provided there is not an over-supply. Easy road access, closeness to motorway junctions or public transport facilities will generally be of benefit to a hotel's trade, but in other instances a remote, attractive countryside location may hold particular appeal. Some hotels may benefit from their proximity to large industrial or commercial establishments or tourist attractions.

The valuer will not normally be required to carry out a structural survey, but as with all property inspections it will be important to note the age, type and quality of construction, and the general condition of the buildings and accommodation both internally and externally. Consideration needs to be given to the fair annual cost of repairs and maintenance, and account taken of any outstanding works or necessary items of expenditure, either current or likely to occur in the near future.

Supply to the property of gas, electric, water and drainage should be confirmed together with the nature and extent of any heating systems, air conditioning etc.

Hotels commonly provide a number of different services within the same building which are often available for the use of both residents of the hotel and non-residents. They include the provision of letting bedrooms, food and beverage services in restaurants, bars and banqueting rooms, conference/meeting rooms and leisure facilities. Full details are required of all the accommodation, facilities and services offered by the hotel, always having regard to the overall layout and ease of operation, staffing and management. The valuer needs to formulate an idea not only of the potential turnover of the hotel but also of the running costs necessary to achieve that turnover.

The letting bedrooms will normally provide the most profitable source of income. It is not necessary to measure all the rooms, but the valuer needs to know the total number of available rooms; the size and specification of the rooms in general terms; the mix between different room types, ie doubles, twins, singles, suites etc; whether the rooms have en-suite bathrooms or showers; the heating or air conditioning systems; the quality, condition and standard of furnishings and fittings; and the facilities provided including such items as televisions, radios, fax machines, modem points, in-house movies, direct dial telephones, hairdryers, mini-bars, safes, tea and coffee making facilities, trouser presses, etc: The general comfort of the rooms in terms of noise, views, etc should also be noted. A certain degree of noise may be expected in city centres for instance, but this can be greatly alleviated by the provision of secondary or double glazing and air-conditioning. Some hotels may charge higher rates for rooms with attractive views, perhaps sea facing. The advertised room rates should be noted, but in the knowledge that these will likely differ significantly from the actual achieved room rates. The provision of lifts and disabled facilities are further important factors.

The net internal floor areas of bars, restaurants, banqueting and conference rooms may be taken, but of greater importance in estimating potential revenue are factors such as the capacities of the facilities, their general layout and location within the hotel, their flexibility and ease of use, modernity and character/ambience. For instance, a bar or restaurant with direct access to the street may attract a good level of non-resident trade, whereas it can be extremely difficult to attract such trade to facilities above ground-floor level. Public rooms situated on more than one floor may necessitate the staffing and management of two kitchens whereas in other circumstances one would easily be able to cope with the same level of trade. Far greater use can be made of large conference rooms that are capable of being subdivided than those that are not. Again, the general comfort of the rooms should be noted in terms of natural light, heating, air conditioning, etc. Record should be made not only of the facilities offered but also of the type of product and services available, functions catered for and pricing structures.

Leisure facilities, as separate profit centres, historically may have added relatively little to the overall profits of an hotel, but their provision is becoming ever more essential to attract trade and improve room rates, particularly in the higher 3 star and above hotels. Many hotels now look to establish viable clubs in their own right in addition to use by hotel guests. Equipment is constantly being updated, and as well as recording the facilities offered and their quality, it is important to determine the level of outside demand, whether there is a separate membership scheme and if so, the joining and annual membership fees, how many members there are, and whether there is a waiting list. The valuer will also need to confirm whether the leisure club is operated directly by the hotel or by a separate operator under a franchise or lease arrangement.

The availability and ease of car parking, preferably secure and directly controlled by the hotel is another important factor. Well tended, landscaped grounds may add to the appeal of an hotel but will also increase running costs. Golf course hotels have become a separate category in their own right, and it is essential to record the quality of the golfing facilities, including the course on offer, separate membership schemes etc.

Note should be made of ancillary accommodation, its suitability, adequacy, and location in the hotel. This will include management and administration offices, kitchens, stores, plant rooms, cellars, staff accommodation, and delivery points.

While not carrying out an inventory valuation, the valuer during his inspection, should ascertain which, if any, items are leased or hired. He should also confirm the existence and validity of all necessary licences and certificates which may include among others a Justices' liquor licence, restaurant licence, and public entertainment licence. The Fire Certificate should be checked, and confirmation sought that there are no outstanding statutory notices.

It is always worth obtaining a copy of the hotel brochure and any available conference packs etc. These will provide a summary of location, style of hotel, facilities offered and pricing structures.

It is essential to meet the manager or proprietor to discuss the hotel, its trading performance and potential. He should be questioned about such factors as occupancy levels; average achieved room rates; the main sources of business and principal business generators; average length of stay; trading profile; reasons for past fluctuations in trade; trade outlook; and any known factors likely to affect the hotel's future trading prospects including new competition. Tenure must be confirmed together with details of any sub-lettings.

A SWOT (strengths, weaknesses, opportunities and threats) analysis is a good means of summarising all relevant factors.

From the inspection and discussions with the manager or proprietor, the valuer will be able to give consideration to the style and class of the hotel, its suitability to the market sector at which it is aimed, and begin to make an assessment of its trading potential and security and the necessary costs involved in running the hotel to achieve that potential.

Definitions

It is worth noting here several terms commonly used in the hotel industry:

Room Occupancy: the total number of letting bedrooms occupied during the year as a percentage of the total number of rooms available.

Bed Occupancy: the total number of bed spaces occupied during the year as a percentage of the total number of bed spaces available.

Double Occupancy: the percentage of total rooms sold occupied by two guests.

Average Achieved Room Rate (AARR): the average rate achieved for every bedroom that is let. AARR is calculated by dividing the rooms revenue by the total number of bedrooms occupied during the year. This will differ from the advertised rack rates because of various discounts offered to guests.

Average Daily Rooms Yield (also referred to as REVPAR – revenue per available room): the average rate achieved for all letting bedrooms available throughout the year. Rooms yield is calculated by dividing the rooms revenue by the total number of rooms available during the year. This is also the product of room occupancy multiplied by the average achieved room rate.

Investigation of Market Sector, Demand and Competition

Having completed the inspection of the particular hotel and on-site enquiries, the valuer will need to investigate the market sector at which the hotel's trade is aimed, demand for the services offered and competition.

It is paramount that an hotel competes in a class of business and market sector to which it is suited. The valuer needs to understand and appreciate the economic profile of the surrounding locality and the sources, nature and level of demand for the hotel's services and facilities, particularly the letting bedrooms as these will normally contribute the greater proportion of profits. Sources of business vary considerably between hotels and can include travelling businessmen, residential conferences, day conferences or meetings, functions such as weddings, dinners etc; tourists, holiday makers, coach parties and leisure breaks. The valuer must be particularly careful if an hotel is heavily reliant upon one client for its trade because any breakdown in that relationship would have a significant effect on the hotel's trade and consequently value. The type of trade coming to the hotel will determine the rates that can be charged for the facilities provided.

The size, nature and location of competitor hotels needs to be established together with the facilities they offer and their relative advantages and disadvantages. Depending upon the type of hotel in question, competitors may be confined to other hotels within a relatively small area, or may be spread throughout the country or indeed internationally. It may be possible to ascertain an idea of the trading performance of competitors because hotels often exchange information on occupancy levels, etc; or conduct their own informal surveys of other competing hotels. The nature of trade may determine that other establishments such as restaurants and leisure clubs are considered to be major competitors. All competition, whether in existence or planned, has to be investigated and any possible effect on the trade of the subject hotel properly taken into account.

The valuer needs to formulate his opinion of the level of trade at the hotel in question, it's security and prospects for improvement.

Trading Potential

From his inspection of the property and other enquiries, the valuer will be able to formulate a reasonable estimate of the trading potential of the hotel by reference to average room occupancy and average achieved room rates. However, wherever possible, the valuer should obtain the actual trading accounts for at least the last three years, the current year and projections for future years. It must be remembered that accounts can be in different formats and may be prepared for a variety of purposes. Hence it is difficult to generalise, but the valuer must carefully analyse whatever accounts are available and be satisfied that the information provided is accurate and reliable. He needs to ensure that all relevant items of income and expenditure are included.

The accounts of a particular property will only show how that property is trading under the particular management at the time, and may or may not represent the market's perception of the fair maintainable level of trade. The task of the valuer is to use his experience, knowledge of the market and analysis of the trading accounts of comparable properties, to form an opinion of the fair maintainable trade achievable not by the best or worst operator, but by the average efficient operator. Assistance may be gained in this respect from the local Tourist Board and statistics published by specialist hotel consultancies such as Pannell Kerr Forster and BDO Hospitality.

Goodwill

This essentially takes two forms:

(i)   Goodwill which attaches to and passes with the property
(ii) Personal goodwill

It must be emphasised that goodwill can only have a market value to the extent that it is transferable. Value which attaches to the building and passes with the property by virtue of circumstances such as its location, design, licences, planning permission and occupation for it's particular use is reflected in its trading potential which can enhance the value of buildings but does not form a separate element in the saleable value of the business. A hotel which is open and trading will probably have a higher value than if it has been closed and empty for some time. A newly built hotel will likely attract a lower value than one with a proven trading history.

This needs to be differentiated from personal goodwill which is attributable to the personal skill, expertise and reputation of the present owner or management. It may also arise as a result of access to a central reservations system or a particular brand name. Personal goodwill attaches to the operator and will disappear when he leaves the property. However, it may be replaced or even enhanced by the personal goodwill of the new operator. The valuer must exclude any element of trade attributable to personal goodwill, but include all or any additional potential that would be realised by an average competent operator. It is therefore, essential to be able to identify the likely potential purchaser (excluding a "special purchaser"). For example, if an hotel is sold by a major chain operator to another similar operator, it may be considered that there will be little or no effect on trading levels. On the other hand, a private purchaser may suffer a drop in trade if he does not have the benefit of national marketing, central reservations etc. Consequently, it is essential for the valuer to have detailed knowledge of the market. He must consider who the potential purchasers are and address the following:

(i)   How, and under what brand, would another operator trade from the hotel?
(ii) What is the anticipated level of turnover and profitability which a purchaser would anticipate?
(iii) What costs would be incurred by the purchaser in re-branding the hotel?

Accounts

Actual accounts may be in different formats, but the valuer must ensure that not only is all income shown, but also that all necessary and relevant items of expenses are included. Items of a non-recurring nature should be discounted as should any items unrelated to the trading activities of the hotel. If for instance, a leisure club is operated separately under a franchise arrangement, it may be appropriate to replace a franchise payment or licence fee shown in the accounts with an estimate of the club's total revenue. Accounts for a new hotel may include joining fees for a leisure club and care needs to be taken to discount these to the extent that they would not be receivable on a year to year basis.

Expenses such as repairs and maintenance may vary significantly from one year to the next, and these will require adjustments to reflect a fair year on year cost. Each item should be considered carefully on its reasonableness and adjustments made if necessary. Head office costs can be substantial and these have to be allowed for where appropriate. Companies sometimes take profits from the bulk purchase of goods at head office level rather than attributing them to individual hotels. The valuer may have to make adjustments for these. Directors' remuneration is not normally allowed as an outgoing when assessing net profit, the directors' return coming from the profit of the business. However, if directors are carrying out duties that would normally be carried out by an employee, then a deduction for appropriate salary costs should be made.

Where actual accounts are not available, the valuer will have to construct his own "shadow" or hypothetical accounts. Trade profiles will vary significantly between hotels and different classes of hotel, and this together with a variety of other factors including, *inter alia*, location, age, size, design, condition, layout of accommodation, facilities and services offered and required staffing levels will affect profitability.

When considering the trading potential of any hotel, emphasis should be given to the rooms revenue because this will normally make a significantly higher contribution to profit than other sales. Typically, about 70% of the rooms revenue will flow to the operated department income or profit line, compared to about 40% of the food and beverage revenue. Total rooms revenue is the product of:

total rooms available × room occupancy × average achieved room rate

or

total rooms available × average daily rooms yield (REVPAR)

For example:

100 bedroom hotel	=	36,500 available rooms pa
× 70% occupancy	=	25,550 let rooms pa
× £40.00 AARR	=	£1,022,000 total rooms revenue

or:

available rooms	=	36,500 pa
70% occupancy × £40.00 AARR	=	£28.00
		£1,022,000 total rooms revenue.

Reference is often made to occupancy levels and average achieved room rates, but it is misleading to consider either in isolation. The product of the two gives the average daily rooms yield (REVPAR), and this is the true indicator of a hotel's performance. For example:

Hotel	Room Occupancy	Average Room Rate	Rooms Yield
A	80%	£42.50	£34.00
B	60%	£56.50	£33.90

If considering these two hotels, either the occupancy levels or average achieved room rates in isolation would give a false comparison. Their true performance, shown by the rooms yield, is virtually identical.

Having considered the potential room revenue, the valuer should turn to the other sources of income. These will vary considerably depending upon the type of hotel in question. For a standard 3/4 star hotel in central London, where most guests tend to go out to eat and drink, rooms revenue will typically account for between 60% and 80% of total revenue. In provincial hotels, where guests are generally more captive, this will usually fall to between 35% and 55% for modern hotels and between 25% and 45% for smaller, older hotels where greater reliance may be placed upon local restaurant and bar trade.

Accounts should always be shown net of VAT, and the following show examples of how hypothetical turnover figures may be calculated:

Study 1

A modern purpose-built 4 star provincial hotel, having 125 double letting bedrooms with well planned, flexible accommodation including restaurant, bar, conference rooms and leisure club with 200 members. Easily accessible just off a motorway junction and with good car parking facilities. Reliant on corporate business and conference trade during the week with some leisure based trade at weekends.

Advertised tariff: Double   £95.00 per night incl. VAT
                     Single   £75.00 per night incl. VAT

Assumptions:

(a) room occupancy   75%
(b) double occupancy   35%
(c) average achieved room rate at 70% of advertised tariff
(d) room revenue 45% of total revenue

1.   Total rooms available = 125 × 365   = 45,625
      75% room occupancy   = 34,219 let rooms
      double occupancy at 35%   = 11,977 rooms let
      single occupancy at 65%   = 22,242 rooms let

2.   Average achieved room rates:
      double room £95.00 at 70% = £66.50 = £54.86 net of VAT
      single room £75.00 at 70% = £52.50 = £43.31 net of VAT

3.   Total room revenue can then be calculated as:

11,977 double rooms at £54.86	=	£657,058
22,242 single rooms at £43.31	=	963,301
Total room revenue		£1,620,359
Say		£1,620,500

      N.B.
      AARR = £47.35
      Room yield = £35.51

4.   Room revenue accounts for 45% of total revenue, which therefore amounts to £3,600,000. Adopting reasonable assumptions for this class of hotel for the other sources of income, this may be made up as follows:

Say:

Rooms	£1,620,500	45.0%
Food	£1,260,000	35.0%
Beverage	£468,000	13.0%
Telephone	£72,000	2.0%
Other	£179,500	5.0%
Total	£3,600,000	100%

Study 2

An old manor house extended and converted to hotel use and providing 65 letting bedrooms mainly in 1970s wings, character restaurant and bar and several small meeting rooms. Set in attractive landscaped gardens in a rural location, and heavily reliant upon non-resident restaurant trade and its local reputation.

Advertised tariff: Double   £75.00 per night incl. VAT
Single   £60.00 per night incl. VAT

Assumptions:
(a)   room occupancy   65%
(b)   double occupancy   40%
(c)   average achieved room rate at 65% of advertised tariff
(d)   room revenue 35% of total revenue

1.   Total rooms available = 65 × 365   = 23,725
       65% room occupancy   = 15,421 let rooms
       double occupancy at 40%   = 6,168 rooms let
       single occupancy at 60%   = 9,253 rooms let

2.   Average achieved room rates:
       double room £75.00 at 65% = 48.75 = £40.22 net of VAT
       single room £60.00 at 65% = £39.00 = £32.18 net of VAT

3.   Total rooms revenue can then be calculated as:
       6,168 double rooms at £40.22   =   £248,077
       9,253 single rooms at £32.18   =   £297,761

       Total room revenue   =   £545,838
       Say   £546,000
       N.B.
       AARR = £35.40
       Room yield = £23.00

4.   Room revenue accounts for 35% of total revenue which therefore amounts to £1,560,000. Adopting reasonable assumptions for this class of hotel for the other sources of income, this may be made up as follows:
       Say:

Rooms	£546,000	35.0%
Food	£624,000	40.0%
Beverage	£320,000	20.5%
Telephone	£23,500	1.5%
Other	£46,500	3.0%
Total	£1,560,000	100.0%

Having arrived at his fair maintainable level of turnover, the valuer needs to turn his attention to the outgoings and expenses necessary to maintain that turnover. Trading projections and actual accounts should ideally be prepared in accordance with the current version of the Uniform System of Accounts. This categorises accounts into:

- Revenues
- Departmental costs and expenses
- Total operated department income
- Undistributed operating expenses
- Income before fixed charges (IBFC)

Revenues and departmental costs and expenses are shown on a department by department basis according to the principal activities of an hotel and, as shown in the above examples, include rooms, food, beverage, telephones, other operated departments, rentals and other income. Departmental costs and expenses only include items which can be specifically identified and related to each department, and include, *inter alia*, wages, costs of food and beverage, licences, travel agents' commissions and central reservations.

Total operated department income is calculated by deducting the total departmental costs and expenses from the total revenues.

Undistributed operating expenses are those outgoings which cannot be specifically related to a particular department, and include administrative and general costs, marketing, property operation and maintenance, energy costs, etc.

The income before fixed charges (IBFC) is calculated by deducting the undistributed operating expenses from the total operated department income and is the operating profit before deducting business rates, rent, interest, insurance, depreciation and amortisation and income taxes. If accounts show a management fee deduction this should be added back before arriving at the IBFC, but it must be ensured that all other expenses are adequate.

An example of the Uniform System of Accounts format is shown below:

Revenues (as calculated in Study 1 above)	£	% of total
Rooms	1,620,500	45.0
Food	1,260,000	35.0
Beverage	468,000	13.0
Telephones	72,000	2.0
Other Departments	179,500	5.0
Total Revenues	3,600,000	100.0
Departmental Costs & Expenses		
Rooms	470,000	29.0
Food & Beverage	1,050,000	60.8
Telephones	30,000	41.6
Other Departments	80,000	44.6

Total Costs & Expenses	1,630,000	45.3
Total Operated Department Income	1,970,000	54.7
Undistributed Operating Expenses		
Administration & General	340,000	9.4
Marketing	90,000	2.5
Property Operation & Maintenance	130,000	3.6
Energy	120,000	3.3
Total Undistributed Expenses	680,000	18.9
Income Before Fixed Charges	1,290,000	35.8

Notes

Figures and percentages shown are for illustration purposes only.

Trading projections, whether purely hypothetical or based on actual accounts should ideally be presented in this format to arrive at the Income Before Fixed Charges.(IBFC).

Further deductions may still need to be made to determine the adjusted valuation net profit, or net free cash flow, which will then be capitalised. Depending upon the nature of the interest being valued, these may include; insurance of the property and it's contents; uniform business rates; finance charges for leased items; a renewals fund to allow a yearly sum for the replacement and maintenance of the furniture, furnishings, fixtures and equipment at a level necessary to maintain the adopted level of trade; rents; and service charges.

Valuations

The valuer must always have regard to the interest being valued, the nature and purpose of the valuation and the appropriate method of valuation to be adopted.

Guidance Note 7.2.2 of the Royal Institution of Chartered Surveyors' *Appraisal and Valuation Manual* 1996 states that the valuation method adopted by the valuer should reflect the approach used by the market for the particular type of property.

Hotels are a type of property which change hands in the open market as fully operational businesses at prices generally based directly on their trading potential for their existing use. Properties will normally be sold inclusive of all trade fixtures, fittings, furniture, furnishings and equipment, with the benefit of all licences, permits and trading potential, but excluding stock in trade, any leased items and equipment and badged items. The whole principle of valuation is therefore based upon potential turnover, net profit and the return on capital required by prospective purchasers bidding against each other in the market.

The primary method of valuation is by capitalisation of income, the two approaches to which are:

- Earnings Multiple
- Discounted Cash Flow (DCF)

In November 1993 the British Association of Hotel Accountants (BAHA) published its *Recommended Practice for the Valuation of Hotels* to which the Royal Institution of Chartered Surveyors (RICS) responded in August 1994.

The relative merits of the two approaches were debated for sometime by the two professional bodies, the debate being highlighted in 1993 as a result of a wide disparity in group valuations of Queens Moat Hotels by two firms of hotel valuers.

The RICS recommends the use of the Earnings Multiple Approach using comparable market transactions, whereas the BAHA recommends the use of Discounted Cash Flow. The two professional bodies sought to prepare an agreed statement of best practice but unfortunately this has not been forthcoming.

Earnings Multiple Approach

This approach, historically known as the "accounts" or "profits" method of valuation, has been traditionally adopted by hotel valuers. The valuer formulates his opinion of the maintainable level of valuation net profit (net free cash flow) after allowing for all items of income and expenditure (excluding finance costs) and taking into account all factors which will impact on future levels of profitability. To this extent the approach mirrors that undertaken during a DCF analysis. The valuation net profit is then capitalised by a multiplier to arrive at the property's capital value to include all trade fixtures, fittings, furnishings and equipment, statutory consents, licences and permits, but excluding stock in trade and any leased or badged items, some of which may be subject to a separate valuation or agreement.

The multiplier to be applied is derived from the rate of return or yield which an hotel operator would require in order to purchase the property taking into account the security and growth prospects of the income. The valuer must assess how the market would view the risk or security associated with the particular hotel and it's profit potential, and establish the level of the multiplier from his experience, regular contact with operators and his analysis of any market transactions on comparable properties.

This approach represents a "mature" situation, and is based upon the maintainable level of net profit in the normal stabilised year of trading. In the case of new hotels for example, this would ignore the build-up of earnings over the first few years of trading, and in such cases a deduction would need to be made from the capital value to reflect any shortfall in earnings prior to the stabilised year. This may commonly take one to three years to reach, but is very dependent on individual circumstances. For example, the rapid growth in budget hotels in recent years has seen many reach their maximum trading potential within a matter of months. Particular care must be taken in the choice of yield or multiplier where a hotel has no proven track record. A deduction would also need to be made to reflect any capital expenditure necessary to maintain the level of income adopted.

Study 3

Income Before Fixed Charges (as calculated above)		£1,290,000
Less:		
(a) Uniform Business Rates Say	£135,000	
(b) Insurance Say	£50,000	
(c) Renewals fund at £10,000 per room = £1,250,000 @ 10% =	£125,000	
Total		£310,000
Valuation Net Profit/Net Free Cash Flow		£980,000 (27.2%)
YP in perp. at 10%		10.00
Capital Value		£9,800,000

Notes
(a) Rates deduction is calculated by multiplying the Rateable Value by the Uniform Business Rate. This may be affected by transitional arrangements.

(c) Renewals Fund: annual sum to provide for the replacement of fixtures, fittings and furnishings necessary to maintain the adopted level of trade.

Discounted Cash Flow Approach (DCF)

Discounted Cash Flow is a well respected financial model used extensively in investment appraisals. It's underlying principle is the "time value of money" – income receivable in the short term is more valuable than that receivable in the longer term. It is a particularly appropriate method in the consideration of new projects or where a property is to be extensively refurbished, extended or improved. It is also a suitable technique to take into account major changes in the market or in the economy as a whole.

DCF involves creating a cash-flow model of a property's projected financial performance over a time scale of typically 5–10 years which is then capitalised by a discount rate which reflects the rate of return an investor would require given the inherent risk attaching to the projected cash-flow and the alternative investment possibilities available. To this is added the residual value where appropriate. The approach simulates a series of cash-flows representing the anticipated receipts, expenses and operating performance of a property together with any required capital expenditure and the effects of inflation.

The discount rate adopted contains three elements:

(i)   Compensation for the effects of inflation.
(ii)   A risk free rate of return.
(iii) A premium compensating for the inherent risk associated with the industry and the individual property.

In normal circumstances, the minimum level of return an investor will expect is the yield obtainable from long-term government bonds. The discount rate is built-up therefore, from the current risk-free rate of return, ie the yield on long-term gilt-edged securities.

Adopting the net free cash-flow calculated above of £980,000 as the starting point, the following shows a DCF model for the purposes of which it is simply assumed that income and expenditure will increase by 5% pa (see table p591):

Comparison of the Two Approaches

Valuations are intended to show the price which an interest in property could be expected to fetch in the market place. The fundamental factor common to the Earnings Multiple Approach and Discounted Cash Flow Approach is that both value hotels by

Year	Total Revenue £	Total Expenses £	Net Free Cash Flow £	% of Revenue	PV of £1 @ 15%	Net present value £
1	3,600,000	2,620,000	980,000	27.2	0.87	852,600
2	3,780,000	2,751,000	1,029,000	27.2	0.76	782,040
3	3,969,000	2,888,550	1,080,450	27.2	0.66	713,097
4	4,167,450	3,032,975	1,134,475	27.2	0.57	646,650
5	4,375,820	3,184,625	1,191,195	27.2	0.50	595,598
6	4,594,611	3,343,855	1,250,756	27.2	0.43	537,825
7	4,824,340	3,511,045	1,313,295	27.2	0.38	499,052
8	5,065,557	3,686,595	1,378,962	27.2	0.33	455,057
9	5,318,835	3,870,925	1,447,910	27.2	0.28	405,415
10	5,584,775	4,064,470	1,520,305	27.2	0.25	380,076
						£5,867,410

Add residual value
YP in perp. @ 10%   1,520,305
15,203,050 × PV of £1 0.25 =   10.00   £3,800,762

                                                  £9,668,172

Say   £9,700,000 reference to recent and current trading performance and future trading potential.

The RICS recommends the use of the Earnings Multiple Approach using comparable market transactions. The valuer must consider the past and anticipated trading performance which gives greater weighting to achieved levels, and thus the method is less reliant on the potential inaccuracy of future projections.

The valuer will have regard to all available market evidence, and the multiplier applied will be derived from the valuer's opinion of how the market will view the subject hotel and its profit potential compared to the multipliers generally applied in the hotel sector. This market led approach results in hotel valuations reflecting the volatility of market conditions. Consequently it can be argued that, at certain times in the economic cycle, these estimations of price may not be supported by the hotel's underlying earnings potential, but they will reflect existing use value.

The discount rate applied in a DCF is intended to approximate the weighted average cost of capital to the likely investor but it is not derived from the hotel market. Many companies use DCF in taking investment and dis-investment decisions but companies have their own fixed rates of return which tend to vary over time and are not necessarily derived from the hotel market. The method is used to calculate worth to the owner and this may vary between owners according to the criteria each company has used in setting its own fixed rate of return.

In many cases the residual value in a DCF may account for up to 50% of the total value. There may be considerable uncertainty as to the accuracy of estimations of profit potential say 10 years ahead, which casts doubt over the accuracy of the valuation as a whole.

BAHA recommends the use of DCF, which it considers is more responsive to anticipated changes in future earnings, expenditure, market conditions etc: and demands a more rigorous, detailed and disciplined approach. The method relies more heavily on general investment yields, inflation, perceived risk and liquidity, and the characteristics and potential of the hotel being valued, than it does on the sentiment influencing the hotel investment market.

BAHA considers the earnings multiple approach to be an indicative method with market transactions as reasonability checks. It is considered a major flaw that while its heavy reliance on market generated yields reflects the market's volatility, it can result in valuations which may be in line with market sentiment but may not be supported by a detailed analysis of the hotel's earnings potential and risk. DCF, it is argued, is a more objective approach which relies on benchmarks set by the financial markets. It must be noted, however, that these are not infallible.

In all cases, it has to be stressed that hotels are a specialist class of property, and it is essential that the valuer has the appropriate experience and expertise to carry out the valuation required. Whichever method of valuation is adopted, it may be wise to carry out a valuation by the other method as a secondary check. The Earnings Multiple Approach is probably more appropriate where a valuation is being carried out of a property with an established trading history, and Discounted Cash Flow where this is not the case – when valuing new hotels/projects, and properties which are going to be subject to major investment in alterations and/or additions.

The end valuation should be checked by reference to analysis of comparable market transactions wherever possible. In some cases this may even provide sufficiently reliable evidence to be adopted as a method of valuation in its own right.

It should also always be pointed out that the value of an hotel could be lower if:

(i)   the business is closed;
(ii) the inventory has been removed;
(iii) the Justices' or other licenses, consents, certificates and/or permits are lost or are in jeopardy; or
(iv) the property has been vandalised.

Evidence of Open Market Transactions

For most property types, valuations are generally based on evidence of open market transactions in similar properties. Valuers must not ignore the evidence of the market. A valuation however, is an exercise in judgment and should represent the valuer's opinion of the price which would have been obtained if the property had been sold at the valuation date on the terms of the appropriate basis of valuation. The valuer is not bound to follow evidence of market transactions unquestioningly, but should take account of trends in value and the market evidence available to him, adjusting such evidence as appropriate, and attaching more weight to some pieces of evidence than others. In a rapidly rising or falling market, undue weight should not be given to historic evidence which may have become outdated, but care is needed where acquisitions may be based purely on speculation and hope value. It is essential that the valuer has the relevant skill and experience to analyse transactions correctly.

For hotels in particular, market knowledge is generally imperfect. The parties to a transaction normally treat the relevant details upon which the sale is based as confidential, and very often sales take place through corporate transfers. Consequently, very rarely will the valuer have available to him all the relevant accounting and other background information to enable a full analysis of other transactions.

Comparison may be made on physical factors, most commonly a price per bedroom. This does allow an analysis to be carried out on information that is readily available, but if adopting such an approach, it is essential that the property being analysed is truly comparable in all respects. This method may be appropriate when considering the valuation of certain categories of hotel such as terraces of privately owned small hotels/guest houses often found in seaside resorts and not necessarily always operated to maximise profits. However, unless all relevant factors are known, care has to be taken because comparison on physical aspects alone can be misleading and unreliable.

In a depressed market, a significant proportion of sales may be by vendors who are obliged to sell, such as liquidators and receivers. The valuer should establish whether or not these sales took place after proper marketing for a reasonable period. Liquidators and receivers are normally under a duty to obtain the best price and their sales may then be regarded as genuine open market transactions. However, there can be a stigma attached to such transactions which can depress the values achieved and they are therefore, not always reliable evidence.

In other instances a purchaser may pay a seemingly high price for a "trophy" property or for a property which will improve the profile of a portfolio as a whole. An operator may be especially keen to gain a presence in a particular locality, and be prepared to pay in excess of what might be considered the norm to secure that presence. Particular care is needed to identify such purchases when analysing transactions.

Often a portfolio of properties may be sold, and when attempting an analysis of such a transaction, it must be remembered that the acquisition may comprise, in broad terms, four elements of value:

* the aggregate of the individual existing use values of the properties.

- a lotting premium reflecting the ability of the purchaser to make a single purchase of a large portfolio of properties with much reduced acquisition costs and a significantly shorter period within which the global acquisition would be earnings enhancing.
- reduced head office costs and possible improvements in buying power having absorbed the acquisition into the estate.
- the acquisition of a strong brand which enables the purchaser to reposition a large part of its existing estate.

Rental Valuations

Relatively few hotels are held leasehold. Generally, considerable investment is required in fixtures, furnishings and equipment, and operators undertaking such expenditure prefer the freedom and flexibility of a freehold interest. From a landlord's point of view, the success of his hotel and consequently his income and the capital value of his asset, is not in his own hands, but is dependent upon the abilities of his tenant, over whom he has little control. As a result, there is a paucity of genuine open market rental evidence, many apparent rents being financing arrangements entered into for taxation or other reasons. Where rents are available for analysis, care has to be taken to ensure that they are genuine open market rents and have not simply been maintained under "upward only" review provisions where the fair rental at the time may actually be lower than the rent passing.

Full consideration must be given to the terms of the lease and particular regard paid to any unusual or onerous terms and conditions. For many property types there is an established pattern for rent reviews, but this is not so evident for hotels, and each has to be considered on its own merits.

Rental values, like capital values, are therefore determined primarily by reference to trading potential, and the amount of profit that will be paid as rent. Whether or not actual accounts are available, the valuer has to go through the same process to arrive at the Income Before Fixed Charges based upon the estimated trading performance of an efficient operator. Deductions are then made to arrive at the level of fair maintainable profit or net free cash flow (often referred to as the Divisible Balance), upon which the tenant will base his rental bid.

Study 4

Income Before Fixed Charges (as calculated previously)			£1,290,000

Less:

(a)	Uniform business rates		£135,000	
(b)	Insurance		£50,000	
(c)	Renewals fund (1) at £10,000 per room			
	= £1,250,000 at 10%		£125,000	
(d)	Interest on tenant's capital:			
	(i) Stock	£50,000		
	(ii) Working capital			
	Say 2 weeks turnover	£138,500		
	(iii) Fixtures and fittings			
	at £10,000 per room	£1,250,000		
	Total	£1,438,500		
	at 10% (2) =		£143,850	
Total				£453,850
Divisible Balance				£836,150
Tenant's Rental Bid (3) at Say 50% =				£418,075
Rental (4), Say				£420,000

Notes

(1) Renewals Fund: annual sum to provide for the replacement of fixtures, fittings and furnishings necessary to maintain the adopted level of trade.

(2) Interest on Tenant's Capital: a debt has to be serviced on working capital in the form of a cash float, food and drink stock and trade creditors. This has traditionally been accepted but in some cases these items may no longer be appropriate. An argument can be made that allowing for a renewals fund and interest on tenant's capital in the form of his outlay on fixtures and fittings is allowing the tenant too much and artificially reduces the calculated rental value. These allowances must, therefore, be considered carefully and in conjunction with the adopted tenant's rental bid.

(3) Tenant's Rental Bid: this will vary depending upon a range of factors that a tenant would take into account including location, the level and security of the income, the lease terms, alternative opportunities, the hotel's design constraints and future developments that might benefit or adversely affect the property. It is generally accepted that the tenant's rental bid will normally fall in the range of 40% to 50% of the Divisible

Balance. Circumstances however, may warrant a bid outside this range, perhaps as low as 35% or as high as 60%. The percentage to be adopted has to be considered in the light of the deductions made to arrive at the Divisible Balance, and it must be ensured that the tenant receives an appropriate return for running the hotel and the risk involved.

(4) When carrying out a rental valuation care must be taken to ensure that the actual rent passing is not included as a deduction before arriving at the Divisible Balance.

Tenant's Improvements

Tenant's improvements may fall to be disregarded for rent purposes at review. This poses the problem of how and where the effect of the improvements is reflected in the rental calculations.

Ideally, the valuer should ignore the improvements when formulating his trading projections and arrive at the net free cash flow for the hotel in its unimproved condition. However, depending upon the nature of the works to be disregarded, this may prove to be an extremely difficult or even impossible task. This being the case, the return that the tenant would expect on his expenditure should be treated as a deduction in the rental calculations This deduction may be made from the calculated rental as above, or as an additional item to take into account before arriving at the Divisible Balance on which the rental bid will be taken. The two approaches may give significantly different end answers as demonstrated below:

(a)	Calculated rental as before		£420,000
	Less:		
	tenant's improvements cost		
	Say £500,000 at 10% return		£50,000
	Rental Value		£370,000
(b)	IBFC as before		£1,290,000
	Less:		
	(i) deductions as before	£453,850	
	(ii) tenant's improvements cost		
	Say £500,000 at 10%	£50,000	£503,850
Divisible Balance			£786,150
Rental Value at 50%			£393,075
Say			£395,000

In such cases the question has to be asked if and why the tenant's improvements should be treated any differently than for instance, the deduction for renewal of fixtures and fittings. Generally, the second approach will be the most appropriate to adopt.

Valuations of leasehold interests will follow the same valuation principles as freehold valuations, except that the rent will be a deduction to make from the income before fixed charges to arrive at the net free cash flow or valuation net profit to be capitalised for the appropriate number of years.

Investment Valuations

Hotel investment valuations will follow normal valuation principles, ie capitalisation of the rental income. It can be seen above that the net free cash flow comprises the two elements of rent and residual profit.

An operator for the hotel will make his bid on both of these and therefore, to arrive at existing use value the two elements are capitalised. However, when considering an investment valuation, it has to be remembered that the residual profit is only available to the tenant or operator, and therefore investment valuations will only capitalise the rental income. Consequently, unlike many other properties, the value of the freehold interest held as an investment will often be significantly lower than the existing use value.

Bricks and Mortar Valuations

Hotels are valued as fully operational business units including trade fixtures, fittings, furniture, furnishings, consents, licences and permits.

Requests may sometimes be made for a "bricks and mortar" or "land and buildings" valuation which is supposed to represent the value of the land and buildings element only, excluding trade fixtures, fittings, furnishings etc.

Practice Statement 2.7.4 of the RICS Appraisal and Valuation Manual 1996, refers to such requests and recommends that the valuer can provide informal apportionments between the different elements contained in the valuation, but must emphasise that it is a hypothetical exercise and that the individual figures do not represent the open market value of the elements involved, since the true valuation can only be the figure taken as a whole.

Estimated Restricted Realisation Price

Valuers in the past were often asked to provide a "forced sale value" but this is no longer a basis of valuation supported by the RICS. It was replaced by "value in default", and now a valuer may be asked to give an estimated restricted realisation price for the existing use as an operational entity valued having regard to trading potential (ERRPEU). The RICS Appraisal and Valuation Manual 1996 defines this as:

> an opinion as to the amount of cash consideration before deduction of costs of sale which the Valuer considers, on the date of valuation, can reasonably be expected to be obtained on future completion of an unconditional sale of the interest in the subject property assuming:
>
> (a)   a willing seller;
>
> (b)   that completion will take place on a future date specified by the client (and recorded in the Valuer's Report) which does not allow a reasonable period for proper marketing (having regard to the nature of the property and the state of the market);
>
> (c)   that no account is taken of any additional bid by a purchaser with a special interest;
>
> (d)   that both parties to the transaction will act knowledgeably, prudently and without compulsion; and
>
> (e)   such of the following as the client requires (and the Valuer shall state in the Report):
>
> > (i)   accounts or records of trade would not be a available to or relied upon by a prospective purchaser;
> >
> > (ii)   the business is open for trade;
> >
> > (iii)   the business is closed;
> >
> > (iv)   the inventory has been removed;
> >
> > (v)   the Justices' or other licences, consents, certificates and/or permits are lost or are in jeopardy;
> >
> > (vi)   the property has been vandalised to a defined extent.

Conclusion

Hotels are a specialist class of property which normally change hands in the open market as fully operational businesses at prices based on the trading potential for their existing use. It is essential to undertake a full study of trading accounts whenever possible and to be knowledgeable of the hotel trade and market.

Chapter 19

Leisure Properties

Introduction

The last decade of the second Millennium marked a time in which the practice relating to the valuation of leisure properties changed enormously as a direct consequence of wider social and economic changes that had gradually been taking place for more than two decades prior to that time.

Earlier decades had witnessed the growth of the service industries and consequent changes in both the specification of, and investor attitudes towards, offices and retail premises; the 1990s saw the manifestation, in property terms, of social and economic changes which resulted in enormous growth of commercial leisure activity. By the end of the century the proportion of average consumer spending devoted to leisure had risen significantly so that, for the average household, it now forms one of the major items of expenditure. The growth in spending on leisure has been disproportionately higher for those people in higher income brackets, for whom it is now the largest item in the weekly bill. As people have devoted more money to leisure so they are expecting better quality facilities and this is very important in value terms. Poorly specified units become obsolete very quickly.

In employment terms, leisure has become of increasing importance with some 20% of all new jobs in recent years being created in the leisure and tourism sectors. All these changes, the product of rising affluence and a shift in economic terms from dependency on primary and secondary industry, have led inevitably towards increases in both the supply of, and demand for, leisure properties.

However, in planning and governmental terms some of these market factors have not yet led to corresponding changes in the statutory framework and this has presented valuers with certain problems. It has also reinforced the necessity for those charged with the valuation of leisure properties to have a good understanding of the market context and the fundamental drivers of value to the leisure operator.

It must also be borne in mind when valuing leisure properties that not all leisure activities are based upon commercial enterprise. Leisure embraces a range of non-commercial activities and this presents a challenge to valuers in terms of methodology as will be outlined briefly below. To illustrate the problem: a multiplex cinema is clearly a commercial enterprise built with the intention of providing profit in the hands of both the operator and the property owner. A community theatre on the other hand is operated for reasons of civic pride and commitment to the enterprise and not for any financial gain to the owner, although the operator will almost certainly require profit. Similarly, and perhaps presenting even more of a difficulty for the valuer, a private sector health and fitness club which is a purely profit driven operation may, at first glance, look very similar to a local authority owned scheme which offers similar facilities but on closer observation the pricing policy may reveal very different ownership agendas.

It can be seen from this introduction that some special issues arise in relation to the valuation of leisure properties. These relate to the wide variety of properties that come under the umbrella heading of "leisure" through to the method of valuation that is appropriate depending on whether the property is, or is not, owned for commercial purposes.

Types of Leisure Property

The question of what are leisure properties is almost unanswerable. To say that they are simply those used for leisure purposes is far too all embracing to be useful for the valuer. Instead it is helpful to consider them under certain categories:

- *Sport and recreation*: including health and fitness, sports stadia and golf;
- *Entertainment*: including cinemas, bingo, bowling and family entertainment centres;
- *Food, drink and hospitality*: including restaurants, bars, public houses, hotels and holiday villages; and
- *Culture and heritage*: including art galleries, museums and historic houses.

This list is certainly not exhaustive and it can be seen that within each of these main categories a wide variation of property types exist, both commercial and non-commercial. Therefore, as well as considering what basic type of property is to be valued, which is necessary if economic analysis is to be undertaken, it is appropriate for the valuer to review also whether the property is owned primarily for:

- *Commercial reasons*: such as a bowling alley or amusement arcade;
- *Social reasons*: such as a local authority swimming pool;
- *Furtherance of a self-help group*: such as a local cricket club; or
- *Free public access*: such as a public park.

By considering both the type of property and ownership motivation the valuer has begun to determine the most appropriate method of valuation.

If the premises are owned commercially, the valuation method chosen will usually be either the profits approach or investment method. If they are held for any other purpose, evidence to support these methods might be limited and a fundamental approach using all available methods, including a cost approach, might be considered.

In this chapter the emphasis is placed on commercial entertainment properties as this sub-category is the one with which the practising valuer is most likely to be concerned. No specific reference is made to public houses and hotels as they form the subject matter of other chapters.

Planning

Under planning legislation, critical factors in relation to the valuation of leisure properties are found not only in the main statutory provisions but also in the statutory instruments and in the policy guidance notes (PPGs). The most important secondary legislation is the Use Classes Order 1987. This must be read in conjunction with the General Development Order. Under these provisions it is clear that there is no one categorisation of leisure property. Some are included within the A (retail) use class as A3 (food and drink) while some are category C (hotels and boarding houses). The majority of entertainment properties, such as cinemas, concert halls, bingo hall, casinos, dance halls and swimming pools come under the assembly class D as D2 (assembly and leisure) but still others (for example theatres and caravan parks) are *sui generis*, that is, considered as individual cases.

Restaurants and food and drink units are often regarded as leisure properties, possibly because they often generate their income from leisure spending and many are located within leisure parks or are regarded as leisure units within shopping schemes. However, for planning purposes strictly they are a sub-group of retail premises and within the Use Class Order any property with A3 consent may, without the need for express consent, be used for A1 (general retail) or A2 (financial services) use, provided no material physical change is in contemplation. One caveat must be added. Where an A3 use has been granted, but not exercised, there is no ability to convert to A2 use.

It is important that the valuer of any leisure property has a good appreciation of the planning situation and, in the case of an A3 unit, an awareness of whether or not a premium value attaches to the A3 use. This will depend on market conditions and in many locations and in certain circumstances an A1 or A2 use may generate a higher level of rental value. If this is the case the alternative use should be considered subject to consideration of, for example, tenure constraints.

Licensing

Many leisure properties rely on a variety of licences in order to trade. Often these will be licences to sell alcohol and for public entertainment but there are also many other licences, such as gaming, upon which the trade depends. Most leisure premises require a Fire Certificate and the valuer should investigate fully to ensure that this and all other appropriate licences are in place. Almost more importantly, it should be ascertained that nothing has occurred, either physically to the building, or by way of management difficulties, that would prevent the renewal of the licences, as many are annually renewable.

Of particular concern will be the question of late night licences. These are often sensitive matters and if there are any objections due to noise or unruly behaviour the police may well object when the licence comes up for renewal. At the time of writing the government are undertaking a review of licensing laws, which it has been argued, are archaic and out of touch with the needs of modern society. The valuer should therefore ensure up-to-date knowledge of both current and proposed licensing legislation which could materially affect the trading potential of the unit.

Tenure Restrictions

At the time of taking instructions the valuer should check what interest is to be valued and how far any investigations as to title should be undertaken. Although the question of tenure will normally be a matter of fact that the valuer can take as given, it is important that appropriate checks as to title are undertaken either before the property is inspected or afterwards. Additionally, when inspecting, the valuer should be alert to any potential problems such as rights of way or access difficulties.

Where the property is subject to a lease this should always be scrutinised carefully and particular note made of any onerous conditions such as restrictions on assignment or change of use. The position in relation to licences and the assumptions at rent review should also be examined carefully.

Until recently, many leisure properties were let on individual styles of leases, with varying lengths. However, with the growth of the leisure parks and the number of multiple operators, so greater standardisation of lease terms has occurred. It is not unusual for leisure properties to be let on terms of 25 years, or even longer, and on standard institutional terms (that is on full repairing and insuring terms and five-yearly upwards only rent reviews). It should be noted that, due to concerns over the availability or otherwise of comparable evidence and of the lack of investment return data, many institutions have insisted on additional protection to the landlord at rent review. Therefore if the property is let on a lease granted in the 1990s it may well have a review clause which ties the reviewed rent to the *higher* of open market rental value or the initial rent plus 3% pa compound increase. Where such a property is being valued, care should be taken to ensure that the onerous nature of the clause is taken into account.

Leisure properties frequently require fitting out to meet operational needs. Depending on the nature of the business this can involve very high capital cost. Accordingly, tenants may require any or all of the following:

- A substantial rent-free period to amortise the costs;
- the right to make subsequent alterations to the premises;
- a lease of length sufficient to allow for full benefit to be gained from the work; and
- reflection of the fit out to be ignored at rent review.

These factors may have significant effect on both rental levels and yields.

RICS Regulatory Framework

All valuers will be aware of the provisions of the RICS *Appraisal and Valuation Manual* and the Practice Statements contained therein. These apply to the valuation of leisure properties as to other valuations and therefore need no general comment. However, in addition to the general provisions, there are provisions that relate specifically to leisure properties and which should be noted by valuers of leisure properties.

The special provisions are included as references to "trading properties" throughout the practice statements in the Red Book. Additionally the provisions of Guidance Note (GN) 7 [Trading Related Valuations and Goodwill] and GN 9 [The Capital and Rental Valuation of Restaurants] should be known.

The particular provisions of the mandatory Practice Statements (PSs) that relate to leisure properties all make the assumption that the property will be valued in relation to trading potential. It lists properties for which it makes that assumption and the list includes, for example, cinemas and bowling alleys. In relation to trading properties the PSs point out that:

- PS 2: care must be taken in taking instructions to clarify the assumptions in respect of fixtures, fittings and licences as these are often important in value terms. It is also pointed out that *personal goodwill* will normally be excluded;
- PS 4: in the case of valuations for Existing Use Value (EUV), although personal goodwill is excluded, the trading potential "which is inseparable from the interest in the property" is included;
- PS 7: when reporting, if the valuation is based on a valuer's estimate of trading revenue and costs and *not* comparable evidence, a statement to that effect must be included;
- PS 12: for valuations for financial accounts purposes, the valuer is instructed that for existing use valuations (EUV) the valuer "will have regard to the trading accounts for previous years, where these are available, and will form an opinion as to the future trading potential and level of turnover likely to be achieved".

The most comprehensive reference to leisure properties occurs in GN 7 which states that most leisure properties are sold as fully operational business units and valued in relation to their trading potential. This is currently open to debate given the rapidly changing market for many types of leisure property, but nevertheless is the extant guidance. It also remains the case that many types of leisure properties do continue, as before, to be traded between operators as occupational units. For these properties the Guidance Note is very informative and provides, in detail, the approach to inspection and analysis of accounts that should be undertaken. It also makes clear that, provided the appropriate information in relation to trade, outgoings, headquarter expenses, competition, etc. is collated and analysed, the valuer is free to choose which *method* of valuation is then considered appropriate to follow. It states that "different methods, including the profits method, Discounted Cash Flow (DCF), analysis of comparable transactions or a combination to these will be appropriate for different types and sizes of property" (GN 7.2.2).

While this advice is *not* mandatory, failure to heed it might, *prima facie*, lead to a claim of negligent behaviour.

The Guidance Note also addresses the question of goodwill. The distinction is made between the value of the trading potential as assessed in accordance with all the factors that determine the trading potential in the hands of an efficient operator and the trade that might attach due to personal reputation, product line or brand success. There is implicit acknowledgement that it is sometimes difficult to differentiate the two, but that this task is within the role of the valuation surveyor.

Where personal goodwill does exist it can be capitalised and transferred – but not included in any valuation for financial statements.

Lastly, GN 9, which relates to restaurants, should be consulted for such valuations.

Collating the Evidence and Referencing

Fundamental to any valuation is appropriate research. PS 4, GN 1 and GN 7 of the RICS *Appraisal and Valuation Manual* give clear guidance on the matters to be investigated when carrying out any valuation. As many of the newer leisure developments have been constructed on "brownfield" sites, particular cognisance should be taken of PS 3.1 and GN 2 regarding investigations concerning possible site contamination.

As well as the standard enquiries two particular sets of factors should be taken into account:

- those which affect profitability of the enterprise in the hands of an operator; and
- those which determine whether the property is likely to be of interest primarily to investors or to owner-occupiers or indeed, whether the nature, layout or location of the property render it of potential interest only to non-profit motivated owners.

Physical Factors

The physical specification of the property is important and the valuer must be aware of the likely operator requirements based on the type of property and location.

Unit Size

The individual requirements of leisure operators are extremely variable and in the past it has been almost impossible to standardise the required specification. However changes during the 1990s meant that some attempt at standard specification can now be made. Figure 1 overleaf sets out an example of the size and location criteria generally sought by operators of commercial leisure properties as at the late 1990s.

It should be noted that operator requirements change rapidly due to the fast evolving nature of the leisure markets. Valuers should ensure that they are fully conversant with latest trends and requirements.

Location

The location must be appropriate for the user. Most forms of commercial leisure property require ease of access and prominent visibility but of equal importance may be proximity of complementary uses. For a commercial leisure unit to operate successfully it is often of paramount importance that it can synergise with other adjacent uses; hence the growth in popularity of the purpose built leisure parks many of which are located adjacent to, or integrated with, retail schemes. Indeed the growth over the last decade and a half of such units has placed great pressure on the viability of "solus" units in terms of market appeal and hence profitability. On the other hand some types of leisure operation can trade successfully in non-purpose built units and, indeed, unusual elements to the fabric can enable a "niche" trade to become established.

Figure 1. Standard Specifications of Leisure Users

Property type	Size (m^2)	Standard specification Location	Building type	Comments
Intensive leisure users				
"Superpub"	3,500–5,000	Town centre or high density suburb	Conversion or new build	Dominated by the major brewers and the "new" multiple chains
Themed restaurants	1,000–1,500	Town centre, strong retail position	Any – unusual buildings liked	Brewer led but with increasing specialist chain operators
Fast food and budget brands	300–1,000	Suburban, provincial or leisure park	Standard retail style unit	Multiples and major brewers setting up national chains
Nightclubs & discotheques	500–3,500	Elite "one-off" location or within leisure complex	New build or warehouse conversion	Mainly independents, few multiples
Family entertainment centres	Varies 500–3,500+	Leisure Parks	New build primarily	Global interest including US, Japanese and UK covenants
Multiplex cinemas	3,000–9,000	Leisure Parks or in-town	New build	Dominated by 8 main chains; site and screen requirements increasing
Bowling	3,000–6,500	Leisure Parks or in-town	New build preferred	Access and grouping with other leisure users important

	Size	Location	Build	Ownership
Bingo	3,000–6,500	Leisure Parks or in-town	New build preferred	Access and grouping with other leisure users important
Health & fitness	1,000–4,500	Leisure Parks or "solus" operation in-town	New build or conversion	Some development of national brand – but nature of use mixes less well with other leisure (Camp 1996)
Amusement centres	150–500	In-town or leisure park or seaside	Any	Some major chains but many independents
Extensive leisure users				
Budget Hotel	3,000+	Motorway or trunk road	New build	Major chains – some international
Leisure/country hotel	9,000+	Ambience, not access, important	Historic/refurb preferred	Mix of independents and chains
Theme parks	20 + ha	Accessible	Requires feature Planning an issue	One main chain, all owner-occupied. Not "speculative"
Golf courts	2.5–6 ha	Adjacent residential or hotel	New build	New concept – as yet little tried – owner occupation
Golf course	60+ ha	Undefined	New build	Held both commercially and municipally

Source: Sayce, 1998

The presence, or otherwise, of car parking should be noted together with proximity and adequacy of public transport facilities. Car parking will be particularly important where either public transport links are weak or where the unit relies on late night trade.

The valuer should also be alert to any proposed changes to traffic planning policies at both national and local level which could adversely affect trade.

Trading Potential

Whether or not the property is to be valued by reference to a full profits approach valuation, any competent leisure valuer should assess the trading potential of the property in the hands of an efficient operator in order to estimate the "fair maintainable trade".

This process begins with the inspection. The valuer should note carefully all details of the existing trade such as admissions price (if applicable) and general tariff together with those factors which affect the ability to build revenue. These include layout, facilities and potential for expanding business within the existing envelope as well as by development. Note should also be taken of any disadvantages such as poor configuration which could adversely affect trade.

Where the property is trading it is normal for access to trading accounts to be supplied but the valuer should always treat these with caution, as, even in the case of correctly audited accounts, levels of trade in the past may not be a good indication of future revenue flows. Additionally, when inspecting accounts, the valuer should take careful note of items that might be allowable for accounting purposes but are personal to the operator and would not necessarily be replicated by other operators. These include items such as depreciation, bank charges and proprietor's drawings. Conversely, the accounts may not show an accurate picture of all items that are necessary to fully cost the business. For example, in the case of family run businesses it is very common that inadequate amounts are entered in the accounts to reflect the true wage costs. At all times, in analysing the accounts the valuer should use experience and judgment to determine whether the current operation is that which could be achieved by an efficient operator. All elements of either under- or over-trading should be disregarded. It is the maintainable trading position and its profitability which is sought.

Increasingly, leisure units are operated by either franchisees or by multiple operators as part of branded operations. In these cases the question of trading potential becomes especially problematic as the trading profile may be closely linked to the quality of the operator and the brand image – rather than the intrinsic property quality– yet it is the latter that the valuer must try to establish by stripping out any element of trade that is deemed to relate to the brand. However, if the valuer deems that the property is one that would be likely to attract a range of different, competing branded operations (as may be the case with cinemas, nightclubs, bingo halls, etc.) then the element of turnover which relates generically to the ability of the *property* to attract a multiple operator can be included. This is, perhaps, one of the most difficult dilemmas facing the valuer in assessing the trading potential and it should not be underestimated.

It follows therefore that when inspection takes place it is important that the valuer obtains as much information as practicable in relation to the trade in order that a true assessment of potential is made.

In addition to careful inspection of the property the valuer ought to take into account any technical changes which could affect the design and trade of the unit in the future. For example, in the case of night-clubs the valuer must be aware of the latest technology used in sound systems so that a judgment can be made as to how the property compares with those with which it is in competition.

It must be remembered that not all leisure properties trade at a profit. There are many units that are owned and/or operated for social reasons. For example, local authorities own many leisure properties, particularly those that fall into the C1 and D2 use classes – such as theatres and leisure centres; also many sporting and recreational clubs operate leisure properties. The common theme for these is that "the bottom line" is not the reason for the ownership and the trading and tariff policies reflect this. Where a valuer is instructed in relation to any non-profit orientated properties, great care must be exercised and consultations with the client should take place in respect of valuation assumptions in relation to trade.

Comparable Evidence

One of the basic tenets of every valuation is that all comparable evidence must be obtained and analysed. This applies to leisure properties as to any other class of property, even if a decision has been made to value by reference to trade potential. However, the collection of comparable evidence may be more difficult than with, for example, a shop or office, as the amount of stock is less and its location more diverse. For example, there are unlikely to be more than one or two multiplex cinemas in any one town – barring the major conurbations. Similarly, bingo halls and nightclubs are not grouped in the way that office blocks are!

Therefore the conventional wisdom has been that comparable evidence must be gathered from a wider geographical base where it can be gleaned at all. This can make analysis difficult as so much of the trading potential of any unit relates to the demographic and economic profile of the area. However, due to the recent growth in the number of commercial leisure schemes the problems in obtaining comparable evidence has declined dramatically for *some*, though not all, types of leisure properties.

In some cases it may be possible to obtain rental evidence from within a single scheme or from other schemes offering a similar range of outlets. A typical leisure park scheme, and there are some 100 in the UK at the end of 1999, would be anchored on a multiplex cinema with bingo, bowling and A3 units dominating the rest of the park. Most recent schemes may also incorporate health and fitness and/or late night venues.

For other types of property there is still a dearth of comparable evidence and for some types of property that appeal to the international market, comparability may take on a cross-national perspective.

Measurement

For most commercial properties, it is usual to measure the property in accordance with the RICS *Code of Measurement Practice* (4th ed). This best practice guide sets out various methods of measurement: gross external (used for insurance purposes), gross internal (used for industrial premises) and net internal (used for retail and offices). The problem with many types of leisure property is that there is no generally agreed basis of measurement. This makes analysis very difficult. Some properties are not normally valued in relation to their floor area but others increasingly are so valued. However, there is no consensus regarding the applicability of zoning or indeed whether gross or net dimensions should be adopted. The practice is evolving fast. At the time of writing commercial leisure units such as cinemas are normally valued on a gross internal basis, but excluding projection rooms and fire exit routes. However, before undertaking any valuation the valuer should ascertain any local practice and ensure that all analysis is carried out on the same basis as is proposed for the valuation.

Appropriate Method

Until very recently the valuation method adopted for leisure properties was almost universally the profits or accounts method. This is referred to in the RICS *Appraisal and Valuation Manual* as a valuation in "relation to trading potential". However, the growth of an investment market in some forms of commercial properties, principally urban entertainment uses such as bingo, cinemas and food and drink units, has shifted the method used towards the investment method in some cases.

This change in context means that before undertaking either a capital or rental valuation the valuer should determine:

- Whether it is operated for profit; and
- Whether the property is likely to be of interest to an investing owner, and let to a major covenant.

If the answer to both these enquiries is "yes" then the most appropriate method is likely to be the investment method but if the answer to only the first is "yes" then a profits valuation should be undertaken.

If the answer to both is "no", then the valuer may, in the absence of any market evidence, have to choose between adopting some form of revenue approach, accepting that a low figure will result, or contemplate use of a cost approach. This however is very much a specialist area and valuers should only attempt such a valuation if they have the requisite experience base.

Accounts Approach to the Valuation of Leisure Properties

The accounts or profits approach is used to establish both rental and capital values for a wide range of trading leisure properties.

Full Accounts Approach to Rental Value

For **rental valuations** the first task is to establish the maintainable trade in the hands of an efficient operator. Often the starting point will be the actual detailed trading accounts. Ideally trading forecast figures will also be available. In the case of properties which are new build, recently refurbished, or vacant due to default or other reasons trading accounts will not be available. The valuer will have to rely on his/her own assessment. Even where figures are available it should be remembered that it is not the *actual* business that is being valued – but the ability of the property to support a business. Therefore great care should be taken at inspection to ensure that all evidence of under- or over-trading is noted (for example, is the tariff appropriate and is the management of high quality?). In forming an opinion as to the maintainable trade every opportunity should be taken to consult staff and managers as well as inspect competing operations.

Once an estimate of the *maintainable* level of trade has been established which the valuer considers to be a fair reflection of all the available evidence, the next step is to examine the expenditure which relates to the estimated turnover. Again the actual accounts can be the best guide to the maintainable level of outgoings, but a critical assessment should be made so that any extraordinary items or items personal to that individual operator are excluded from the calculation and should be "added back" (see below).

The accounts should also be adjusted, if necessary, to allow for a realistic amount for wages in the case of, for example, a family run operation. It is helpful in estimating this to consider the expenditure under two main headings:

- The variable costs; and
- The fixed costs.

The **variable costs** are those items which will vary directly with the level of trade achieved, such as the cost of food and drinks purchases in the case of a restaurant. Therefore where the valuer has made adjustments to the level of turnover recorded in the accounts care should be taken to adjust the variable costs accordingly. It is also important that unit costs are appropriate to the style of operation being undertaken.

The **fixed costs** are those which the business must carry whatever the level of trade, such as rates, heating, repairs etc. In examining the fixed costs, the valuer should ensure that the costs that are likely to be incurred in an average year are used and that any extraordinary items, such as periodic major repairs, are excluded from the calculation but are included, nominally, in the annual average.

Two other major adjustments should be undertaken by the valuer. These are: adding back and adjustment for sinking fund.

Adding back

To the notional accounts a series of "add back" adjustments are normally made. These include any personal items related to the particular operator and not connected with the notional business – such as interest, loan and bank charges, depreciation costs, director's remuneration and, in the case of multiple operations, head office costs – provided that the equivalent "solus operation" costs are included. Wages and salaries should also be adjusted – either up or down – to provide figures which would be appropriate to the operation of the business on a purely commercial basis. This can often be difficult as many small operations would not be viable if full labour costs were included. In this case, the valuer must exercise judgment; if a likely hypothetical tenant would bid for the premises working on the assumption that family members would help to run the business unpaid, then it is reasonable to build this assumption into the figures. However, if it is necessary so to do to reach a figure of viability then the long term future of the enterprise must be seriously questioned.

Adjustment for sinking fund

The notional accounts should be further adjusted to include a reasonable amount to cover:

- A sinking fund to allow for the periodic replacement of trade fixtures, fittings and equipment; and
- A return on the notional capital invested in the business (that is to allow for the acquisition of the fixtures and fittings and to supply working capital).

The fully adjusted accounts will reveal a level of **adjusted net profit**. This figure is that which a reasonable competent operator could reasonably expect year on year to be sustained as at today's trading performance and as at today's prices. It represents the amount that is available for:

- Rent; and
- Profit in the hands of the operator.

If the valuer is seeking to establish a rental value the next step is to decide how much of the **adjusted net profit** would be bid away as rent. Typically this may be in the order of between 35% and 60% of the profit level. In determining what percentage to be adopted as the rental value the valuer should consider:

- The likely level of demand for the unit from other potential operators;
- The future prospects, based on trends in performance;
- The risk of profit being eroded (note where the fixed costs are a high percentage of total costs, the profit is vulnerable in a downturn but potential is great in an upswing); and
- The total level of profit and its relationship to turnover.

In relation to the last point it should be borne in mind that different types of operation have different "norms" of level of net profit as a percentage of turnover. Any operation where the net profit falls below say 15% of the turnover could be regarded as highly risky.

Alternative Approaches to establishing Rental Value

In many ways the estimation of a rental value from a fully adjusted set of accounts is the truest approach to establishing a rental value as the relationship between the occupier's benefit from use and the value ascribed is transparent. However, it is very subjective and, in the hands of an inexperienced valuer, can be misleading. Therefore where it is possible to establish rental value by analysing comparable transactions, this will normally be preferred. At the time of writing the growth of leisure parks development and urban leisure schemes has created a whole range of new stock that has been let on lease, so obtaining evidence for standard commercial leisure premises is now much easier than it was only five years ago.

Even though evidence may be found great care should be taken to establish that true comparability does exist in terms of specification, location, access and in lease details. If comparables are used the valuer should make every effort to read the lease as there are considerable variations in the types of leases used for leisure properties.

The following clauses are critical:

- *Term length*
  Currently it is not unusual for leisure properties to be let on terms of 25 or 30 years. Given that practice for other commercial properties is in the order of 15 years, this represents a significant advantage to the landlord and potential burden to the tenant, especially if assignment requirements are onerous. If they are, the rental level could well be depressed.

- *Repair*
  Modern practice will normally dictate that the lessee is responsible for all repairs and redecoration work but the valuer should carefully examine the clause and ensure that no unduly onerous provision is included.

- *Improvements and fit-out*
  For many leisure properties the cost of fitting out can be substantial; there is also a need to refurbish more frequently than with many other types of property. Not only will the fabric require on-going commitment to alter and update as fashion trends demand, but also the operator will require to spend on fixtures and fittings. For this reason the well-drafted leisure lease should make adequate provision for the lessee to carry out works of improvement and alteration to the premises as trading conditions require, but care should be taken to ensure that the value of such improvements are excluded at rent review. Therefore the valuer should check the provisions regarding alterations and improvements carefully and note any inter-relationship with the rent review clause. Due to the costs of fitting out it is common practice for lessees to be granted either rent-free periods or capital contributions at the commencement of the term. In such cases the value of the work may well be included at rent review and the valuer should check this position.

- *User clause*
  As stated above, leisure comes under a number of different use classes. Care should always be taken when assessing a rental value to check the permitted planning use and note whether or not this is related to any particular planning use class.

- *Licensing and other consents*
  As with user clauses, the valuer should check whether the lease places any special requirements on the lessee in relation to the gaining and keeping of licences or other consents as these can materially affect value.

- *Rent review provisions*
  The rent review provisions in any lease will be critical in establishing rental value, both initially and at review. Over recent years the use of standard lease drafting has led to many review clauses being five yearly upward only to Open Market Rental Value. With many commercial leisure properties concerns about the likely availability of comparable evidence at review has led to the inclusion of many special provisions

being included. This is especially so with D2 and *sui generis* properties. In these cases the review clause may include reference to the greater of *either* Open Market Rent Review *or* a prescribed increase, typically 3% pa compound. Other formulae sometimes used are to: open (A1) retail use; a percentage of turnover some other prescribed use or by reference to other retail or office rents in the locality. All these formulae, of which the escalator clause appears to be the most common, are reflective of concerns about lack of evidence. If any of these are present the valuer should consider whether they are sufficiently onerous as to reduce the rental value.

Where a rental value has been established by reference to comparable letting evidence it is still advisable to check that the rent is sustainable by reference to profits – if it is feasible so to do. Another approach sometimes adopted to find rental value is by reference to gross turnover, with the rental value being expressed simply as a percentage of take. This approach has several advantages. It

- is simple to ascertain;
- is not open to argument as it can be related to audited accounts and it is difficult to manipulate in a non-transparent way; and
- can form a basis for rent review which encourages the lessee to maximise profit – as the containment of costs will not lead to higher rent.

However set against these advantages are several disadvantages:

- If costs escalate due to inflation but without profitability improving, this will act against the tenant;
- The relationship between profit and turnover is not static over time or between types of operation and thus it is at best a crude measure; and
- It does not take into account issues of improvements or lease terms generally.

Establishing Capital Value

There are several approaches to establishing the capital value of commercial leisure units. These are:

- The profits approach;
- The investment method; or
- Capital comparison.

The profits approach

The profits approach can be used either where the property is owner occupied or likely to transfer between occupying operators or in the event of let premises. It is most commonly used when the premises are owner occupied.

There are three variants in use:

- The single earnings multiplier;
- The dual capitalisation approach;
- Discounted Cash Flow.

Single Earnings Multiplier

The single earnings multiplier is the approach most commonly adopted for commercial leisure properties that are sold to an operator rather than an investor. It assumes that the capital value is related directly to the profitability of the premises in the hands of an efficient operator. It builds on the full profits approach to finding rental value (see page 613). The starting point is the adjusted net profit.

The method appears deceptively simple as in essence it is the net profit multiplied by a single figure to find a capital value, as set out below:

Maintainable Net Operating Profit (NOP) Say	£300,000
(arrived at after full adjustment of accounts)	
YP @ Say	6
Capital Value	£1,800,000

It is obvious from this valuation that the choice of YP is extremely critical as a slight error at this stage can lead to enormous error in the capital valuation. This is probably the most important single reason that leisure valuations are regarded by many as being the province of specialist valuers.

It follows that it is important to understand what drives the choice of YP. First it must be stressed that research has revealed that there is no consensus among valuers as to the nature of the YP in the case of the single earnings multiplier. To some, it is a YP in perpetuity at a property yield (in the case above approximately 16.5%); to others it is the YP for an assumed term of the life of the business (for example 6 is approximately the YP for 10 years at

10%). To yet others it is a simple mathematical device. For many experienced valuers dealing with this type of property the tendency is to go straight to the capitalisation factor, which will typically be in the range of 5 to 9.

The valuer, in choosing the multiplier, should have regard to a number of factors:

- **Comparable evidence**
  Due note should be taken of any comparable market evidence, although unless the valuer has first hand information of the deal he or she is unlikely to be aware of the level of adjusted net profit to which a capitalisation factor was applied. This limits the use of comparables.
- **Likelihood of sustainability of the profit**
  To establish this a careful note should be taken of the profit trends and analysis of competition and other external factors. It is also important that the valuer assesses any issues of branding and considers whether the level of maintainable profit is independent of any brand imaging considerations.
- **Level of demand**
  The valuer should consider the state of the market and the availability of other premises on the market.
- **Development potential**
  The valuer should assess whether the possibility exists to further exploit the trading potential of the unit either with, or without, major capital expenditure and adjust the multiplier accordingly.
- **Profitability of the underlying operation**
  This is very important as the value of any trading leisure property relates to the profitability of the underlying activity. For example at the time of writing cinema admissions have undergone very significant growth over a fairly long period. This has reflected both in rentals achievable and in capital values. However for profitability to continue, not only must the property specification be right but also the type of films screened must be capable of attracting audiences. A downturn in the film industry and shortage of "blockbusters" could quickly adversely affect property confidence and, hence, capital values.

In making the judgment of what capitalisation rate to use the valuer should, at all times, ensure that no element of double counting with adjustments made to the accounts, has taken place.

For example, if the level of turnover has been adjusted upwards to reflect the presence of an inefficient operator and the potential for increased trade, it would be important that the capitalisation factor is not increased – or yield decreased – to reflect this fact. Similarly where the expectations of revenue have been muted due to the likelihood of a new competing development depressing trade, care should be taken before this factor is built into the capitalisation. At all times it is important to take a *reflection* on the trading figures and ensure that they are used to guide the valuer to an appropriate figure and not used as the only evidence of value.

Although hotels do not specifically fall within the remit of leisure properties the debate that has taken place in recent years within the profession regarding the most appropriate method of valuation to be applied is equally applicable to leisure properties. This debate, led by BAHA (the British Association of Hotel Accountants) in the early 1990s as a call for the adoption of DCF, led the RICS to conclude that for many units the single earnings multiplier was the most appropriate method of valuation. There is acknowledgement that for city centre hotels and, by implication, many large-scale leisure units, DCF may become the preferred method (see below).

For leisure properties where the single earnings multiplier is used it is important that valuation checks (see below) are carried out to add confidence to the resultant figure.

The Dual Capitalisation Approach

The dual capitalisation builds directly on the full profits approach to finding rental value. It is favoured by some valuers as providing the "truest" method of establishing the market value as it separates out the value of the purely property element and the value attributable to the trade. It is used normally where the market is dominated by owner operators because the very construction of the model implies that there is an addition to capital value over and above that which an investing owner would pay for the premises.

The starting point is to take the two elements of the divisible balance (rent and residual profit) and to capitalise each separately. The resultant figures are then added to give the capital value. So, taking the example used above, the calculation becomes:

Stage 1: the divisible balance

Maintainable Net Operating Profit (NOP) Say		£300,000
(arrived at after full adjustment of accounts)		
Annual rental value at say 55% of NOP		£165,000
Deduct from NOP		
Therefore residual profit		£135,000

Stage 2: capitalisation

Annual rental value	£165,000	
YP in perpetuity @ Say 11%	9.091	
Value of rental slice		1,500,015
Residual profit	£135,000	
YP @ Say	2	
Value of residual profit slice		£270,000
Capital Value		£1,800,000

The rationale for splitting the two elements is that the rental income slice is more secure and is capable of being capitalised at a property yield, whereas the residual profit is less secure. In assessing the yield to be applied to the bottom tranche of income, the valuer should take cognisance of all the factors that have been detailed above in relation to the quality of the property as an investment, but excluding consideration of the strength of the covenant must be. This method of valuation is used *only* where the anticipation is that the most likely purchaser in the open market would be an operator. If an investment sale of the freehold interest is in contemplation then the use of the dual capitalisation approach is inappropriate. An investment method must be adopted.

In assessing the "top slice" multiplier (not yield) which is typically in the region of 2 to 5 YP the factors listed above become paramount in relation to the potential for growth of the profit stream and the likelihood of sustainability of profit in the hands of an efficient operator. It must be stressed that the residual value in the hands of the operator does *not* represent personal goodwill; it is the residue that an efficient operator would expect to make over and above the amount "bid away" as rent. As such the valuer should acknowledge that it is part of the property valuation and not something that adheres to the occupier – it runs with the land.

There is one potential situation in which the valuer may need to make an adjustment to the valuation. In the event that the premises to be valued are new or for some other reason have no recent trading record, the value of the projected residual value in the hands of the operator may be depressed, as upon transaction the operator will have to establish the trading reputation of the premises. This additional risk factor can be taken into account by making an adjustment to the multiplier used to capitalise the residual profit. In these circumstances, the use of the dual capitalisation approach is more flexible than the single earnings multiplier as it enables the valuer to reflect accurately within the valuation both the quantum and the nature of the trading risk.

Theoretically the capital valuation arrived at by use of the dual capitalisation approach should be reconciled with that achieved by use of the single earnings multiplier.

The use of the dual capitalisation approach brings into sharp focus the notion that such a unit is potentially worth more to an occupying owner than a property investor, to whom the top slice of residual profit is irrelevant. This point is important because, as has been pointed out earlier, a strong investment market in leisure properties has emerged over recent years and there are now a significant number of investment transactions. The implication of this is that the use of the dual capitalisation method is likely to fall into disuse for establishing open market value of many types of urban leisure properties – although the approach remains relevant to establishing the existing use value. In any event the *principle* of deriving rental value by reference to a hypothetical tenant's ability to pay remains sound.

Discounted Cash Flow

Valuers are making increasing use of full DCF approaches to the capital valuation of leisure properties facilitated by have access to either industry standard software (such as Excel) or bespoke property programmes (such as KEL). Whenever a DCF approach is used two important principles should be recognised:

- The quality of a DCF calculation is only as good as the inputs on which it is based. If these are objective and market based it is simply a more flexible method by which capital value can be derived; and
- If the inputs are taken with reference to specific client requirements and without full reference to the market and to the factors set out above in relation to the profit's approach, what will be produced will be in the nature of a Calculation of Worth, as defined by the RICS *Appraisal and Valuation Manual*,

and not an Open Market Valuation. That is, it becomes a calculation that represents an evaluation of the net monetary worth of the asset to the individual client, based on *their* requirements and not necessarily reflective of market opinion.

In order to carry out a DCF appraisal not only must the NOP be calculated but also it must be calculated over a specific time period. The example below shows an annual basis but it can easily be adjusted to allow for monthly or quarterly cash flow periods. The cash flow should build in specific projections for changes in profit levels. Where an OMV is in contemplation (not a Calculation of Worth) any growth built in to reflect, for example, inflation must be related to the discount rate which is chosen, by analysis of external objective indicators, and represents the valuer's best opinion of the overall return relating to that class of property required by investing owners active in the market.

Having set the discount rate and made the explicit assumptions for cash flow the valuer should then determine the period over which the calculation will be run. In general, due to market uncertainty, it is unusual to take a time frame longer than say 10 years. This is because the specific projections on which a DCF calculation is based are difficult to establish with any degree of accuracy for periods beyond such a time period. At the end of the time period for discount it is necessary to establish an "end value" either by taking a capital sum direct or by calculating the last estimated profit at a realistic market yield or "exit" yield.

Once all the inputs have been established the discounted cash flow for each year (or period) can be computed and the sum of all of these will represent the net present value or capital value of the asset.

To illustrate the methodology the same example as used above is set out below to show a DCF approach, although in this case, for simplicity, the time frame has been reduced to six years. It is accepted that in practice this is likely to be longer, although seldom will it exceed 15 years.

Year	1	2	3	4	5	End
NOP allowing for growth @ 2.5% pa (000s)	300	307.5	315.2	323.1	331.1	339*
Capitalisation at exit yield of 12%						8.3

PV£1@ 18%	0.847	0.718	0.609	0.516	0.437	0.370
DCF	254.1	220.8	192.0	166.7	144.7	1,041.1

Total Present Value:	£2,018,569
Capital Value, Say:	£2,000,000

In the example above the "exit" yield has been taken as slightly above that which was applied as a market capitalisation rate for the dual income approach. This is because at the end of the set period the unit may be suffering from obsolescence. However such an adjustment is not always justified and the valuer should exercise due judgment in establishing the appropriate yield.

The DCF approach, as with any variation of the profits approach requires great care in execution, particularly as the very nature of the calculation means that it is difficult to obtain good objective comparable evidence. It therefore should not be undertaken without much experience. In many cases its use is best suited to the establishment of a Calculation of Worth. It is however useful to run a DCF calculation alongside another approach to gain a better overall picture of value.

Investment Method

The use of the investment method to establish the capital value of a leisure unit has become widespread over the last three years as institutional investors have bought into the market. In general their interest has been restricted to

- High street A3 users;
- Well specified and let leisure parks – preferably integrated with or adjacent to retail schemes;
- Well let multiplexes; and
- Other large commercial D2 units with strong covenants and tried and tested brand product.

Where such a unit is to be valued for capital purposes, the usual approach will be to simply capitalise the rental value, in the case of a rack rented unit, at an appropriate all-risks yield or, where a reversionary value is to be established, to capitalise using a hardcore approach and equivalent yield.

The yields that might be appropriate for prime leisure units have declined from around 9% in the mid 1990s to approaching 6.25% by the end of 1999. This places them on the hierarchy of yields at a little over good retail units and similar to retail warehousing and shopping centres. The very significant yield shift that took place demonstrates the realignment of some types of commercial leisure from a non-institutionally attractive sector, to one which can compete, as of right, for such funds. It must however be stressed that it is only the very best well let schemes that will be of attraction to investors.

In assessing the attractiveness to investors the valuer should be aware of the factors that will determine the attractiveness of the property to an investor. These principal factors can be listed as follows:

- **Depth of the investment market**
  The investor will wish to be assured that there are sufficient units of this type and within similar localities to ensure that comparability exists to support valuations and to ensure liquidity in the event that they wish to sell.
- **Quality of the covenant and potential occupational demand**
  Not only will the investor need to establish that the present or, in the case of a pre-let scheme, prospective tenant is a satisfactory covenant but also they will require assurance that there is a depth in tenant demand in the event of a proposed assignment to ensure long-term performance. Given that the last two years of the 1990s have seen enormous consolid-ation of leisure operators through mergers and acquisitions and that process is ongoing, this is a real investor concern.
- **Planning considerations**
  The fast moving nature of the leisure market means that there may be a need for the occupier to "refresh" the image and brand during the life of the lease. It is important that, as far as possible, the planning considerations are not likely to hinder this process. At a macro-level consideration of potential policy decisions (for example in relation to transport policy) should be "factored in". At the time of writing, concerns about the introduction of local transport plans is leading to investor preference for in-town schemes over car dependent edge-of-town or out-of-town schemes.
- **Design and obsolescence factors**
  Many early leisure schemes are showing signs of obsolescence; this in turn will reflect in poor rental growth prospects. The valuer should be aware that the very nature of leisure means that functional and design obsolescence are real problems that

can lead to quickly falling values. However, for some other types of leisure property, such as heritage schemes, obsolescence is not an issue and may add value! Currently however heritage schemes are not usually of any attraction to institutional investors who seek prime modern stock.

- **Location, accessibility and footfall**

  The location, accessibility (including visibility) and ability to generate footfall are all issues which will underpin the long-term viability of a commercial leisure property. Therefore these considerations and analysis of economic and demographic trends should be undertaken before a yield is chosen.

- **Brand image and tenant mix**

  The brand image can be important to an investor as it will be influential in creating an image for the whole investment. The presence of recognised brand names in a leisure scheme is just as important as in a shopping centre where the brand image of the anchor tenant and the appropriate trading mix are crucial factors to securing a good long-term investment. For a leisure scheme the mix of users that will create synergy is constantly changing. At the time of writing emphasis is shifting from purely entertainment uses (cinema and bingo and food and drink) to a mix of entertainment and health and fitness, reflecting the changing lifestyles of the affluent sectors of the population.

- **Lease terms**

  Imperative to institutional investors is a lease which provides the terms that have become established over three decades for other commercial premises. The nature of these has been discussed above and can be summarised as 20 to 25 years in length with five yearly rent reviews, upward only and possibly with a "ratchet" and on full repairing and insuring terms. It is sufficient here to emphasise that a property which is not so let is likely to be valued at a considerably higher yield and lie in the portfolio of a small or specialist investor.

It must be pointed out that a valuation carried out on the investment method will not normally be prepared for existing use value.

Capital comparison

Most commercial leisure properties will be valued using either the profits approach or, in the case of let schemes, an investment method. To an extent both of these rely on some measure of comparable evidence. It is unusual that a leisure property will be valued directly to capital comparison completely. This is because, either it will be let, in which the investment method is the preferred method or, in the case of owner occupied properties, the trading potential sooner than the physical specification will drive the valuation. However, there are instances where capital comparison can be used. If evidence does exist of transactions of similar properties with comparable trading potential then a simple analysis and comparison should be used although a check against a profits approach is advisable.

Valuation for Various Purposes

General

The valuation methods outlined above are those which are usually employed in the preparation of open market valuations (OMV) or existing use valuations (EUV) as defined in PS 4 of the RICS *Appraisal and Valuation Manual.* In many cases these valuations will be the same as a valuation for estimated realisation price (ERP) but valuers should note the definitions and adjust their calculations in accordance with the details of the definitions. It has been pointed out that adherence to the statements contained in the manual are mandatory and must be observed by valuers, unless there are specific, justified and clearly articulated reasons for a departure to take place.

In preparing a valuation, the valuer should also take account of the provisions of PS 7 which states the reporting and other requirements. This statement makes cross reference to PS 2 which sets out the requirements for the taking of instructions. Among these requirements is that the valuer should state both the purpose and the basis of valuation.

The most common purpose for which a valuation of leisure property is required is for open market transaction, but frequently valuations are required for either loan security or balance sheet.

In all these cases the appropriate basis is normally OMV. For open market transfer and for secured lending an additional valuation provided on the basis of ERP may be requested and in some cases, particularly where the purpose is for loan security an estimated restricted realisation price (ERRP) may also be specified. This basis, defined in the *Appraisal and Valuation Manual*, requires the valuer to assume a constrained time frame for the projected sale as set by the instructing client.

Valuations for Loan Security

There are no specific rules contained in the practice statements in relation to loan security valuations of properties that are valued as trading operational entities. However, GN 7 does address some particular concerns although it accepts that properties valued on a trading basis are valid for loan purposes.

The value of a trading leisure property is very dependent upon the availability (or otherwise) of trading records and it follows that there is greater risk attaching to the security offered by a property so valued as compared with say, an office investment property where the value is less dependent on the nature and trade of the occupier. Therefore, due to these inherent risks included in trading properties, the guidance note (GN 7.8) makes it clear that the valuer should include within the report all "significant" matters which may "assist the Lender in his assessment of risk". . . . Further, it states that it "would be helpful" for the valuer to indicate "whether the valuation figure incorporates a multiplier or discount rate which reflects the Valuer's opinion of the market perception of risk associated with the income stream from the property".

The purpose of including this assessment is to give information to the lender which will assist in determining the amount of loan that will be granted against the security of the property. It follows that the valuer is instructed to give the lender information concerning the sources of trading evidence and in particular whether the assessment of value is founded on *actual, estimated* or *projected* accounts.

It may sometimes be the case that the nature of the operation is one for which the potential demand is particularly fickle. Where this is the case and the valuer considers that the market demand for the property in its existing use may change significantly over the period of the projected loan, the matter should be reported to the client. If appropriate, a valuation on the basis of alternative use may be advised.

From time to time the valuer may be asked to produce a "bricks and mortar" or "land and buildings" valuation. This presents a conceptual difficulty in that a valuation produced on the trading basis is a global figure that does not relate to the summation of a trade element and a "bricks and mortar" element. However, there is recognition by the RICS that such figures are requested by clients from time to time. If such an request is received PS 2.7.4 instructs the valuer to exclude from the valuation a figure for the fixtures and fittings, the value of all licences and consents and any goodwill attaching to the property. It also instructs the valuer to make clear to the client that any such apportionment is informal only. This must also be stated in the Valuation Certificate.

Where a loan security valuation is required of a property that is let as an investment the usual considerations for any investment property apply.

Valuations for Balance Sheet

The specific requirements for valuations for balance sheets and other financial statements is covered in PS 12. This statement makes clear that three bases are recognised:

- Open Market Value (OMV);
- Existing Use Value (EUV); and
- Depreciated Replacement Cost.

In all cases the choice of appropriate basis should be determined by the valuer in agreement with the client. OMV is the basis used where the property is either let as an investment or is operational land but is surplus to requirements. EUV is only used where the property is operational land "occupied for the purposes of the business" and for which "there would be a market for sale to a single occupier". Such properties are defined in the Appraisal and Valuation Manual as "non-specialised" properties.

Those properties that are tenanted will be valued on the basis of OMV, normally using the investment method. However, for those leisure properties that are owner-occupied the basis will normally be EUV in which case the trading approach will normally be utilised, but the practice statement does refer to the situation where a business has been closed down and the fixtures and fittings removed. In this case, PS 12.6 indicates that the normal approach will be that either:

- The Directors will declare it surplus to requirements and an OMV will be prepared – possibly for alternative use; or
- The intention will be that the property will be re-opened. In this case the costs of re-fitting must be included in the valuation and this must be explained in the report.

Non-Commercial Leisure Properties

The valuation methods and considerations detailed above relate to trading leisure properties that are owned and operated primarily for commercial purposes. There are, however, very many leisure properties that are not so held. For example, local authorities normally hold within their portfolios a whole range of leisure properties. Typical local authority portfolios contain leisure and sports centres, municipal golf courses, theatres, community centres, museums, playing fields, parks and gardens. Many of these are revenue producing and, indeed, some have very high turnovers; others are free access. The common thread is that they are provided as a service to the local populace in furtherance of political and social ambitions.

In general such properties have seldom been sold on the open market and until the introduction of changed accounting procedures in the mid 1990s there was no general requirement to place a monetary valuation on them. Where valuations were prepared (for example for rating purposes) the general approach was to use a cost-based method. For rating this was the Contractor's Test and for other purposes the Depreciation Replacement Cost (DRC). The underlying principle prompting these decisions rested on the assumption that the properties were "specialised" within the definition in the *Appraisal and Valuation Manual* and that the owner (ie the public body) would be the only possible purchaser or tenant for the unit.

It follows that until the mid 1990s the question of the basis of valuation was largely academic as properties were seldom valued other than for rating; even here the valuations did not carry real financial significance to the owners. However, changes to accounting policies and local government finance has meant that the valuation of such units is no longer academic. In rating terms the decisions matter and local authorities are now required to produce capital valuations of most of their assets.

At the time of writing the position is that many publicly owned leisure properties are still valued on the basis of DRC for capital accounts purposes and the Contractor's Test for rating. There is pressure to effect a change of methodology, due to both market changes and changes in the philosophy of public provision. The changes to the market, as detailed above, have created a more transparent market and there is now more evidence on which to base valuations of leisure properties than ever before. This means that the argument that no market exists and that they are "specialised" becomes increasingly less sustainable.

Additionally, over the same time period there has been a major shift of social policy and public funding the results of which have seen a general increase in capital moneys available for projects (for example the National Lottery) but a pressure on revenue, leading in many cases to private operators being called in to run the facility, albeit not for profit, but on cost efficiency principles. The nature of the provision has also changed with many public sector properties mirroring quite closely the style and facilities that are offered in the private sector. The drive has been towards higher grade provision and in some cases joint venturing with the private sector. These moves collectively led to the development of new properties which, although publicly owned, are similar to those in private sector hands. This, in turn, is prompting a move away from the use of cost approaches and towards market methodologies.

In addition to local authorities, many other public and private sector bodies hold leisure properties for non-commercial purposes. For example the National Trust and other charities own many trading heritage properties and visitor attractions. Additionally any number of local and national sporting and cultural organisations own leisure properties ranging from large stadia to small museums. For these similar considerations exist to those which apply to local authority provision. In deciding what method to use, the valuer should consider:

- The purpose of the valuation;
- The attraction of the property to other potential occupiers or owners;
- Discounting the effect on value of any particular trading policy introduced for social or political reasons; and
- The effect of grant funding on both the original capital provision of the property and its revenue expenditure.

It must be stressed that the methods to be adopted in relation to the type of properties detailed above is open to debate as at the time of writing. Many rating disputes have occurred under both the 1990 and 1995 lists and several are still outstanding. The area is dynamic and currently such valuations lie firmly within the realm of the specialist valuer.

Case Study Examples

Case Study 1: Freehold Valuation of a Leisure Park Investment

You are instructed to value the freehold interest of a leisure park investment for possible sale. The scheme is situated in the south-east of England and the vendors have been the owners since its inception some five years ago. The scheme itself abuts an edge-of-town retail scheme and is fully let to mainly multiple tenants. It has adequate on-site parking and there is good visibility and vehicular access from a main road. A motorway junction is within 2 miles but public transport links are limited. The scheme is anchored on an 8 screen multiplex cinema and has two other D2 users (bingo and bowling) and has four A3 units (a "branded" public house a restaurant and 2 fast food units).

The leases are as follows:

- **Cinema**
  Let on a 30 year term with a five-yearly upward only rent review clause to Open Market Rental Value (OMRV) or 3% pa compound. At the time of letting the tenant received a rent-free period of six months to allow for fit out. The user clause restricts the use to a cinema.
- **Other D2 units**
  Let on 25 year terms with five-yearly upward only rent review clause to Open Market Rental Value (OMRV) or 3% pa compound. The user clauses permit any D2 use.
- **Food and Drink units**
  Let on 20 year terms with 5 yearly upward only rent review clause to Open Market Rental Value (OMRV). The user clauses are restricted to any A3 use.

There is evidence from recent lettings of other leisure units in the region to show that since the leases were originally granted rental growth has occurred and the rents have all just been reviewed. The rental growth of the cinema has exceeded 3% pa as did that of the other D2 units, so in every case evidence supported a review to OMRV.

There is also evidence of yields from other investment sales. The adequacy of the location and access combined with the quality of the covenants, which are all multiple traders, lead you to rank this as semi-prime. The age of the scheme, although only five years, means that the design is not "state of the art" and the poor public transport links lead you to question its ability to achieve rental growth in line with the best of new schemes.

As all units have been reviewed recently they are now fully let and a straight capitalisation approach can be adopted. (NB: if the evidence had not supported the minimum level of uplift prescribed, effectively the units would have been over-rented, in which an adjustment in yield would have been factored in or an equated yield approach to the valuation adopted).

Valuation
8 screen multiplex
Rent passing (OMRV) (3,000 m² @ 130 per m²)   £390,000

D2 Units
Unit 1
Rent passing (OMRV)   £64,000
(800 m² @£80 per m² – all ground floor)

Unit 2
Rent passing (OMRV)   £60,000
(600 m² @£80 per m² – ground floor
300 m² @ £40 per m² – first floor)

A3 Units
Unit 1
Rent passing (OMRV) (500 m² @£160 per m²)   £80,000

Unit 2
Rent passing (OMRV) (640 m² @£150 per m²)   £96,000

Unit 3
Rent passing (OMRV) (200 m² @£140 per m²)   £28,000

Unit 4
Rent passing (OMRV) (180 m² @£140 per m²)   £25,200

Total Rent Roll (OMRV)		£743,200
YP in perp at 7%		14.2857
Gross Capital Value		£10,617,143
less Purchasers Costs		
Stamp Duty @ 3.5%	£371,600	
Agents and legal fees @ 2%	£212,342	£583,942
Net Capital Value		£10,033,201
Say		£10,000,000

Case Study 2

You are instructed to act on behalf of a tenant in a rent review negotiation of a 20-lane bowling alley. The property is located in a town centre and it is let to an independent trader. It is not is a prime pitch. The building, of which it comprises part of the ground floor, is now some 30 years old. The remainder of the building comprises a restaurant unit and there are four floors of offices over. The unit still trades well but due to the pitch and the age of the building it is unlikely to attract a multiple operator. The premises have an internal net area of 2,100 m^2.

The lease is for a term of 20 years and this is the second rent review. The lease is on the equivalent of full repairing terms as there is a full service charge and the rent review clause specifies a review to open market rental value.

After inspection you consider that the tenant is trading to the full potential of the unit. In addition to the income derived from the lane hire the tenant has maximised the opportunities for food and drink sales and from machine revenue.

The audited accounts show that the trade for the last three years has increased slightly above inflation but you are also aware that a new leisure park located some one mile away is due to open in a few months. This includes a 36-lane bowling alley. You therefore consider it appropriate to reflect the potential competition in preparing adjusted accounts on which to advise a rental value.

The accounts show a current turnover of just over £900,000 and a profit of £400,000. However you consider that the turnover is unsustainable in view of the impending competition and examination of the accounts reveals that a large sum is included for annual depreciation. The tenant is also showing in the books costs of running a small catering business which is run from the premises. These items you exclude by the add back process to produce an estimated maintainable trade.

There is little rental evidence of bowling alleys in the immediate area as this unit has a local monopoly position. However you are aware that in neighbouring towns bowling alleys have commanded rents in the region of £80 per m^2.

Estimated Maintainable Trade
Income

Admissions	£50,000
Lane hire	£560,000

Shoe hire	£55,000	
Food sales	£50,000	
Drinks	£70,000	
Merchandising	£10,000	
Machine revenue	£40,000	
Total Revenue		£835,000
Costs		
Wages	£350,000	
Heating and lighting	£30,000	
Promotion	£10,000	
Operating expenses	£40,000	
Food and drink purchases	£55,000	
Service Charge	£17,000	
Total outgoings		£502,000
Adjusted Net Profit		£333,000 (40%)
Rental bid at say 42% of profit		£139,860
Say		£140,000

This bid represents some 16.7% of turnover.

To check the valuation an analysis is carried out against floor area. The suggested rental bid represents a figure of £66.70 per m². This is lower than the comparable evidence but appears reasonable given the poor quality of the building and the future competition.

The Use of Valuation Checks

The use of valuation checks is widespread in practice but there is a danger in their use by those not practiced in such properties. In a sense every rental value found by reference to a per m² is a check method in that underlying the rate adopted will be the ability of the premises to be profitable at that figure in the hands of an efficient operator.

Common units of comparison are per seat (theatres and cinemas), per court (sports such as squash) per cover (restaurants) and per bedroom (hotels). However it is advocated that, however experienced the valuer, the use of checks are used for just that purpose only – as a check against a more detailed calculation.

Summary

The leisure property sector is extremely disparate. It comprises both commercial and non-commercial property; it includes properties located in urban positions and those in rural locations. Leisure properties are found in the occupational portfolios of both public sector bodies and private sector organisations and increasingly they are acceptable within institutional investment portfolios.

All these facts are reflected in the wide range of valuation approaches that may be appropriate. Recent changes to the market for leisure properties and the increasingly commercialism in sport and leisure provision has increased the applicability of the investment method. Rental values of such investment properties are found by analysing comparable transactions on a square metre basis, although the method of measurement is in need of clarification.

However, a great number of leisure properties are still primarily traded for owner-occupation. For these, the profits or trading approach continues to be the most frequently adopted method and the use of the full profits approach does provide the best method of establishing the maximum rental value that an operator could afford to pay.

Within the field of social provision, the use of cost approaches to valuation is still widespread whenever it can be argued that the property is not one for which there would be a ready market. With the growth of private sector provision and consequent market evidence this method is becoming less relevant.

© Sarah L Sayce 2000

Chapter 20

Easements and Wayleaves for Sewers, Pipelines and Cables

Introduction

While the underlying principles of easement and wayleave valuations and the legislative background have remained largely the same, one or two points of detail have been altered or clarified since the last edition of this publication.

Practitioners are becoming ever more aware of the number of promoters who are entitled to acquire such rights, with the established sewer, water, gas, electricity, telephone and oil companies being joined by cellular telephone, cable TV and other telecommunications promoters, and by competitors such as Mercury and private gas transmission companies, and with fibre optic telecommunications cables being strung between existing electricity poles and included with many new pipelines. Indeed, because the scope of such organisations is now so wide, going far beyond the usual concept of an acquiring authority for compulsory purchase and even, in some cases, down to private individuals, they will be referred to throughout as "the promoter". Similarly, those from whom such rights are being acquired will be described as "the claimant". Where land is let the owner and each of the tenants or lessees will usually be independent claimants and will have to be treated separately by the promoter.

The majority of easements and wayleaves involve agricultural land and, except where otherwise indicated in relation to sewers in particular, the practical implications are considered from an agricultural point of view.

Legal background

In the majority of cases it is unnecessary for the valuer to study the detail of the statute, if any, under which a particular scheme is being promoted. A brief background to legislation can be found in the Royal Institution of Chartered Surveyors *Guidance Notes on*

Agricultural Wayleaves and Easement (RICS Publications, December 1985*), in the *Cables and Wires* (CLA A4/97, September 1997) and *Pipes and Sewers* (CLA A6/97, December 1997) and with a more detailed consideration appearing in *Compensation for Compulsory Acquisition of Agricultural Land* by R.N.D. Hamilton LLB (4th ed, RICS Publications, 1988*).

In general, most sewers and some water mains are laid under statutory powers without any further legal documentation beyond the original notice and without any easement being acquired. Others, such as gas, oil and other pipelines, are installed by voluntary deed of grant with compulsory powers being used only rarely. The electricity companies install apparatus under a combination of voluntary deeds and statutory powers and the burgeoning telecommunications industry tends to do likewise, although, in the case of many telegraph lines, no deed of grant is entered into and under the Telecommunications code there are certain circumstances in which equipment can be installed with the agreement of the occupier and without obtaining the consent of the owner although this does involve the promoter making an application to the county court.

The valuer will need to give consideration to the terms of any deed concerned, to ensure that it suits the parties' circumstances with or without amendment, and because the terms will have a bearing upon the valuation. The claimant's solicitor should be consulted on the wording of the deed at as early a stage as possible.

Exceptionally, in the case of sewers, private individuals may have indirect access to statutory powers. Generally, however, such individuals can install equipment in the land of another only if the owner and any occupier agree when, in the interests of all parties, a formal deed of grant should be prepared by the respective solicitors to be attached to the title deeds of each. This should be a comprehensive document incorporating not only the terms commonly used by industrial promoters but also the additional undertakings that such promoters usually volunteer, including an arbitration or expert procedure in the event of disputes and confirmation that the promoter will reimburse professional fees incurred by the claimant.

---

*Ed. both are currently out of print.

Role of the valuer

Paradoxically, the initial role of the valuer instructed on an easement or wayleave scheme, whether for promoter or claimant, is usually not valuing but preparing for the scheme and the implications that it will have on the claimant's land, particularly during the construction period. Such preparations have a bearing on the subsequent compensation claims and in particular that for surface damage.

The valuations themselves fall under two heads:

1. Compensation for permanent damage, where the basis of calculation tends to differ slightly with the different rights being granted.
2. Compensation for surface damage, where the principles of calculation tend to be the same regardless of the rights actually being granted.

Where the claimant is an owner-occupier the two heads will be included in the same claim.

The valuer might have further roles: in deciding whether the compensation attracts interest and, if so, how it is calculated; in the assessment of the promoter's contribution towards fees; and in the disputes procedure if compensation cannot be agreed.

On the larger schemes in particular, the valuer's role might continue for several years after the equipment is installed as certain items of damage do not reveal themselves until a later stage and others take time to restore.

In the case of surface wayleaves, particularly for electrical apparatus, the annual payments should be checked from time to time by the claimant or his valuer to take account of changes to field boundaries and land use.

In certain circumstances the valuer with the claimant's accountant might give advice on the taxation of alternative types of compensation.

Preparing for the scheme

General

The valuers for both the promoter and the claimant have roles to play during the inception and planning stages.

In the case of the water, sewerage and telecommunications industries, codes of conduct or practice are required by statute and the gas industry, in particular, negotiates such an agreement every three years with the Country Landowners Association and National Farmers Union. On larger schemes the promoter will invariably have agreed a specific code for the project concerned with those same organisations. These codes will cover most of the points described below, but the valuer must still have regard to the particular circumstances of each claimant and to any amendments that are required to the relevant code as a result.

While such preparations are necessary to mitigate the physical damage resulting, their primary purpose is to anticipate and avoid problems to the mutual benefit of both promoter and claimant. The promoter's valuer will need to guide his engineers most of whom will have little agricultural knowledge. The claimant's valuer will have a similar role with the occupier who will have little experience of civil engineering schemes. Close liaison between the valuers is essential from the outset and continuing through the construction period.

The most important points that valuers need to consider are listed below. On major schemes great attention will have to be paid to all or the majority of these together with others which might be peculiar to that particular scheme or the land affected. On others certain points may not be relevant but the valuer should be wary as small schemes can give rise to major problems.

This aspect of the valuer's role should not be underestimated, for it is generally far better to anticipate and prevent a problem than to pay compensation for something which could have been prevented. For instance, most farmers would prefer a live cow rather than compensation for its death following a fall into an unfenced pipeline trench.

In addition to the specific preparations listed, the valuer must bear in mind those items which he will be including in his compensation claim at a later date for certain of those will require monitoring as the project proceeds. For instance, the difficulties in extra costs involved in farming small areas of land which are wholly or partly cut off from the remainder of the farm.

Integrated Administration and Control System (IACS)

Pipeline and cable projects can have implications for the Arable Area Payments Scheme and for forage area declarations under the Beef Premium Scheme. The Ministry of Agriculture have produced a guidance note on this subject which the parties should obtain at an early stage.

Route

The route to be taken by the scheme is one of the first considerations and both promoter and claimant will want the shortest route possible. However, variations may be necessary for a variety of reasons.

Even on the smallest schemes it is necessary to avoid dwellings, buildings and other structures and also locations where such development is expected in the future. On the larger schemes this achieves greater significance, as the deed of grant may impose a limitation, if not a total prohibition, on building within the easement width. For safety reasons promoters are tending to increase the easement width where dangerous materials are being carried through the pipeline and the Health and Safety Executive may oppose future development on an even greater width.

If possible, woodland, fruit orchards and trees, particularly if subject to a preservation order, should be avoided and care should be taken to lay pipelines at sufficient distance to avoid damage to root systems. The deed of grant may limit the size of trees that can be replanted or prohibit such replanting altogether.

A claimant may persuade a promoter to change the route for other reasons, such as the need to avoid tile drains, but this is less likely on the larger schemes, where the costs of increasing the length, or even installing bends to avoid such obstacles, can be considerable. Also it must be remembered that gravity sewers have to be laid to a gradient and that promoters prefer to avoid laying pipes in public roads because of the costs involved.

A large number of underground pipelines are now constructed of plastic which is difficult to detect thereby giving rise to problems if the precise position needs to be determined at a later date. To avoid this problem it is advisable that triangulation points be used to identify the precise location in the deed of grant plan . This is no less important on the smaller schemes where a leaking joint between two lengths of alkathene water pipe can be hard to find.

Surface structures

Consideration needs to be given to the location of surface structures such as manholes, marker posts, aerial marker posts, stiles and gates.

Sewer manholes and surface chambers on water pipelines have to be provided at regular intervals or at particular locations determined by contours and need to be borne in mind when the route is under discussion. Ideally they should be located in farm roadways, tracks, hedge or fence lines. However, if a claimant is faced with a significant length of pipeline he may have to become resigned to a number within farmable land. While such structures appear intrusive when positioned in the middle of a field, they can cause less hindrance there than in headlands, and particularly in field corners, where agricultural machinery passes and turns and especially if insufficient width is left for machinery to pass between the structure concerned and the field edge.

Some promoters are happy to bury manholes which is of obvious benefit to claimants, although triangulation points should be considered to enable easy relocation. Otherwise there is merit in surface structures being provided with tall markers which can be spotted by the operators of farm machinery when the field is cropped, thereby avoiding damage by or to the machinery concerned.

Marker posts, whether for ground or aerial detection, are more easily accommodated as are the stiles and gates which some promoters use to indicate the presence of underground pipelines. These can be located within field boundaries, although consideration must be given to the possibility of such boundaries changing in the future.

Trial borings

Before the scheme begins, it may be necessary for the promoter to carry out trial borings to determine the type of underground strata or to locate existing pipelines or underground equipment. Such borings should be carried out only after consultation between the parties, with a record of condition being prepared and with compensation being paid for all damage caused. Such compensation is usually confined to surface damage.

Timing of the work

The valuer should establish the likely date of entry and the probable duration of the work. Provision should be made for the promoter to serve a notice of intention to commence work, which should be of not less than 14 days duration with the ideal being nearer 28 days. On the larger schemes the land occupier should be made aware of the promoter's detailed programme, which should refer to all movements of equipment, machinery and vehicles required for construction purposes, and should be kept in touch with progress and changes.

Ideally, schemes should be carried out when the weather is likely to be dry, between April and October, thereby minimising the risk of damage to the soil.

Depth

It is obviously important that pipelines, buried manholes, cables and other underground equipment should be laid at sufficient depth to avoid damage to or by normal agricultural operations. This is generally regarded as being a minimum of 900 mm cover measured from ground level.

Both parties should have regard to the possibility of ground level changing with erosion and the claimant should take care to avoid subsequent operations which materially affect the cover.

In rare situations, where a claimant has installed or is planning an unusually deep tile drainage or irrigation system, a greater depth might be necessary, but generally an excessive depth should be avoided because of the greater upheaval should it become necessary to locate or excavate the pipeline.

Ditches

Where underground equipment passes beneath a ditch problems can arise if the ditch is cleaned or enlarged subsequently. The sound bottom of the ditch, which is often below the apparent bottom, should be located. The equipment should be protected by a concrete pad at least 300 mm thick with the top of that pad being a minimum of 300 mm below the sound bottom of the ditch, thereby giving a minimum 600 mm of cover.

Protection of the soil

Protection of top soil is a fundamental consideration, especially following the publication by the Ministry of Agriculture of *Modern Farming and the Soil* in 1970 and the *Code of Good Agricultural Practice for the Protection of the Soil* in 1993. Those emphasise the extent to which top soil is a living medium in which water, air and particles are in balance and stress the long-term damage to soil structure which can result from compaction by heavy equipment, particularly in wet weather. If such damage occurs the possibility of remedial action is limited and losses of crop can result over many years.

Accordingly it is important to ensure that top soil is protected from compaction, that it is handled only when it is dry and that it is kept separate from subsoil and from any base rock such as chalk.

On all but the smallest schemes, top soil should be stripped from the entire width of the working area including all access points, so that the promoter's vehicles and machinery are working on subsoil, should be stockpiled alongside the original position and should be protected from compaction, in some cases being fenced off from the remainder of the workings. The subsoil should be kept in a separate stockpile with any base rock being kept separate.

On the larger schemes, great attention needs to be paid to this point. Typically, trial holes should be dug to determine the depth of the top soil; next the top soil should be excavated and stockpiled; after that the claimant should be allowed to inspect to ensure that the top soil has been properly stripped; and only then should the subsoil be excavated. At the end of the scheme, any base rock should be placed in the bottom of the trench and should be compacted. The subsoil should then be replaced and similarly compacted, being ripped or subsoiled to a depth of at least 450 mm. The claimant should be allowed to inspect the reinstated subsoil before any top soil is replaced, which should be the last activity as the promoter leaves site, other than the removal of loose stones and litter. Ideally the length of trench open at any time should be kept to a minimum with top soil being reinstated as soon as practicable.

Even on smaller schemes, such as the flying of a private electricity cable over a neighbour's land, the risk of soil damage is considerable. If the weather is fine, the forecast good and the soil dry, smaller machines can be permitted on top soil particularly in pasture land and especially if fitted with wide tyres or tracks, provided the project can be completed within the forecast spell of good weather. However, such an arrangement should be on the understanding that if rain falls work will cease until conditions improve.

If small schemes involve excavation of a trench, for instance for a private water pipe, the top soil and subsoil should be kept separate, ideally on different sides of the trench, and should be replaced in the correct sequence. The only exception is when a narrow chain and bucket trencher can be used where the width of top soil and subsoil mixing will be very narrow.

Frequently promoters are left with surplus soil, partly as a result of the space taken up by the pipeline itself and partly of the difficulty of compacting the replaced material in the trench line. Only in exceptional circumstances should such soil be removed from the site. On most schemes, particularly the larger ones, subsidence will occur subsequently over the pipeline area. In anticipation of this, it is sometimes wise to mound the subsoil slightly before reinstating the top soil and to remove any surplus material as subsoil rather than top soil. If subsidence does occur top soil may have to be imported. This can be variable in quality and arrangements should be made to enable the claimant to inspect each load before tipping, which should be carried out only in dry conditions.

Protection of the soil is so important that, in extreme cases, specialist advice should be taken from a soil surveyor even before the scheme begins.

Wet weather

The arrangements with the promoters should make provision for wet weather during the course of the scheme and for steps to be taken to avoid any damage to top soil as a result.

Promoters are unlikely to agree to halt a scheme, or to delay its commencement, if conditions are too wet but once their equipment is installed they are usually ready to agree that reinstatement of the top soil be postponed until the land has dried. In practice this can mean that the top soil is left mounded throughout the winter months until it is fit for handling again in the following spring or summer.

Stacking and protecting turf and plants

Where a scheme involves non-agricultural land, particularly domestic gardens and sports fields, it is usually best for the turf and as many of the plants as possible to be dug up beforehand to be stored on one side on plastic sheeting, tended and watered, and replaced after the promoter has finished. Such action is possible only if the scheme can be carried out swiftly but it does enable lawns to be left with a more uniform appearance than if turfs are brought in and enables gardens to be left with their original plants rather than less mature replacements, although some losses may occur particularly in dry weather.

Working hours

Usually it is in the interests of both promoter and claimant that the scheme be completed and the land reinstated as quickly as possible even when this involves work at weekends and after dusk. However, where all or part of a scheme is within sight or sound of domestic dwellings, the promoter should be encouraged to cease or curtail its activities during unsocial hours.

Land drainage

As natural or tile land drainage is susceptible to severe damage by compaction or excavation the claimant should give the promoter copies of any available land drainage plans at as early a stage as possible.

The most obvious problem will occur in land where tile drain runs will be cut by the promoter's trench. Each drain so cut should be marked prominently within the trench and should be recorded and noted on a plan by the promoter. These records should be available to the claimant, who should be given an opportunity of inspecting to ensure that all such drains have been detected.

In extreme cases, depending on the number of drains and their angle to the scheme trench, it may be necessary to employ a specialist land drainage contractor to lay an additional header tile drain parallel to the trench line and connecting into each of the severed drain runs on the uphill side.

More commonly, each land drain is reinstated individually, with the subsoil first being replaced to the level of the drain and compacted and with the tile drain run across the trench being replaced with a rigid porous pipe. This pipe should be cut into undisturbed land for a distance of at least 300 mm on either side of the trench line to give it a sound bedding and should be supported beneath by a stout timber or concrete bridge lintel, similarly cut into the sides of the trench, to reduce the risk of the reinstated land drain being rendered ineffective by sagging when subsidence occurs. The claimant should be given an opportunity of inspecting all such reinstated drains before they are covered with any further soil.

Problems can occur in poorly drained land when soil water follows the promoter's trench line and collects at low points. Additional tile drainage runs will have to be provided to take this water away.

Because of the weight of the promoter's machinery, damage by crushing or compaction might be caused to tile drains or natural drainage within the working area but outside any excavations. Tile drains damaged in this way will have to be excavated and repaired. Sometimes damaged natural drainage can be restored by mole ploughing, subsoiling or ripping but tile drains might be required to deal with problem areas.

The valuer should be aware that drainage problems might not become apparent until after the scheme has been completed. Accordingly it is important that the acceptance of reinstatement should be qualified by a reservation in respect of drainage defects and that surface damage compensation settlements be subject to a similar reservation.

Land drainage problems can be so severe that the valuers find it necessary to take expert advice thereon from a soil surveyor even before the scheme begins.

Working width

It is important that adequate width be granted to enable the scheme to be executed properly and rarely in either the promoter's or claimant's interest for it to be reduced. This width must accommodate separately the excavated top soil, the sub soil, any base rock, access for the promoter's machinery, space to assemble the pipes, cables or other equipment, and space for the machinery to lower the equipment into the ground or raise it on to poles. If insufficient width is allowed, there is a risk of damage to crops and soil outside the width granted and to the excavated and stored top soil through compaction or mixing with subsoil.

In certain places, particularly adjoining major roads and railways, additional working areas might be required. These should be negotiated beforehand and are sometimes conditional upon a payment by the promoter in the nature of a rent in addition to compensation for permanent and surface damage.

Fences

The limits of the working area and any access points should be fenced to a specification agreed between the parties beforehand. This is necessary particularly when the land on either side is grazed by livestock.

Fences are less important in arable land but should be provided to ensure that the promoter's employees stay within the working area. While a lesser specification will suffice it should comprise post and wire rather than plastic bunting.

Where the scheme severs existing field boundary fences provision must be made for straining posts on either side of the breach to maintain tension. Again, this is important in fields grazed by livestock but it applies none the less in arable fields, as fences deteriorate once tension is lost.

Access points for promoter

On most projects the promoter will have sufficient access within the working area. Any additional access points should be the subject of prior agreement incorporating the same terms as apply to the working width. Access beyond what is agreed should be prohibited because of the risk of damage occurring.

Access points for claimant

Where a field is severed by a scheme, the parties must decide how the cut-off portion is to be dealt with if there is no alternative access. If the cut-off area is small, the promoter might agree to pay compensation for the loss of the cut-off crops. Alternatively, access points might be left to enable the claimant to cross the working area so that he can continue to farm the cut-off portion. These points need careful planning, particularly if they are to be used by livestock when a system of gates will be required, or by unusually large or heavy machinery, such as combine harvesters, where an appreciable width and substantial bridging of the trench might be necessary. Additionally, there will be times when even these accesses need to be closed and it is important for the parties to liaise over such closure to prevent substantial losses occurring.

Continuity of supplies

If the scheme crosses the line of any services, such as water, electricity, gas, drainage or telephone, arrangements must be made for these to be protected or reinstated without delay. Their precise position needs to be located beforehand whenever possible. The existence of certain water supply pipes in particular is often unknown and their severance can pass unnoticed. To guard against this, claimants should be encouraged to read their water meters at frequent intervals as excavations continue.

Particular problems can occur in areas covered by irrigation abstraction licences if the source of the water is isolated from the remainder of the farm.

Trees

Trees should be removed only after prior consultation with the claimant, where both parties should ensure that there is no breach of any preservation order nor of the legislation regarding trees in conservation areas. Prior agreement should be reached as to whether the promoter is to remove the tree completely, paying compensation for its timber value, or is to leave it for the claimant to sell.

Reinstatement of roads and paths

Generally the excavated width across roads and paths should be reinstated. However, a precise match of materials is rarely possible and in particularly sensitive areas of high amenity value, usually involving residential rather than agricultural property, it might be necessary to resurface the entire roadway or path concerned.

Control of dust

In dry periods, the promoter's machinery may raise dust which can have a deleterious effect on domestic property, on market garden crops including fruit and on grass for grazing by livestock. On larger schemes the promoter should provide water bowsers and sprinklers to damp down any such dust.

Noise vibration and explosives

The noise and vibration caused by the promoter's machinery can be a problem, especially during unsocial hours, when close to residential property and when livestock is involved. In such cases the promoter should reduce noise and vibration by restricting working hours and by the use of alternative machinery. Explosives should be used only when the parties have agreed beforehand.

Infectious diseases

Both parties should guard against an infected area being declared on account of foot and mouth disease, fowl pest, swine fever, rhizomania or other notifiable disease. In such circumstances the project should continue only with the approval of the Ministry of Agriculture.

Cathodic protection

Pipelines constructed of metal frequently need cathodic protection to avoid decay caused by ionisation. The equipment required should be the subject of separate negotiations and should be referred to in the deed of grant. The promoter should take steps to ensure that existing buildings, structures and other installations are protected in accordance with any British Standard Code of Practice.

Pollution of water supplies

Promoters should be wary of polluting water supplies, particularly in water catchment areas and in marshy and drained land. Noxious liquids deposited on the land can often seep through the soil into water and be carried long distances by drains. Such pollution can spread over a wide area posing a threat particularly to livestock and to a wide variety of arable crops through spray irrigation.

Soil-borne pests and diseases

The promoter should be required to take care to avoid the spread of soil-borne pests and diseases such as potato cyst eelworm and verticillium wilt and should be alerted by claimants where such a risk exists.

Fossils and artefacts

Prior agreement should be reached that any fossils, coins or other objects of value discovered by the promoter will remain the property of the claimant unless found to be treasure as defined by the Treasure Act 1996.

Security and the prohibition of certain activities

The promoter should be prohibited from using the working area for purposes or activities other than the scheme itself. Caravans or site offices should not be permitted. Dogs should not be allowed except in the case of properly controlled guard dogs. Poaching should be prohibited and firearms or shotguns should not be allowed on the working area for any reason. The use of metal detectors should be confined to the locating of underground services. On larger schemes, where numerous people will be employed by the promoter, adequate means of identification should be carried at all times.

Abandonment

Both parties need to agree the action to be taken in the event of the equipment being abandoned in the future.

Promoter's representative

Before the project starts, claimants should be provided with the name, address and telephone number of the promoter's representative. On larger schemes liaison officers may be provided but otherwise the representative is usually the resident engineer. Provision should be made for resolving problems that arise overnight and at weekends.

Record of condition

Shortly before the scheme commences, the promoter's valuer should make a record of condition of the working width, access points and any features nearby which risk being damaged. This record should be as full as possible and should include photographs. The record and photographs should be sent to the claimant's valuer before work commences for approval with or without amendments and further photographs.

Residential or other buildings which lie close to the scheme should be included within the record of condition and the valuer should arrange a building survey where such buildings are old, of great value or have a history of structural problems.

Diary

The promoter's representative and the claimant should keep diaries and photographs of problems that arise as the project proceeds. The diary should include a detailed record of incidents as they occur, including dates, times, names of people and the description and registration numbers of vehicles involved, to form the evidence for complaints or compensation items. The claimant should report more serious problems immediately to the promoter's representative and to his own valuer.

Compensation for permanent damage

General

The first of the two categories of compensation is that for the permanent or long-term presence of the equipment in or on the land and is referred to variously as easement consideration, as the claim for freehold damage, as a recognition payment, as the freehold claim or as the landlord's claim.

The amount of compensation for permanent damage depends in part upon the restrictions imposed on the claimant by the presence of the equipment and the rights granted to the promoter for inspection and subsequent re-entry on to the land.

This head should include compensation for surface structures such as manholes, even if buried, for cathodic protection equipment and for injurious affection to any other property belonging to the claimant.

Where the deed includes the right to install further equipment at a later date, care should be taken that the relevant clauses are worded so as to entitle the claimant to full compensation at values then current.

Specific provisions should apply to land containing minerals where the precise legal position must be clarified.

In common with the general law of compulsory purchase, promoters using statutory powers can deduct betterment from any compensation that would otherwise be due.

In *Mercury Communications Ltd* v *London & India Dock Investments Ltd* [1993] 1 EGLR 229 the court declined to award a ransom element within the compensation for freehold damage and in *Northern Electric plc* v *Addison* [1997] 39 EG 175 a similar attitude was taken by the Court of Appeal on a rent review of an electricity sub-station.

Advance payment

Generally the claimant will be entitled to an advance payment of
90% of the compensation as agreed or estimated, either on the date
notice is served, date of entry or on proof of title. The claimant's
valuer should apply for such advance payment as soon as a
relevant notice is served or at least three months before the expected
date of entry and instruct solicitors to prove title without delay, as
outstanding compensation does not always attract interest.

On some schemes the final amount of compensation will not be
assessed until detailed "as laid" plans have been produced. These
should be checked by the valuers and should form part of the deed
of grant and be attached to the title deeds.

Sewers and water pipes

Despite occasional urging by valuers, the Lands Tribunal have
refused to lay down rigid formulae for the calculation of
compensation for permanent damage although such formulae are
widely used in practice. This is because of the need to judge each
individual case on its merits.

In *St John's College, Oxford* v *Thames Water Authority* [1990] 1
EGLR 229, the Tribunal expressed an opinion that such valuations
should be made ideally on a "before and after" basis but conceded
that no such evidence was available.

In *Felthouse* v *Cannock Rural District Council* and *Markland* v
Cannock Rural District Council (1973) 227 EG 1173 the Tribunal
referred to four main factors:

1. The trend of payments for sewer pipelines.
2. The relationship of sewer pipeline payments to capital land
    values.
3. The trend of capital land values.
4. The trend of payment for other types of pipeline.

The common formula for agricultural land involves three elements
and is calculated per metre run of pipe:

1. The value of the land.
2. A percentage of that land value.
3. A notional width.

The land value used should be the value to the owner of the farm
as equipped as a working farm but should exclude any uplift for

amenity or non-agricultural value unless such uplift is itself affected by the scheme. Promoters tend to adopt a standard value for land of a similar quality within each scheme and to vary that figure only where circumstances are markedly different. Traditionally the vacant possession value of the land has been taken, even where it is let on an agricultural tenancy, although the Lands Tribunal adopted the tenanted value in the *St John's College* case.

Almost universally for sewers and water pipelines, a 50% reduction in value is used and has been adhered to generally by the Lands Tribunal, despite other promoters using higher percentages. However, where the route of a pipeline was altered at the request of a claimant so that it followed a longer route around the edge of a field, the Lands Tribunal reduced this to 40%; see *Abercrombie* v *Derwent Rural District Council* (1971) 219 EG 1397.

Usually the greatest scope for discussion concerns the width affected. To a certain extent this is determined by the size and depth of the sewer but typically widths of 6 m to 8 m are common. The width used in such calculations tends to be less than the working width employed by the promoters, and sometimes a multiplier of two-thirds of that working width is used, but in the St John's College case the Lands Tribunal award was calculated on the full 18.3 m working width necessary to lay a 750 mm water pipe at a depth of 2 m to 2.5 m. An argument in favour of a larger width is that, on the rare occasions when major problems occur subsequently, necessitating the excavation of a sewer or water pipe, a far greater width is required than that when the pipe was first laid. On the other hand, a direct link with the working width might encourage promoters to reduce that width, which is in neither party's interests for reasons explained previously.

Besides agricultural land, sewers in particular and water pipes are laid most commonly through dwelling-house gardens. Compensation is paid rarely for the permanent presence of such pipes because the dwelling concerned invariably benefits from the scheme through the provision of a connection to the public sewer instead of a cesspool or septic tank: see *Collins and Collins* v *Thames Water Utilities Ltd* [1994] 9 RVR 13. Otherwise there is no formula and much will depend upon the skill and experience of the valuers. Obviously the rights of inspection and re-entry assume great significance in such circumstances where they involve loss of privacy within a domestic garden, although these rights are rarely exercised.

If the pipeline interferes with a building plot, by preventing or reducing the size of the potential development, compensation will be based on the usual "before and after" valuations.

In addition to the payment for the pipeline, separate payments are made for surface chambers. The amount will depend upon the precise location, with more being paid for chambers within the cultivated area than for those within tracks or fence lines. Typical payments in agricultural land range around £100–£250 for structures within the cultivated area. Compensation is payable also but at a lesser rate for buried manholes: see *Felthouse* v *Cannock Rural District Council* and *Markland* v *Cannock Rural District Council* (1973) 227 EG 1173.

No form of payment is made to agricultural tenants to cover the permanent presence of sewers and water mains, despite such payments being common for pipelines carrying other materials. Additionally some tenancy agreements except and reserve all wayleave and easement payments to the landlord.

Special provisions apply to land containing minerals and are set out in Schedule 14 to the Water Industry Act 1991.

Gas and other pipelines

The valuer's role in the calculation of compensation for permanent damage for gas and other pipelines tends to be more limited than for sewers and water pipes. The same type of formula is used but often the land value is subject to a generous minimum level. The percentage is usually around 80% and the width is specified in the deed of grant, so there is less scope for disagreement, especially as the promoters of such schemes invariably agree such terms beforehand with the Country Landowners Association and National Farmers Union, on an industry-wide basis in the case of gas, and for each individual scheme in the case of oil and chemical pipelines.

Nevertheless, the valuer should consider whether the terms are adequate for his client and should pay particular regard to the possible loss of development potential and safety restrictions. Most promoters include the right for the claimant to be awarded additional compensation in the future if development, including mineral extraction, is prevented by the pipeline but such clauses need careful drafting. For safety reasons, some development is being prevented and additional restrictions imposed by the Health and Safety Executive over a greater width even than that envisaged by the promoter. In cases where such additional restrictions are possible, the parties should seek the views of the Health and Safety Executive at an early stage and should base the compensation and the restrictions in the deed of grant thereon.

Surface chambers are less common on gas and other pipelines than on sewers and water pipes. Areas required for pressure-reducing stations or valve chambers tend to be purchased on a freehold basis.

Frequently the promoters of gas and other pipelines make a payment per metre run to the land occupier, although usually this is expressed to be in return for prompt co-operation with the project rather than as compensation. In the case of an owner-occupier, this further payment is in addition to the compensation. Where land is tenanted, the tenant will receive the occupier's payment.

Electricity pylons, poles and fibre optic cables

While the promoters of electricity projects have extensive compulsory powers they prefer to negotiate private agreements with claimants on either an annual or permanent basis.

Annual wayleaves are more common, are the norm for smaller apparatus and are in exchange for annual payments of rent and compensation which are revised each year by agreement between the electricity industry and the County Landowners Association and National Farmers Union. The rental payments are made to the landowner and depend upon the size and type of apparatus concerned. The annual compensation is paid to the occupier and is dependent, not only on the size and type of apparatus, but also on the type of land, with different rates for apparatus in hedges/ditches, for permanent pasture/long leys and for arable, with a further uplift for various specialist categories such as paddock/strip grazing, orchards and hop gardens and double-cropped market garden land. Owner-occupiers receive both rental and compensation payments and certain agricultural landlords receive the compensation in addition to the rent under the terms of their tenancy agreement.

Permanent or capital wayleaves apply where the parties agree that the equipment shall remain on the land in perpetuity in exchange for a once-and-for-all capital payment.

In *Clouds Estate Trustees* v *Southern Electricity Board* [1983] 2 EGLR 186, it was accepted that the annual payments agreed with the

Country Landowners Association and National Farmers Union were based on agricultural values with no allowance for any loss of amenity. Accordingly where the claimant's property suffers loss of amenity, such as that caused by a large pylon close to a dwelling, the valuer should negotiate higher rental payments to reflect such loss. If agreement cannot be reached, the claimant should consider serving notice terminating the annual wayleave, whereupon the promoter will have to apply for compulsory powers and the usual compensation will be payable for injurious affection.

The annual nature of such wayleaves is relevant when the apparatus is found to interfere with subsequent proposals for development. Again the claimant should consider serving a termination notice, whereupon the promoter either will have to move the apparatus so that it does not interfere with the proposed development or will have to pay compensation for the loss in development value. As an alternative, a permanent wayleave can incorporate a clause providing for future compensation for loss of development value but such clauses need careful drafting: see *Turris Investments Ltd* v *Central Electricity Generating Board* [1981] 1 EGLR 186 and *Mayclose Ltd* v *Central Electricity Generating Board* [1987] 2 EGLR 18.

While the capital sum obtained for a permanent wayleave is attractive to claimants, it will be seen from the foregoing that annual wayleaves offer flexibility which can be valuable, particularly where there is the possibility of future development. However a capital claim may be more appropriate where amenity has been damaged.

Because the annual compensation payments are based on the use to which the land is put, from time to time they need to be checked by the valuer who should notify the promoter of changes. For instance, the annual compensation due to the claimant should be increased: where hedges have been removed or ditches filled so that apparatus located previously in a hedge/ditch line is in the centre of an enlarged field; where grassland has been ploughed and converted to arable; or where arable land has been planted with orchards or double cropped with market-garden produce.

Telecommunications

In general, telecommunications equipment is installed by agreement rather than under compulsory powers on the basis of either permanent or annual wayleaves similar to those for electricity.

Annual wayleaves are again far more common than permanent agreements and enable the claimant to serve a notice of termination if he requires the equipment to be removed under the Telecommunications Act 1984.

Most telecommunications operators negotiate standard annual payments with the Country Landowners Association and National Farmers Union which are revised each year and some have agreed a model deed of grant and letter of undertaking. Nevertheless the valuer must be satisfied that these are adequate to the particular circumstances of each case.

In *Mercury Communications Ltd* v *London & India Dock Investments Ltd* [1991] 1 EGLR 229 it was held that compensation for a wayleave agreement under the telecommunications code should be fair and reasonable which is not necessarily the same as the result in the market and that a ransom value approach was inappropriate.

Private easements and wayleaves

There are a wide variety of circumstances in which a private individual or company might require an easement or wayleave over the land of another ranging from a private individual wishing to lay a small water pipe through a neighbour's property to a large company requiring an extensive range of services.

Where a private individual is able to requisition a public sewer, compensation for permanent damage will be assessed on the basis described previously. Otherwise compensation for permanent damage is a matter between the parties. In certain cases, for smaller projects, the claimant might settle on a neighbourly basis for 50% of the land value as for public sewers and water pipes, or even less. In others the valuer should consider a "ransom strip" valuation similar to that in *Stokes* v *Cambridge Corporation* (1961) 180 EG 839. In all cases the valuer must consider the precise terms of the deed of grant to ensure that the rights and liabilities granted and imposed are reflected in the figures.

Compensation for surface damage

General

The second category of compensation is that for the damage caused during the installation of the apparatus, which is described variously as surface damage, temporary damage or the occupier's or tenant's claim.

A very wide variety of items can be included within this category, some being described in more detail below, and the valuer should take care to ensure that no item of loss or damage is omitted.

The claimant is under an obligation to mitigate his loss and the scope for such mitigation can be considerable. However, care needs to be taken, particularly during the initial planning of a project, and steps towards mitigation should not be taken until the promoter has given written confirmation of a precise date for entry. If a claimant adjusts his cropping in anticipation of a scheme which is then cancelled or postponed, he could be left with losses for which he is not entitled to compensation.

In certain circumstances compensation is payable where losses result from construction works not on the claimant's land but in an adjoining highway; see *George Whitehouse Ltd (trading as Clarke Bros (Services))* v *Anglian Water Authority* [1978] 2 EGLR 168 and *Leonidis* v *Thames Water Authority* (1979) 251 EG 669.

Advance payments

It is unwise for the claimant's valuer to submit a final surface damage compensation claim until the project is completed and the promoter has moved off the claimant's land. Even then haste may be inadvisable as evidence for items of claim may be delayed, for instance, where a crop is sold into a co-operative which reimburses its members from a pool price at a later date.

In certain circumstances, interest may be payable on outstanding compensation as soon as a notice is served or at least three months before entry but the rate allowed is likely to be less than the claimant is paying on an overdraft.

Accordingly the claimant's valuer should apply for 90% advance payments of compensation at least three months before entry and at appropriate intervals thereafter so that his client is not out of pocket owing to such delays.

Monitoring

The valuer should inspect the progress of the scheme from time to time, particularly in the early days after entry is taken, and should keep in touch with his client so that he is aware of problems that arise. Notes should be taken during these inspections, together with photographs, and problems should be communicated immediately to the other party's valuer by telephone and letter. Joint inspections by the two valuers might be necessary to agree the extent of losses before the evidence is lost and such agreement should be confirmed in writing. For instance, damage to standing crops should be agreed before those crops are harvested. If agreement cannot be reached, the claimant's valuer must assemble sufficient evidence to enable him to prove the loss subsequently. This should include photographs and might necessitate the appointment of a further valuer so that a second opinion is available.

Loss of growing crops

Loss of growing crops on entry is an obvious head of claim. The valuer will need to measure the precise area lost, which may exceed the working width in arable fields by at least a metre owing to the difficulty of harvesting close to the fences enclosing the working area. The yield and price of each crop from the remainder of the fields concerned must be recorded and, where appropriate, an allowance made for costs saved.

The extent to which costs are saved must be considered, especially where entry is made just before or soon after a crop is sown. An allowance should be included for direct or variable costs such as seed, fertiliser and sprays saved. It is debatable whether deductions should be made to reflect notional savings in regular labour, machinery costs and overheads when such fixed costs are not related directly to output. However, the valuer should aim for consistency and may decide, on balance, that notional costs should be deducted under this head so that increases in such costs resulting from the scheme can be claimed under other heads. Usually fixed costs calculations are based on the costings produced annually by the Central Association of Agricultural Valuers (CAAV).

Some loss of crop will occur outside the working area, in particular, in cereals owing to the need for the claimant to set out fresh tram lines and turning areas where the originals are cut by the project and additional headlands created.

In domestic gardens some items of plants and vegetables will be lost. In the case of perennial plants, the cost of purchasing and planting replacements is a measure of loss but, in certain circumstances, it might be preferable to allow instead for removing, storing and replanting. This procedure should be agreed beforehand with the promoter's valuer because the promoter might prefer to carry out such work or because the cost of such double handling might exceed the cost of purchasing replacements and be disallowed or reduced on the mitigation argument. Furthermore, some plants will die as a result of being disturbed twice and a right should be reserved to claim for these.

Loss of grazing

The claim for loss of grazing is similar to that for loss of arable crops in that the precise area of loss needs to be measured. However, the calculation can be more complicated because stock numbers in the field are rarely reduced on account of part being lost to the project. For simple grazing systems involving sheep or beef cattle, the going rate for hiring alternative grazing in the area is a guide as is the gross margin per acre for the enterprise concerned. Where the field affected is intensively grazed under a strip or paddock grazing system or by dairy cows a more detailed calculation based on hay equivalents might be necessary. In extreme cases extensive supplementary feeding might be required and the cost included in the claim but the promoter's valuer should be informed in advance.

Reseeding grassland

At the conclusion of the project the working area through grass fields will need to be reseeded and the parties must agree whether this is the responsibility of the promoter or the claimant. Promoters rarely have the expertise to carry out such work but on larger projects may employ agricultural contractors when the claimant's valuer will have to ensure that the correct seed mix, fertiliser and sprays are used.

It is more usual for claimant to reseed and to claim for the costs incurred being the materials used and operations based on CAAV costings or the cost of employing an agricultural contractor.

Extra costs

However well the promoter executes the project, invariably the claimant is involved in additional work to restore the working area at the end of the scheme and incurs extra costs as a result. In many cases subsoiling is necessary, even when similar operations have been carried out by the promoter, as are a number of extra cultivations to restore the land to a consistent level and tilth with the remainder of the field. Further work may be required to control weeds which have sprung up and reseeded during the construction period and to restore soil structure by the incorporation of farmyard manure, shoddy or similar organic material. All such costs should be included in the claim.

Additional cost of working severed areas

Fields which are divided into two or more parts by a project invariably take more time to drill, cultivate, tend and harvest than the original undivided field. This is due partly to the extra turning time required, especially for larger machinery, and is increased where the only access to severed portions is over the working width of the project. The extra costs based on the time involved should be included in the claim. Where particular difficulties are envisaged the claimant's valuer must consider whether he should mitigate his claim by abandoning the crop in the severed areas, where its value is less than the extra costs involved; close liaison with the promoter's valuer is essential.

Future losses

However well a project is executed and the land reinstated, some shortfall is likely in succeeding crops and, where soil structure has been badly affected, these have been recorded for up to 10 years. Also a weed problem, such as wild oats, can occur. The two valuers should inspect each year before harvest so that the extent of the losses can be agreed.

Most promoters are willing to pay compensation in succeeding years as losses occur but equally are amenable to agreeing a lump sum in advance to cover future losses. The claimant's valuer needs to decide which alternative to pursue but should clarify the taxation position with his client's accountant before agreeing a lump sum.

Where land has suffered no undue damage and has been properly reinstated, a rule of thumb for the calculation of such a lump sum is a 50% loss in the first year after reinstatement, 25% in the second year and 10% in the third year.

In *Smith, Stone & Knight Ltd* v *Birmingham City District Council* [1988] EGCS 49 in Official Referee's Business in the High Court it was held that damage caused by subsidence occurring more than six years after the initial laying of a sewer was not statute-barred.

Fences and hedges

At the conclusion of the project fences cut by the working width will require reinstatement and, where the original fence was strained, as will lengths of fence outside the working width from which tension has been lost. The cost of such remedial action should be claimed based upon the claimant doing the work himself or employing a contractor. On major projects promoters may employ fencing subcontractors when the claimant's valuer must see that the work is to a sufficient standard.

Hedges can be replanted only during the winter months so this task often is left to the claimant. Again it is an item for which the claimant may employ a contractor. Once planted, the new length of hedge should be protected by fences along each side against damage by cattle and rabbit netting might be necessary. Such hedges will require maintenance for up to five years and the cost thereof should be included in the claim.

Where a hedge has served as a windbreak for orchards or market garden crops, an item should be included for plastic or netting windbreak until the hedge is re-established.

In domestic gardens, hedges can provide a degree of privacy. In such cases, an item should be claimed for the cost of erecting a timber screen fence to restore privacy until the hedge is sufficiently mature.

Disturbance to animals

Certain types of livestock, most notably poultry and dairy cows, are susceptible to disturbance caused by noise and unfamiliar activities. Losses in egg and milk production can result and, in extreme cases of panic, livestock can be injured or killed. Such disturbance may be difficult to prove so care must be taken in collecting evidence from experts such as vets and in advising the promoter's valuer. The quantum of such loss is often hard to calculate without a detailed study of production over the months and years prior to and following the project.

Deficiency of top soil

Top soil can be lost as a result of removal by the promoter, mixing with subsoil or subsidence in the trench line. Usually supplies of top soil for making good are limited and of suspect quality but such supply and spreading is an item of claim none the less. In extreme cases, where large amounts are required, the promoter's valuer might argue that the cost of such work exceeds the amount of the claimant's loss. Such an argument should be resisted as, invariably, such problems arise from subsidence or promoters failing to keep to their undertakings.

In *Lucey (Personal Representative of)* v *Harrogate Corporation* (1963) 186 EG 745, where a deficiency of top soil was alleged, the Lands Tribunal held that the occupier might please himself whether he restored the top soil, claiming the cost thereof, or whether he tolerated the physical damage, claiming compensation for the resultant losses.

Claimant's time

Even where a claimant employs a valuer, he will spend time on matters related to the project, and on major schemes this will be considerable. On properly managed projects the promoter and claimant will establish a close working relationship which will operate to their mutual benefit because it will reduce the number and size of problems that arise and the amount of the surface damage compensation claim. Accordingly it is in neither party's interest for such time to be stinted.

During the preparation for the scheme the claimant will attend meetings to consider the points referred to above. Once work has begun, he will need to inspect the working fences at frequent intervals and keep an eye on the promoter's activities. Furthermore time will be taken on matters such as checking the stripping of soil, gaining access across the working width, checking land drains and their reinstatement and reading water meters. At the end of the scheme further checks will be needed to top soil and land drains and discussions will take place concerning items of reinstatement. Throughout this period the claimant will have to liaise with his valuer and provide much of the evidence on which the surface compensation claim is based.

An item should be included for such extra effort on the claimant's part in accordance with the decision of the Court of Appeal in *Minister of Transport* v *Pettit* (1968) 209 EG 349. Ideally, the basis of calculation should be agreed with the promoter at the outset. Wherever possible, it should be supported by evidence such as photographs, diaries, time sheets and similar records: see *Rush & Tompkins Ltd* v *West Kent Main Sewerage Board* (1963) 187 EG 47.

Miscellaneous

The claimant's valuer must ensure that a claim is submitted for every item of loss resulting from the scheme. Certain of these might be unusual or unique to the particular claimant concerned. In the past, compensation has been paid for loss of shooting days and a lack of pheasants owing to the construction of a pipeline; for loss of fruit and vegetables rendered unmarketable by a coating of dust; for a weed infestation in severed land owing to access being unavailable during the crucial spraying period; for additional travelling time when the main access to a farm was blocked necessitating a longer alternative; for the signing of footpaths temporarily rerouted during the construction of a pipeline; for a reduction in crop resulting from an inability to spray irrigate because of pollution of the water course concerned by a promoter; for loss of profits in a public house during a period when much of the car park was excavated for a pipeline scheme; and for loss of privacy within a dwelling-house garden crossed by a scheme.

Full and final settlement

Claimants should be wary of agreeing compensation in full and final settlement of all claims. Particularly on larger schemes it might be advisable to delay submitting the compensation claim until sufficient time has elapsed after the completion of the project for evidence to become available. Even then a full and final settlement should be confined to the losses claimed and reservations should be included in respect of subsidence, land drainage and other items not apparent at the settlement date.

Contractor's liability

Invariably promoters employ civil engineering contractors to install their equipment and it is not uncommon for such contractors to cause damage to the claimant's property by negligence, nuisance or trespass and in breach of their contract with the promoter.

While they may have a claim against the contractors concerned, claimants are advised to pursue all such claims against the promoter alone, for in *Smith, Stone & Knight Ltd* v *Birmingham City*

District Council [1988] EGCS 49 in Official Referee's Business in the High Court, it was held that the promoter could not deny the validity of a statutory claim by seeking to establish that the damage was outside the statute because it was attributable to the negligence of its contractors.

Claimants and their valuers should avoid any dealings with the contractors in matters directly related to the scheme. Otherwise there is a risk of them making an inadvertent contractual arrangement under which their claim will lie against the contractor rather than the promoter. Specifically, complaints arising during the scheme should be directed to, and any meetings arranged with, the promoter's representative and not the contractors. By the same token, requests from the contractors for additional working areas or additional access points should be refused unless redirected through the promoter.

It might be to the claimant's financial advantage to negotiate a site office, pipe dump or similar facilities for the sole use of the contractor, but such arrangements should be regarded as quite separate from the scheme. The claimant should ensure that his position is protected as he will not have any claim against the promoter for losses that result. Ideally a legal agreement should be signed before entry is taken, all rents or similar moneys together with fees paid in advance and a bond against claims held by a stakeholder.

Betterment

Where schemes are executed under compulsory powers, betterment can be set off against any compensation that would otherwise be payable whether for permanent or for surface damage. Strictly calculated, this can result in a nil figure and frequently does in the case of sewer schemes affecting residential property. Nevertheless, most promoters are willing to pay some compensation towards surface damage. The Lands Tribunal took a similarly benevolent view towards limited compensation for permanent damage in the case of *Raper* v *Atcham Rural District Council* (1971) 219 EG 1397 and the promoter did not pursue such an argument in the case of *Rush & Tompkins Ltd* v *West Kent Main Sewerage Board* (1963) 187 EG 47.

Interest

Whether this applies and the date it is payable from varies depending on the type of claim as detailed in the Planning and Compensation Act 1991, section 80, Part 1 of Schedule 18 to that Act.

For most schemes provision is made for interest at the statutory rate on any outstanding compensation for permanent damage calculated from the date of entry. The claimant's valuer should confirm that this is the case, particularly where private easements and wayleaves are being negotiated, and should consider the insertion of a commercial rate of interest in place of the statutory rate wherever possible.

Generally interest is not paid on compensation for surface damage from the date of entry, even on schemes other than sewers and water pipes, but from the date of submission of the claim. If interest is not applicable by statute valuers should consider the inclusion of such an item to run from the date of submission of the claim or from a reasonable date thereafter, to encourage prompt negotiation and settlement by promoters. Where possible the final settlement of compensation should include an agreement that interest will run if payment is not made by a specified date.

Fees

Valuers are advised to agree fees with their client when taking instructions rather than relying on any contribution that the claimant might receive from the promoter.

For schemes carried out under certain statutory powers a contribution to valuer's fees will be paid to the claimant calculated on Ryde's Scale (1996).

Several other promoters make contributions to valuer's fees on Ryde's Scale or some variation thereof but the electricity industry has a scale of its own and promoters sometimes introduce special scales for particular schemes. Wherever possible the claimant's valuer should ensure that the promoter reimburses valuer's and any solicitor's fees paid by the claimant in full and a clause to that effect incorporated in the deed of grant, particularly for private schemes.

Within Ryde's Scale there is provision for fees to exceed the scale in exceptional circumstances as explained in the general note to the scale.

Where the claimant is registered for value added tax (VAT) the promoter's contribution should be to the fee net of tax which the claimant will be able to reclaim from Customs and Excise. Where the claimant is not so registered the promoter's contribution should include VAT.

Disputes

Where valuers cannot agree the compensation payable, the usual recourse is to the Lands Tribunal for schemes executed under statutory powers. Otherwise most promoters provide for disputes on compensation for to be resolved by arbitration. The claimant's valuer should ensure that such a provision is included within the deed of grant, although on smaller private schemes an independent expert might be more appropriate than an arbitrator.

Claimants should be advised that the larger commercial promoters offer compensation for permanent damage on a higher basis than that awarded by the Lands Tribunal: see *Wells-Kendrew* v *British Gas Corporation* (1974) 229 EG 272 and *Sanders* v *Esso Petroleum Co Ltd* (1962) 182 EG 257.

The claimant's valuer should be aware that the onus of proof in a dispute lies with his client. Furthermore, a higher standard of proof is required by the Lands Tribunal and arbitrators than is common when valuers are negotiating privately, with considerable detail being required to prove items such as crop yields if these cannot be agreed.

The Lands Tribunal can award only one lump sum to cover past and future losses which may be a disadvantage if surface damage is likely to continue.

Study 1

This illustrates the calculation of a claim for compensation for permanent damage after the laying of a sewer through a tenanted farm.

Date of entry February 17. Date of claim July 31.

Compensation for pipeline			(£)
Length of pipeline (m) (See over, p670)	1155		
Value of land with vacant possession (£ per ha)	6000		
Width of land affected by pipeline (m)	7		
Depreciation in freehold value (%)	50		
Therefore depreciation per m run equals (£)		2.10	
Compensation for pipeline			2425.50

Extent of scheme

OS field number	Description	Length of pipe (m)	Manholes and references	Location of manholes
4724	Arable	366	–	
1557	Recreation ground	60	1 (A)	Verge
8993	Arable	348	–	
9756	Track and shave	13	–	
5899	Track and shave	11	–	
5198	Pasture	112	2 (B)	Centre of field which could be ploughed in future
			(C)	Centre of field which could be ploughed in future
3892	Farmyard	155	3 (D)	Beside fence line. Some betterment to farmhouse
			(E)	Close to building by gateway preventing extension of building concerned
			(F)	In area used for storage of machinery
5348	Pasture	48	–	
8721	Farmyard	42	2 (G)	Beside buildings
			(H)	Waste area
Total		1155	8	

Compensation for manholes

No	Manhole reference	per man-hole (£)	total (£)
2	B and C	150.00	300.00
2	D and E	125.00	250.00
1	G	85.00	85.00
1	F	50.00	50.00
2	A and H	35.00	70.00

Compensation for manholes		755.00
Compensation for permanent damage		3,180.50

Fees

Fee on Rydes Scale (1996) scale 2.4 table E	710.00	
Travelling and out of pocket expenses	7.50	
	717.50	
VAT @ 17.5%	125.56	
Fees		843.06

Note: The claimant is not registered for VAT

Grand total	4,023.56
Deduct advance payment requested on November 1 and received on February 20	2,700.00
Balance outstanding	£1,323.56

Further advance payment. Settlement of this claim is requested, or a further advance payment demanded, in accordance with the Water Industry Act 1991 before August 31.

Study 2

This illustrates the calculation of a claim for compensation for surface damage after the laying of a sewer through a tenanted farm.
    Date of entry May 18. Date of claim October 15.

Field OS No 3421

(Crop – winter wheat)		£
Loss of grain		
Length of area lost (m)	384.00	
Width of area lost (m)	7.62	
Yield from remainder of field (tonnes per ha)	7.78	
Price (£ per tonne)	85.00	
Therefore value of grain lost		193.5

Loss of straw
Area lost	as for grain	
Yield from remainder of field (tonnes per ha)	2.50	
Price (£ per tonne)	12.00	
Therefore value of straw lost		8.78

Extra tramlines
Length of area affected (m)	384.00	
Width of area affected (m)		
12 metre strip each side of pipeline	24.00	
Yield of grain from remainder of field	as above	
Extent of grain loss within area affected (%)	7.50	
Price	as above	
Therefore value of grain lost		45.71
Yield of straw from remainder of field	as above	
Extent of straw loss within area affected (%)	7.50	
Price	as above	
Therefore value of straw lost	2.07	

Removing corn. The promoter asked the claimant to mow the immature crop from the working area at an agreed price of 50.00

Extra time taken to work the field in two parts (based on Central Association of Agricultural Valuers (CAAV) costings)

	extra time for each operation (hours)	
2 sprays (225 litres per ha)	0.5	16.86
Combining	1	81.71
Baling and straw carting		
Say		10.00

Loss of crop in cut off area
(by agreement of the parties an area of severed crop was left without access from May until August. This could not be sprayed and suffered from disease as a result)

Area affected (hectares)	0.35	
Yield of grain from remainder of field (tonnes per ha)	7.78	
Extent of grain loss within area affected (%)	10.00	
Price	as above	
Therefore value of grain lost		23.15

	extra time for each operation (hours)	
Extra cultivations to restore working area (based on CAAV costings)		
1 cultivation	1.5	25.52
1 disc harrow	1.5	25.52
1 power harrow	1.5	61.14

Future losses of crop
(The project was carried out in a satisfactory manner by the promoter in reasonably dry weather. No unusual long term damage is envisaged and the parties have agreed to settle a figure now)

Area affected (ha)	0.29	
First year after reinstatement. Crop – winter barley		
Estimated yield of grain from remainder of field (tonnes per ha)	7.50	
Extent of grain loss within area affected (%)	50.00	
Price (£ per tonne)	80.00	
Therefore value of grain lost		87.00

Second year after reinstatement. Crop – oil seed rape		
Estimated yield from remainder of field (tonnes per ha)	3.71	
Extent of grain loss within area affected (%)	25.00	
Price (£ per tonne)	155.00	
Therefore value of oil seed rape lost		41.69

Third year after reinstatement. Crop –winter wheat		
Estimated yield of grain from remainder of field (tonnes per ha)	7.78	
Extent of grain loss within area affected (%)	10.00	
Price (£ per tonne)	85.00	
Therefore value of grain lost		19.18

Fence repairs
(Further re-inforcing needed of promoter's reinstatement)
North fence. Extra strand of wire, one stake and labour at CAAV rates
Say 7.50
South fence. Extra strand of wire, one stake and labour at CAAV rates
Say 7.50

Field OS No 5198
(Crop – long ley grazed by sheep)

Length of area lost (m)	119.00	
Width of area lost (m)	6.40	
Value of grazing at gross margin of (£ per ha)	500.00	
Therefore value of grazing lost		38.08

Extra cultivations to restore working area
(based on CAAV costings to include time taken
to attach implement, travel to field, undertake
task, return and remove implement)

	extra time for each operation (hours)	
1 power harrow	0.75	30.57
1 man, tractor and trailer stone picking	4	60.74

Reseeding
(based on CAAV costings to include time
taken to attach implement, travel to field,
undertake task, return and remove implement.
Note: 2.5 hours is the minimum time needed
to assemble, adjust, use and dismantle the seed
drill)

	time for each operation (hours)	
1 power harrow	1	40.76
1 fertilizer application (250 kg per ha)	0.5	9.85
1 seed drill	2.5	59.34

Cost of fertilizer 250kg per ha @ (£)125 per tonne		2.38
Cost of seed		
Say		5.00

Future loss of grazing
(see comments on future losses of crop in
OS 3421 above)

Area affected (ha)	0.08	
Value of grazing at gross margin of (£ per ha)	500.00	

First year after reinstatement

Extent of loss within area affected (%)	50.00	
Therefore value of grazing lost		20.00

Second year after reinstatement

Extent of loss within area affected (%)	25.00	
Therefore value of grazing lost		10.00

Third year after reinstatement		
Extent of loss within area affected (%)	10.00	
Therefore value of grazing lost		4.00

Fence repairs
(Further re-inforcing needed of promoter's reinstatement)

North fence.	@ (£)	
19 m netting	0.85	16.15
38 m barbed wire	0.09	3.42
1 driver fence post		5.00
1 strainer fence post		3.50
10 chestnut stakes	1.00	10.00

Labour at CAAV rates		
6 man hours	7.21	43.26
3 tractor hours	5.79	17.37
3 fencing equipment hours	2.50	7.50

South fence.		
12 m netting	0.85	10.20
24 m barbed wire	0.09	2.16
1 driver fence post		5.00
1 strainer fence post		3.50
6 chestnut stakes	1.00	6.00

Labour at CAAV rates		
6 man hours	7.21	43.26
3 tractor hours	5.79	17.37
3 fencing equipment hours	2.50	7.50
Occupier's time spent reorganising affairs		150.00
Compensation for surface damage		1336.67

Fees	(£)	
Fee on Ryde's scale 2.4 table E	625.00	
Travelling and out of pocket expenses	15.00	640.00
Note: The claimant is registered for VAT		

Grand total		1,976.67
Deduct advance payment requested on		
February 2 and received on June 1		200.00
Balance outstanding		1,776.67

Further advance payment. Settlement of this claim is required or a further advance payment demanded in accordance with the Water Industry Act 1991 before November 15
Reservations. The right is reserved to claim additional compensation for items not included above

Study 3

This illustrates the calculation of compensation for permanent and surface damage after the laying of an 18-inch gas main through an owner-occupied arable farm.

Date of entry May 17. Date of claim October 24.

Compensation for permanent damage			(£)

(The gas company have acquired a 6 m wide easement and have an agreement with the Country Landowners Association (CLA) and National Farmers Union (NFU) to pay compensation for permanent damage based on 80% of the freehold value)

Length of pipeline (m)	163		
Value of land with vacant possession (£ per ha)	7500		
Width of easement (m)	6		
Depreciation in freehold value (%)	80		
Therefore depreciation per metre run equals (£)		3.60	
Compensation for permanent damage			586.80

Note

(1) The gas company will make an advance payment of (£) 528.12 being 90% of the compensation for permanent damage on the date of entry or on proof of title, whichever is the later

(2) The gas company will pay interest at the statutory rate on all outstanding compensation for permanent damage from the date of entry

(3) The balance of compensation for permanent damage, together with interest due thereon, is based on the final length of the pipeline measured after it has been laid

Occupier's payment		
Length of pipeline (m)	163	
Rate per m (£)	1.50	
Occupier's payment (£)		244.50

Note

(4) This payment, which is calculated as agreed between the gas company and the CLA and NFU, is made direct to the occupier on or about the date of entry

Compensation for surface damage		(£)
Field OS No 3421 (Crop – bulb onions)		
Length of area lost (m)	163.00	
Width of area lost (m)	9.00	

Yield from remainder of field (tonnes per ha)	17.91	
Price (£ per tonne)	100.00	
Therefore value of crop lost		262.74

Extra headland west
(An extra 9 m headland had to be created to the
west of the pipeline. The crop in this area was
severely damaged by wheelings as the claimant
carried out normal husbandry operations.)

Length of area damaged (m)	163.00	
Width of area damaged (m)	9.00	
Yield from remainder of field (tonnes per ha)	as above	
Price (£ per tonne)	as above	
Amount of crop loss (%)	85	
Therefore value of crop lost		223.33

Extra headland east and cut off portion
(The cut off area could not be sprayed resulting
in a substantial shortfall in crop yield)

Area damaged (ha)	0.29	
Yield from remainder of field (tonnes per ha)	as above	
Price (£ per tonne)	as above	
Amount of crop loss (%)	75	
Therefore value of crop lost		389.54

**Extra time taken in working the field in two
parts** (based on Central Association of
Agricultural Valuers (CAAV) costings)

	extra time for each operation(hours)	
4 sprays (225 litres per hectare)	0.5	34.63
Harvesting by hand in cut off portion	24	165.12

Reinstatement of hedges. Lowest of three estimates from landscape gardeners for replanting hedges damaged by scheme, maintaining them for three years and replacing losses	1200.00
Extra top soil. Cost of purchasing and spreading two loads in north east corner	65.00
Additional weed killing. An area of cultivated horseradish was crossed by the scheme and roots have spread as a result. Allow to spray resulting volunteers	65.00

Claimant's time	(hours)		
January 28 letter to valuer	0.30		
March 20 dealing with letters, pre-entry consents solicitors etc	1.00		
March 21 producing 5 years cropping records	1.50		
May 10 meeting gas company's representative	2.00		
May 24 meeting gas company's representative	0.25		
June 15 inspecting working area	0.30		
June 29 inspecting working area	0.25		
July 13 inspecting working area	0.25		
July 27 inspecting working area	0.25		
September 8 final meeting with gas company's representative	0.75		
October 20 providing valuer with information for compensation claim	1.00		
Total	7.85		
Say 8.50 hours until November 23		300.00	
Compensation for surface damage			2,705.36
Total compensation for permanent and surface damage			3,292.16
Fees			
Fee on Ryde's scale 2.4 table E		710.00	
Travelling and out of pocket expenses		25.00	735.00

Note

(5) The claimant is registered for VAT
(6) The occupier's payment is regarded by the gas company as being ex gratia and accordingly is excluded from the fee calculation

Grand total	4,027.16
Deduct advance payment received on May 17	500.00
Balance outstanding	£3,527.16

Interest. Interest is claimed on the outstanding compensation for permanent damage from the date of entry

Further advance payment. Settlement is requested, or a further advance payment demanded, before November 23rd

Reservations. The right is reserved to claim for additional losses not included above

Study 4

This illustrates the calculation of compensation for surface damage following the laying of a water main through an intensive tenanted dairy farm.

Date of entry June 1. Date of claim September 4.

Growing crop of intensive grass for silage and aftermath grazing		(£)
Reduction in yield of first-cut silage taken 3 weeks early to allow entry by promoter		
Area affected. Whole field of (ha)	8.50	
Reduction in yield (tonnes per ha)	5.00	
Value (£ per tonne)		
33% of a hay price of (£ per tonne) 80	26.40	
Therefore value of silage lost		1,122.00
Total loss of second cut silage		
Length of area lost (m)	1,012.00	
Width of area lost (m)	15.00	
Therefore area affected (ha)	1.52	
Yield from remainder of field (tonnes per ha)	22.25	
Value (£ per tonne)	26.40	
Therefore value of silage lost		891.67
Loss of aftermath grazing		
Area affected (ha)	1.52	
Value of grazing at gross margin of (£ per ha)	125.00	
Therefore value of grazing lost		189.75
Loss of second cut silage from inaccessible cut off area		
Area affected (ha)	0.32	
Yield from remainder of field (tonnes per ha)	22.25	
Value (£ per tonne)	26.40	
Therefore value of silage lost		187.97
Loss of aftermath grazing from inaccessible cut off area		
Area affected (ha)	0.32	
Value of grazing at gross margin of (£ per ha)	125.00	
Therefore value of grazing lost		40.00
Cost of reseeding		
Area affected (ha)	1.52	

Cost based on lowest of three quotations from local agricultural contractors (£ per ha)	350.00	
Therefore total cost of reseeding		531.30
Extra sprays to prevent weed infestation in reseeded area		
Area affected (ha)	1.52	
Cost (£ per ha)	47.85	
Therefore total cost of spraying		72.64
Allowance for patching up and chickweed sprays after first cut of silage from new sowing		
Area affected (ha)	1.52	
Cost (£ per ha)	43.50	
Therefore total cost		66.03
Allowance for managing severed areas		50.00
Compensation for surface damage		3151.36
Fees	(£)	
Fee on Ryde's scale 2.4 table E	710.00	
Traveling and out of pocket expenses	15.00	
Note: The claimant is registered for VAT		725.00
Grand total		3,876.36
Deduct advance payment received on June 1		2,150.00
Balance outstanding		£1,726.36

Further advance payment. Settlement of this claim is required or a further advance payment demanded in accordance with the Water Industry Act 1991 before October 3

Reservations. The right is reserved to claim additional compensation for items not included above.

© Exors. of DC Elgar (Dec'd) 2000

Chapter 21

Assessments for Insurance Purposes

Insurance is essentially a device by which the losses of the few are made to fall as lightly as possible upon the many. It is a social arrangement which provides financial compensation for the effects of misfortune, the payments being made from the accumulated contributions of all parties participating in the scheme.

While insurance assessment is primarily concerned with cost and value, the surveyor's knowledge cannot be restricted solely to this area. He must be clearly aware of what is to be insured, what risks or perils can be covered, what provisions a policy contains and how appropriate insurance can be arranged.

What is to be insured?

Apart from its human resources there is no doubt that the principal assets of any company will be the buildings it occupies to carry on its business, the machinery it utilises to manufacture its goods for sale and its stock and work in progress. Such assets will be protected in terms of material security by wall, gates, roofs and alarms but they should also be protected in terms of financial security by insurance. A prudent businessman will also insure probably by separate policies a potential loss of profit, legal liability to the public or to his employees and such other risks as are thought appropriate.

A prudent householder will ensure that his major investment, the fabric of his house, is protected by insurance and will in the same way protect house contents, liability and money.

What risks or perils can be covered?

The standard fire policy

The standard fire policy for a commercial concern would cover the following perils:

(a) FIRE, subject to certain exclusions and provisions regarding origin and ignition.

(b) LIGHTNING, whether fire results or not.
(c) EXPLOSION,
    (i)   Of boilers used for domestic purposes;
    (ii) In a building not forming part of a gas works, of gas used for domestic purposes or for lighting or heating the building.

The exclusions referred to under section (a) above are where fire is occasioned by or happens through:

(i)   its own spontaneous fermentation or heating or its undergoing any process involving the application of heat;
(ii)   earthquake and subterranean fire;
(iii) riot and civil commotion, and
(iv) insurrection, rebellion, war, invasion and other acts of foreign enemy.

Special perils extension

There are times, usually when the economy is affected by recession, when businesses might for reasons of cost restrict their insurance cover to that provided by the standard fire policy. There are other times when insurers refuse to provide an optimum level of cover or any cover when they consider the risk of a claim being made to be high. This could be because the business to be insured is situated in a high risk inner city area, where there are frequent malicious attacks or where there is a high risk process being carried out.

Most modern businesses would regard the cover provided by a standard fire policy to be insufficient to meet their requirements. For example if the building from which they trade was damaged by storm, flood or burst pipe the standard fire policy would not provide any insurance cover. In such circumstances if cover for these additional risks is required they would add "special perils" insurance to the benefits provided by a standard fire policy.

The special perils for which insurance cover is normally provided can be separated into four main groups:

(a) Perils of a chemical type:
    Explosion
    Spontaneous combustion, heating or fermentation.

(b) Social perils:
    Riot, civil commotion, strikers, locked-out workers or persons taking part in labour disturbances, or malicious persons and theft.

(c)   Perils of nature:
Storm and tempest
Flood, hail and thunderbolt
Earthquake
Subterranean fire
Subsidence and landslip.

(d)   Miscellaneous perils:
Bursting or overflowing of water tanks, apparatus or pipes
Aircraft
Impact by vehicles, horses or cattle.

Cover for special perils is evidenced not necessarily by a separate policy but by an endorsement to a fire policy.

Insurable interest

All property whether material or represented as a right is insurable provided that the party seeking insurance has an insurable interest in the property or right for which insurance is sought. The insured need not, however, have an insurable interest at the time insurance is effected but he must have such an interest at the time of the loss. For example, a freeholder has an insurable interest in the property which he owns because if it is destroyed he stands to sustain a loss. However, he will not have an insurable interest in an adjoining property which he does not own or lease even though its destruction could cause him loss in terms of, say, loss of support for his own buildings.

Utmost good faith

Insurance is based on the doctrine of *uberrima fides* or utmost good faith. This means that the insured, the policyholder, is required to disclose to the insurer any information which is likely to cause the insurer to review the basis upon which he is insuring a property or refuse to insure it at all. Failure to disclose material facts would make the policy voidable. For example, failure by a policyholder to declare that a building, for which insurance is sought, shows cracks in the brickwork, even though they may be old, could make the policy voidable.

Calculation of rebuilding costs

Method of measurement

To enable a client to fix a sum insured, the method to be adopted in assessing building costs for insurance purposes requires that industrial and commercial properties be measured to obtain the gross internal floor area. This is attained by taking measurements from inside external walls, including all areas within the envelope of the building, and multiplying the result by the number of floors in the building. It is not unusual to obtain the cube measurements of the property by taking heights of individual floors.

Houses are measured to attain the gross external area by taking measurements from outside external walls and multiplying the result by the number of floors.

The RICS/ISVA Code of Measuring Practice provides guidelines for the measurement of properties for various valuation purposes.

Property type

The unit cost of a building will depend on many factors but the surveyor will need to establish a basic description of the property, whether it is single- or multi-storey, a factory, warehouse or office building and, in the case of a house, detached, semi-detached, terraced or bungalow.

Construction type

As complex foundations can add considerably to the cost of reconstruction, it will be necessary to ascertain details of the site and any particular problems which it might present. Principal construction details will be required such as whether the building to be assessed is of traditional brick and slate construction, a light steel-framed metal-clad or a reinforced-concrete-framed brick building. A surveyor will need to establish such internal features of the building as partition walls, suspended ceilings and such services provided as lighting, heating, fire and intruder protections. The survey will encompass details of thickness of external walls, quality of floor and roof finishes, types of windows and doors, quality of internal and external finishes and the extent of internal and external services.

Method of assessment

The calculation of rebuilding costs to fix a sum insured is generally by applying unit cost rates to superficial or cube measurements of a building, adjusting the rates to particular types of building. In certain cases it will be preferable to produce priced full or elemental bills of quantities but these are labour intensive exercises and therefore costly. However, following destruction of a building which is to be rebuilt there is probably no alternative to the provision of a specification and full bill of quantities using the services of a quantity surveyor and others.

Sources of cost information

Surveying practices involved with building projects will be able to analyse details of costs and record them for future assessment use. Other sources of building costs are the Building Cost Information Service of the RICS, which provides, *inter alia*, quarterly reviews of building prices taken from actual tenders, and technical books such as Spons and Laxtons.

The *Guide to House Rebuilding Costs* published by the Building Cost Information Service and by the Association of British Insurers is a useful guide to the cost of domestic construction costs.

Additions to the basic assessment

So far we have reached the stage of establishing the cost of rebuilding a particular property but there are several additions required to complete an assessment suitable for insurance purposes.

In assessing the replacement cost of a particular structure a valuer needs to consider what affect current regulation would have on design, materials and methods if the building were to be destroyed by a fire. For example, it is possible that an existing old factory with a steel frame and corrugated iron walls and roof, without proper insulation, could continue in use indefinitely. If, however, it were seriously damaged by fire it would be subject to current planning policy and building regulations. If the owner of the property intends to rebuild it in a modern form using modern materials (and perhaps altering the size and height) a new planning permission would be required. If the owner's new plans do not conform with current planning policy, permission could be refused.

This could force the owner to move to a different, more suitable site which, if ground conditions are poor, could increase the cost of building the sub-structure compared with the original site. The valuer needs to be aware of such matters so that the sum insured set for the existing building is sufficient to cover the higher sub-structure costs and the cost of complying with current regulations.

While the replacement of an old factory on an existing site following a fire may be acceptable to the planning authority, it is almost certain that building regulations will force the owner to improve the replacement building, requiring possibly more substantial foundations, better insulation, lighting and means of escape. This too will impact on cost and the pre-fire cost assessment must take account of such requirements.

A valuer preparing a cost assessment for insurance purposes needs a crystal ball to foresee the future and to reflect in his assessment the cost of local and statutory authorities' requirements. However in commercial property insurance to ensure that the insurance policy will compensate the factory owner for increased construction costs, it needs to be specifically amended to incorporate a number of additional valuable clauses. (Such clauses are usually automatically included in a policy insuring a domestic property.)

Further additions to the basic assessment of building costs will be required. First, the damaged building will require demolition and the valuer needs to add for this and site clearance. Second, when the insured contemplates a new factory to replace the one destroyed he will do so probably with the assistance of a construction team composed of one or all of the following: architect, quantity surveyor and engineer. He will also be required to appoint a planning supervisor under the Construction (Design and Management) Regulations 1994 which impose legal obligations on the construction team in the area of health and safety. As will be seen later a surveyor will also make allowances for debris removal cost and professional fees.

Public authorities clause

This clause, which must be specifically included in the policy for the insured to gain its benefits, covers the additional cost incurred in reinstatement of damage to comply with building or other regulations made in pursuance of any Act of Parliament or bye laws. The costs to be incurred under this item are not separately insured but are to be included in the overall sum insured covering the buildings.

Debris removal clause

As with the public authorities clause it is necessary to extend the policy to include for debris removal. By removal of debris it is usually meant to refer to the cost of removing debris, demolishing, shoring or propping up necessarily incurred by the insured with the consent of the insurer. As far as buildings and machinery are concerned the cost of removal of debris should be included within the relevant cost assessment which forms the sum insured. However, the removal of debris left from destroyed stock and materials-in-trade has to be insured as a separate item with a separate sum insured.

Architects', surveyors' and consulting engineers' fees clause

Professional fees are not automatically insured and special provision by means of a relevant clause in the policy must be made for them. They will normally form part of the overall sum insured on buildings and relate to fees necessarily incurred in reinstating or repairing the damage, not for preparing a claim. Professional fees will normally include the cost of appointing a planning supervisor in accordance with the Construction (Design & Management) Regulations 1994.

Fees for building regulation and planning applications

As the cost of fees for building regulations approval and for planning applications can amount to several thousand pounds in the case of industrial and commercial building projects, allowance should be made for them in the assessment at appropriate levels.

Value Added Tax

There are two reasons why VAT may not need to be taken into account when assessing value for insurance purposes. Firstly, the building works may be zero-rated (VAT free) and secondly, if VAT is chargeable on the building works, it may be reclaimable from Customs & Excise by the insured business using the property.

1. *Zero-rating is available for certain works to the following types of buildings:*
(a) new dwellings
(b) charitable buildings (for non-business use or as a village hall or similar)
(c) residential buildings (communal residences e.g. nursing homes, student accommodation, hospice)

For these buildings, new building works in the event of complete reconstruction following a fire or other damage are zero-rated. Repair work to such buildings will be standard-rated. It has been accepted by insurance companies that the sum insured on the building of a dwelling does not need to include VAT, despite the fact that claims for partial damage will attract VAT. In such cases insurers will reimburse the cost of repairs plus VAT.

2. *Recovery of VAT on building works*
An insured business that is registered for VAT and owns a property will be able to reclaim VAT incurred on works to the building if it is occupied by the business for the purposes of making taxable supplies (zero-rated, lower-rated or standard-rated). For example, a VAT registered manufacturing business will be able to reclaim VAT on works to the factory that it owns and occupies to manufacture its products for sale. Consequently, insurers exclude VAT payments from claims by such businesses. Assessments of buildings and plant and machinery belonging to such VAT registered businesses need not, therefore, include VAT.

However, certain businesses e.g. financial institutions and private schools, which are exempt for VAT purposes, will not be able to reclaim VAT on works to the buildings that they own and occupy. Exempt businesses do not have to charge VAT to their customers, and are not able to reclaim VAT on their expenses. This means that the complete reconstruction of buildings owned and occupied by such organisations will attract VAT that they will not be able to recover from Customs. In these circumstances, the assessment of the cost of buildings and contents of such organisations for insurance purposes must include VAT.

In some instances businesses may have some income which is liable to VAT and some which is exempt. In these cases, the business is partly exempt and a proportion of the VAT on the works will be recoverable if the business is registered for VAT.

Where property is held as an investment to provide rental

income, the VAT situation will depend on the type of property and whether or not the investor has elected to waive exemption. The letting of dwellings, charitable property and communal residential property referred to in (1) above is exempt from VAT and therefore no VAT is recoverable on any works to such property.

The letting of commercial buildings is exempt unless the investor has elected to waive the exemption. If an election has not been made, no VAT is recoverable on works to the building and this must be taken into account when assessing the cost of building for insurance purposes. However, if the investor has elected to waive exemption, this usually means that VAT is chargeable on the rent and therefore VAT is recoverable on works to the building. Where an effective election is in place, no VAT need be taken into account when assessing the cost of the buildings for insurance purposes. There are some instances however, when even though an election has been made, the election is disapplied (e.g. for supplies between connected parties when certain anti-avoidance provisions may apply). In these circumstance VAT on works to the building will not be recoverable. If a building is completely destroyed the investor's election on the original building will end and it will be necessary for the investor to make an election in respect of the new building if VAT is to be recoverable on the construction of the new building.

Insurance premium tax

Since October 1994 the government has levied a tax on insurance premiums which is presently set at 5% of the premium total. Unlike VAT there are no circumstances in which it can be avoided or recovered.

Insurance of buildings

Valuers will be aware of the two main bases of cover available under which buildings can be insured, namely indemnity and reinstatement, and perhaps be familiar also with three little used special types of cover for buildings, namely "first loss", "obsolete buildings" and "modern materials".

Indemnity

The standard fire policy allows the insurer to:

pay to the insured the value of the property at the time of the happening of its destruction or the amount of such damage or at its option reinstate or replace such property or any part thereof.

Insurers rarely ever take the option to reinstate or repair an insured's property but it is a device available to them if they are dealing with a difficult or intransigent insured. It is an option, however, which is of doubtful advantage to the insurer since there would be the possibility of a continuing dispute with a dissatisfied claimant.

The insurers have agreed by the above wording to pay "the value of the property" but the policy does not define "value". It is reasonable to assume, however, that its normal tort definition would apply where value would be that amount which would place the insured by payment or otherwise in the same position as he was immediately before the fire, neither better nor worse. In these circumstances indemnity value would probably be the cost of reinstatement or repair less a deduction for depreciation, since the replacement of a fire damaged roof or walls has provided something better because they are newer rather than superior or more extensive. If the reinstated building were to be more extensive or superior than the original then the additional costs involved would not be recoverable from insurers. Indemnity value could be, however, diminution in open market value.

Study 1

The insured owns an old four-storey factory in Birmingham with a gross internal floor area of 4,000 m². The building has brick walls, timber floors and a slated roof on timber trusses. It is in reasonable condition, is heated by warm air units and is protected by a sprinkler installation. Because of the recession the insured's business is failing and occupies only two floors of the premises. He has been advised to insure the buildings on an indemnity basis. What sum insured should he select?

Gross internal area (m²)	4,000
Reinstatement cost per m²	£400
	£1,600,000
Add:	
Professional fees and debris removal costs, Say 15%	240,000
	£1,840,000

Deduct:

Depreciation, age, wear and tear, Say 40%	736,000
Indemnity value	£1,104,000

There are at least two important points to be raised in connection with study 1. First, there can be no set rule as to the amount of the deduction for depreciation and each case must be dealt with on its merits. If the building in question had been erected only one year ago and it could be expected to last for 40 years, the annual depreciation could be as little as 2.5%. Second, it is important to realise that the reinstatement cost of the building which is used to arrive at the indemnity value should be that prevailing at the time of reinstatement not that which would be appropriate at the time of the loss.

In deciding what should be the correct value for insurance purposes, the surveyor might need to consider the merits of a sum insured based on a market value. There are many cases where the cost of reinstating a building which has been badly damaged or destroyed will be far greater than the market value of the reinstated building and the insured might decide that it would be pointless to reinstate. In the case of *Reynolds & Anderson* v *Phoenix Assurance Co Ltd* [1978] 2 EGLR 38, the judge believed the evidence of the insured that they intended to reinstate and awarded them £343,320, which was the estimated cost of reinstatement duly adjusted for betterment. The insurers' initial calculation of the loss was on a diminution in market value basis of £5,000 against a total sum insured of £550,000. They later increased their offer to £55,000, arguing that this would be a fair settlement representing, as it did, the cost of a modern equivalent building, since no commercial man in his senses would think of spending so much on an obsolete building if he could buy a modern structure for so much less.

However, settlement of a claim in the case of *Leppard* v *Excess Insurance Co Ltd* [1979] 1 EGLR 50 was found to be correct on a market value basis where the insured had placed his house on the market for sale shortly before it was destroyed by fire. The insured argued that an indemnity was £8,694, the cost of reinstatement less an allowance for betterment, despite the fact that he expected to receive £4,500 from the proceeds of sale.

If the calculation in study 1 is revised so that the indemnity value is on a market value basis, the sum insured would be radically different.

Study 2

The insured has decided that in the event of serious damage to the four-storey factory in Birmingham (see study 1) he would not intend to repair or rebuild the damaged premises. In such a case what are the insurers likely to offer in settlement of a claim for total destruction?

Gross internal area (m²)	4,000
Capital value per m²	£80
	£320,000
Add:	
The cost of demolishing and removing building debris	20,000
	£340,000
Deduct:	
Market value of the site which will remain in the insured's ownership	80,000
Indemnity value (diminution in market value basis)	£260,000

It will be apparent that there is a considerable difference between the indemnity value of £1,104,000, calculated by reference to depreciated building costs, and the indemnity value of £260,000 calculated on the basis of diminution in market value. The criterion to be adopted must be this. If there is an intention to rebuild, the basis of insurance must be depreciated replacement cost at £1,104,000 but it does not follow that if the insured expresses an intention not to rebuild, in the event of a serious loss, he should insure the building on a market value basis. What if the property is only partially damaged to the extent of repair costs of £10,000? He will most certainly wish to repair it and he will expect insurers to pay in full. As the basis of the payment of £10,000 will be depreciated replacement cost, the sum insured must also be on this basis at £1,104,000, otherwise the insured's claim will be reduced substantially by average (see under "Average conditions" later).

In short, the indemnity value of a building for insurance purposes should be based on reinstatement cost less an allowance for betterment. But if the building is destroyed or badly damaged and the insured does not reinstate he can expect insurers to propose to settle the claim on the basis of diminution in market value, which could be at a much lower figure than the sum insured.

Insurers have devised other methods of insuring buildings which allow the insured to select a lower sum insured than that provided by reducing the reinstatement cost for depreciation without suffering the penalties for under-insurance. These will be considered later.

Reinstatement

Under statute

It will be recalled that under the standard fire policy the insurers have an option to reinstate a building if they so wish, but the option is seldom used. However, insurers can be required by statute to reinstate or replace any property. The Fires Prevention (Metropolis) Act 1774 contains provisions which relate to the reinstatement of houses and other buildings and which require insurers "to cause the insurance money to be laid out and expended, as far as the same will go, towards rebuilding, reinstating or repairing . . ." unless within a specific period the insured gives an assurance that he will himself expend the money or unless the insurance money is laid out within the same period to the satisfaction of all parties. The most usual circumstances where the provisions of the Act are invoked is where a tenant gives notice under the Act to the insurer of buildings so that the tenant can be sure that the landlord will be required to expend moneys in the reinstatement of the building.

Reinstatement memorandum

Insurance arranged on a reinstatement basis provides the optimum level of cover but, to be effective, the policy must contain the "reinstatement memorandum". The memorandum provides that the amount payable in respect of any destroyed item shall be the reinstatement of the item destroyed or damaged subject to certain special conditions.

Reinstatement is defined in the memorandum as:

The carrying out of the aftermentioned work, namely:
(a) Where property is destroyed the rebuilding of the property, if a building, or, in the case of other property, its replacement by similar property, in either case in a condition equal to but not better or more extensive than its condition when new.
(b) Where property is damaged, the repair of the damage and the restoration of the damaged portion of the property to a condition

substantially the same as but not better or more extensive than its condition when new.

It will be apparent that a reinstatement policy goes further than an indemnity policy, which places the insured in the same position as he was before the fire, since it allows him to rebuild, replace or repair a property to a condition equal to or the same as "its condition when new".

It is often overlooked at the time of a valuation that the reinstatement memorandum does not allow an insured to obtain settlement of a claim for replacement or repair on a reinstatement basis if the act of replacement or repair is not carried out. This and other controls on the insured are exercised by the "special provisions" which form part of the reinstatement memorandum and which, because of their importance, deserve some mention.

The work of reinstatement may be carried out on another site provided that the liability of the insurers is not increased. This means that the insurers would not object to the insured's decision to relocate to London a factory destroyed in Glasgow provided that the cost to insurers would not be greater. However, inclusion in the policy of the Public Authorities Clause would allow for increased costs at the new site if it were a requirement of a statutory body that the insured should vacate the present site following damage to buildings and insurers accept the arguments.

The special provisions provide that the work of reinstatement must be carried out with reasonable despatch, which is taken to mean that there must be no unreasonable delays in the rebuilding or repair of a property which would result in insurers being asked to pay a higher amount.

Special provision 3 is of vital importance and knowledge of it should prevent a surveyor from valuing a building on a reinstatement basis which he knows his client will not intend to rebuild or repair. It states:

> 3. No payment beyond the amount which would have been payable under the policy if this memorandum had not been incorporated therein shall be made until the cost of reinstatement shall have been actually incurred.

If the cost of reinstatement is not incurred, the insured will only be able to recover settlement of the claim on an indemnity basis.

If the insured does intend to reinstate but it will be a long time before completion of the reinstatement or replacement he can apply for a payment on account equal to an indemnity settlement (probably on a diminution in market value basis) pending completion of the reinstatement and payment therefor.

At this point it would be useful to consider a basic reinstatement assessment and to extend the calculation later to show the effect of the average clause.

Study 3

The gross internal floor area of six small, modern factories in one building on an open site in Edinburgh is 1,079 m^2. The construction consists of a steel portal frame, brick facings and shallow pitched roof with roof sheeting. The services provided are basic, each unit having toilet facilities, nominal power and lighting installations but no heating. Define the reinstatement cost for insurance purposes including professional fees and debris removal costs. Make allowance for inflation bearing in mind that it would take six months to rebuild the property.

Total gross internal floor area (m^2)	1,079
Reinstatement cost per m^2	£320
	£345,000
Architects', surveyors' and engineers' fees at Say 12.5%	43,000
	£388,000
Debris removal costs	10,000
Reinstatement cost (base cost)	£398,000
Inflation provision, insurance year plus 5%	19,900
	£417,900
Inflation provision, building period 6 months plus 2.5%	10,447
Reinstatement cost for insurance purposes	£428,347
Say	£430,000

A note about inflation

Study 3 shows not only a base reinstatement cost of £398,000 but also additions to this cost to cover inflation. The danger of fixing the sum insured at £398,000 without any inflation provision are easily explained. On January 1 the sum insured of £398,000 would represent the cost of reinstating the six small factories in Edinburgh but one month later, with inflation running at 5% pa, the cost of reinstatement would already have increased to £400,000. By June 30 the sum insured of £398,000 would be under-

Valuation: Principles into Practice

assessed by 2.5% and the cost of reinstatement would have increased to £408,000.

By November 30 the reinstatement cost of the buildings in question will have increased by 4.6% since January to £416,000. If the buildings were destroyed by fire on November 30 the insured would receive insufficient funds to reinstate them. Consider the following facts:

Event: The total destruction of factory units on November 30

Sum insured	£398,000
Cost of reinstatement on January 1	£398,000
Inflation January 1 to November 30 plus 4.6%	18,308
Cost of reinstatement on November 30	£416,308
Inflation during the rebuilding period, 6 months, plus 2.5%	10,408
Cost of reinstatement in the following year when the buildings have been replaced	£426,716
Say	£427,000

Provided that the insured reinstates the factory units in question and incurs £427,000 he will receive £398,000 being his own insurer for £29,000.

The practice of building inflation provision into the sum insured is one method of dealing with the problem of under-insurance. There are others including the most important and popular device "Day-One Reinstatement" which will be considered later.

Average conditions

Ordinary pro rata average

Most valuers will be aware of the punitive effect of the average clause which is now incorporated into most classes of fire and special perils insurance. The pro rata condition of average operates where there is under-insurance and provides that the insurer will be required to pay only that proportion of the loss which the sum insured bears to the true indemnity value of the property at the time of the fire. The pro rata condition of average affects assessments calculated on an indemnity basis. It is an attempt to ensure that the indemnity value is correct at the time of the fire, not at the time of reinstatement.

In study 1 the indemnity value of the factory in Birmingham was calculated to be £1,104,000. Let us assume that the insured decided not to insure the property to its full value but fixed the sum insured at £500,000. Shortly after the valuation the property was damaged by fire and the cost of reinstatement was estimated to be £12,000. The repairs would provide some improvement, so that an equitable assessment of the loss on an indemnity basis, ignoring under-insurance, would be £10,000. However, it is clear that there is under-insurance and that average will apply on the following basis:

Pro rata condition of average formula:

$$\frac{\text{Sum insured}}{\text{True indemnity value at the time of the fire}} \times \text{Loss}$$

$$\frac{£500,000}{£1,104,000} \times £10,000$$

Insurer's liability £4,528

Reinstatement average

Policies containing the reinstatement memorandum are also subject to average but its application is slightly different from that of pro rata average. As we have seen, pro rata average requires that the sum insured should be equal to the value of the property at the time of the fire. Reinstatement average determines that the sum insured should be equal to the value of the property at the time of reinstatement which, in a large building contract, could be two or three years after the destruction.

Combating inflation

During periods of high inflation it is difficult for an insured to ensure that the level of insurance to cover his assets remains at adequate levels. If it does not, a claim could be reduced by the application of the average clause, as described earlier. In the last few years building and other costs have risen less steeply so that some of the devices to combat inflation are not as common. The surveyor dealing with insurance assessment should be aware of the following, all of which must be specifically added to the policy for their benefits to be gained.

Day-one reinstatement scheme

With the day-one scheme, which is now the most popular anti-inflation device, the insured declares the total rebuilding cost at prices ruling at renewal each year. Cover for inflation is provided by reason of the fact that the sum insured is stated to be a certain percentage more than the declared value at the beginning of the insurance year. However, the declared value must be accurate because it will be used as the basis for the average calculation at the time of the loss. An additional premium is levied by insurers to provide payment for the inflation element.

Escalator clause

This clause can be incorporated into the policy at extra cost and it provides that the sum insured at the renewal date will be increased each day automatically by 1/365 of the specified percentage increase. The specified percentage increase is the insured's estimate of the annual increase in inflation. Its disadvantage is that the addition stops at the time of the loss and because of this there will be no insurance cover for any cost increases which take place after the fire before rebuilding is completed.

85% Reinstatement memorandum

As a temporary amendment to the reinstatement memorandum insurers decided to restrict the operation of the average clause to those cases where, at the time of reinstatement, the sum insured is less than 85% of the full reinstatement cost. For example, if the sum insured represents 90% of the reinstatement cost the average clause will not be applied to reduce the amount of the claim (unless of course the property is totally destroyed, in which case the insured will have to stand 10% of the loss himself). Where, however, the sum insured at the time of reinstatement represents less than 85% of the full reinstatement cost, any claim will be subject to average as if the 85% concession did not exist.

Other anti-inflation devices such as the Notional Reinstatement Value Scheme and the Valuation Linked Scheme are now seldom used.

Inflation indices for building cost assessments

Every valuer needs to have access to inflation indices and there are several good sources of information.

The Building Cost Information Service calculates, *inter alia*, the Tender Price Index, which indicates the movement of tender prices for new building work in the United Kingdom. Because the data is derived from actual tenders, the indices reflect the conditions prevailing at the time of pricing. The indices measure not only changes in building costs such as labour rates and materials prices but also the influence of economic conditions operating at the time in a competitive market.

The index gives details of actual changes in tender prices and also forecasts likely changes in the immediate future, the latter being of assistance to valuers in giving recommendations on escalator clauses and inflation projections.

Alternative methods of insuring buildings

Obsolete buildings insurance

For many years it has been possible to insure obsolete industrial and commercial buildings on a special basis of cover. If the design and construction of the building is such that it would be impracticable to rebuild it in a like manner and the building is not subject to an obligation to reinstate in the existing style and materials, the basis of insurance could be the cost of purchasing a similar building in the open market or the cost of erecting a modern building providing comparable facilities. In each case it would be necessary to add an amount to the chosen sum insured to cover debris removal costs and professional fees as appropriate. Although this type of cover is technically still available it is apparently little used.

First loss insurance

A policy of this type is one in which the sum insured is deliberately restricted to a figure which is less than the full value of the property. In a case where the building has been destroyed the insurers will pay losses up to the sum insured without the application of average. First loss insurance is commonly found and is sometimes used to insure buildings and plant and machinery belonging to large corporations where a single sum insured "floats" over all the assets of the company on a world wide basis.

Insurance of machinery and chattels

The methods of insuring machinery and chattels are almost identical to those used for insuring buildings in that a decision is required as to whether a reinstatement or indemnity sum insured would be appropriate. Where there is no intention to reinstate or repair damaged machinery, insurers will wish to pay no more than a market value and a plant and machinery valuer will need to bear this in mind in fixing appropriate values.

In many instances a valuer will be asked by his client not only to produce a report giving recommendations as to the correct sum insured on plant and machinery but also to prepare a detailed inventory of the contents of the factory. The inventory will describe individual machines, giving type and serial numbers and separate reinstatement or indemnity values. The benefits of an inventory are at least two fold as it provides the insured with a detailed record of the plant, machinery, fixtures and fittings which can be revised annually and it is an invaluable record of these should the premises be destroyed.

Study 4

Your client has taken delivery of a second-hand Komori Lithrone model 544, five-colour sheet fed offset printing press and is unsure whether to insure it on a reinstatement or indemnity basis. Calculate the appropriate values allowing for delivery and installation costs.

Reinstatement cost	
New replacement cost of a similar machine	£1,150,000
Add:	
Delivery installation and commissioning	50,000
Machine foundations	50,000
Cost of reinstatement with new	£1,250,000
Indemnity value	
Either:	
Depreciated replacement cost, as follows:	
Reinstatement cost as above	£1,250,000
Less:	
Allowance for age, wear and tear, functional and technical obsolescence say 45%	562,500
Indemnity value, depreciated replacement cost basis	£687,500

OR

Open market value, as follows:

Cost of acquiring a similar machine of a similar age and condition	£200,000
Add:	
Delivery, installation and commissioning costs	50,000
Machine foundations	50,000
Indemnity value, open market basis	£300,000

A sum insured calculated on a reinstatement basis provides the optimum level of cover and, should the Komori printer be destroyed, the insured would be able to replace it and recover the full cost from the insurers.

There is disagreement as to whether an indemnity calculated by reference to a depreciated replacement cost or by reference to market value provides the second best level of cover. If the insured is advised to insure his machinery on a market value plus delivery and installations cost basis, he can expect to be paid on this basis in the event of total destruction. If the insured's machinery is seriously damaged and no second-hand replacements are available, he might be forced to repair the damaged machines and the cost of so doing could be considerably greater than the market value. In the circumstances it would seem prudent to insure the machines on a depreciated replacement cost basis, if the insured is unable to take the reinstatement option.

Installation costs

Care needs to be taken in assessing the cost of installing machinery since such costs can vary substantially from industry to industry. For example, the cost of installing engineering machinery such as lathes is likely to be significantly less than the cost of installing chemical plant.

Debris removal costs

Whatever value is chosen for the insurance of machinery, valuers frequently overlook the fact that to the basic costs of the machine and installation needs to be added the cost of removing machinery debris from the premises after the fire.

Engineers' fees

If it is a complex plant which is to be insured it is almost certain that the services of qualified design and consulting engineers will be required in the event of its replacement. In these circumstances the assessment should be increased to cover such fees.

Inflation indices for machinery cost assessments

Capital Replacement Costs of Twickenham provide indices covering changes in annual average prices for a wide range of plant and machinery. There is a composite index-covering chemical and allied plant, food manufacturing machinery, machine tools and office equipment which are described as Industrial Plant. The indices are available on a quarterly basis.

Other indices of various types can be obtained from the Office for National Statistics from their headquarters in London.

Sources of cost information

Details of machinery replacement cost can be obtained from manufacturers or their agents or from the valuer's own records. Market values can be obtained by discussion with machinery dealers and careful consideration of information contained in various publications and in auction records.

Other forms of insurance

Loss of rent

A landlord who has let his property will probably wish to insure the potential loss of rent which he would incur if the property were damaged or destroyed. He can do this under a fire policy by means of a separate item specifying the period and the amount of the rent to be covered. The rent clause, which will be added to the policy, defines that loss of rent will be paid during the period the property is unfit for occupation, not during the period it remains unoccupied after repairs have been completed, when new tenants might be sought.

Business interruption

To avoid a loss of rent which might be incurred in any subsequent period while new tenants are found, a property owner could decide not to extend his policy by the addition of an item on loss of rent but to take out a separate "business interruption" policy, previously called a consequential loss policy. This type of policy covers all losses incurred during a specific period including expenditure incurred to minimise a loss of rent such as additional travelling expenses to alternative accommodation.

In a more conventional sense a business interruption policy provides the support to a fire policy which covers the material losses to buildings and machinery etc. A business interruption policy has two sections one to cover loss of gross profit and the other to indemnify a claimant for the increased cost of working following damage. The gross profit section covers the profit on lost sales as a result of an insured loss during a specified period, usually 12 months, when the company has been unable to trade at all or only partially. The increase in cost of working section provides cover for the insured to incur expenditure on measures which will minimise a loss of sales. Such measures might include the setting up of a temporary production plant or the increased cost of having goods made elsewhere, the total cost of the measures being limited to economic levels.

Some business interruption policies are subject to an average clause when there is under-insurance. But in most cases insurers provide declaration-linked non-average policies where the insured declares the estimated gross profit for the current period, which he increases proportionately if the indemnity period is for more than one year. The estimated gross profit can be increased by a percentage, usually 33.3%, to cover inflation. If the policy is arranged in this way, average will not be applied in the event of a claim.

Employers' liability policy

Employers' liability insurance, which is compulsory, covers the employer's liability to his employees for injury or death arising out of the negligence of the employer or persons for whose negligence he is responsible arising out of and in the course of the employees' employment.

Public liability policy

A public liability policy covers the liability of an individual or firm to pay damages to a third party, members of the public, for accidents caused by the negligence of the insured or his or their employees or defects in the insured's premises, machinery or plant. It is not compulsory.

Loss adjusters and assessors

Loss adjusters

A loss adjuster will normally act for insurers and will be paid by them although, under the code of conduct of the Chartered Institute of Loss Adjusters, they are to act impartially. The adjuster's role is to investigate the cause of the damage, comment on the infringement or otherwise of policy conditions, examine and approve costs and recommend to insurers amounts to be paid.

Loss assessors

Loss assessors generally act for the insured, who is responsible for the payment of their fees which cannot be recovered as part of the payment of a claim. Assessors are not organised within any particular professional body. Their main function is to assess the amount of the loss to buildings, machinery and stock and to submit the details to the loss adjuster for discussion. Often in major losses the assessor will liaise with the reconstruction team consisting of architect, quantity surveyor or engineer. A loss assessor requires a detailed knowledge of assessment and valuation techniques and insurance principles and practice.

Chapter 22

The Valuation of Care Homes

During the second half of the 1980s and on through the early and mid 1990s there was substantial growth in the provision of private residential and nursing homes. This is evidenced by the fact that at the beginning of 1985 there were 189,000 beds in private care homes. This figure had grown to around 380,600 in 1999, out of a total of 554,100 care beds in a residential setting (inclusive of voluntary and public sector provision).

As care homes are essentially a property-backed asset, there simultaneously evolved a need for professional valuations for sale, loan or security purposes. Prior to this a limited number of private homes changed hands very much on a local basis and the involvement of business transfer agents and professional valuers has emerged as this sector has evolved.

Originally the sector was traditionally a "cottage industry" dominated by small owner-operated homes, often a husband and wife team where the wife, a former nurse, acted as matron while her husband provided the limited management function. Some of these operators sought to increase the size of their facilities or move to larger homes. Others "colonised" their area by setting up, or acquiring further homes in their locality. At the end of 1991 there were 9,000 residential homes, 4,000 nursing homes and 550 dual registered homes, a total of 296,400 beds estimated to be worth about £3.3 billion.

Steady growth occurred in the late 1980s as groups of homes were acquired by small companies, which in turn, became a target for acquisition by larger corporate grouping. In late 1998 these corporate operators provided around 26% of all private care beds (only 5% in 1989). The majority of these beds are in purpose-built facilities developed in the early 1990's or substantially extended and upgraded former residential properties.

Naturally the majority of these corporate operators were looking to acquire homes which had a high level of registration, with, say, 40 or more beds. However, since the majority of the then existing homes were of a smaller size, with average nursing home then only having 27 beds, many of these corporate operators pursued the "new build" options.

The tightening of standards following the introduction of the 1984 Registered Homes Act helped to promote this course of action, as a set of minimum standards could be taken into account when designing a home. Additionally the "new build" option enabled a practical and ergonomically viable building to be created which resulted in substantial savings in operating and staffing costs. There was also the opportunity to provide better facilities in the way of en-suite cloakrooms or bathrooms.

Relatively few of these homes have been built in the south east, where operators compete against the residential housing market for development land. Provided the land could be acquired at a reasonable level, particularly in the Midlands or Northern England, it was more attractive to develop homes in those areas. However, it should be borne in mind that while land values were cheaper in these areas, the resources of potential residents in these localities are also significantly less. Therefore few if any will be private pay residents able to afford the higher level of fees that a home which offers a higher level of facilities should be able to command

The establishment of "new build" homes also offers the opportunity to develop "close care units". These are a new departure from established care homes – a sort of halfway house where perhaps rather younger, physically disabled persons can live, cooking and fending for themselves but having nursing care if it is needed. Alternatively, this type of accommodation might be suitable for a couple where one partner needs nursing care while the other is still able.

As a result of recent market consolidation by acquisition and merger, there were in December 1998 34 corporate operators who owned, leased, or managed over 500 beds. BUPA Care serviced the largest, then operating 217 facilities providing 15,967 beds, which was 4.2% of total private market share. Ashbourne plc the second largest had 142 homes with 8,180 beds.

The pace of new build development in the early 1990s had resulted by late 1998 in 7,998 residential homes providing 156,313 beds with an average size of 20 beds, 3,856 nursing homes providing 145,118 beds with an average size of 38 beds, and 1,728 dual registered homes providing 80,258 beds with an average size of 46 beds, resulting a total of 381,700 beds estimated to be worth about £5.90 billion.

Significant Legislation

Registered Homes Act 1984

Two major influences have seriously affected the growth of the residential and nursing home market. The first of these was the coming into effect on January 1 1985 of the Registered Homes Act 1984, which governs the operation of both residential and nursing homes. This legislation sought to eradicate abuses among a minority of disreputable operators, to enhance standards generally and to fund regulation from fees payable by proprietors.

The principal changes in the Act were:

1. Any residential home offering accommodation to four or more persons must register with the local authority, while nursing homes must register with the local health authority (homes of fewer than four beds became registerable from 1991).
2. Homes can apply for dual registration if they wish to offer both residential and nursing care.
3. An initial registration fee is charged, together with a per-resident annual re-registration fee to cover annual inspection costs.
4. Appeals against refusal of registration or de-registration are heard by a Registered Homes Tribunal.

The Act also sought to distinguish between residential and nursing homes, debarring the former from purporting to offer, or actually offering, nursing care. Residential homes are defined as those providing board and personal care as distinct from nursing care. In practice, the principle applied is that in a residential home the level of care provided should be no more than that which a caring relative would provide to an elderly person in his or her own home.

Accompanying the new legislation a pair of codes of practice were published, one for nursing homes and the other for residential homes. The fairly conservative code for nursing homes was drawn up by the National Association of Health Authorities (NAHA, subsequently re-named NAHAT as NHS Trusts became members) to harmonise standards applied by the 200 health authorities in England and Wales.

However, Home Life, the code for residential homes drawn up by the Centre for Policy on Ageing, proposed a number of fairly radical and idealistic changes which the local Registration Authorities, whose powers and funding were greatly increased by the Act, initially seized upon and implemented owing to the lack of any actual legislation. There was considerable variation in the way in which the code was applied and this led to anomalies in standards between adjoining registration authorities, however Home Life was subsequently revised in 1993.

It should be appreciated that historically increasing levels of competition, (due to the growth in the number of homes with a consequent overall reduction in occupancy levels) has had a greater influence on improving the quality of accommodation offered, rather than the upgrading of standards by the Registration Authorities following the Registered Homes Act.

The introduction of funding limits and the changes enacted through the 1993 Community Care Act.

The other significant factor, which influenced the early evolution of the residential and nursing home market, was the imposition of funding limits for those residents receiving supplementary benefits. Prior to the limiting of funding in November 1985 the DSS were prepared to pay the going rate in their area to place a resident in a private residential or nursing home.

The funding levels are reviewed annually and from April 1998 to April 1999 the rates were £213 for a resident in a residential home and £318 per week for a resident in a nursing home. These fees are inclusive of the attendance allowance and the only additional funding is an extra £44 for rest home residents in Greater London areas and £49 for nursing home residents. A person becomes eligible for such supplementary benefit funding provided his or her assets are less than £8,000. Between £8,000; and £12,000 there is a sliding scale.

In the late 1980s the Government became increasingly concerned over the rapidly escalating expenditure on Income Support for residents of private homes, which rose from £12m in 1984 to £200m in 1989. Consequently Sir Roy Griffiths was commissioned by the government to investigate alternative arrangements. He subsequently submitted a report recommending "Care in the Community". This report was primarily based on a trial research project carried out in Kent, which appeared to indicate that this would be a cheaper option than residential care.

In 1989 a White Paper entitled *Caring for People* was published containing proposals which were originally due to be enacted in April 1991. The main provisions of the legislation were that all those unable to fund their own care should to be assessed by the social services to establish whether they could remain at home with community services provided to them. Alternatively, whether day care, respite care, or residential care would be more appropriate, according to their dependency level, with an emphasis naturally being placed upon community care. Additionally, where persons require full nursing care the health authority needs to participate within the assessment procedure.

Health authorities and social services departments were encouraged to co-operate and plan the provision of services within their localities and are required to prepare Community Care Plans outlining their proposals.

While funding previously provided via the income support system was diverted to local authorities, it was only "ring-fenced" for the first three years. Thereafter money required to fund placements would be within the local authorities' block grant allocation from the Treasury (their Standard Spending Assessment SSA). Thus social services departments now need to compete against the budgetary demands of the other departments of the local authority.

Transitional arrangements continued for those already in receipt of Income Support funding until their demise. Social services were empowered to negotiate with the private sector to contract for beds, and services and moved slowly towards becoming enablers rather than providers of care services. Local authority-run homes (Part III Homes) which had previously been unregistered were required to be inspected by a newly formed "arm's-length" Inspectorate.

Given the size of the task it soon became clear that the April 1991 implementation date was too optimistic and so there was a phased implementation until April 1993.

Initially the private sector viewed the prospect of the local social services departments being "fund holder, contractor, gate-guardian and inspector" with some trepidation, especially as some officers felt the need to be responsive to local political leanings. Initially little progress was made in consulting with the private sector on such key areas as contracts, specifications, approved provider status etc.

Post April 1993 there was an almost immediate initial contraction of demand when the new system came fully into effect, particularly for residential beds, as many of the local authorities took the opportunity to fill there Part III facilities to there full capacity.

The more enlightened were conscious that due to high wage and benefit costs in the public sector, these facilities would never be as cost effective as the private sector. They took the opportunity to move their Part III facilities into the not-for-profit sector, often run by their former officers with the benefit of generous dowries, with the staff still benefiting from the same terms and conditions. The Conservative government subsequently blocked this practice.

Some local authorities initially funded community-based packages of care schemes with scant regard to evaluating their true cost. Gradually the financial reality began to impact, and they cautiously utilised the private sector with government encouragement and directives.

Initially the majority of local authorities (with the exception of some of the prime Home Counties), set the fee levels for residential or nursing care, at or around the same rate as the on-going Income Support rates as determined by the DHSS for clients whose residency pre-dated the Community Care Act legislation which are revised each April broadly in line with inflation. Subsequently many local authorities have failed to increase fee rates in line with the revised DHSS rate or inflation.

Of even greater significance to the private sector has been the practice of local authorities social services departments managing their limited budgets by dragging out the assessment procedures, thereby delaying resident's discharges from hospitals, and only allowing a limited number of referrals each week into the private sector. This has naturally had a profound effect on the occupancy levels of facilities in certain areas.

The other suspected practice of local authorities which has had a significant effect on levels of profitability, is the seemingly high preponderance of assessments which result in the need for only residential care (which is around £100 a week cheaper than nursing care). This has resulted in many former nursing homes re-registering as dual registered facilities in order to try to maintain flagging occupancy levels. This change of registration has not necessarily automatically resulted in reductions in staffing levels and consequent costs. This relates particularly to trained staff since these are typically banded at one per each 20 clients

Valuation of residential and nursing homes

There is no such thing as a typical residential or nursing home. The majority of these homes have been set up in converted Victorian and Edwardian properties. They are generally detached and stand in their own grounds, although in coastal locations it is quite common to find three- or four-storey Victorian semi-detached properties which have been converted into homes and in some cases subsequently linked with the adjoining semi to make a larger home. The growth in the corporate operator sector in recent years has primarily been by the development of green-field sites, and in 1999 they provided approximately 26% of registered beds throughout the country.

During my time in private practice with the leading business transfer agents and valuers of the care home market in the early and mid 1980s, I received numerous calls from surveyors asking for guidance and, more particularly, information as to recent sales of rest and nursing homes which they could use to aid them for a valuation they were undertaking. Invariably they would admit, if pressed, that they had carried out few, if any, such valuations, since the advent of residential homes and nursing homes in many areas were then a recent development.

Many of their instructions emanated from their friendly high street bank, which wanted a simple valuation for mortgage security purposes. Some of these valuations were concerned only with the bricks and mortar value, as this provided sufficient security for the loan.

However some of these valuers, who had been asked to quote the business value or the fully traded value as an operational business, had not been given, nor had they asked for, details of the current trading position of the home, the fees charged, current occupancy levels and, most important of all, the certified trading accounts.

Additionally, few were aware of local Registration Authority regulations and whether the home in question met these criteria, particularly with regard to losses in bed spaces, which would naturally have a dramatic effect upon income and profitability.

Residential homes and nursing homes are most appropriately valued by the Profit's Method of valuation: see Studies 1(A) and 2(A). The use of comparable evidence from recent sales in the locality or the application of a per-registered bed multiplier for an area should be used only as a check, as no two businesses are ever the same and so much depends upon the business acumen of the proprietors, the goodwill they have built up, the demand for beds in the area and a host of other factors.

As with other profit's method valuations, the most important figure to determine is the "adjusted net profit" that the home is capable of producing. To determine the adjusted net profit the valuer needs to take the net profit as shown in the certified accounts and then add back any items that are personal to the individual operator of the home (including finance and depreciation charges) to arrive at the level of profit any operator of that home might expect to generate.

In the case of a smaller home, which may well be a family-run operation where the proprietor may even live in self-contained accommodation within the home in order to save on staffing costs, there may be a high level of expenditure on such items as provisions, telephones and motor costs as well as high mortgage repayment charges.

Similarly, in those homes owned by a company, there may be a central or head office management charge levied in the accounts. Consequently, the valuer needs to make adjustments to the profits shown in order to produce a figure that truly represents the home's profitability.

One attraction of dealing with this type of business is that, with very few exceptions, the trading accounts are usually a fairly reliable record of the business. Generally most private residents pay by cheque or banker's order, through their family, solicitors, or trustees, and the great majority are funded by the local authority under the Community Care arrangement, with a dwindling number supported by the DSS funding.

Having determined the adjusted net profit, the years' purchase multiplier to be applied will need to be considered. At the current time this can vary from 4.75 to 6.25 depending upon the location of the home and the level and quality of accommodation it offers. A 4.75 to 5YP figure might be used for a below average home in, for example, the oversupplied South Coast or for homes in Northern England which generally command lesser fee levels, with many operating a basic local authority/ DSS fee level; whereas 6 to 6.25 YP might be utilised for a prime Home Counties location, or a purpose built facility. It is the valuer's experience and expertise, which will lead him to choose a particular Years' Purchase multiplier. There is no precise determinant or true rule of thumb: at the end of the day it comes down to the individual valuer's judgment and relevant experience.

A further refinement of the profit's method, which may be applicable, particularly where there are known variations in the future adjusted net profit, is the utilisation of a Discounted Cash Flow approach.

Alternative methods of valuation

There are two alternative, less reliable, approaches to the assessment of the value of a residential or nursing home. Specifically for a facility, which is the conversion of former large domestic property, the valuer can determine the bricks and mortar value by comparison with other properties in the locality, and then add in the value of the goodwill. This will normally be taken to be one to one and a half years' profits plus the value of the fixtures and fittings: see studies 1(B) and 2(B).

This approach is not very relevant to recent large purpose built facilities, where there are no local comparable properties, and additionally such facilities have few alternative uses, other than as a care home. In such circumstances the actual original construction cost of developing the facility, or other similar facilities, may be utilised as the base value when combined with the site value.

The other method, which is really to be utilised only as a check, is a per-bed multiplier: see studies 1 and 2. The figures quoted vary considerably, anything from as low as £15,000 to £18,000 for a failing care home in a poor area charging low fees, with low occupancy and resultant low levels of profitability to £25,000 to £30,000 or more for a successful high-quality facility located in the prime Home Counties, charging high fees, with consistently high occupancy levels, and resultant high operating margins. Again it comes down to the question of expertise as to what multiplier is chosen.

Others factors affecting the valuation

Naturally when the 1984 Act first came into effect there were a number of homes, which did not comply with the new requirements. Originally some were so far out of line that it proved uneconomic to adapt them and many others were given periods of grace of up to three years to be put in order. However valuers should still be wary that a home, which they are to market or value does comply with all the requirements of the original Act, and more relevantly the ever changing and increasing standards of the relevant local Registration Authorities.

The Registration Authorities may have been prepared to be a little accommodating with an existing home owner on the annual inspections, will inevitably take a much tougher line when they are asked to re-register a home for a new owner. It must not be automatically assumed that because a home has been registered for a certain number of residents this will automatically continue. Enquiries will need to be made to the relevant Registering Authority, and their published standards and guidelines studied in depth.

The most common cause of a reduction in registration will be the registration authority's refusal to re-register a bedroom owing to its failure to meet the requirements with regard to size. While there is currently no national standard, most Registration Authorities require an absolute minimum of 10 m² for a single bedroom and 15 m² for a double.

A major change introduced by the 1984 Act was that care homes were only able to offer accommodation in either single or double rooms: rooms which were formerly registered as triples or more have to be reduced to doubles whatever their size.

Many Registration Authorities also limit the number of double bedrooms that can be registered in a facility, and apply a set a ratio which is typically 80% of bedrooms to be singles, and 20% as doubles, arguing that the majority of elderly persons would prefer a room to themselves. On a change of registration this ratio may be more rigorously applied, resulting in a reduction in capacity. From a practical point of view, operators are finding it increasingly difficult to let double bedrooms, as the great majority of the elderly are single persons rather than couples.

The Registration Authorities also set standards with regard to the provision of day-space per resident and most require a minimum of 2.5 m² per resident of lounge space and 1.5 m² of dining space in care homes.

Other factors, which may result in additional expenditure upon re-registration, may be associated with the fire precaution requirements. While rest homes and nursing homes are not issued with a fire certificate, the Registration Authority will consult with the fire officer to ensure that the building meets the regulations. Many home owners operating in, or setting up homes in, old Victorian and Edwardian properties have been put to considerable expense in fire-proofing ceilings and stairways.

Similarly, the Registration Authorities will consult with the local Environmental Health Authority to ensure that a home meets the requirements of the various Health Acts.

A reduction in registration may also result from the fact that the registration authority will register upper floors only if a suitable lift is installed. Obviously the lift cannot be used in times of fire, so a secondary means of escape and fully protected stairwells must be provided to all floors.

Some older homes were fitted with stair-lifts which climb and descend on a track running up the stairs, these are now becoming increasingly unacceptable to a number of Fire Authorities, who consider that the stairway may be too restricted in width by the tracking. There have also been fatalities in rest homes in vertical home lifts. These were due to bad maintenance and the fact that these simple lifts were designed for limited domestic use rather than constant use in the care home environment.

Valuers may come across a vast variety of equipment within residential and nursing homes. The basic standards applied by the great majority of Registration Authorities are that there must be a WC for every six persons, a bath to every eight persons and a bath to each floor where bedrooms are located. Additionally, nursing homes are required to provide sluices to each floor.

All homes must be equipped with a nurse call system with call points beside each bed, in bathrooms, cloakrooms, lounges and common areas etc. These nurse call systems may be a simple bell system rigged up by a local electrician through to a full intercom system with call logging facilities, and many homes have been equipped with elaborate wireless systems or paging systems which indicate the resident's room number. Similarly, all homes must have a modern, zoned fire alarm system fitted throughout the home with a smoke detection and emergency lighting system.

When setting up their care home, some operators are tempted to lease equipment, particularly nurse call and fire alarm systems, lifts and even assisted baths and laundry equipment. The leases may be difficult to break upon a sale. The new purchaser will be required to take over the lease contract and such annual premiums must be allowed for in any assessment of the operating costs of homes and resultant profitability.

Catering costs

The catering costs within a residential or nursing home may be considerably lower than might be envisaged, typically around £15, per person, per week, including some provision for feeding staff. Some corporate operators achieve around £12 with their greater purchasing power. It should be born in mind that while the elderly residents within homes eagerly await their meals; their appetites are limited, so the food needs to be presented attractively but not in large quantities.

Kitchens within homes may vary quite considerably in their

levels of equipment. Many smaller residential homes may still have a kitchen partially fitted out with domestic equipment, especially if the home has operated on a limited budget. Alternatively, the valuer may find a kitchen fitted out with the full range of catering equipment and valuers may need to make adjustments to the valuation if it is felt that the home's kitchen is deficient.

Staffing costs

This is a labour intensive business operation, so staffing costs are by far the highest of all operating costs. This is particularly true in nursing homes, as there is an ever-increasing national shortage of trained nursing staff. In some areas this is now reaching crisis proportions, with both operators and Trusts having to resort to extensive agency usage, and even needing to recruit trained staff from overseas. Naturally increasing competition between homes to recruit and retain these essential trained staff has resulted in substantial increases in rates of pay.

Consequently, wage costs in a managed rather than an owner-operated nursing home can easily be between 55% and 65% of turnover. In recent years this has resulted in a significant drop in the profitability of nursing homes.

In a smaller home with a qualified owner-operator, especially one who lives in and who is therefore effectively on call 24 hours a day, some savings are possible. However, valuers should be careful to add back the true costs of the owner-operator's input should they be acting for a client who intends to run the home under management. Valuers should also be careful to ensure that the home is staffed to the correct level in accordance with the requirements of the health authority. Nursing homes are issued by the Health Authority with a section 25 notice (1984 Registered Homes Act) which details the numbers of both trained staff and nursing auxiliaries who must be on duty at any one time.

The majority of Health Authority section 25 notices require one person to care for five residents during the day, and one for every 10 at night, with around 35% of these staff being trained nurses. Thus in a typical 40 bed home there would be a requirement for eight staff on from eight in the morning until two in the afternoon, night comprising three trained nurses and five nursing auxiliaries. From two in the afternoon until eight at night seven staff comprising three trained nurses and four nursing auxiliaries. From eight at night until eight in the morning there would be a requirement for two trained and three nursing auxiliaries. This equates to 26.25 care hours per client per week. The Matron would normally be fully super-numerary in a facility of this size, below 30 beds some of her/his hours may be within the requirement.

Should an operator consistently fail to roster sufficient staff to meet the section 25 notice, the Registration Authority may then seek to de-register the facility, which would then result in closure, and the removal of the residents.

In addition there will be around six hours per resident per week of ancillary staff comprising catering staff, domestic staff, laundry staff, reception and administrative staff, and maintenance man.

The social services who register residential homes do not issue such precise staffing requirements however typically they require around 15 hours of care staff per resident per week. There is no requirement for trained staff. Similar levels of ancillary staff will be required.

The problems of staffing are not so acute in the residential home environment. The majority of Registration Authorities do not require the matron of a home to be a trained nurse. The care assistants within a residential home are usually recruited from the local labour market and given the necessary basic induction training before moving on to study for their National Vocational Training qualifications. Residential homes may have an extensive staff roster with a number of local staff working just a few hours a week. While it is difficult to generalise, wage costs in a residential home are usually within the range of 40% to 50% of the turnover.

Another factor, which will inevitably affect staff levels, particularly in a nursing home, is the type of residents in the home and their level of dependence. Homes may vary quite considerably as to the type of resident they are having to accept in order to maintain occupancy levels and staffing levels may need to be increased to deal with intensive nursing care cases, such as incontinence, double incontinence and heavy disabled patients.

Other outgoings

The relevant Registration Authority in order to ensure a warm, cosy environment lays down minimum levels of heating of residential and nursing homes. Thus heating bills, particularly in older style homes, may be quite substantial when compared with typical domestic running costs. Other outgoings include items such as insurance, sundries, vehicle operating costs, repairs, replace-ments and, finally, depreciation.

Levels of profitability

While many of the cost-centres detailed above can be fairly rigorously controlled, staffing costs are by far the greatest operating cost. Failure to control this will seriously affect the profitability of a home. Generally speaking, one would expect the adjusted net profit to be in the region of 25%–30% of the turnover in a nursing home, while in the residential home, where staffing costs are considerably less, this will generally be around 35% of turnover.

Operators may quote a range of fees for their private residents as depending on the duration of their stay with them, these may vary quite considerably. Operators naturally find it difficult to increase their fee rates to existing private residents other than on an annual increase in line with inflation. However, when a vacancy occurs, operators naturally seek to get as high a fee as possible, depending on the level of demand for private beds in their area.

Owing to the substantial increase in the number of residential homes and nursing homes which opened throughout the country in the late 1980s and early 1990s, together with a lack of Community Care referrals from local authorities, in some areas operators have in some areas experienced a significant reduction in occupancy levels. In many cases the percentage has dropped to the low 80%s, rather than the low 90%s previously achievable.

Recent and future trends, the effects of external factors, and their affect on the valuation of care homes

Evolution of the Sale and Leaseback market

The Sale and Leaseback market began to evolve in the mid 1990s with the emergence of Principle Healthcare Finance (advised by Omega (UK) Ltd), and Nursing Home Properties plc, as the two key players in this new and creative source of funding for corporate operators, which has grown to over one billion million pounds invested by the end of 1999.

These transactions either involve the refinancing of operator's existing portfolios thereby providing repayment of traditional debt and mortgage funding, or more frequently the acquisition of single facilities or groups of care homes. A simultaneous legal transaction takes place whereby the operator buys the facilities from the existing vendor, immediately sells the Freehold interest of the property on to the sale and leaseback funder for the same figure, (thus providing the funds to satisfy the original vendor).

Simultaneously a wholly owned subsidiary company of the operating company is granted a 25 or 30 year full repairing and insuring lease.

The sale and leaseback funder is prepared to pay the full "going concern" value for the Freehold interest, because the operating company guarantees payment of the rent, and more significantly the shares of subsidiary company have been pledged. Thus in a default situation (such as non-payment of rent, or failure to maintain) the pledge can be exercised by the sale and leaseback funder, thereby regaining control and full ownership of the facility, it's fixtures, fittings, equipment, and its goodwill.

Additionally, as the subsidiary company is the required registered operator of the facility, registration is maintained, and there is no disruption to the residents or trading, as the Registration Authorities are merely advised of a change of control, by the appointment of new directors of the subsidiary company.

An intriguing development of sale and leaseback funding is the valuation of the sale and leaseback funder's freehold interest and the operator's interest. As the operator guarantees payment of the rents, and provided it has significant net worth, this provides a very secure income stream to the funder, thereby justifying a low yield expectation, resulting in a valuation of around 10 Year's Purchase.

Turning to the valuation of the operator's leasehold interest, provided it is commercial and efficient, it will enjoy a surplus after payment of the rent out of the adjusted net profit derived from operating the facility. This surplus will be partially depleted by the operator's internal management operating costs, however it is a form of "profit rent" which, subject to the success of the facility and potential erosion by rent reviews, is enjoyed for the full term of the lease.

To date there have been very few transactions involving the sale of operating leases between operators, however this is likely to become a developing market as sale and leaseback funding grows. In the USA where sale and leaseback originated and now dominates the funding of corporate operator activity, operating leases regularly change hands at multiples of around five to six times the post rent profit. Arguably this is due in part to the profitability of care home operations being more predictable and assured in the USA, due to restriction in the licensing of new care facilities. A UK valuer may consider a multiple of two and a half to three times to be appropriate at the present time.

Minimum Wage and EC Working time Directive

The introduction of a minimum wage in April 1999 at £3.60 per hour inevitably had a significant effect of the operating costs of the majority of care homes throughout the UK. While there were exceptions, particularly the Home Counties, London and the south east, there were still many parts of the country where care staff, domestic, catering and ancillary staff were paid below the minimum wage rate. Consequently this inflated wage costs typically around 3%–4.5% (of turnover), with a consequent reduction in profitability, when taken in conjunction with the introduction in the Autumn of 1998 of an EC requirement that all workers were henceforth entitled to three weeks paid holiday, rising to four weeks in October 1999.

Royal Commission

The Royal Commission appointed in 1997 by the then new Labour Government to review long term care with particular regard to the financial aspects and reported in January 1999. The government having published the Royal Commission's report entitled *With Respect to Old Age* has to date done nothing to implement the main recommendation that the costs of long term care should be split into personal care costs (to be paid by the government out of general taxation), and living and housing costs to be paid by the individual (subject to a means test). No doubt their lack of enthusiasm for the findings relates to the substantial cost implications.

Proposed Legislation

At the time of writing the government is proposing by the drafting of new legislation the introduction of National Required Standards (NRS) to be administered by a National Commission for Care Standards (CCSs) to take on responsibility for the regulation of all care services (administered by eight regional divisions). This will go some way to reducing anomalies, as this will replace the 150 local authorities and 100 health authorities, who currently undertake the role.

In April 2000 the Local Government Bill which enacts best value in local authority purchasing should hasten the demise of Part III local authority residential homes.

These two initiatives are clearly seen as beneficial to the private sector, however the current debate revolves around just how onerous the national required standards are likely to be and whether they will be only be applied to new registrations or applied retrospectively to existing homes.

Draft proposals circulated for discussion have included such measures as a maximum of 20% of beds in shared rooms, all single rooms to be 10 m^2, and all new homes including conversions to have single rooms of 12 m^2 excluding en-suite facilities. Many small homes would not meet such a standard, even if periods of grace were permitted to carry out the necessary re-modelling. It might prove un-economic resulting in the facility closing, and the property returning to the domestic residential market.

Study 1

(A) Profit's method valuation of a small residential home
Griffiths Lodge Residential Care Home situated on the South Coast and registered for 26 beds. Owned by Mr and Mrs A Milburn who live on the premises in the second-floor flat.

	£	£	£
Income			
Fees (1)			239,033
Less:			
Catering (2)		21,780	
Staff costs			
Salaries (3)	70.980		
NI & holiday pay	7,800		
Agency charges	4,161		
Total staff costs (4)		82,941	
Establishment costs			
Water & rates	1,874		
Property insurance	663		
Repairs & renewals (5)	9,430		
Electricity	2,520		
Gas	3,642		
Total		18,129	
General overheads			
Telephone (6)	1,501		
Postage, stationery & printing	247		
Laundry & cleaning	3,788		
Advertising	1,596		

Equipment rental	1,208	
Television rental	806	
Equipment charges	525	
Medical supplies	1,459	
Depreciation	3,526	
Sundry expenses	2,419	
Motor expenses	3,445	
Bank charges	575	
Legal & audit	2,750	
Mortgage interest	45,325	
Total general overheads		69,170
Total expenditure		192,020
Profit		47,013
Pension	5,000	
Directors' remuneration	20,000	25,000
Net profit		£22,013

Notes

(1) Average occupancy achieved over the 12-month period was 85% at an average fee of £208 per week. The low occupancy was in part due to the fact that three of the bedrooms are doubles, and have proved difficult to let other than as singles.

(2) Catering costs includes private consumption of, say £1,500.

(3) Excludes owner's remuneration and no cost for a matron.

(4) This is 35% of turnover.

(5) Repairs and renewals includes repair to roof of £5,500, a non-recurring expense.

(6) Telephone costs include private usage of £500.

Adjusted net profit for valuation purposes	£	£
Net profit		22,013
Add back:		
Depreciation	3,526	
Repairs & renewals	5,500	
Catering	1,500	
Telephone	500	
Mortgage interest	45,325	
Bank charges	575	
Pension	5,000	
Directors' remuneration	20,000	81,926
Less: Provision for matron's salary	18,000	63,926
Adjusted net profit		£85,939 (36% of turnover)

Therefore, using a YP multiplier of 5, this would provide a capital value of £429,695, which equates to £15,526 a bed.

Note: it was to be noted on inspection of the property that one single bedroom opened directly off the dining room. Therefore enquiries should be made to the Registration Authority to ensure that the existing registration for 26 beds will be maintained or, accordingly, the valuation may need to be re-valued pro rata to 25 beds.

(B) Bricks and mortar valuation approach
As discussed in the foregoing text this method of valuation involves three elements, namely:

(a) An estimate of the value of the bricks and mortar by comparison with similar adjoining properties including consideration of alternative use value and taking into account the state of repair of the property.
(b) A figure for goodwill, which is determined by applying a multiplier of 1 to 1.5 times the adjusted net profit.
(c) An estimate of the value of the trade fixtures, fittings and furnishings etc in situ. This should be at current value, taking into account present condition, wear and tear etc, and not be taken as the replacement costs.

In the case of Griffiths Lodge the valuer, having studied the local property market and taking into account the value of adjoining properties, alternative use value and the state and condition of the property, considered the bricks and mortar element was fairly represented in the sum of £275,000.

In determining the goodwill element he decided to use a multiplier of 1.5, having noted, by the accounts provided, that the facility had been trading for a number of years and had been achieving consistent occupancy levels. Through enquiries made to the inspection and registrations unit of the social services he had established that the home enjoyed a good local reputation and saw no reason for this not to continue following a change of ownership.

Having inspected the trade fixtures, fittings, furnishings and equipment and having taken due regard of their present condition and the need for some replacements, the valuer estimated their worth to be £30,000.

	£
The bricks and mortar element	275,000
Goodwill (1.5 × £85,939)	128,908
Fixtures, fittings and furnishings	30,000
The value by this method is estimated to be	433,908

Comparison per bed-space valuation approach
From his knowledge of the locality and sales achieved etc, the valuer is aware that small residential care homes are achieving values in the region of £16,000–£18,000 per bed.

Method 1 resulted in a capital value of £16,526 per bed.
Method 2 resulted in a capital value of £16,688 per bed.

Both approaches produced a capital value per bed within the expected range.

However, this method should really be used only as a check, since the trading acumen of operators varies widely even the most luxurious home which is trading at only low occupancy will not be able to service the capital repayments of a mortgage.

Study 2

(A) Profits method valuation of a large Dual Registered Home
Blair Lodge is a recently constructed purpose built dual registered residential and nursing home situated on the outer fringes of London and registered for a total of 50 beds with a maximum of 20 nursing residents. Owned by a limited company with two directors Mr G Brown and Mrs R Cook a RGN who currently acts as Home Manager, while Mr Brown's accountancy firm provides the bookkeeping service, at a cost of £18,000 pa.

Income	£	£	£
Fees (1)			676,260
Less:			
Catering		32,292	
Staff costs			
Salaries (2)	291,876		
NI & holiday pay	32,106		
Agency charges	7,325		
Total staff costs (3)		331,307	

Establishment costs

Water & rates	3,764	
Property insurance	1,823	
Repairs & renewals	7,426	
Electricity	5,530	
Gas	4,645	
Total		23,188

General overheads

Telephone	1,231	
Postage, stationery & printing	763	
Laundry & cleaning supplies	5,464	
Advertising	1,823	
Equipment maintenance charges	1,845	
Medical supplies	2,432	
Depreciation	5,762	
Sundry expenses	3,212	
Motor expenses (4)	8,721	
Bank charges	875	
Legal & audit	2,250	
Mortgage interest	96,000	
Accountancy charges	18,000	
Total general overheads		148,378
Total expenditure		535,165
Profit		141,095
Pensions	25,000	
Directors' remuneration	50,000	
		75,000
Net profit		£66,095

Notes

(1) Average occupancy achieved over the 12-month period was 90% at an average fee of £289 per week.
(2) Excludes the cost for the home manager/ matron.
(3) This is 49% of turnover.
(4) Mrs Cook runs a luxury car.

Adjusted net profit for valuation purposes	£	£
Net profit		66,095
Add back:		
Depreciation	5,762	
Mortgage interest	96,000	

Motor expenses reduction to a more modest vehicle	5,000		
Bank charges	875		
Accountancy charges	18,000		
Pension	25,000		
Directors' remuneration	50,000		
	200,637		
Less:			
Provision for a home manager /matron's salary	23,000		
Provision for an administrator/ bookkeeper based at the facility	12,000		
		165,637	
Adjusted net profit		£231,73	(34.2% of turnover)

Therefore, using a YP multiplier of 6, this would provide a capital value of £1,390,392 which equates to £27,808 a bed, alternatively using a YP multiplier of 6.25, this would provide a capital value of £1,448,325 which equates to £28,996 a bed

(B) Bricks and mortar valuation approach
As set out previously this method of valuation involves three elements, namely:

(a) An estimate of the value of the bricks and mortar by comparison with similar adjoining properties including consideration of alternative use value and taking into account the state of repair of the property.
(b) A figure for goodwill, which is determined by applying a multiplier of 1 to 1.5 times the adjusted net profit.
(c) An estimate of the value of the trade fixtures, fittings and furnishings etc in situ. This should be at current value, taking into account present condition, wear and tear etc, and not be taken as the replacement costs.

Therefore, in the case of Blair Lodge the valuer is aware that there are no other direct local comparable properties and this purpose built care home has little alternative use value. However, he is aware that the three-quarter acre site was acquired two years before for £225,000, and that the advised construction cost of £900,000 which equates to £18,000 per bed space with 32.5 m^2 of built area per bed space is a similar cost of construction to other facilities.

Given the quality and condition of the property, he therefore considers the bricks and mortar element is fairly represented in the sum of £1,125,000.

In determining the goodwill element he decides to use a multiplier of 1.25, having noted, by the accounts provided, that the property quickly filled upon opening and thereafter has achieved consistent high occupancy levels. Through enquiries made to the inspection and registrations unit of the social services he had established that the home enjoyed a good local reputation, and a high level of referrals, and saw no reason for this not to continue following a change of ownership.

Having inspected the trade fixtures, fittings, furnishings and equipment and having taken due regard of their present condition and the need for some replacements, the valuer estimated their worth to be £70,000.

	£
The bricks and mortar element	1,125,000
Goodwill (1.25 × £231,732)	289,665
Fixtures, fittings and furnishings	70,000
The value by this method is estimated to be	1,484,665
(which equates to £29,693 per bed space)	

Comparison per bed-space valuation approach
From his knowledge of the locality and sales achieved etc, the valuer is aware that good quality purpose built care homes can achieve values approaching £30,000 per bed.

Method 1 resulted in a capital value of between £27,808 and £28,996 dependent on the YP multiplier per bed.
Method 2 resulted in a capital value of £29,693 per bed.

Both approaches produced a capital value per bed within the expected range, however the profits method approach is more accurate, as it is based on the actual level of profitability achieved at the facility.

Chapter 23

Negligence and Valuations

"Quidquid agas, prudenter agas, et respice finem."
Gesta Romanorum

"Whatever you do, do cautiously, and look to the end."

One of the great perils facing all professions today is the threat of an action for negligence. The current opinion that the valuer has more claims than the other professions is mainly due to reading selected texts. Every profession tends to read more works about its own subject than general works on law and tends to see only the cases affecting their own colleagues. The same fears now beset all the major professions: solicitors, architects, engineers are all in the same position as valuers.

It is true that some of the cases involving valuation, such as *Smith v Eric S Bush* [1989] 1 EGLR 169, have made headlines, but this is only because they affect many people in their everyday affairs. Anything that is news for house-buyers is bound to make headlines in all the media.

Liability for Negligence

Liability in negligence may arise as the result of the breach of a contract, where the valuer fails, contrary to his obligations under the contract, to provide his client with a reliable opinion because of his negligence. It may also arise, independently of contract, from the tort of negligence, where third parties claim to have been injured as a result of the valuer's negligent opinion. Occasionally, but now rarely, liability may arise out of both concurrently, ie the party injured may be able to sue in both, or either, contract and tort.

The same act, omission or opinion may give rise to suits by separate parties, one of whom claims in tort and the other in contract. For example, if a valuer acting for a mortgage lender is negligent in the performance of that duty there may be a liability to those who gave the instructions (who sue in contract) and those who rely upon that opinion (who sue in tort). In this case, the claim in tort amounts (in fact, if not in strict law) to a third party being able to sue upon the contract. In these cases, of which *Yianni* v *Edwin Evans & Sons* [1981] 2 EGLR 118 is a leading example, the plaintiff claimed that he had been injured because the valuer was negligent in the performance of his contract to value the property (in that case, for the Halifax Building Society). It is difficult to imagine circumstances in which the valuer who has complied with the instructions given, and not been negligent to the contractual client, could be held liable to the third party.

It is not necessary, or always even possible, for both such parties to sue. In *Yianni* (*supra*) the Halifax Building Society could not sue the valuer since it had suffered no loss. The borrower was making his mortgage payments; there was no question of the exercise of the power of sale; and no financial harm befell the lenders.

Generally speaking, the same principles apply to determine liability, and loss, whether the action is brought in tort or in contract. There are some differences in the time during which the action may be brought (but these rarely affect cases of valuation) and differences in assessing the losses. In the case of a breach of contract, there is no need for the plaintiff to have to prove the proximity of the parties which features largely in the tort cases. The contract itself establishes the connection between them.

Required Standard

The standard of professional behaviour which determines whether a valuer has been negligent is the same as that applied to all professions: "Has the defendant met the requirements of the average, competent practitioner?" If that standard has been achieved the defendant has not been negligent. The same standard applies to all members of a profession. The defendant may not plead that he was just starting out in practice and was less well qualified. By the same token, highly experienced practitioners are under no greater duty. However, those who claim to be specialists, or experts, in any particular field are, however, to be judged by the standard of the "average, competent specialist (or expert)".

The unqualified who seek to stray into the activities of the professionals are to be judged by the same standards. In *Kenney* v *Hall, Pain & Foster* [1976] 2 EGLR 29 an unqualified negotiator informed a would-be seller that his house would fetch between £90,000 and £100,000. On the strength of this opinion, the vendor went ahead and bought another property, using bridging finance until his house was sold. The opinion was negligent and the house was sold ultimately at about £50,000. The firm was liable for the losses incurred. Also in *Freeman* v *Marshall & Co* (1976) 239 EG 777 a person who advertised himself as a "surveyor" could not escape his liability in negligence simply on the basis that he was "an unqualified person doing his best".

The standard of competence does not vary with the fee charged. If no fee is payable, much the same standard applies. The law on gratuitous obligations requires that the "volunteer" carries out the task to the same standard as would a reasonable person in the conduct of his own affairs. With professional people, this seems to indicate that they would use all their skills, as they would if acting for themselves. The standard would therefore seem to be the same as if the work were to be paid for. In *Roberts* v *J Hampson & Co* [1988] 2 EGLR 181, the negligent valuer pleaded, as part of his defence, that the fee for the carrying out of a valuation for mortgage purposes was very small, and as a result, the time available to perform the task was limited. The court was not impressed. Ian Kennedy J stated that in all forms of work there were some winners and some losers and that if a particular valuation took disproportionately longer than normal it had to be put in the balance with the others. This poses a particular problem for those who engage in the assessment of reinstatement costs for insurance purposes – a task which frequently is associated with valuation. Some buildings take much more work in measurement and calculation yet the fee will usually be based on the total figure supplied.

Reassessment of building costs has it own perils as was shown in *Beaumont* v *Humberts* [1990] 2 EGLR 166. The valuers were concerned with a large listed building. They did not quote the actual rebuilding cost. Instead, they gave the cost of reinstatement of a building which would have complied with current Building Regulations. This amounted to rather less than half the cost of *actual* reinstatement. The valuers were held not to have been negligent. Care should be taken against using this as a precedent. The RICS *Appraisal and Valuation Manual* (the "Red Book") should be consulted. In any event, if the calculation is made on any basis other than exact reinstatement the basis should be spelled out, in advance, so that the parties are aware of the situation before instructions are confirmed.

Basis of Negligence

As in all cases of negligence, tortious liability stems today from the principles laid down by the House of Lords in *Donoghue* v *Stevenson* [1932] AC 562. The plaintiff must prove:

(i)   That the defendant owes the plaintiff a duty of care;
(ii) That the defendant has broken that duty; and
(iii) That the plaintiff has suffered loss as a result.

To this must be added comment about *Hedley Byrne & Co Ltd* v *Heller & Partners Ltd* [1964] AC 465, which extends the duty of care from the previous classifications of "negligent acts" and "negligent omissions" to "harm caused by negligent words".

For the valuer, the extension to "mere words" is significant, since the valuer's opinion is expressed in words only, figures being the equivalent of words. Because words are different from "acts" or "omissions", the House of Lords, in *Hedley Byrne* (*supra*), was careful to lay down certain special rules in cases stemming from negligent words. They appreciated that words could be overheard, be misconstrued, be passed to persons not intended to rely upon them, etc. They therefore limited liability for negligent words to those cases where the defendant has "accepted a special responsibility" to the plaintiff. This has been reaffirmed recently by the House of Lords in *Henderson* v *Merret Syndicates Ltd* [1995] 2 AC 145 and *Marc Rich & Co* v *Bishop Rock Marine Co* [1995] 3 WLR 227.

Sometimes it is difficult to know whether such responsibility has been accepted or not. In *Smith* v *Eric S Bush* (*supra*) and in *Harris* v *Wyre Forest District Council* [1989] 1 EGLR 169 (both cases are in the same report) the House of Lords found that the standard form of denial of liability on the part of the valuer, in performing a valuation for mortgage purposes, was not a denial of the existence of such a special responsibility. It appears that the acceptance of a special responsibility is not so much a matter for the parties as it is to be found by the courts. The parties may appear to deny such a responsibility but the courts may spell it out from their conduct. Denial alone may not be enough, since it falls to be considered under the Unfair Contract Terms Act 1977. That Act (which also applies to liability in tort) requires that all such exclusions of liability be determined by the test of "reasonableness".

Other cases since *Smith* v *Bush* (*supra*) have established that liability is unlikely to extend to a "class of persons" such as the shareholders of a company *Caparo Industries plc* v *Dickman* [1990] 1

All ER 568; or the readers of a published work *Mariola Marine Corporation* v *Lloyd's Register of Shipping – "The Morning Watch"* [19901 1 Lloyd's Rep 54. It would appear, therefore, that liability is limited to cases where the plaintiff is a named, or identifiable, person or body.

The liability for negligent words extends to cases where, although the plaintiff has not seen or heard the words, he knows, by necessary implication, what they were. Before the practice of disclosing the reports in mortgage valuations became general, the borrower knew that the valuer had formed an opinion on which a building society might lend money. In *Yianni's* case (*supra*) the plaintiff did not see the negligent report and valuation. He was able to deduce that the property was valued, at the very least, at the amount which the building society advanced by way of mortgage. The situation may be different if the lender is not a building society or a local authority. Only those bodies are under a statutory duty of obtaining a valuation before making an advance of money. For reasons of commercial prudence, trustees, banks, and other corporations usually have a valuation before lending money, but they are, generally, not so obliged. The valuer acting for a bank may have suggested one figure but the bank, for good commercial reasons, may have lent more. In such cases, it would appear to be more difficult for the plaintiff to rely on any implication of value if the report has not been disclosed.

What are the Main Causes of Negligence?

It appears from the cases that valuers are liable to be found negligent in their work if they have not followed correct procedures. This is not to say that an incorrect figure is irrelevant, it is the very core of liability, but does acknowledge that the courts are aware that valuations are matters of opinion. They are, however matters of informed opinion. If the valuer has so acted that the opinion is not based upon proper referencing of the site, knowledge of legal principles affecting the valuation, available comparable evidence and the like, then the valuation may be called into question. The investigation, which leads to this evidence being adduced, commences with the plaintiff claiming that the figure is, in his opinion, incorrect. It may be found that the structure valued is ruinous; that the expected development potential is non-existent; that the property may not lawfully be used for the purpose required; or, it has been found that the property is worth a lot less on trying to realise the security. It is only then that the absence of the correct procedures becomes apparent.

The defendant valuer is put to question as to what information was taken into account, what procedures were followed, what amount of the structure was examined etc. The plaintiff's expert valuer, with the inestimable benefit of hindsight, will demonstrate the way in which (in his opinion) the valuation should have been performed and the defendant will be cross-examined as to why the procedures in question were not the same.

In *Corisand Investments Ltd* v *Druce & Co* [1978] 2 EGLR 86 the valuer defendants had omitted to enquire whether an hotel had a fire certificate. That requirement had been law for some three years by the time of the valuation. The fact that they had neglected to ask for a sight of this important document then opened up an enquiry into the whole of their valuation. The difference between the two sets of expert valuers (one for the plaintiff and one for the defendant) on each of the parameters going to make up the valuation was very small. There were small differences as to the appropriate yield to be adopted, as to the occupation rate, the occupation charge etc. The sum of these differences amounted to, what was considered by the court to have been, an overvaluation by 100%. The defendants were found to have been negligent and the difference between the two valuations was awarded as damages. On reading the case one cannot but feel that, but for the absence of the fire certificate, there would have been no inquiry into the valuation. The valuers were particularly unfortunate in that they were required to value the hotel one week before the property crash of 1974. The judge held that a competent valuer, with a proper feel for the market, should have known that the market was about to crash!

In *Singer & Friedlander Ltd* v *John D Wood & Co* [1977] 2 EGLR 84 the defendants had been asked to value some development land in Gloucestershire for the purpose of using the value of the land as collateral for the loan made by the plaintiffs. It transpired that the valuer for the defendants had not seen the land; had not made inquiries of the planning officers; but stated in evidence that he had done so. This is a classic example of a case where a valuation is prepared in response to a contractual obligation made to the borrower in the knowledge that a lender would see it and rely upon it. The plaintiff's action was in tort – in fact, alleging a breach of the contract with the borrower – which spilled over and affected them. (But see later in *Banque Bruxelles* etc where doubt is cast on this proposition.)

To seek to overcome the problem of liability attaching to a third party, the Conditions of Engagement prepared by the Royal Institution of Chartered Surveyors contained a clause to limit the ambit of the report to the client and his/her/its professional advisers. Any other party relying on the report does so at his own risk. Good as this practice is, it will not prevent liability attaching to the valuer when it is known that the purpose of the report is to secure a loan. Either the potential lender will require to become a party to the report, when it was thought that he was linked to the valuer in contract, or, if the identity of the lender is known to the valuer, the court will find that the valuer had accepted a special responsibility to the lender. (But see later the opinion of Phillips J in *Banque Bruxelles Lambert SA* v *Eagle Star Assurance* (*infra*) where doubt was cast on the respective position of the parties.) The valuer may well be liable to an unknown party if he is aware that the report will be used by some third party as the basis of a known transaction. If the contract seeks to exclude that sort of liability it may be that such a clause will be held to be unreasonable.

In the field of valuation for mortgage purposes there have been many cases where valuers acting for the lender have been held liable for negligence towards the purchaser. These started *Yianni* v *Edwin Evans & Sons* (*supra*) in 1981. Since that time there have been many other cases before the courts. It is pertinent to point out that none of those cases has involved a valuation which was carried out after the *Yianni* case was reported. In other words, it would seem that the professional mortgage valuer has learned the lesson spelled out in that case. The theory of the law is that judges interpret the law; they do not make it. Any set of circumstances which arose before 1981 is, therefore, judged on the basis of *Yianni*. It is reasonably clear that many valuers have changed their practice since the law was "declared" in 1981. The Royal Institution of Chartered Surveyors and the Incorporated Society of Valuers and Auctioneers collaborated in the production of a Guidance Note for Mortgage Valuers – now in the Red Book. This spells out what is the duty of the valuer in carrying out this class of work. It also spells out what is not expected of the valuer.

Despite the furore which greeted the publication of the original Guidance Note, it has already acted to protect valuers from some claims. Although the first edition was not produced until December 1985 its contents have been accepted by the courts as being the distillation of good practice and in establishing what was the level of competence of the average competent practitioner. In *Whalley* v

Roberts & Roberts [1990] 1 EGLR 164 the court accepted that a valuer would not normally carry a spirit level in performing a valuation for mortgage purposes. That case concerned a property which was constructed so as to incorporate a slope of some 2.5 inches across the main room. There were no "defects" in the property; there was no cracking; no sign of doors jamming. The property was not unsound, it was very unusual. The judge accepted the limits of inspection as laid down in the guidance note.

In *Gibbs* v *Arnold, Son & Hockley* [1989] 2 EGLR 154 the judge accepted that the duty of a valuer was limited to carrying out a "head and shoulders" inspection of the roof unless other defects indicated the necessity for a further inspection. There was a crack in the chimney breast within the roof space which, the court held, could not have been observed from the roof hatch. Although the property was defective the valuer had not been negligent. He had performed his duty as laid down by the profession itself.

These cases contrast with the finding of negligence in *Smith* v *Eric S Bush, (supra)* where the valuer had not noted that, although all of the chimney breasts had been removed from the lower floors, the chimney stack remained and was unsupported. It collapsed through the house. The valuers sought to rely on the standard exclusion clause limiting the valuers' liability to the lender client only. The Court of Appeal and the House of Lords found such a clause unreasonable and therefore unenforceable. Their lordships said that such exclusions might be acceptable in the cases of "commercial property, very expensive houses and in the purchase of blocks of flats." It is clear from these words and other *dicta* in the judgments that the courts will look very hard at any exclusion clause where they feel that the parties are not of equal bargaining power. They explicitly cited the fact that the valuer was insured, thus casting the loss over a wide range of people rather than have it fall on the unfortunate Mrs Smith.

At the same time, the House of Lords heard the appeal in *Harris* v *Wyre Forest District Council, (supra)* where the same principle was applied to a loan made by a local authority. The authority, and not the valuer, was sued because the valuer was a member of authority's staff. It was sued as an employer, and not as a lender, under the principle of vicarious liability. The house in question had been mortgaged to the District Council by the plaintiff some years before the *Yianni (supra)* case. Later, when the owners tried to sell it, the same authority refused to accept the property as suitable security. Partly, this was because, by the time of the second inspection, the Yianni principle was well known. The house was found to be ruinous and was demolished. In the Court of Appeal the District Council had been successful in claiming that the standard exclusion clause operated, not as an exclusion, but as a denial of the acceptance of a special responsibility – the core of the decision in *Hedley Byrne & Co Ltd* v *Heller & Partners Ltd (infra)*. This argument found no favour in the House of Lords, where the clause was held to be an exclusion – and an unreasonable one, too.

Terms of Engagement

Apart from the failure to take the necessary steps in the collation of information likely to affect the valuation, one of the principal areas of difficulty between the valuer and the client is the lack of communication of ideas. Frequently it appears that the client thought that every valuation entailed a survey, if only for the over-simplistic view "How can you value something if you don't know it is sound?" Far too few valuers take the trouble of agreeing the terms and conditions on which they are to work. They fill their reports with all the exclusions about high alumina cement and asbestos, un-discoverable dry rot and the like, but they fail in the simple matter of agreeing such exclusions with the client before starting. All such exclusions attached to a report which have not been agreed with the client beforehand are so much waste paper. Furthermore, it is dangerous waste paper, for it indicates that the valuer knew what he was going to do before he started and did not inform the client.

In order to put this right, and to give the institution's *imprimatur* to the concept, the Royal Institution of Chartered Surveyors produced two pamphlets on conditions of engagement; one for residential property and the other for commercial property. They form a simple contract between the valuer and the and inform the client as to the nature of the task which the valuer undertakes to perform. They also include all the requisite forms of exclusion clauses. If such a pamphlet is sent to the client, accompanied by letter saying that the valuer will be pleased to carry out the client's instructions "in accordance with the Conditions of Engagement which are enclosed herewith" then the situation becomes completely different. The client knows, in advance, what will be included and what excluded. The identical terms of the exclusions in the report have been conveyed to the client, who has agreed to a valuation in accordance with those terms. Those words, which previously were so useless and dangerous, are now the essence of the contract. This has an additional great advantage. Any third party (eg a borrower) would find it very difficult to maintain an action in tort claiming that the valuer's liability to him is greater than the contractual obligation to the client. That is not to say that such a claim would be impossible; it would, however, be difficult to sustain such an argument. If the valuer is carrying out a Home Buyer Survey and Valuation he may still find that differing responsibilities are owed to each party.

Any particular practice whose needs are not met by the RICS conditions is free to draw up its own. There is no obligation to use the printed form. What is important is that there are some agreed conditions of engagement between client and valuer. Who wrote them is less important. The advantage of using the RICS form is that it is becoming well known, and almost expected. The disadvantage is that that form may not cover all the eventualities of a particular practice. It is for the practice to decide which course to follow.

The Red Book

The terms of the RICS pamphlets originally formed part of the text of both the *Manual of Asset Valuation Guidance Notes* (the "Old" Red Book) and the *Manual of Valuation Guidance Notes* ("the White Book"). Those two volumes were, largely, designed to try to ensure that valuers did not carry out valuations without the assembly of sufficient information. Within the parameters of each guidance note it ought now to be assumed that the guidance notes (now incorporated in the new "Red Book – RICS *Appraisal and Valuation Manual*) constitute the minimum standard of the "average, competent practitioner". It does not follow that every valuer who follows what the guidance notes say will automatically escape any liability for negligence. There will be times, as the guidance notes point out, when further investigations and procedures are necessary. It is for the valuer to decide, in each case, what must be done. Further, the guidance notes rarely suggest the form of calculation required to meet the circumstances. Generally, they do not dictate the mathematical processes to be employed. Thus a valuer may use "DCF", "hard core", "dual rate", "tax adjusted" or any other reliable form of calculation, including the "Spot Valuation" which is the only method available for residential mortgage valuations.

What a valuer may not do, with impunity, is ignore the guidance notes completely or deviate substantially from its procedures. If a valuation which is called into question discloses methods which differ markedly from the guidance notes, the burden of proof will undoubtedly fall on the defendant (instead of the plaintiff) to justify the figures. But, even more than the figures, the valuer will have to justify the method used.

Even given compliance with the guidance notes, the resultant figure must be "correct". That is to say, it must be arithmetically correct. There is still room for professional dispute as to the correctness of the ultimate figure as representing the value of the interest in question, but the figures on each page must add up correctly. If a computer is used, the data entered must correspond to the findings of the valuer in the field. Put in its worst form, the date must not have been added in with the data, (or, as has happened, with the file number!)

Some valuers are notorious for arriving at a figure and then justifying it on paper. There is nothing wrong with this in principle, and many valuers are better at valuation by this method than any other. However, if such a valuation becomes the subject of a negligence claim the former standard response, "I have been doing this, man and boy, since I left school", no longer meets the requirements of the profession or the courts. Provided that the paper proof meets the figure which the valuer has thought up first; provided that the forms of inquiry and investigation, the comparative evidence and the systems of the guidance notes all appear to have been followed, then such a valuation may stand a judicial inquiry.

Measuring the Difference between Valuers

There are judicial *dicta* which suggest tests as to the competence of the valuer. It was suggested in *Singer & Friedlander Ltd* v *John D Wood & Co* (*supra*) that two valuers should be within 10% of each other. Other cases have suggested a wider margin. 15% is now often regarded as a tolerable margin. However, I find this unhelpful. If valuers are not within these tolerances, how do you judge which one has been negligent? I know of many occasions when parties have, by reason of different interpretation of the documents, justifiably been much further apart, certainly at the initial stages negotiation. I doubt whether any such figure is the best test of negligence. One may have been totally negligent and hit on the right answer. Fortunately, it is unlikely that anyone will find out. If they did, there could still be no action since there would have been no loss.

The essence of negligence in valuation is that the valuer has failed to provide the correct figure. But, frequently, it is other matters which first give rise to the allegation of negligence; in *Yianni* (*infra*) the ruinous condition of the property; in *Corisand* (*infra*) the absence of the fire certificate; in *Singer & Friedlander* (*infra*) the absence of the expected development value. Despite these facts, it is the valuation which is the key to the question of negligence. If the figure provided is correct the valuer cannot be found guilty of negligence. It follows that the valuer must have taken into account the impact of these factors and adjusted what would have been the value if he found otherwise.

The unreported case of *Ralphs* v *Francis Hornor & Sons* (1987)* illustrates the point. An architect valued property at £60,000 and informed a client that expenditure of a further £60,000 would make a valuable property worth more than the total outlay. The client bought the property and then found that the lowest estimate for the works was £93,000. He sued the architect for negligence. The plaintiff lost his action on production of evidence that the property purchased was worth the £60,000 which had been paid for it. The architect would have had to have acted according to the standard of the average competent practitioner in *valuation*. (Since he had taken upon himself the task of a valuer, he had to be judged as a valuer.) The fact that the plaintiff could not realise his bargain and that he had insufficient funds to carry the project through were both irrelevant. He was advised to pay £60,000 and he got £60,000's worth of property.

Importance of accurate records

In almost all of the other cases cited, the purchaser had to prove that, because the valuer had omitted to take into account matters which affected the value of the property, more had been paid than was justified. From this it follows that the valuer seeking to keep away from claims for negligence will keep accurate, systematic notes of what he finds. The courts acknowledge that a valuation does not involve a survey (*Gibbs* v *Arnold Son & Hockley* and

---

* I am indebted to John Murdoch for the information on this case.

Whalley v *Roberts & Roberts (supra)*. However, failure to record gross deficiencies in the structure; absence of material facts; absence of comparable evidence; etc. will make for difficulties in proving that all the right considerations were borne in mind in arriving at the figure. If the valuer always follows the same course of conduct in the work performed, always records the information in the same manner, always sums up his valuation (so as to show what amounts have been allowed for in arriving at the total), his position will be the stronger if the valuation comes under attack: see *Mount Banking Corporation* v *Brian Cooper & Co (infra)*. The valuer who says he "took these matters into account" but fails to record the manner of so doing will find life much more difficult if the valuation becomes the subject of a testing cross-examination. The decision in *Bere* v *Slades* [1989] 2 EGLR 160 illustrates this well. A valuer acting for a mortgage lender reported a number of defects but the plaintiff maintained that there were other substantial matters of which he was not advised. The judge, in finding in favour of the valuer, was impressed by the quality of his site notes, his evidence and his presentation. He found that the valuer had exercised a proper degree of care.

Type of property being valued

In most of the cases cited above the valuation in question has been a valuation for mortgage purposes. This is because, for some years, it was the area where the actions have been fought. Latterly the courts became inundated with a number of commercial valuations. It is clear that the cycle follows closely on the repetitive Boom and Bust under which the British economy has for so long suffered. Valuations performed in the height of a Boom tend to fall for scrutiny in the depths of the subsequent depression of the market.

It must be borne in mind that all the principles set out in the residential market are of equal validity in all forms of valuation work; save only that in commercial work the ability to exclude responsibility to third parties is more readily available. It is important to recognise gross defects if valuing for any purpose. The question of the small fee is not limited to the mortgage valuer. Gratuitous advice need not be only that given over the garden fence or in the saloon bar of the local. Much free advice is given during the everyday course of business relationships. There are clients who will ring up with a query where there can be no question of submitting a fee account. Yet, it is in this very area where greatest danger can lie. Free advice is often given without proper consideration. The valuer knows this – but does the recipient? It is probably easier to limit liability for free advice, but there still remains the question of proving that such limitations were imposed.

Valuers will be aware that, over recent times, there have been considerable disputes involving commercial valuations of great amounts. In addition to those which have been before the courts there are some which have been settled by the parties; some have gone to arbitration. The problem for the rest of the professional community is that we do not know on what terms these matters were settled or on what basis liability was established. Neither do we know how we may avoid some of the same pitfalls ourselves.

Measure of damages

If negligence has been established, there remains the question of assessing the amount of damages to be paid. In residential property, commonly the plaintiff claims for all the costs incurred putting the property into the condition it was expected to have been in had there been no negligence. The claim might also include the cost of alternative accommodation while the works are carried out and a further sum for anxiety and distress caused by the events. In the great majority of cases this is not the proper method of assessment of damages for a faulty valuation. The approved method of fixing the *quantum* is to find the difference between the value of the property in its actual state and the value in the state in which it was assumed to have been at the date of the valuation report. Such a difference is frequently a great deal less than the cost of execution of the works.

The leading authority for this position is *Perry* v *Sidney Phillips & Son* [1982] 2 EGLR 135. This was a case of a "structural survey" which proved to be negligently performed. One might have thought that, in the case of a survey report, there was a greater chance of the plaintiffs recovering the cost of the works. The court held, following *Dodd Properties (Kent) Ltd* v *Canterbury City Council* [1980] 1 EGLR 15 that the difference in value was the true measure of damages. Some judges tried to circumvent this rule by finding that the correct method of assessing the "difference in value" is by finding out the cost of works and deducting that from the price paid. Recently, however, the Court of Appeal has made harsh criticism of Official Referees following such a practice. It would appear that, in some of these cases, insufficient evidence has been adduced to show what the actual market differences would have been. If no such evidence is made available then the court will have no yardstick to apply.

In cases involving domestic premises it is normal for the court to award some extra small sum for the anxiety and worry caused to the plaintiff, particularly at the lower end of the housing market. This is usually of the order of £1,000–£2,000. This might rise as prices etc rise.

In commercial property the normal claim is for the difference in value between what was paid and the true value. This figures largely in the ensuing part of this chapter.

More Important Recent Decisions – Stage 1

We must now consider a number of more recent cases affecting the commercial property market. This list comprising, principally:

- *Mount Banking Corporation* v *Brian Cooper Co* [1992] 2 EGLR 142
- *Banque Bruxelles Lambert SA* v *Eagle Star Assurance* [1004] 2 EGLR 108
- *Nyckeln Finance Co Ltd* v *Stumpbrook Continuation Ltd* [1994] 2 EGLR 143
- *Mortgage Express Ltd* v *Bowerman & Partners* [1995] 1 EGLR 129
- *United Bank of Kuwait* v *Prudential Property Services Ltd* [1996] 2 EGLR 93

The "Boom & Bust" of the 1970s had caused a number of well known cases and this had already lead to the creation of the Red Book for Asset Valuations. The profession, rather late in the day, was beginning to put its house in order. The various books have now been consolidated into the RICS *Manual of Appraisal and Valuation* ("The Red Book") covering all aspects of valuation. Disciplinary procedures within the profession have been strengthened and failure to adopt the correct procedures can result in action by the professional bodies. This may take place where no action has been taken against the member in the courts, or after such an action, if further measures are considered necessary.

Such measures were far from popular with many members of the profession. The original fear that the Institution was making "whips and scorpions for learned counsel to use to flay us in the witness box" has receded and there is now more general acceptance of the need for guidance and control by the professional bodies.

The work of Michael Mallinson, and his team, in producing the working paper at great speed has been continued by those responsible for various sectors bringing out papers at a furious rate. Although the "paper chase" might be considered, by some, to be too fast, it merely reflects haw far behind the situation the Institutions had been.

If the profession has been bad at anything which it knows it ought to do, it is the taking of formal instructions. This is one of the prime factors contributing to negligence claims. RICS Insurance Services estimated that 58% of their case load would never have arisen had the parties agreed terms and conditions before undertaking the work.

The more recent crop of cases commenced with one which produced a sigh of relief throughout the profession.

The case was: *Mount Banking Corporation Ltd* v *Brian Cooper & Co* (*supra*).

In this case the valuer had been asked to provide a number of valuations for the purpose of securing a bank loan on a relatively small development in South London. The valuer's instructions, in fact, came from the developer. He, however, acted in the knowledge that the defendant valuers were normally retained by that bank. There was considerable confusion about the manner and dates in which the instructions were given. In addition to the instructions from the developer, there was correspondence between the plaintiff and the defendant. The time scale was very compressed and some of the letters appear to have been answered before they were written! However, the judge found, as a fact, that the defendants were employed in a contractual relationship to the bank who could expect competent valuation advice from the defendants in the four matters in dispute.

The case was heard by Deputy Judge R M Stewart QC, who heard very complex arguments involving *Singer & Friedlander* v *John D Wood* (*supra*); *Corisand Investments* v *Druce & Co* (*supra*); *Bolam* v *Friern Hospital Management Committee* (*infra*). The first two arose from the boom of the early seventies; the latter, which dates from 1957, is an important case involving allegations of medical negligence illustrating that:

(i)   Law can be derived from any source and,
(ii) The concepts can be applied generally.

These commercial property cases were judged after the spectacular crash of 1974.

Valuers were instructed to prepare the following valuations of a south London property:

1. Open Market Value as existing.
2. Forced sale as existing.
3. Open Market as extended and refurbished.
4. Forced sale as extended and refurbished.

They gave the figures as follows:

1. £530,000
2. £477,000
3. £765,000
4. £688,500

The valuer did not know, and had not asked for, the actual purchase price at the time of the valuation. It was £440,000.

The purchaser took the property with a 70% loan of the "existing value" (as defined by the defendants) with a promise of 70% of redevelopment finance to be provided by the plaintiffs. The purchaser bought the property. He paid nothing to the lenders! The property was sold for £125,000. The plaintiff sued the valuers in negligence and lost the action.

The action concentrated on the open market valuation of the premises in the sum of £530,000. The plaintiffs conceded that the 10% margin, (*Singer & Friedlander* v *John D Wood & Co* (*supra*)) was not a correct guide and adopted a margin of 17.5% as being the acceptable margin of error.

The valuation was dated August 18 1989. The borrower first approached the lender on August 31 which was the actual date fixed for completion! On September 1 Mr Shah, of the plaintiffs, spoke to the defendants and Mr Cohen, of the defendants, gave him a figure of £530,000 as open market value in existing use. Mr Shah confirmed his instructions that day but this crossed with a letter from the defendants, also dated September 1 containing their report.

The judge found, as facts:

1. That despite the original contact made to the defendants by the borrower, there was a contractual relationship between the parties under which the plaintiff could expect "competent professional valuations in the four respects",
2. That the plaintiffs relied upon the advice given by the defendants,

3.   That, had the valuation shown a lower figure the loan would have been limited to 70% of that figure, and

4.   That, had that figure been anything over £450,000 the amount of the loan would have been sufficient and the deal would have proceeded.

There was criticism of the defendant in that he had made no allowance for interest on capital during the refurbishment period. Since that capital amounted only to some £48,000 (and for only a few months) out of a project valued at £765,000 at completion, I doubt whether I would have worried over much. However, the judge found this to be a fault in the valuation method.

When asked to consider the individual items of the valuation and to judge whether each was reached properly or negligently the judge said that that was not within the *Bolam (infra)* principle and neither was it to be found in *Corisand (supra)*. He stated the law, as he saw it, thus:

> If the valuation that has been reached cannot be impeached as a total, then, however erroneous the method or its application by which the valuation has been reached, no loss has been sustained because, within the *Bolam* principle, it was a proper valuation. I do, however, accept that if, and where, errors are demonstrated either in the approach, or in the application of the approach, then any judge should look carefully at whether that valuation is, despite those errors, none the less an acceptable value within the *Bolam* principle.

Many points emerged from the *Mount Banking* case (though some are already out of date):

1.   If the valuation is correct in total it is not negligent albeit that there may be some rational dispute about some of the constituent parts.

2.   Less than total acceptance of Watkin J's comment in *Singer & Friedlander* to the effect that valuers must be within 10% of each other. The Deputy Judge said that 10% should not all apply to all cases. Even the pleadings accepted 17.5% as a tolerable margin of error.

> I think the judge must approach the question, first, by asking where the proper valuation or bracket of valuation lies. Then, if the defendant is more than the permitted margin outside that proper figure, the inference should be drawn.

3.   The defendant valuer's case notes demonstrated that he had considered all the correct matters in arriving at his valuation.

The fact that other valuers would have given different values to some, or all, of those elements does not amount to negligence. He considered the following matters:

(a) The size, condition, tenure and location;
(b) Planning permission and permitted use;
(c) The value and use of adjoining properties;
(d) The value (both rental and capital) of other offices in the locality and their comparability;
(e) The state of the property market (given as "buoyant demand for this type of B1 office refurbishment at the time");
(f) The effect of the rise in interest rates from 7.5% in May 1988 to 14% in August 1989;
(g) The realism of the project;
(h) The defendant had carried out a number cross check valuations to verify his answers before making his report. As the judge recorded in the judgement:

> Mr Cohen was doing exactly what a competent valuer ought to do, in the sense of going through the right processes . . . it should be made clear, and I so find, that these notes themselves demonstrate that Mr. Cohen approached his valuation, sought evidence and prepared test calculations in precisely the manner to be expected of a competent valuer using proper skill and care.

He did not consider the purchase price. Evidence given showed that a body of professional opinion considered it proper to make a valuation without knowing the price. (That same evidence was given, by the same expert, in the later *Banque Bruxelles* case (*supra*) where his submission that the purchase price was irrelevant was heavily criticised.)

4. "Valuation by trained, competent and careful professional men is a task which rarely, if ever, admits of precise conclusion" (from Watkins J in *Singer & Friedlander*) cited and approved.

5. The *Bolam* principle. ". . . Where you get a situation which involves the use of some special skill or competence, then the test as to whether there has been negligence . . . is the standard of the ordinary skilled man exercising and professing to have that special skill. A man need not possess the highest expert skill; it is well established law that it is sufficient if he exercises the ordinary skill of an ordinary competent man exercising that particular art" *per* McNair J in *Bolam* v *Friern Hospital Management Committee* [1957] 2 All ER 118.

6. One of the expert witnesses was heavily criticised for being unbending in his opinion. An expert's prime duty is to aid the court. Being totally "right" on all matters, without acknowledging that there could possibly be a range of figures, does not assist the court. His evidence was ignored.
7. Failure to take account of the purchase price itself was not considered to amount to negligence.
8. The defendant's main expert witness had come to the conclusion that a correct figure might have been about £500,000. Mr Cohen's figure was within 10% of that and so the judge, in spite of having denied the validity of the 10% margin, was happy to find that the valuation fell within it and so the defendants were not negligent!

It is important to point out at this stage that this was a remarkable file. It was the very precision of the file which largely led to the finding that the defendant was not guilty of negligence. Particularly, as it appears that Mr Cohen had, unfortunately, died before the case was commenced.

More Important Recent Decisions – Stage 2

These commence with the monumental judgement of Mr Justice Phillips in:

*   *Banque Bruxelles Lambert SA* v *Eagle Star Insurance Co Ltd* and others [1994] 2 EGLR 108

The "Others" referred to in the title are:

Maurice Markowitz: Allied Dunbar Assurance plc

Lewis & Tucker Ltd: John D Wood Commercial Ltd

This case demonstrates, most emphatically, that the valuer's duty is one of reflecting the market and not trying to make it. Contrary to *Mount Banking Corporation* v *Brian Cooper* (*supra*), this case illustrates just how much reliance is to be put on market transactions; transactions which are to be ignored at the valuer's peril.

In a case of daunting complexity, the transcript runs to 37 pages of the *Estates Gazette Law Reports*, Phillips decided that John D Wood Commercial Ltd (JDW) were negligent in valuing properties at figures far in excess of those only recently achieved in the open market. Of the six properties involved in the action, four had been valued by Lewis & Tucker Ltd, (of which one was also confirmed by JDW) and two were valued solely by JDW. (For the perils of giving an overview of a valuation see *Charterhouse Bank Ltd* v *Rose* (*infra*).)

The case concerned the history, over a very short period, of six properties. All of them had been properly exposed in the market before sales were agreed. The properties fetched large sums of money.

The properties had been acquired by companies under the control of Mr Markowitz. The purchaser companies were newly formed companies (often called "Shelf Companies" because most firms of commercial solicitors have a number of such companies ready formed and available for clients immediately – they are reached "off the shelf"). They are also referred to in the judgement as SPV or SPC ("Single Purpose Vehicle" or "Single Purpose Company"). They are intended to own only one property each.

Having acquired the properties, Mr Markowitz sought valuations from his professional advisors. Each was asked for an "armchair valuation", which, more or less, means what it says. The valuation was to be done without inspection and on the basis of information provided by Mr Markowitz. The firm of valuers who came up with the highest armchair valuation was then asked to prepare a formal valuation.

The total purchase price paid for the six properties was £243.775m. The total of the valuations, (and the latest was prepared three months after the sale and purchase while the average period was about six weeks) was £346.1m – an increase of £102.325m.

Three of the properties had been "sold on" since acquisition from the original vendors. The purchasers were, in each ease, nominee companies of Mr Markowitz.

Mr Markowitz used the valuations for the purpose of raising money from the Belgian Bank. That bank took a deal of persuading that the London property market was a good place to invest. However, the London office convinced head office to let the loans go ahead. In one of the cases, the explanation given for the discrepancy between the market prices and the valuers' figures was "that the owners, Legal & General, had sold the property below its value because LPT (the purchasers) had offered a prompt cash payment." Adding that "the management of Legal & General were unimaginative and unaware of the true value of their property".

In total the Belgian Bank had advanced £307.935m against properties bought for £243.775m making an immediate cash sum available amounting to £64.16m.

Mr Markowitz sought and obtained a Mortgage Interest Cap Policy – Mr Markowitz also took commission from the sponsors, the borrowers, the lenders and the insurers. He placed cover for key personnel with Allied Dunbar (a sister company of Eagle Star). Sun Alliance were also involved in insuring the risks (their contribution giving BBL cover up to 100% of the sums advanced) but did not participate in the action.

One of the things which the bank had "learned" was that valuations, made for Mr Markowitz, should be rewritten addressed to the bank. This procedure was common and intended to ensure that the connection between the valuer and the lender is one in contract: see the views expressed by Phillips J on this point later on.

BBL had taken the view that, with the insurances in place, they were not dealing with property – that all of their risk was an insurance risk and stated in one memorandum:

> We are not interested in the valuation of the property *per se* as a matter of credit consideration. It is the cover that we are concerned with in establishing the amount advanced. Sun Alliance will determine their level of cover on the basis of their due diligence.

Clearly, nobody in BBL imagined that anything could happen to the property market which would be outside the extensive insurance cover which they had arranged.

The original defendants had joined each of the other co-defendants. Eagle Star maintained that Mr Markowitz was an agent of BBL and that his knowledge was therefore theirs. Whereas BBL claimed exactly the opposite – that Mr Markowitz was an agent for Eagle Star. Eagle Star alleged that Mr Markowitz knew that the valuations were excessive and claimed damages. So also did BBL.

Eagle Star claimed that BBL had not acted as prudent bankers and had failed to disclose this fact. The valuers also claimed that BBL had acted imprudently and should be liable for some of the losses themselves.

Settlements were reached between BBL and Eagle Star; between Mr Markowitz and BBL; between Mr Markowitz and Eagle Star; between Lewis & Tucker and BBL.

This left John D Wood Commercial Ltd as the sole defendants. JDW was in considerable difficulty. Its insurers of professional indemnity refused to assist it even to the extent of providing solicitors and counsel. It was only after some weeks into the trial that some of the "second line insurers" paid for JDW to be represented. They were rewarded for their investment in securing a finding of contributory negligence against the bank.

Lewis & Tucker having made an agreed settlement, the action related to the three valuations of JDW, two original and one confirmatory of a valuation made by Lewis & Tucker. BBL lost the action in respect of the confirmatory valuation because it had not relied on it – it still relied on the valuation of that property provided Lewis & Tucker Ltd. (See the later case of *Charterhouse Bank Ltd* v *Rose* for a considered opinion on "taking a view" on another valuer's valuation.) For the other two, JDW were found to be negligent – but the plaintiff bank was found to have been guilty of contributory negligence to the extent of 30%.

The result was that JDW were found liable to pay the £10.8m to the plaintiffs and £5.3m to Eagle Star. John D Wood Commercial Ltd went into immediate liquidation.

The court found that, contrary to the views expressed in *Swingcastle* v *Alastair Gibson (infra)*, the valuers were not liable for the losses caused by the fall in market values.

Also of great significance is the finding that the valuations, made for the purchasers but addressed to the bank, were not enough to base a claim in contract by the bank – such liability as there was, was found to lie in tort only. This again contradicts the findings in *Mount Banking Corporation* v *Brian Cooper & Co (supra)*. Without this there would normally have been no counterclaim for contributory negligence.

It has been the normal practice for lenders to require such letters and valuations to be addressed to them. Phillips J found it improbable that the bank had paid for such highly expensive valuations upon which the transaction was meant to depend. Without a favourable report, the loan would not go ahead. It is normal for the borrower to commission such a report. It is his money which is at risk in the payment of fees; he has to set up the transaction as to tempt the lender to make the advance; he commissions and pays for the report. The fact that a copy of that report is addressed to the lender does not make him party to the contract for the valuer's services.

Contrast this with the somewhat cavalier way in which the judge in *Mount Banking (supra)* dismissed the manner in which the instructions were given. In that case, where clearly the borrower had made the first contact with the defendant valuers, the judge had no difficulty in finding that there was a contractual relationship between the lender and the valuer. It was not even treated as a point of law; it was found as a fact. This does not imply criticism of Stewart QC because it may well turn on the manner in which the case was argued before him.

In the *BBL* case the relationship between the parties became a major issue. The mere fact that there were so many of them would indicate a necessity to isolate the nature of the relationship between each of them. The point therefore emerged with great force. In my view, *BBL* should be regarded as being more authoritative on this issue than *Mount Banking*.

Measure of Damages

This now becomes the major issue of the day. In *Banque Bruxelles* there was considerable argument as to the amount of the losses to be attributed to the negligent valuation. Was it to be the entire loss – the difference between the amount loaned and the amount of subsequent realisation of the assets, or was some other measure appropriate? The real problem lies in the fact that, in addition to the errors of valuation there had been a collapse in the market for London offices (and a lesser one for the provincial ones). After considerable argument and an in-depth review of the cases Phillips J held that the loss(es) attributable to the defective valuation was the amount of the difference between the values given by the valuers and the "true value" – which in this case was the original sale price.

In *Nyckeln Finance* v *Stumpbrook Continuation Ltd*,* which also concerned a valuation of property for mortgage purposes, a property was originally sold for £23.5m; valued at £30.5m and ultimately realised £3.1m. The fact that the valuation was negligent was admitted. The major questions arising were:

(a) What was the true value?
(b) Was the plaintiff guilty of contributory negligence?
(c) If so, to what degree?
(d) Had the plaintiff done everything possible to mitigate the losses?

The facts are very confusing (and the case is not easy to read since the major personnel were Norwegian, Swedish and Dutch). However, we get a similar scenario of a foreign bank, reluctant to

---

*This appears to be a subsidiary of Jackson Stops & Staff.

lend on the London property market being persuaded so to do, largely because the parties are good friends and clients of each other.

On the issue of damages, it was suggested that all that the plaintiff could recover was the difference between the true value and the "highest non-negligent valuation"((this was given in evidence by expert witnesses) which was held to be £23.5m. The judge chose to follow Phillips J in *BBL*, (which had been decided a few days earlier) and stick with the actual sale price. He decided that this bank was also guilty of contributory negligence which, in the circumstances of that case, he assessed at 20%.

Interest on Damages

There then remained the question of interest. Successful plaintiffs are entitled to interest on the award of damages and this will normally run from the date of the "injury" to the date of judgement (after which any further interest runs at the statutory rate). But, in this case, the plaintiff, largely because of the goodwill which still existed between the borrower and the lender, was dilatory in exercising its right to sell the property. It had a good opportunity of a sale at £17m in July 1990. Due to lack of response it lost that sale and finished with a sale at £3.1m. Interest was held not to run after August 13 1990 being a date at which they might have mitigated loss at £17m. The judge added that, if he were held to be wrong in denying the plaintiff's claim for the losses caused by the fall in the market the figure of £17m, which would have been available to them, was to form the "bottom line".

In both of these cases the judges were able to "distinguish" the cases from earlier precedents – notably *Swingcastle Ltd* v *Alastair Gibson* [1991] 1 EGLR 157, a decision of the House of Lords.

In that case, a valuer negligently produced a valuation of £18,000 as a forced sale value for a property. A loan of £10,000 was made on the basis of the valuation. There was a default almost immediately following which a sale at £12,000 was achieved. Had the lender known that the value achieved would have been so small, the loan would never have been made. The lender claimed the loss of the default interest (at 45% pa!) which had been contracted to be paid

---

*See the later comments by Lord Hoffmann in *South Australia Asset Management Corporation* v *York Montague Ltd* (*infra*).

under the loan should there be a default in performance. The House of Lords held that the correct measure of damages was the actual loss rather than the difference in value because the transaction would not have occurred at all had there been no negligence.

This followed the pre-war case of: *Baxter* v *FW Gapp & Co Ltd* [1939] 2 All ER 752.

However, it would seem from the judgement of Phillips J in *Banque Bruxelles* that there is doubt about both the relevance of this case and the accuracy of the report. But the House of Lords in the *South Australia Case* (*infra*) made no comment on the accuracy of this case.

No such doubts troubled Gage J in *United Bank of Kuwait* v *Prudential Property Services Ltd* [1994] 2 EGLR 100. (A further appeal in this case was held with *South Australia Asset Management Corporation* v *York Montague Ltd*. Both appeals are referred to later.) Another defective valuation of commercial property for mortgage purposes. The plaintiff bank agreed to advance £1.75m or 70% of a professional valuation, whichever was the lower, on the security of property owned by Sallows Development Ltd. The valuation turned out at £2.5m (containing a considerable error in the matter of the over-renting of some of the units). The loan was made and the owners defaulted. A receiver was appointed and the property sold for £950,000.

The "correct" figure for valuation was between £1.8m and £1.85m. The defendants were found not guilty of contributory negligence, as had been alleged. Damages were awarded on the basis of *Swingcastle* v *Alastair Gibson*, a distinction being drawn between "transactional" and "no-transactional" negligence.

"No-transactional" negligence is defined as applying to cases where, if the negligence had been apparent, the transaction would never have taken place. It follows from this that the losses sustained by the lender are not limited to the difference between the true market value and that ascribed by the valuer, but include all the subsequent losses, including (in particular) losses attributable to the fall in the market. Unfortunately, we do not know what this actually amounted to in this case. It was agreed between the parties.

The difficulty is that some of the "no-transaction" cases appear to differentiate between "Negligence" and "Gross Negligence." If this is a true interpretation of the distinction it is hard to find a case of negligence more "Gross" than *Banque Bruxelles*.

First National Commercial Bank plc v *Humberts* [1995] 1 EGLR 142. In that case the Court of Appeal considered the liability of valuers following negligent valuations. The Court of Appeal, following *Swingcastle Ltd* v *Alastair Gibson*, held that the negligent valuer was liable for all of the losses. The land had been valued by Humberts at £4.4m against a true value of £2.7m. The Court of Appeal held Humberts liable for all of the advance (the borrowers having become insolvent); any other losses which the plaintiff had suffered; what it might have made on other deals; or the interest which the money would have made if left in the bank.

The note of the case is very short indeed. It looks as if the Court of Appeal considered itself bound by the decision of the House of Lords and looked to see whether the negligence was "transactional" or "non-transactional".

Banque Bruxelles and four other cases were later heard by the Court of Appeal which decided that the cases were "non-transactional". Accordingly, the valuers were found liable for all of the losses occasioned, including the losses attributable to the fall in the market.

There was no comment made about:

(a)   The liability being in tort not in contract;
(b)   Accordingly, there is still the offset for contributory negligence.

Charterhouse Bank Ltd v *Rose* [1996] 3 WLR 87 brought another blow to the valuers. In this case, which would have been against the directors of John D. Wood Commercial (and against the company itself), the only defendants left were Savills and Mr Marland of that firm who had been asked to provide an "overview" on the valuations of John D Wood Commercial Ltd.

The plaintiff bankers were concerned that John D Wood (and in particular Mr Browne, who had figured so prominently in the *Banque Bruxelles* case) might have given a valuation which was too high for the security offered. The case was similar in that the property had only recently been sold by an insurance company after considerable exposure to the market. John D Wood's valuation was very considerably higher than the sale figure. Savills were asked for an overview of the valuation.

Mr Marland provided a draft which was not sympathetic to JDW's views. This fell foul of the wishes of the bankers to make the loan. Mr Marland changed his opinion to suit the bank. It also transpired that Savills acted for the tenant of the building concerned in the matter of rent reviews. The Court, having decided where the liability lay, left it to the parties to work out the amount of damages. Accordingly we do not know the amount finally agreed but it would be several millions of pounds.

The case is remarkable for several points:

(a) Giving a "view" of somebody else's valuation can be just as dangerous as giving your own original figure.
(b) Valuers should not be swayed by the opinions of others against their better judgment. The bankers in *Banque Bruxelles* were as eager to make the loan as were Charterhouse Bank in this case.
(c) Savills should have declared a strong conflict of interest and declined the instructions. It is totally humiliating to be confronted with evidence, given in an arbitration to the effect that the premises are poor, out of date, off site etc. and have the court compare those views with views given in the trial giving a contrary opinion!

Decision of the House of Lords

At last these matters went to the House of Lords. Earlier courts had come to differing positions and the Court of Appeal held itself bound by the House of Lords decision in *Swingcastle Ltd v Alastair Gibson (infra)*. The cases submitted to their Lordships house were:

(a) *South Australia Asset Management Corporation* v *York Montague Ltd*
(b) *United Bank of Kuwait* v *Prudential Property Services Ltd.*
(c) *Nykredit Mortgage Bank plc* v *Edward Erdman Group Ltd.*

All at [1996] 2 EGLR 93.

The case note reads [On appeal from *Banque Lambert SA v Eagle Star Insurance Co Ltd*] so we get the surprising situation where the case under appeal is one which is not actually before the court.

We now have a clear decision from Lord Hoffmann, with all of the other Law Lords in agreement.

In a lucid judgement Lord Hoffmann destroyed the concept of "transactional" and "non-transactional" cases. *Swingcastle Ltd v Alastair Gibson (supra)* was demolished. He instead used the preferred terminology of "no-transaction" and "successful transaction" which more clearly indicates the division. In his view, such matters are quite irrelevant to the scope of the duty of care.

He said:

The distinction is not based on any principle and should . . . be abandoned.

Every transaction induced by a negligent valuation is a "no-transaction" case in the sense that *ex hypothesi* the transaction which actually happened would not have happened.

On the subject of the amount of the damages to be awarded once negligence has been established Lord Hoffmann would calculate the difference between the valuation relied on and the mean of the other figures produced in evidence by expert witnesses. He found that it would be wrong to take the highest, or the lowest, alternative*. He quoted, again, the *dictum* that valuation is seldom an exact science. He thus rejects the view that the measure of damages is the difference between the valuation in issue and the highest "non-negligent" valuation. Such a comparison might indicate whether or not there has been negligence (eg by examining the bracket of possibilities) but is too crude a measure for the assessment of damages.

What can we learn from the cases put together?

1. Valuers must agree terms and conditions with their clients before undertaking any work. If this causes delay the valuer can post terms to the client while proceeding with the valuation – provided that no indication of the result is allowed before the client returns the signed agreement. The former course is much to be preferred.
2. There is no such thing as an over-full file. Reliable records must be kept of everything. Those records must be clear – not just clear at the time, but clear when you are asked to explain them – perhaps seven years later.
3. There is no substitute for diligence such as that displayed by Mr Cohen of Brian Cooper. The more information collected, and recorded, about the property and the project, the comparables, the degree of comparability and the state of the market at that time, the better. (You will be unlikely to remember in six years time what the state of the market is at the moment at which you are now reading this. You will if you

---

* We still do not know whether Phillips J was correct in holding to the market price – but I would suggest that that is right. Valuations only become necessary where there is no market transaction.

make a marginal note in your papers – and if you can find them again in six years time).

4. The valuer's job is, was, and, so far as we can tell, always will be, to reflect the market as it is at the date of the valuation.

5. The prime information about any property is its own history particularly its market history. Agents should be prepared to discuss with their valuation brethren not only comparable evidence available today, but also the marketing history of property recently sold or let.

6. Despite some comments by Phillips J against using yields and rents to arrive at property value, this method remains sound and valid, unless the property has recently been marketed. The judge had had an arduous trial which had a number of unpleasant overtones to it. He handled the evidence of valuer's practices extremely well, but he only heard part of the story. When property has recently been sold you may choose to throw away your *Modern Methods of Valuation* and your *Parry's Valuation Tables* if you will.

   But the fact remains that it is quite rare for valuers to be called upon to value property immediately after it has been sold. We are more often asked to value it just before it is to be sold.

   This might be to recommend a sale price; to suggest a buyer's bid price; to assess the value for loan purposes etc etc. Heaven forbid that we should value all of these by reference to its recent sale. To rely on that method alone would be to open the door to malpractice on a scale undreamed of. If the maxim is "I've just bought it. It must be the right price and value" what need would there be for the statutory provision for Building Societies to have an independent valuation?

   In most cases, there will have been no recent sale. In such circumstances we must continue to use the tried and tested methods of analysis of recent transactions; breaking that transaction down to constituent parts (per square foot, per square metre, per acre etc) assess the instant property in terms of comparability; determine, by similar processes what the yield rate should be (and we *are* free to use Term & Reversion, Equated Yield, Equivalent Yield, DCF or whatever arithmetical method best suits our use and our practice).

7. Valuers are not liable for the overall fall in market prices. Previously I had found the decision in *Swingcastle* v *Alastair Gibson* to be one which is almost to be confined to its facts. In

the *BBL* case (and the later cases referred to the House of Lords) the lenders did know the true facts, and as a result, were found guilty of contributory negligence.

8.   Do not let the above lead you to thinking that the price paid at a collective auction of repossessed residential property is a true market figure on which you can rely. You may have strong reasons to doubt any sale which is patently a repossession.

The terror of the moment is that there are still firms working on the back of an envelope with an undated jotting. The Conditions of Engagement for Commercial Valuations have been available since 1984. Many valuers have never seen them; fewer of them use them. If there is no contract for your services, how do you, or your clients, know if you have performed them?

Perhaps I may pass on a comment made to me by the senior valuation partner of a major West End firm who prefers to remain anonymous. He says that, in his firm, no valuation may be signed by anybody other than a full partner (usually himself). Each file must be presented to the partner together with the valuation. The valuation(s) are made on coloured paper so as to be readily identifiable. The order in which the contents of a valuation file is to be kept are laid down. So that, for example, the photographs of the property are filed adjacent to the site plan. (I do not know the actual order and it does not matter – what does matter is that every file in the firm has the same information in the same place and the same order.) My friend admits that this looks like sheer bureaucracy. But, he adds, this system has kept that firm out of the courts for all of the years since he introduced it. Perhaps even Brian Cooper & Co could learn from this example, excellent though their file proved to be.

In my own firm, small though it was, I insisted on all valuations being made on coloured paper which was ruled with vertical lines on the right hand edge. This was intended to keep rental elements from being added into the column for capital sums – and it also kept the date and the data well clear of one another when arriving at the answer.

I am indebted to Professor Keith Davies for reading and correcting this chapter. However, if there are any faults: *Mea culpa; mea maxima culpa.* VWT

Index

A

D

E

M

O

Q

R

W